Advanced Series in Agricultural Sciences 18

Advanced Series in Agricultural Sciences

Volumes already published in the series

Volume 1: *A. P. A. Vink*
Land Use in Advancing Agriculture

Volume 2: *H. Wheeler*
Plant Pathogenesis

Volume 3: *R. A. Robinson*
Plant Pathosystems

Volume 4: *H. C. Coppel, J. W. Mertins*
Biological Insect Pest Suppression

Volume 5: *J. J. Hanan, W. D. Holley, K. L. Goldsberry*
Greenhouse Management

Volume 6: *J. E. Vanderplank*
Genetic and Molecular Basis of Plant Pathogenesis

Volume 7: *J. K. Matsushima*
Feeding Beef Cattle

Volume 8: *R. J. Hanks, G. L. Ashcroft*
Applied Soil Physics

Volume 9: *J. Palti*
Cultural Practices and Infectious Crop Diseases

Volume 10: *E. Bresler, B. L. McNeal, D. L. Carter*
Saline and Sodic Soils

Volume 11: *J. R. Parks*
A Theory of Feeding and Growth of Animals

Volume 12: *J. Hagin, B. Tucker*
Fertilization of Dryland and Irrigated Soils

Volume 13: *A. J. Koolen, H. Kuipers*
Agricultural Soil Mechanics

Volume 14: *G. Stanhill*
Energy and Agriculture

Volume 15: *E. A. Curl, B. Truelove*
The Rhizosphere

Volume 16: *D. P. Doolittle*
Population Genetics: Basic Principles

Volume 17: *A. Feigin, J. Ravina, J. Shalhevet*
Irrigation with Treated Sewage Effluent (in preparation)

Volume 18: *D. Gianola, K. Hammond*
Advances in Statistical Methods for Genetic Improvement of Livestock

Daniel Gianola · Keith Hammond (Eds.)

Advances in Statistical Methods for Genetic Improvement of Livestock

With 5 Figures

Springer-Verlag
Berlin Heidelberg New York
London Paris Tokyo
Hong Kong

Prof. Dr. DANIEL GIANOLA
University of Illinois
Department of Animal Sciences
126 Animal Sciences Laboratory
1207 West Gregory Drive
Urbana, Illinois 61801
USA

Dr. KEITH HAMMOND
Animal Genetics and
Breeding Unit (AGBU)
University of New England and
NSW Agriculture and Fisheries
Armidale
New South Wales, 2351
Australia

ISBN 3-540-50809-0 Springer-Verlag Berlin Heidelberg New York
ISBN 0-387-50809-0 Springer-Verlag New York Berlin Heidelberg

Library of Congress Cataloging-in-Publication Data. Advances in statistical methods for genetic improvement of livestock / Daniel Gianola, Keith Hammond, eds. p. cm. -- (Advanced series in agricultural sciences ; 18) Based on an international symposium held in Armidale, Australia, Feb. 16-20, 1987. 1. Livestock--Breeding--Statistical methods. 2. Livestock--Genetics--Statistical methods. I. Gianola, Daniel, 1947- . II. Hammond, Keith, 1942- . III. Series. SF 105.A48 1990 636.08'21--dc20 90-9586

Printed in Germany

2131/3145(3011)-543210 – Printed on acid-free paper

First row: (left to right): Robert A. Young, Peter Smith, Dr. Thorvaldur Arnason, Gerard Davis, Dr. Jean-Louis Foulley, Prof. Brian W. Kennedy, Dr. Peter Parnell, Dr. Neal Fogarty, Dr. Vijay Chauhan, Dr. Stephen Bishop, David Klassen, Dr. Rohan L. Fernando, Dr. Scott Newman, Gisela Ahlborn-Breier.
Second row: Dr. John James, Dr. Kerry Rathie, Mary Rose, Prof. W. (Bill) G. Hill, Dr. Laurie Piper, Graham W.M. Kirby, Dr. Hans-Ulrich Graser, Dr. Chris Hagger, Dr. Markus Schneeberger, Bruce Tier, Ross Shepherd.
Third row: The late Prof. Charles R. Henderson, Dr. Karin Meyer, Margaret Mackinnon, Alicia L. Carriquiry, Dr. Robert C. Elston, Dr. Raul W. Ponzoni, Dr. Horacio Raul Guitou, Dr. A.S. Del-Bosque-Gonzalez, Dr. Öje Danell, Dr. Nan Laird, Prof. Leo Dempfle, Dr. R. (Bob) Scarth.
Fourth row: Dr. Robin Thompson, Andrew Swan, Dr. Kevin D. Atkins, Dr. Sue Mortimer, Dr. Ian R. Franklin, Dr. Alan Stark, Dr. Geoff K. Robinson, Prof. Dan Gianola, Dr. Les P. Jones, Dr. Arthur Gilmour, Prof. Robert Anderson, Maurie Josey, Dr. David Johnson.
Fifth row: Dr. Keith Hammond, Dr. Siva S. Sivarajasingam, Dr. Stephen Smith, Paul Nicholls, Dr. David A. Harville, Dr. John Lax, Dr. Mick Tierney, Dr. Mike Goddard.
Absent: Prof. J. Stuart F. Barker.

Contributors

J. Bouix
T.P. Callanan
L. Dempfle*
J.M. Elsen
R.C. Elston*
R.L. Fernando*
J.L. Foulley*
D. Gianola*
B. Goffinet
D.A. Harville*
C.R. Henderson*
W.G. Hill*
S. Im
B.W. Kennedy*
S. Knott
M.W. Knuiman
N.M. Laird*
F.W. Macedo
L.R. Schaeffer
S.P. Smith*
R. Thompson*

* Invited speaker

Discussion Summaries

R.D. Anderson
J.S.F. Barker
A.R. Gilmour
M. Goddard
H.-U. Graser
J.W. James
R. Thompson

Preface

Developments in statistics and computing and their application to the genetic improvement of livestock have gained momentum during the past 20 years. In particular, best linear unbiased prediction and associated mixed linear model methodology are now considered to be standard procedures for identifying genetically superior animals and estimating genetic trends in breeding programs. Nevertheless, research in this field continues to be very active, to: (1) develop improved statistical and computing strategies, (2) assess robustness to departures from ideal conditions, and (3) obtain maximum economic benefit from information in data sets resulting from recording the performance of animals. Ideally, it would be desirable to integrate these statistical procedures with the designs of the breeding programs.

This volume grew out of an international symposium which took place in Armidale, Australia, February 16-20, 1987. The idea of organizing such a symposium evolved during a visit of Keith Hammond to the University of Illinois in November, 1984. We felt that it was time to review and consolidate the underlying statistical foundations of animal breeding, in the light of sweeping developments in techniques for genetic evaluation taking place in the 1970's and 1980's. We considered it important to include both Bayesian and frequentist approaches.

The 12 main speakers and 7 moderators invited from 6 different countries represented well "the state of the art" as well as areas which merit further research and development.

This 23-chapter volume is organized into seven main sections: I General, II Design of Experiments and Breeding Programs, III Estimation of Genetic Parameters, IV Prediction and Estimation of Genetic Merit, V Prediction and Estimation in Non-Linear Models, VI Selection and Non-Random Mating, and VII Statistics and New Genetic Technology. Each of the sections contains three or four "main" chapters plus a summary written by the corresponding moderator; the summaries reflect the moderator's viewpoint and the main points discussed when the presentations took place. The volume is, therefore, lengthy but complete, and we feel that it "cuts the work" needed in the next 20 years or so.

In editing this book, no attempt was made to unify notation and terminology because it would have been difficult to achieve this objective. We read the original manuscripts as carefully as feasible, and attempted to clarify the message of the authors when this was needed. The authors of individual chapters are responsible for the substance of the contribution, but we are responsible for typographical or grammatical errors remaining. We worked hard to catch obvious mistakes and would very much appreciate it if the readers bring additional ones to our attention, for future correction. No book is entirely free of errors (even after several printings!) and this will probably be no exception.

The book should be useful as a reference source to animal breeders, quantitative geneticists, and statisticians working in these areas. Alternatively, it could be used as a text in graduate courses in animal breeding methodology with prerequisite courses in linear models, statistical inference, and quantitative genetics. Many universities in North America and other countries now offer post-graduate courses in statistical aspects of animal breeding. Several chapters in this book could be used in such courses to provide excellent supplementary reading; however, intense instructor guidance will be needed because of the depth and difficulty of the material.

The symposium was sponsored by the A.S. Nivison Trust, Australian Association of Animal Breeding and Genetics, Australian Meat and Livestock Research and Development Corporation, Devon Cattle Breeders' Society of Australia Ltd, Holstein-Friesian Association of Australia, Pig Research Council, Reserve Bank of Australia, The Australian Poll Hereford Society Ltd, Mr. Dugald Mactaggart of "Waterloo", Glen Innes and Mr. Jock Nivison, of "Yalgoo", Walcha NSW.

We thank Mrs Coral Rogers for assisting with the Symposium, Ms. Glen Andrews for carrying the brunt of the typing, Mrs Elaine Farrell for much of the proofing, and Mr. Bruce Tier for technical assistance with word processing. We also appreciate additional editorial assistance provided by Drs. R.W. Everett, S. Sivarajasingam, S.P. Smith and R.D. Scarth and Mr. A.A. Swan.

We hope the material in this book will be useful to stimulate further research needed in this area of animal breeding, so vital to scientific animal production.

Daniel Gianola, Urbana
Keith Hammond, Armidale

September 1989

Acknowledgements

A.S. Nivison Trust
Australian Association of Animal Breeding and Genetics
Australian Meat and Livestock Research and Development Corporation
Devon Cattle Breeders' Society of Australia Ltd.
Holstein-Friesian Association of Australia
Pig Research Council
Reserve Bank of Australia
The Australian Poll Hereford Society Ltd.
Waterloo, Glen Innes
Yalgoo, Walcha

Contributors

Numbers in parentheses indicate the pages on which the authors' contributions begin.

Addresses

Bouix, J. (277), INRA, Station d'Amélioration Génétique des Animaux, Toulouse, France

Callanan, T.P. (136), Applied Statistics, Management Services Division, Eastman Kodak Company, Rochester, New York, USA

Dempfle, L. (98,454), Institut für Tierwissenschaften, TU München, D-8050 Freising Weihenstephan, Federal Republic of Germany

Elsen, J.M. (277), INRA, Station d'Amélioration Génétique des Animaux, Toulouse, France

Elston, R.C. (41,495), Department of Biometry and Genetics, LSU Medical Centre, 1901 Perdido Street, New Orleans LA 70112, USA

Fernando, R.L. (15,118,437), Department of Animal Sciences, 1207 West Gregory Drive, University of Illinois, Urbana, Illinois, 61801, USA

Foulley, J.L. (15,277,361), INRA, Station de Genetique Quantitative et Appliquee, 78350 Jouy-en-Josas, France

Gianola, D (15,118,210,361,437), Department of Animal Sciences, 1207 West Gregory Drive, University of Illinois, Urbana, Illinois, 61801, USA

Goffinet, B. (277), INRA, Laboratoire de Biométrie, BP 27, Auzeville, 31326 Castanet-Tolosan Cedex, France

Harville, D.A. (136,239), Department of Statistics, Iowa State University, Ames Iowa 50011, USA

Henderson, C.R. (2,413), Department of Animal Science, Cornell University, Ithaca NY 14853 USA and Department of Animal Sciences, University of Illinois, Urbana, Illinois, 61801, USA

Hill, W.G. (59,477), Institute of Animal Genetics, University of Edinburgh, West Mains Road, Edinburgh EH9 3JN, Scotland

Im, S. (15,210,361), Laboratoire de Biométrie, Centre de Recherches de Toulouse, Institut National de la Recherche Agronomique, 31326 Castanet-Tolosan Cedex, France

Kennedy, B.W. (77,507),Centre for Genetic Improvement of Livestock, University of Guelph, Guelph Ontario N1G 2W1, Canada

Knott, S. (477), AFRC Institute of Animal Physiology and Genetics Research, West Mains Road, Edinburgh EH9 3JQ, Scotland

Knuiman, M.W. (177), Department of Biostatistics, School of Public Health, Harvard University, 677 Huntington Avenue, Boston, Massachusetts, 02115, USA

Laird, N. (177,329), Department of Biostatistics, School of Public Health, Harvard University, 677 Huntington Avenue, Boston, Massachusetts, 02115, USA

Macedo, F.W. (210), Department of Mathematics, University of Tras-os-Montes e Alto Douro, Vila Real, Portugal

Schaeffer, L.R. (507), Centre for Genetic Improvement of Livestock, University of Guelph, Guelph Ontario N1G 2W1, Canada

Smith, S.P. (190,344), Animal Genetics and Breeding Unit, University of New England, Armidale. NSW 2351, Australia

Thompson, R. (312), Institute of Animal Physiology and Genetics Research, West Mains Road, Edinburgh EH9 3JQ, Scotland

Discussion and Summaries

Anderson, R.D. (56), Department of Animal Science, Massey University, Palmerston North, New Zealand

Barker, J.S.F. (533), Department of Animal Science, University of New England, Armidale NSW 2351, Australia

Gilmour, A.R. (410), Agricultural Research and Veterinary Centre, Forest Road, Orange NSW 2800, Australia

Goddard, M. (474), Department of Agriculture and Rural Affairs, P.O. Box 500, East Melbourne VIC 3002, Australia

Graser, H.-U. (309), Animal Genetics and Breeding Unit, University of New England, Armidale NSW 2351, Australia

James, J.W. (132), Department of Wool Science, University of New South Wales, P.O. Box 1, Kensington NSW 2033, Australia

Thompson, R. (207), Institute of Animal Physiology and Genetics Research, Edinburgh Research Station, Roslin, Midlothian EH25 9PS, United Kingdom

Contents

❋ ❋ ❋ ❋ ❋ ❋ ❋

Part I: General

1 Statistical Methods In Animal Improvement: Historical Overview

C.R. Henderson[1]

This review is restricted primarily to some history of the development of methods for evaluation of animals. It deals with linear models with emphasis on missing subclass problems. Some topics covered are selection index and its pioneers including Wright and Lush; analysis of variance particularly as related to variance estimation and maximum likelihood, both due to Fisher; formalized selection index of Smith and Hazel; unequal numbers analysis due greatly to Fisher, Yates and Iowa State University.

Variance and covariance estimation has advanced rapidly with the development of ML, REML, and MIVQUE. The invention of BLUP, the advances in computing, and the discovery of a rapid method for computing $\mathbf{A}^{-1}$ have resulted in rapid adoption of mixed model BLUP for evaluation of animals in many nations. Rules for optimum selection have been derived in recent years, with notable work by Bulmer, Goffinet, Fernando and Gianola.

1.1 Introduction

Acceptance of the assignment to review statistical methods in animal improvement was probably foolhardy on my part. The topic is much too large for a comprehensive review. Consequently, this paper is restricted to a subset of the entire problem. I shall be concerned only with the multivariate normal distribution and with application to prediction of breeding values and the related problems of estimation of fixed effects and covariance components. Certainly not all problems permit multivariate normality assumptions, and the questions of how to design and analyze breeding experiments and how to plan breeding programs on the farm are of equal importance.

Nevertheless, the rapid expansion of artificial insemination (AI) and field testing programs, the advances in statistics and computing, and the entry into animal breeding of many workers well trained in statistics have undoubtedly contributed remarkably to animal improvement using multivariate normal theory and employing only field data.

[1] Cornell University, Ithaca and University of Illinois, Urbana, Illinois, USA.

1.2 Pearson's Pioneering Work

Selection index and later improvements in selection methods owe much to the work of K. Pearson (1903) who derived conditional means and variances for the multivariate normal distribution. Selection index can be perceived as computation of conditional means of breeding values to use as selection criteria. Best linear unbiased prediction can be defined as using for the predictors breeding values conditional on a set of n-r linearly independent, translation invariant, linear functions of the data, where n is the number of observations and r is the rank of **X**, a matrix defined later.

To describe Pearson's results we use the standard notation of animal breeding in which the linear model is

$$\mathbf{y} = \mathbf{X}\boldsymbol{\beta}+\mathbf{Z}\mathbf{u}+\mathbf{e}, \tag{1.1}$$

where **X**, **Z** are fixed and known, $\boldsymbol{\beta}$ is fixed and unknown, and **u**, **e** are random vectors with null means and var(**u**)=**G**, var(**e**)=**R**, and cov(**u**,**e**')=**0**. Pearson's (1903) result is:

$$E(\mathbf{u}|\mathbf{y}) = \mathbf{G}\mathbf{Z}'\mathbf{V}^{-1}(\mathbf{y} -\mathbf{X}\boldsymbol{\beta}), \tag{1.2}$$

where **V**=var(**y**)=**ZGZ'**+**R**. Then it follows that

$$E(\mathbf{m}'\mathbf{u}|\mathbf{y}) = \mathbf{m}'\mathbf{G}\mathbf{Z}'\mathbf{V}^{-1}(\mathbf{y}-\mathbf{X}\boldsymbol{\beta}). \tag{1.3}$$

We show later how these results are related to the selection index.

Pearson also dealt with the problem of means and variances, conditional on some selection function. Let **w** be some vector variable correlated with [**u**, **e**, **y**]. The unconditional means of

$$\begin{bmatrix} \mathbf{y} \\ \mathbf{u} \\ \mathbf{e} \\ \mathbf{w} \end{bmatrix} \text{ are } \begin{bmatrix} \mathbf{X}\boldsymbol{\beta} \\ \mathbf{0} \\ \mathbf{0} \\ \mathbf{W} \end{bmatrix} \text{ and the unconditional covariance matrix is}$$

$$\begin{bmatrix} \mathbf{ZGZ'+R} & \mathbf{ZG} & \mathbf{R} & \mathbf{B}_y \\ \mathbf{GZ'} & \mathbf{G} & \mathbf{0} & \mathbf{B}_u \\ \mathbf{R} & \mathbf{0} & \mathbf{R} & \mathbf{B}_e \\ \mathbf{B}'_y & \mathbf{B}'_u & \mathbf{B}'_e & \mathbf{H} \end{bmatrix} \tag{1.4}$$

Now suppose that $\mathbf{w}$ is selected in such a manner that in repeated sampling its mean is $\mathbf{s}\neq\mathbf{d}$, and its variance is $\mathbf{H}_s\neq\mathbf{H}$. Then Pearson derived the conditional means and covariances of $[\mathbf{y}, \mathbf{u}, \mathbf{e}]$. The conditional means are

$$\begin{bmatrix} \mathbf{X}\boldsymbol{\beta}+\mathbf{B}_y\mathbf{H}^{-1}(\mathbf{s}-\mathbf{d}) \\ \mathbf{B}_u\mathbf{H}^{-1}(\mathbf{s}-\mathbf{d}) \\ \mathbf{B}_e\mathbf{H}^{-1}(\mathbf{s}-\mathbf{d}) \end{bmatrix}, \tag{1.5}$$

and the conditional variances are

$$\begin{bmatrix} \mathbf{ZGZ'}+\mathbf{R}-\mathbf{B}_y\mathbf{H}_o\mathbf{B}_y' & \mathbf{ZG}-\mathbf{B}_y\mathbf{H}_o\mathbf{B}_u' & \mathbf{R}-\mathbf{B}_y\mathbf{H}_o\mathbf{B}_e' \\ & \mathbf{G}-\mathbf{B}_u\mathbf{H}_o\mathbf{B}_u' & -\mathbf{B}_u\mathbf{H}_o\mathbf{B}_e' \\ \text{symmetric} & & \mathbf{R}-\mathbf{B}_e\mathbf{H}_o\mathbf{B}_e' \end{bmatrix} \tag{1.6}$$

where $\mathbf{H}_o=\mathbf{H}^{-1}(\mathbf{H}-\mathbf{H}_s)\mathbf{H}^{-1}$.

These results have been used to derive prediction methods utilizing data in which selection has occurred (Henderson 1975). Pearson's contributions to methodology used in animal breeding are enormous.

1.3 Fisher's Work of the Late Teens and the Twenties

Many of the methods used in animal breeding trace to Fisher's work of the late teens and early twenties. We need mention only his invention of maximum likelihood (1922), analysis of variance with implications for variance component estimation (1925), and his influence on the pioneering work of Brandt (1933) and of Yates (1934) on the analysis of data with unequal cell numbers. Fisher had a profound influence upon the development of statistics at Iowa State University and the founding of modern animal breeding at that same institution due to Snedecor and Lush, respectively. My first course in statistics taught by Snedecor in 1933-34 utilized the first edition of Fisher's book. Also, I found that Lush used this same book extensively. Linear model methods used in animal breeding derive very much from Fisher's influence. See for example a recent review by Herr (1986).

1.4 Wright's Work of the Teens and Twenties

It is hardly disputable that the most important single contributor to statistical methods used in animal evaluation is S. Wright through his path coefficient techniques that quantified the notions of inbreeding and genetic relationships. His **A** matrix is widely used in evaluations by best linear unbiased prediction (BLUP). A few of his publications that influenced our field are Wright (1921a,b, 1931). The great interpreter of Wright's work was J.L. Lush, who almost single-handedly introduced these concepts into animal breeding practice.

1.5 Lush and Wright - Early Prediction Methods

I am reasonably certain that Lush and Wright were the first practitioners of selection index and would be more widely recognized for this had they given it a name. The naming of a procedure seems to be an important prerequisite for recognition in science. Lush used such terms as "most probable producing ability" and "most probable breeding value" to describe his prediction techniques, and Wright simply described the techniques. Although their derivations may not have been stated completely, I have no doubt that they are based on Pearson's conditional mean in the multivariate normal distribution. Since Lush's computing procedures looked very much like those for computing standard partial regression coefficients, some animal breeders may have failed to notice the important differences in assumptions. Lush's assumptions clearly were that **y** and **u** have a multivariate normal distribution and null means. This implied that the **y**'s used in the analysis were actually $\mathbf{y}-\mathbf{X}\boldsymbol{\beta}$, with $\boldsymbol{\beta}$ presumably "known". Also, Lush routinely transformed all variables to variance=1 and, therefore, dealt with correlations rather than with variances and covariances as is now a much more common practice. This is not surprising because he drew heavily upon Wright's path coefficient methods, which used standardized variables. On a personal note, I think that I was the first student of Lush to write a Ph.D. thesis with no path coefficients. A long break between my M.S. and Ph.D. with 4 years of army service may have given me the temerity to do this.

Lush's assumption of known $\mathbf{X}\boldsymbol{\beta}$ still persists as evidenced by continued use of selection index by many animal breeders. More will be said about this deficiency of selection index when BLUP is discussed. Early examples of prediction using conditional means are Lush (1931, 1933) and Wright (1931).

1.6 Selection Index

Formalization of selection index is due to Smith (1936) and Hazel (1943). The former's work was applied to plants and the latter's to animals. Hazel probably deserves the credit for clarifying the issue of multiple trait selection. He defined genetic and environmental correlations and suggested how they could be estimated. He also emphasized the importance of weighting by economic values. Henderson (1963) gave a simple proof that economic weights could be applied to the $\hat{\mathbf{u}}$ computed for single traits, but utilizing correlated records coming from measurements on other traits either on the individual or its relatives. That is, if the total genetic merit of an individual is defined as $\mathbf{m'u}$, then the selection index prediction is $\mathbf{m}'\hat{\mathbf{u}}$, where $\hat{\mathbf{u}}$ is the selection index prediction of the individual elements of $\mathbf{u}$. This procedure has the advantage of permitting each producer to provide his own economic weights, and these certainly differ among herds, and also permits economic weights changing with time.

The motivations of both Smith and Hazel were to maximize the mean of breeding values of individuals selected by a linear index. Their proofs as well as the one taught by Lush (1948) assumed multivariate normality, known $\mathbf{X\beta}$, all candidates for selection have the same amount of information, the breeding values and records uncorrelated between any pair of animals, and selection is by truncation on the computed indexes. Users of the index usually had animals that were related and with unequal information, but apparently thought that optimality applied to that situation. The proof that this was true came about 30-40 years later. Meanwhile it was proved (Henderson 1963) that selection index maximized the probability of correct pairwise ranking in spite of correlated information and unequal information. This gave some comfort to users of the index, but this was not proof of "global" optimality. A far more serious problem was the assumption of known $\mathbf{X\beta}$. What was done in practice was to estimate this and then to treat estimated $\mathbf{X\beta}$ as the parameter value. Of course ($\mathbf{y}$-$\mathbf{X\beta}^*$, $\mathbf{u}$) does not have the same distribution as ($\mathbf{y}$-$\mathbf{X\beta}$, $\mathbf{u}$) so the selection index was not E[$\mathbf{u}$|($\mathbf{y}$-$\mathbf{X\beta}^*$)], where $\mathbf{X\beta}^*$ was some unbiased estimator of $\mathbf{X\beta}$. Further, selection index, in practice, seldom used all information. Generally, the index was constructed using a selected subset of $\mathbf{y}$ for each individual. When that is done the index is not E($\mathbf{u}$|$\mathbf{y}$) even with $\mathbf{X\beta}$ known, but rather E(u_i|y_i), where $\mathbf{y}_i$ is the subset used to predict u_i. Had my mixed model equations been generally recognized earlier than 1973, the date of publication of the Lush symposium, and had the simple method for computing $\mathbf{A}^{-1}$ been available, E($\mathbf{u}$|$\mathbf{y}$) could have been computed by solving for $\hat{\mathbf{u}}$ in these equations

$$(\mathbf{Z'R^{-1}Z+G^{-1}})\hat{\mathbf{u}} = \mathbf{Z'R^{-1}}(\mathbf{y}-\mathbf{X\beta}).$$

The properties of this $\hat{\mathbf{u}}$ as a selection criterion are discussed later. In spite of its deficiencies, selection index has been very useful in animal breeding selection problems to discourage over-emphasis upon traits with low heritabilities or with small economic values. It has also been helpful in ranking animals with markedly different amounts of information, notably in AI sire evaluations.

Cochran (1951) extended to any distribution the selection index result regarding optimality. That is, with truncation selection, use of E($\mathbf{u}$|$\mathbf{y}$) as the selection criterion maximizes the expected mean of the selected individuals. As required in the proofs of Smith,

Hazel, and Lush for the multivariate normal distribution, the candidates must have equal information and be uncorrelated.

1.7 Early Development of Linear Model Methods for Unbalanced Data

The preceding section emphasizes that application of selection index requires that $\mathbf{X\beta}$ be estimated. The $\boldsymbol{\beta}$ vector represents a host of fixed environmental effects such as ages, years, seasons, experimental treatments, inbreeding, etc. Seldom, if ever, were levels of cross-classified factors observed in a design with equal cell numbers, thus permitting estimates from simple means. Animal breeders were among some of the pioneers attempting to solve the problem of estimation of $\boldsymbol{\beta}$ in such situations, for example, Hazel and Terrill (1945) and much earlier Brandt (1933), who had a minor with Lush in his graduate program. The really definitive work was by Yates (1934) who presented least-squares solutions for two different two-way cross-classified models. These papers assumed a model, $\mathbf{y}=\mathbf{X\beta}+\mathbf{e}$, where $\text{var}(\mathbf{e})=\mathbf{I}\sigma_e^2$. Animal breeders such as Hazel and Terrill clearly understood that their models contained a random vector, **u**, but they estimated fixed effects by treating **u** as fixed in the computation. Some animal breeders still follow this practice even though the resulting estimators of estimable functions of $\boldsymbol{\beta}$ are not BLUE. Of course, these same types of computations are used in many variance component estimation methods, e.g., balanced ANOVA, unweighted means ANOVA, weighted squares of means ANOVA, and Henderson's Method 3 (1953).

Although these new methods for estimating fixed effects were an important breakthrough as compared to earlier use of raw means with obvious biases due to unequal subclass numbers, the methods were not at all good for prediction of breeding values, when the least-squares solutions to **u** were used as selection criteria. As was pointed out by Lush in an Iowa State extension publication entitled **Out on first record?**, culling cows on the basis of the mean of their lactations (essentially a least-squares method) causes too many first lactation animals to be culled. Conversely, selecting a small fraction of progeny tested sires on their progeny means selects mostly sires with few progeny. "Shrinkage" based on numbers of records or progeny corrects this deficiency.

1.8 Derivation of Best Linear Unbiased Prediction

The development of my mixed model equations for BLUP was the consequence of an attempt to combine the power of least-squares (or better, generalized least-squares, GLS) to estimate fixed effects, with the appealing features of selection index such as shrinkage, combining information from relatives, multiple trait evaluation, and optimizing the mean of selected individuals under certain restrictions. These equations were first reported at the 1949

American Dairy Science Association Meeting (Henderson 1949), and the method of derivation was reported by Henderson (1950). The proof that the solution to $\boldsymbol{\beta}$ is GLS is in Henderson et al. (1959), and that $\hat{\mathbf{u}}$ is BLUP in Henderson (1963). I attempted without much apparent success to persuade animal breeders that the method was nothing more than selection index using $\mathbf{y}\text{-}\mathbf{X}\boldsymbol{\beta}^o$ in place of $\mathbf{y}\text{-}\mathbf{X}\boldsymbol{\beta}$, and that this leads to smaller prediction error variances than selection index using any other $\mathbf{y}\text{-}\mathbf{X}\boldsymbol{\beta}^*$, where $\boldsymbol{\beta}^o$ is a GLS solution, and $\mathbf{X}\boldsymbol{\beta}^*$ is some other unbiased estimator of $\mathbf{X}\boldsymbol{\beta}$.

Of course, it should be recognized that few animal breeders were very competent in linear algebra and, particularly, matrix algebra in 1949 and, further, none had computing facilities that could deal with the large matrices required. In fact, the method was not used in AI sire evaluation at Cornell until 1970 even though we had been evaluating sires since the beginning of AI in New York in the early 1940's.

BLUP by mixed model methods is now the standard method for evaluation in dairy cattle, beef cattle, and swine in most nations.

1.9 The Development of Methods for Estimation of Variances and Covariances

Animal breeders have certainly been in the forefront of use of variance component estimates as well as in the development of methods of estimation. An excellent review of methods of estimation and, particularly, some early history has been written by Anderson (1979). The earliest methods involved equating the mean squares of balanced ANOVA to their expectations. Many of the early animal breeders knew how to do this but, unfortunately, they seldom had balanced data available. Consequently, they used various approximations including treating two-way cross-classified designs with interaction as a two-way nested design. Thus, for example, in a model with herd by sire interactions there would be an estimation of σ^2_s and $\sigma^2_{h/s}$, and then estimation of σ^2_h and $\sigma^2_{s/h}$. Lush did this routinely in the 1940's. Snedecor (1946) described in his book how to take expectations in nested models. Many animal breeders also knew how to deal with expectations in a one-way random model with unequal numbers of observations per level. Hazel and Terrill (1945) extended this idea to a mixed model with one random factor. They estimated the fixed effects by least-squares, treating the random factor as fixed, and then did a one-way ANOVA on $\mathbf{y}\text{-}\mathbf{X}\boldsymbol{\beta}^*$, where $\boldsymbol{\beta}^*$ was the OLS solution. Of course, taking expectations as though they had $\mathbf{y}\text{-}\mathbf{X}\boldsymbol{\beta}$ resulted in biased estimators, but probably not seriously biased in large data sets. The correct expectations in this case were presented as Method 2 (Henderson 1953). This paper also presented Method 1, an easy method for the strictly random model, and Method 3, a more general method than 2 for the mixed model. Methods 1 and 3 have been used very widely in animal breeding as well as other fields plagued by unequal numbers of observations in the cells. In fact, these probably are still the favorite methods. Their popularity has been due in no small measure to the availability of two computer packages, in particular, that have implemented them, namely SAS and Harvey. Also, Searle's journal publications and books have had much influence.

These and related methods for unequal numbers, are largely in the class of translation invariant, quadratic, unbiased estimators. Since it is possible to derive a host of such estimators, it is difficult to make any claims regarding their relative merits. Presumably most would choose that one, if it existed, which minimized sampling variance. This is impossible to attain except in the balanced case, in which ANOVA is the optimum in this sense.

The first general solution to the problem of minimization of sampling variances in the unequal number case was Rao's MIVQUE (1971). Minimum variance quadratic unbiased estimation is, of course, a misnomer because minimization can be obtained only if the variances and covariances are known to proportionality, which is all we really need for most animal breeding applications. If we know this, why bother to compute MIVQUE? But to be fair, if we have a "reasonably" good idea about the proportionalities we can probably obtain better estimates by MIVQUE than by computing other estimates in the class of translation invariant, quadratic, unbiased estimators.

Another approach to estimation was maximum likelihood, and the early promising work in this area was by Crump (1947), who unfortunately was unable because of health problems to continue with this research. The first general solution to the problem of maximum likelihood estimation in the unbalanced case was presented by Hartley and Rao (1967). We at Cornell (Henderson 1973) were able to obtain this result by iterating on $\hat{\mathbf{u}}$ and $\hat{\mathbf{e}}$ of the mixed model equations. It turned out that both Hartley and Rao's and our solution gave the same result as EM of Dempster et al. (1977). Patterson and Thompson (1971) presented REML, which got around the bias in ML. This bias can be severe when the number of elements in $\boldsymbol{\beta}$ is large relative to the number of observations. In recent years REML has been rather widely conceded to be the method of choice in animal breeding, provided that it can be computed. Gianola et al. (1986) provide a further rationale for using REML in this situation based on a Bayesian argument. Rapid advances in computing strategies and in development of computer hardware made computation of REML feasible in problems of moderate size. However, all of the supercomputers in the world working together would be unable to handle large problems such as the entire milk record data bank of the USDA or the multiple trait analysis of Simmental data currently being conducted by Quaas and Pollak at Cornell.

REML does have the most important property of estimating the base population parameters even though only data arising from selection are available. As pointed out in my second paper in this conference (Henderson Chap.18, this Vol.) these are the parameters that should be used in mixed model equations with data from selected animals.

1.10 Some Recent Developments in Computing Strategies

One of the important findings that enabled large mixed model solutions was the recognition that Gauss-Seidel iteration could be used. The first large-scale application of this was probably the Northeast sire evaluation. We discovered rather quickly that we could solve to suitable accuracy equations of order approximately 6000 in about 10 rounds requiring only a few minutes on a computer that now would be easily beaten by an IBM/AT or clone. With the advent of multiple trait BLUP, a logical extension was to perform block iteration. Various relaxation methods have also been tried. Very recently, Schaeffer and Kennedy (1986) proposed a method of solution that does not require setting up the mixed model equations.

A remarkable advancement in REML for multiple trait problems with no missing data has been the canonical transformation suggested by Thompson (1977). This method is used in multiple trait sire evaluation at Cornell where the traits are milk yield and composition in which we have a subset of data with such observations on all progeny.

One of the useful tricks that adds little to computing costs was the discovery of a very rapid method to compute $\mathbf{A}^{-1}$ (Henderson 1976). This has enabled the routine use of all relationships among AI sires in sire evaluation in almost all BLUP evaluation schemes. It has also encouraged the development of programs utilizing the animal model with $\mathbf{A}^{-1}$. As is mentioned in my later paper at this symposium, $\mathbf{A}^{-1}$ can be used effectively to control bias in breeding value predictions that can be caused by selection.

The implementation of the animal model on a regional or national scale results in a spectacularly large number of equations, as a minimum nt, where n is the number of animals and t the number of traits. The reduced animal model (RAM) of Quaas and Pollak (1980) is a remarkable example of the use of equivalent linear models to reduce computational labor, particularly in those species where the number of progeny that are not parents greatly exceeds the number of parents, e.g., pigs and poultry.

If evaluation of non-additive effects is desired, the computing strategy for estimating variances and for solving mixed model equations of Henderson (1985) can be used to markedly reduce computational labor. These methods apply only to non-inbred progeny and use the result of Cockerham (1954).

1.11 Recent Work in Optimum Selection Criteria

Selection index and BLUP are the two most widely used methods for computing selection criteria. Earlier work showed that selection index is optimal for selection under rather restricted conditions. These restrictions are described above in this paper. Recently Bulmer (1980), Goffinet (1983), and Fernando and Gianola (1986) have shown that if k individuals are to be selected from n candidates, selection on $E(\mathbf{u}|\mathbf{y})$ maximizes the expected value of the mean of breeding values of the selected individuals. Note that the previous restriction on the form of distribution in Cochran's result is eliminated by the simple change from truncation selection to selection of a fixed number from a fixed number of candidates. Of course, selection index is a particular case of E(**u**|**y**) for the multivariate normal distribution. This is a very nice result but unfortunately is not very useful for classical statisticians who assume a fixed and unknown $\boldsymbol{\beta}$. What do we do in this case? Goldberger (1962) and Henderson (1963) restricted the predictors to linear, translation invariant functions of **y** and, in the class of such predictors, minimized the prediction error variances. This of course is BLUP. Under multivariate normality these predictors have the same optimal properties as selection index, but only in the class of translation invariant functions.

Bayesians get around the problem of unknown $\boldsymbol{\beta}$ by declaring it to have a prior distribution. Then assuming that this is correct, E(**u**|**y**) can be obtained. However, in a classical framework this cannot be proved to be an optimum selection criterion. In fact, in the classical context, it seems obvious that we can seldom, if ever, know what is optimum. Presumably with quadratic loss functions, minimization of mean square error (MSE) of prediction is optimum. If that is true, then it must be obvious that when $\boldsymbol{\beta}$ is unknown there must exist a biased predictor that is better than BLUP. The difficulty is that with absence of knowledge about $\boldsymbol{\beta}$ we do not know what it is. Consider some practical problems. Should we modify mixed model equations to eliminate biases due to certain types of selection as described in my later paper at this symposium? In particular, should we modify for group+sire selection in sire evaluation that results in $\mathbf{L'X} \neq \mathbf{0}$, or should we use regular mixed model equations as is now done everywhere? Also, should we treat herds (or herd-year-seasons) as fixed to eliminate bias due to association between sire and herd values, or should we treat herds as random? Another case in which biased predictors probably beat BLUP is the situation when $\mathbf{X'X}$ is an ill-conditioned matrix. For example, we could have three fixed factors cross-classified with all possible interactions and with many missing subclasses. In that case BLUP could have very large prediction error variances. The predictors might be improved markedly in the MSE sense by dropping certain interactions from the model or by computing as though they are random.

As I simulated examples for my second paper at this symposium I found that by altering parameter values and/or selection intensity I could for the same incidence matrix find a biased predictor with smaller MSE than BLUP. Should we then continue to use BLUP? I rather think we should, but I am open-minded on the question. Every situation is different and there needs to be much research done before implementing a program for evaluation on a national or international scale. What are some things that can be done by research workers?

1. Characterize the population as accurately as possible.
2. Study the goals of producers. They often differ dramatically, especially in the different segments of the beef industry. Also, should we be concerned more with the producers or with the consumers?

3. To accomplish the first of these two, animal breeders need to know a great deal about the industries with which they work.
4. With adequate knowledge can we solve our thus far unsolved problems by analytical methods? We should go as far as possible in this direction, but probably simulation will be required for most problems. Rapid advances in computing make this more feasible than formerly.

To conclude, we animal breeders have our work cut out for us. Can we persuade funding agencies that we have something to offer to agriculture and society in general?

References

Anderson RD (1979) On the history of variance component estimation. In: Van Vleck LD, Searle SR (eds) Variance components and animal breeding. Cornell Univ, Ithaca, New York, pp 19-42

Brandt AE (1933) The analysis of variance in a 2xs table with disproportionate frequencies. J Am Stat Assoc 28:164-173

Bulmer MG (1980) The mathematical theory of quantitative genetics. Clarendon Press, Oxford

Cochran WG (1951) Improvement by means of selection. In: Neyman J (ed) Proc Second Berkeley Symp Math Stat and Prob, pp 449-470

Cockerham CC (1954) An extension of the concept of partitioning hereditary variance for analysis of covariances among relatives when epistasis is present. Genetics 39:859-882

Crump SL (1947) The estimation of components of variance in multiple classifications. PhD Thesis, Iowa State Univ

Dempster AP, Laird NM, Rubin DB (1977) Maximum likelihood from incomplete data via the EM algorithm. J R Stat Soc B 39:1-38

Fernando RL, Gianola D (1986) Optimal properties of the conditional mean as a selection criterion. Theor Appl Genet 72:822-825

Fisher RA (1922) On the mathematical foundations of theoretical statistics. Phil Trans R Soc Lond Ser A 222:309-368

Fisher RA (1925) Statistical methods for research workers. Oliver and Boyd, London

Gianola D, Foulley JL, Fernando RL (1986) Prediction of breeding values when variances are not known. In: Dickerson GE, Johnson RD (eds) Proc 3rd World Congr Genet Appl Livest Prod. Agric Commun, Univ Nebraska, Lincoln, Nebraska XII:356-370

Goffinet B (1983) Selection on selected records. Genet Sel Evol 15:91-97

Goldberger AS (1962) Best linear unbiased prediction in the generalized linear regression model. J Am Stat Assoc 57:369-375

Hartley HO, Rao JNK (1967) Maximum likelihood estimation for the mixed analysis of variance model. Biometrika 54:93-108

Hazel LN (1943) The genetic basis for constructing selection indexes. Genetics 28:476-490

Hazel LN, Terril CE (1945) Effects of some environmental factors on weanling traits of range Rambouillet lambs. J Anim Sci 4:331-341

Henderson CR (1949) Estimates of changes in herd environment. J Dairy Sci (Abstr) 32:706

Henderson CR (1950) Estimation of genetic parameters. Ann Math Stat (Abstr) 21:309

Henderson CR (1953) Estimation of variance and covariance components. Biometrics 9:226-252

Henderson CR (1963) Selection index and expected genetic advance. In: Statistical genetics and plant breeding. Nat Acad Sci - Nat Res Counc Publ No 982, Washington DC, pp 141-163

Henderson CR (1973) Sire evaluation and genetic trends. In: Proc Anim Breed Genet Symp in Honor of Dr. J.L. Lush. ASAS and ADSA, Champaign, Illinois, pp 10-41

Henderson CR (1975) Best linear unbiased estimation and prediction under a selection model. Biometrics 31:423-447

Henderson CR (1976) A simple method for computing the inverse of a numerator relationship matrix used in prediction of breeding values. Biometrics 32:69-83

Henderson CR (1985) MIVQUE and REML estimation of additive and nonadditive genetic variances. J Anim Sci 61:113-121

Henderson CR, Kempthorne O, Searle SR, von Krosigk CM (1959) The estimation of environmental and genetic trends from records subject to culling. Biometrics 15:192-218

Herr DG (1986) On the history of ANOVA in unbalanced, factorial designs: the first 30 years. Am Stat 40:265-270

Lush JL (1931) The number of daughters necessary to prove a sire. J Dairy Sci 14:209-220

Lush JL (1933) The bull index problem in the light of modern genetics. J Dairy Sci 16:501-522

Lush JL (1948) The genetics of populations. Animal breeding graduate students, Iowa State Univ, Ames, Iowa Mimeo, p 381

Patterson HD, Thompson R (1971) Recovery of interblock information when block sizes are unequal. Biometrika 58:545-554

Pearson K (1903) Mathematical contributions to the theory of evolution. XI. On the influence of natural selection on the variability of organs. Phil Trans of the R Soc London A 200:1-66

Quaas RL, Pollak EJ (1980) Mixed model methodology for farm and ranch beef cattle testing programs. J Anim Sci 51:1277-1287

Rao CR (1971) Minimum variance quadratic unbiased estimation of variance components. J Multiv Anal 1:445-456

Schaeffer LR, Kennedy BW (1986) Computing solutions to mixed model equations. In: Dickerson GE, Johnson RK (eds) Proc 3rd World Congr Genet Appl Livest Prod. Agric Commun, Univ Nebraska, Lincoln, Nebraska XII:382-393

Smith HF (1936) A discriminant function for plant selection. Ann Eugen 7:240-250

Snedecor GW (1946) Statistical methods. First Edition. Iowa State College Press, Ames

Thompson R (1977) Estimation of quantitative genetic parameters. In: Pollak E, Kempthorne O, Bailey TB (eds) Proc Int Conf Quant Genet. Iowa State Univ Press, Ames, pp 639-657

Wright S (1921a) Correlation and causation. J Agr Res 20:557-585

Wright S (1921b) Systems of mating. Genetics 6:111-178

Wright S (1931) On the evaluation of dairy sires. Proc Am Soc Anim Prod (Abstr) 71

Yates F (1934) The analysis of multiple classifications with unequal numbers in the different subclasses. J Am Stat Assoc 29:51-66

2 Mixed Model Methodology and the Box-Cox Theory of Transformations: A Bayesian Approach

D. Gianola[1], S. Im[2], R.L. Fernando[1], and J.L. Foulley[3]

It is often assumed in animal breeding theory that models used for data analysis are "correct" with respect to functional form and distributional assumptions. However, a transformation may be needed to achieve this. An extension of the Box-Cox theory of transformations to univariate mixed linear models is presented. The discussion includes estimation of the transformation and of the required variance components, including computing algorithms. An analysis of fixed effects and breeding values after the transformation involves the following steps: (1) estimate ratios of variance components and the transformation parameter from their joint posterior distribution; (2) conditionally on these values, integrate out the residual variance (σ_e^2) from the joint posterior distribution of fixed, random effects and σ_e^2, and (3) complete inferences using a multivariate-t distribution.

2.1 Introduction

Many programs of genetic improvement of farm animals rely on mixed linear model techniques (Henderson 1973, 1984) for assessing merit of candidates for selection. These techniques rely on best linear unbiased estimation and prediction, and do not require assumptions other than the existence of finite first and second moments (Searle 1974; Harville 1985), and known dispersion parameters. Given certain conditions (Kackar and Harville; 1981, Gianola et al. 1986), it is possible to relax this latter requirement. Mixed model methodology provides a flexible, versatile and often computable statistical tool for enhancing productivity of livestock. For example, if the data comprise subpopulations that differ in location and dispersion characteristics, a multivariate mixed model analysis can be conducted (Henderson and Quaas 1976). Production testing often occurs in a range of environments, e.g., herds, with heterogeneous genetic and residual variances, as well as covariances. In this

1 Department of Animal Sciences, University of Illinois, Urbana, Illinois, USA
2 Laboratoire de Biometrie, Institut National de la Recherche Agronomique, Castanet-Tolosan, France
3 Station de Génétique Quantitative et Appliquée, Institut National de la Recherche Agronomique, Jouy-en-Josas, France

situation, across-herd selection can be viewed as a problem of choosing candidates from several distributions. As shown by Haldane (1930) and Hill (1984) assuming normality, the proportion of individuals selected from each distribution can be markedly affected by differences in variance among such distributions. Heterogeneity of variance has been found for growth traits in pigs (Rook 1982) and milk yield in dairy cattle. Cows with extreme levels of production tend to be found in herds with large variance (Everett et al. 1982; Powell et al. 1983; Lofgren et al. 1984). If residual variance is larger at lower levels of production (Mirande and Van Vleck 1985) and this is ignored in statistical analyses used to evaluate animals, a large proportion of selected individuals will come from strata of the population having lower production. This difficulty can be overcome with a multivariate mixed model analysis in which measurements in each of the strata are treated as different traits, so a completely general covariance structure can be specified (Gianola 1986). However, this analysis has high computing requirements and all necessary dispersion parameters need to be identified. Hill (1984) described procedures for simple selection schemes but it is not clear how these generalize to more complex settings.

From an animal breeding perspective, the most appealing property of genetic evaluations based on best linear unbiased prediction is the maximization of expected merit in the selected sample provided that a fixed number of candidates is to be selected (Bulmer 1980; Goffinet 1983; Fernando and Gianola 1986); this property requires normality. In the absence of homogeneous variances and normality, it would be useful to find a metric in which the observations follow an approximately normal, linear model with constant variance. Logarithmic transformations have been employed in dairy cattle breeding following empirical considerations (Hill et al. 1983; Mirande and Van Vleck 1985; Brotherstone and Hill 1986) but the general problem of finding a transformation for data which can be reasonably described by a mixed model has not been studied.

In a classical paper, Box and Cox (1964) described a general family of transformations for producing a metric under which the classical assumptions of the fixed linear model, e.g., additivity, constancy of variance and normality, hold. The theory has been expanded in a number of directions (Gaudry and Dagenais 1979; Pericchi 1981; Spitzer 1982; Seaks and Layson 1983), including estimation of variance and covariance components by maximum likelihood in one-way balanced models (Solomon 1985). We describe an extension of the Box-Cox theory for univariate models of the type arising in animal breeding applications, using Bayesian methods. The discussion includes the estimation of the transformation and of the required variance components, with computing algorithms, and the analysis of fixed effects and breeding values after the transformation.

2.2 Motivation: A Simple Sire Evaluation Model

We consider a hypothetical example patterned after Box and Tiao (1973) but using a mixed rather than a fixed linear model. Suppose there is interest in ranking sires based on testicular traits obtained in their male progeny with the hope that these measurements are genetically correlated with reproductive efficiency in females. If this is so, indirect selection may be worthwhile. In this hypothetical scheme, testicular area is estimated using a complex procedure and used to calculate sire evaluations. Suppose further that testicular diameter, rather than area, can be reasonably described by the model

$$y_{ijk} = h_i+s_j+e_{ijk} = h_i+\varepsilon_{ijk}, \tag{2.1}$$

where y_{ijk} is a testicular measurement on son k of sire j in herd i. Let h_i be fixed, $s_j \sim NIID(0,\sigma_s^2)$ and $e_{ijk} \sim NIID(0,\sigma_e^2)$, where NIID stands for "normal, independent and identically distributed"; further, assume that all possible pairs of s_j and e_{ijk} variables are independent. In (2.1), $\varepsilon_{ijk}=s_j+e_{ijk}$, so $\sigma_\varepsilon^2=\sigma_s^2+\sigma_e^2$. Now, the sire analyst is unaware of the linear relationship holding for testicular diameter and works with the area instead, which is proportional to the square of the diameter. Thus, provided (2.1) holds:

$$y_{ijk}^2 = h_i^2+\varepsilon_{ijk}^2+2h_i\varepsilon_{ijk},$$

so

$$E(y_{ijk}^2) = h_i^2+\sigma_s^2+\sigma_e^2.$$

One can write

$$y_{ijk}^2 = E(y_{ijk}^2)+w_{ijk}, \tag{2.2}$$

where

$$w_{ijk} = \varepsilon_{ijk}^2+2h_i\varepsilon_{ijk}-\sigma_s^2-\sigma_e^2.$$

Clearly, $E(w_{ijk})=0$, and

$$Var(w_{ijk}) = Var(\varepsilon_{ijk}^2)+4h_i^2(\sigma_s^2+\sigma_e^2),$$

because the expectation of the cubic term in ε_{ijk} vanishes in the normal distribution. Also,

$$\mathrm{Var}(\varepsilon^2_{ijk}) = (\sigma^2_s+\sigma^2_e)^2\mathrm{Var}\{[\varepsilon^2_{ijk}/(\sigma^2_s+\sigma^2_e)]\}$$

$$= 2(\sigma^2_s+\sigma^2_e)^2,$$

because the variance is taken over a chi-square variate with one degree of freedom. In summary, if (2.1) holds, the consequences of working with (2.2) are the following:

1. the mean of y^2_{ijk} is a linear function of the square of the herd effect and of the variances σ^2_s and σ^2_e;
2. $\mathrm{Var}(y^2_{ijk})=2(\sigma^2_s+\sigma^2_e)[h^2_i+E(y^2_{ijk})]$ so the variance clearly depends on the mean, and
3. the residual in (2.2) cannot have a normal distribution because, as shown above, it is not a linear function of ε_{ijk}.

It should be noted that if (2.1) had additional fixed effects, e.g., years, the mean of y^2_{ijk} would contain a cross-product (joint effect) term between herd and year effects. This implies that an interaction term would appear in (2.2). This illustrates that the choice of an inappropriate metric can complicate the analysis, or potentially render it meaningless. For example, estimation of parameters and variances by maximum likelihood assuming normality would be sensible in (2.1) but not in (2.2), where the distributional assumption is manifestly violated. If the sire analyst would apply the square root transformation to (2.2), then (2.1) would hold and the "standard" analysis could be carried out. Unfortunately, the required transformation is not always known on theoretical grounds so it must be estimated using the available data and prior knowledge. As pointed out by Box and Tiao (1973), a transformation can be found when the problem arises because of an unsuitable choice of metric (in the example, testicular area rather than diameter), and this may not always be the case. Further, transformation may not eliminate simultaneously all difficulties, e.g., induce normality and remove heterogeneity of variance.

2.3 Family of Transformations

In this paper we consider transformation of the response variable only, but the theory is more general (Box and Cox 1964; Box and Tiao 1973; Judge et al. 1985). The family of power transformations can be represented as

$$y^{[\lambda]}\begin{cases}(y^{\lambda}-1)/\lambda & (\lambda \text{ different from } 0)\\ \ell n\ y & (\lambda=0)\end{cases}$$

for y>0. It can be verified that this function is continuous at $\lambda=0$. Also, it can accommodate several commonly used transformations. For example $\lambda=1/2$ gives the square root transformation, and $\lambda=-1$ yields the reciprocal transformation. It is assumed that for a certain λ, the model

$$\mathbf{y}^{[\lambda]} = \mathbf{X}\boldsymbol{\beta}+\mathbf{Z}\mathbf{u}+\mathbf{e} \tag{2.3}$$

holds such that the distribution of the vector of residuals is $N(\mathbf{0},\mathbf{I}\sigma_e^2)$. In (2.3), $\boldsymbol{\beta}$ and $\mathbf{u}$ are vectors of "fixed" and "random" effects, respectively. The probability density function of the untransformed observations $\mathbf{y}$ is

$$p(\mathbf{y}|\boldsymbol{\beta},\mathbf{u},\sigma_e^2,\lambda) = p(\mathbf{y}^{[\lambda]}|\boldsymbol{\beta},\mathbf{u},\sigma_e^2,\lambda).|J(\lambda,\mathbf{y})|, \tag{2.4}$$

where

$$|J(\lambda,\mathbf{y})| = \prod_{i=1}^{n} |\delta y_i^{[\lambda]} / \delta y_i| = \prod_{i=1}^{n} y_i^{\lambda-1}$$

is the Jacobian of the transformation. The likelihood function is then

$$p(\mathbf{y}|\boldsymbol{\beta},\mathbf{u},\sigma_e^2,\lambda) \propto \sigma_e^{-n} \exp[-(\mathbf{x}-\mathbf{X}\boldsymbol{\beta}-\mathbf{Z}\mathbf{u})'(\mathbf{x}-\mathbf{X}\boldsymbol{\beta}-\mathbf{Z}\mathbf{u})/2\sigma_e^2]|J(\lambda,\mathbf{y})|, \tag{2.5}$$

where from now on $\mathbf{x}$ and $\mathbf{y}^{[\lambda]}$ will be used interchangeably.

2.3.1 Prior Distributions

The unknown parameters are $\boldsymbol{\beta},\mathbf{u},\sigma_e^2,\lambda$ and the variance of the random effects which will be designated as σ_u^2. The joint prior density function of the parameters can be written as:

$$p(\boldsymbol{\beta},\mathbf{u},\sigma_e^2,\sigma_u^2,\lambda) = p(\boldsymbol{\beta},\mathbf{u},\sigma_e^2,\sigma_u^2|\lambda).p(\lambda),$$

since it is reasonable to expect that the magnitudes of the effects of $\boldsymbol{\beta}$ and $\mathbf{u}$ and of the variances will be affected by λ. We assume that $p(\lambda)$ is "flat" and, further, that the above joint prior density can be written as:

$$p(\boldsymbol{\beta},\mathbf{u},\sigma_e^2,\sigma_u^2,\lambda) \propto p(\mathbf{u}|\sigma_u^2).p(\boldsymbol{\beta},\sigma_u^2,\sigma_e^2|\lambda) \tag{2.6}$$

with $\boldsymbol{\beta}$ and $\mathbf{u}$ conditionally independent. The dependence of the distribution of $\mathbf{u}$ on λ is mediated through the variance σ_u^2. Using the traditional model of quantitative genetics, we take

$$\mathbf{u}|\sigma_u^2 \sim N(\mathbf{0},\mathbf{A}\sigma_u^2), \tag{2.7}$$

where $\mathbf{A}$ is a known, positive-definite, matrix. For example, when $\mathbf{u}$ is a vector of transmitting abilities of sires, the matrix $\mathbf{A}$ may contain additive genetic relationships among the sires in question. It will also be assumed in (2.6) that given λ, the two variance components are conditionally independent, a priori.

We now consider identification of the functional forms of the prior densities in (2.6). Following Box and Cox (1964), an approximate relationship between the transformed and raw observations is given by

$$y_i^{[\lambda]} \simeq a_\lambda + l_\lambda y_i, \tag{2.8}$$

where a_λ is a constant dependent on λ, and l_λ is some "representative" value of the first derivative of $y_i^{[\lambda]}$ with respect to y_i. Box and Cox (1964) take

$$l_\lambda = [\prod_{i=1}^{n} |(\delta y_i^{[\lambda]}/\delta y_i)|]^{1/n} = |J(\lambda,y)|^{1/n} = J(\lambda)^{1/n}, \tag{2.9}$$

which is the geometric mean of the absolute value of the derivatives. It should be noted in (2.9) that the l_λ chosen depends on the observations. As shown later, this creates a dependence of the prior on the observations, which has been criticized (e.g., Pericchi 1981). However, this dependence is only "mild", and this is illustrated by writing

$$\ell n l_\lambda = (1/n)\sum_i(\lambda-1)\ \ell n y_i = (\lambda-1)\ \overline{\ell n\ y}\ .$$

As n tends to ∞, $\ell n l_\lambda$ tends to $(\lambda-1)E(\ell n y)$; this shows that for large n, the dependence of l_λ on y is nil. Now, we want to find a relationship between the parameters of $\mathbf{y}^{[\lambda]]}$ and those of $\mathbf{y}$. Consider first the identity

$$\boldsymbol{\beta} = (\mathbf{X'X})^{-1}\mathbf{X'X}\boldsymbol{\beta} = (\mathbf{X'X})^{-1}\mathbf{X'}E(\mathbf{y}^{[\lambda]}),$$

where it is assumed without loss of generality that $\mathbf{X}$ has full-column rank p. Using (2.8)

$$\boldsymbol{\beta} \cong (\mathbf{X'X})^{-1}\mathbf{X'1}a_\lambda + l_\lambda\boldsymbol{\phi}, \tag{2.10}$$

where **1** is a vector of ones, and $\phi=(\mathbf{X}'\mathbf{X})^{-1}\mathbf{X}'E(\mathbf{y})$. Likewise, from (2.8)

$$Var(y_i^{[\lambda]}) \cong l_\lambda^2 Var(y_i).$$

Letting y_{ij} and $y_{ij'}$ be any two observations in level i of random factor u

$$Cov(y_{ij}^{[\lambda]},y_{ij'}^{[\lambda]}) = \sigma_u^2 \simeq l_\lambda^2 Cov(y_{ij},y_{ij'}) = l_\lambda^2\sigma_u^{2*} \tag{2.11}$$

and

$$\sigma_e^2 = Var(y_{ij}^{[\lambda]})-\sigma_u^2 \simeq l_\lambda^2[var(y_{ij})-\sigma_u^{2*}] = l_\lambda^2\sigma_e^{2*}. \tag{2.12}$$

It is assumed that prior knowledge about ϕ, σ_u^{2*} and σ_e^{2*} is vague, so using the theory of non-informative priors (Zellner 1971; Box and Tiao 1973) it is reasonable to take:

$$p(\phi,\ell n\sigma_u^{2*}, \ell n\ \sigma_e^{2*}) \propto \text{constant}$$

as prior density. Using (2.10), (2.11) and (2.12), the change of variables $\phi\rightarrow\boldsymbol{\beta}$, $\ell n\sigma_u^{2*}\rightarrow\sigma_u^2$ and $\ell n\sigma_e^{2*}\rightarrow\sigma_e^2$ yields as prior density

$$p(\boldsymbol{\beta},\sigma_u^2,\sigma_e^2|\lambda) \propto |l_\lambda|^{-p}.\sigma_u^{-2}.\sigma_e^{-2} \propto J_\lambda^{-p/n}.\sigma_u^{-2}.\sigma_e^{-2}. \tag{2.13}$$

Collecting (2.7) and (2.13) in (2.6), the joint prior distribution of all parameters is:

$$p(\boldsymbol{\beta},\mathbf{u},\sigma_u^2,\sigma_e^2,\lambda) \propto J_\lambda^{-p/n}.(\sigma_u^2)^{-(q+2)/2}.\sigma_e^{-2}.\exp[-1/2\mathbf{u}'\mathbf{A}^{-1}\mathbf{u}/\sigma_u^2], \tag{2.14}$$

because it is assumed that $p(\lambda)$ is uniform over the real line.

2.4 Some Posterior Distributions

2.4.1 Joint Posterior Distribution of all Parameters

The joint posterior distribution can be written as

$$p(\boldsymbol{\beta},\mathbf{u},\sigma_u^2,\sigma_e^2,\lambda|\mathbf{y}) \propto p(\mathbf{y}|\boldsymbol{\beta},\mathbf{u},\sigma_e^2,\lambda).p(\boldsymbol{\beta},\mathbf{u},\sigma_u^2,\sigma_e^2,\lambda).$$

Collecting now (2.5) and (2.14), one obtains

$$p(\boldsymbol{\beta},\mathbf{u},\sigma_u^2,\sigma_e^2,\lambda|\mathbf{y}) \propto (\sigma_e^2)^{-(n+2)/2}.\exp[-1/2(\mathbf{x}-\mathbf{X}\boldsymbol{\beta}-\mathbf{Z}\mathbf{u})'(\mathbf{x}-\mathbf{X}\boldsymbol{\beta}-\mathbf{Z}\mathbf{u})/\sigma_e^2]$$

$$.(\sigma_u^2)^{-(q+2)/2}.\exp[-1/2\mathbf{u}'\mathbf{A}^{-1}\mathbf{u}/\sigma_u^2].J_\lambda^{(n-p)/n} \qquad (2.15)$$

with

$-\infty<\beta_i<\infty$, $i=1, ..., p$

$-\infty<u_j<\infty$, $j=1, ..., q$

$\sigma_u^2>0$

$\sigma_e^2>0$

$-\infty<\lambda<\infty$.

In (2.15), the power to which J_λ is raised stems from (2.14) and from the fact that $|J(\lambda,\mathbf{y})|$ in (2.5) is also J_λ.

Let $\mathbf{W}=[\mathbf{X}\ \mathbf{Z}]$, $\theta'=[\boldsymbol{\beta}'\mathbf{u}']$ and define $\hat{\theta}$ such that $(\mathbf{W}'\mathbf{W}+\Sigma)\hat{\theta}=\mathbf{W}'\mathbf{x}$ where

$$\Sigma = \begin{bmatrix} 0 & 0 \\ 0 & \mathbf{A}^{-1}\sigma_e^2/\sigma_u^2 \end{bmatrix}$$

so $\hat{\theta}$ is a function of λ and of the ratio of variance components. Then we have the equality

$$(\mathbf{x}-\mathbf{X}\boldsymbol{\beta}-\mathbf{Z}\mathbf{u})'(\mathbf{x}-\mathbf{X}\boldsymbol{\beta}-\mathbf{Z}\mathbf{u})+\mathbf{u}'\mathbf{A}^{-1}\mathbf{u}\sigma_e^2/\sigma_u^2$$

$$= (\mathbf{x}-\mathbf{W}\theta)'(\mathbf{x}-\mathbf{W}\theta)+\theta'\Sigma\theta$$

$$= \mathbf{x}'\mathbf{x}-2\theta'\mathbf{W}'\mathbf{x}+\theta'(\mathbf{W}'\mathbf{W}+\Sigma)\theta$$

$$= \mathbf{x}'\mathbf{x}-2\theta'(\mathbf{W}'\mathbf{W}+\Sigma)\hat{\theta}+\theta'(\mathbf{W}'\mathbf{W}+\Sigma)\theta$$

$$= \mathbf{x}'\mathbf{x}-\hat{\theta}'(\mathbf{W}'\mathbf{W}+\Sigma)\hat{\theta}+(\theta-\hat{\theta})'(\mathbf{W}'\mathbf{W}+\Sigma)(\theta-\hat{\theta}). \qquad (2.16)$$

Note that $\mathbf{x}'\mathbf{x}-\hat{\theta}'(\mathbf{W}'\mathbf{W}+\Sigma)\hat{\theta}=\mathbf{x}'\mathbf{x}-\hat{\theta}'\mathbf{W}'\mathbf{x}$ in (2.16) can be interpreted as a mixed model residual sum of squares. Using (2.16) in (2.15), the joint posterior distribution becomes:

$$p(\boldsymbol{\beta},\mathbf{u},\sigma_u^2,\sigma_e^2,\lambda|\mathbf{y}) \propto (\sigma_e^2)^{-(n+2)/2}.(\sigma_u^2)^{-(q+2)/2}.J_\lambda{}^{(n-p)/n}$$

$$.\exp[-1/2(\mathbf{x}'\mathbf{x}-\hat{\theta}'\mathbf{W}'\mathbf{x})/\sigma_e^2]$$

$$.\exp[-1/2(\theta-\hat{\theta})'(\mathbf{W}'\mathbf{W}+\Sigma)(\theta-\hat{\theta})/\sigma_e^2]. \tag{2.17}$$

2.4.2 Posterior Distribution of the Variance Components and of λ

The marginal posterior distribution of σ_u^2,σ_e^2 and λ is obtained integrating θ out of (2.17). Observe that the only part of (2.17) that is a function of θ is the second exponential expression. This is the kernel of the distribution $\theta|\sigma_u^2,\sigma_e^2,\lambda,\mathbf{x}\sim N[\hat{\theta},(\mathbf{W}'\mathbf{W}+\Sigma)^{-1}\sigma_e^2]$, of order p+q, and non-singular because **X** has full-column rank. Also

$$\int_{R_\theta} \exp[-1/2(\theta-\hat{\theta})'(\mathbf{W}'\mathbf{W}+\Sigma)(\theta-\hat{\theta})/\sigma_e^2]\, d\theta=(2\pi.\sigma_e^2)^{(p+q)/2}|\mathbf{W}'\mathbf{W}+\Sigma|^{-1/2}$$

using properties of the normal distribution. Then

$$p(\sigma_e^2,\sigma_u^2,\lambda|\mathbf{y}) \propto (\sigma_e^2)^{-(n+2-p-q)/2}.(\sigma_u^2)^{-(q+2)/2}.|\mathbf{W}'\mathbf{W}+\Sigma|^{-1/2}$$

$$.\exp[-1/2(\mathbf{x}'\mathbf{x}-\hat{\theta}'\mathbf{W}'\mathbf{x})/\sigma_e^2].J_\lambda{}^{(n-p)/n} \tag{2.18}$$

gives the marginal posterior distribution of the variance components and of λ. Some simplification can be obtained using results in Searle (1979):

$$|\mathbf{W}'\mathbf{W}+\Sigma| = |\mathbf{X}'\mathbf{X}|.|\mathbf{Z}'\mathbf{M}\mathbf{Z}+\mathbf{A}^{-1}\alpha|,$$

where $\alpha=\sigma_e^2/\sigma_u^2$ and $\mathbf{M}=\mathbf{I}-\mathbf{X}(\mathbf{X}'\mathbf{X})^{-1}\mathbf{X}'$. Now, **A** is a positive-definite matrix so it can be decomposed as **A**=**LL**' with **L** non-singular. Thus:

$$|\mathbf{Z}'\mathbf{M}\mathbf{Z}+\mathbf{A}^{-1}\alpha| = |(\mathbf{L}')^{-1}(\mathbf{L}'\mathbf{Z}'\mathbf{M}\mathbf{Z}\mathbf{L}+\mathbf{I}\alpha)\mathbf{L}^{-1}|$$

$$= |(\mathbf{L})^{-1}|^2|\mathbf{P}+\mathbf{I}\alpha|,$$

where **P**=**L'Z'MZL**; note that **P**=**Z'MZ** when **A**=**I**. Now, **P** can be written as **P**=**U'DU** where $\mathbf{D}=\{\gamma_i\}$ is a diagonal matrix containing the eigenvalues of **P**, and **U** is a qxq matrix of eigenvectors. Because **U** is orthogonal **U'U**=**UU'**=**I** and so $\mathbf{U}^{-1}(\mathbf{U}')^{-1}=(\mathbf{U}')^{-1}\mathbf{U}^{-1}=\mathbf{I}$. Hence

$$|\mathbf{Z'MZ}+\mathbf{A}^{-1}\alpha| = |\mathbf{L}|^{-2}|\mathbf{D}+\mathbf{I}\alpha|,$$

and the only part of this determinant that depends on σ_e^2 and σ_u^2 is:

$$|\mathbf{D}+\mathbf{I}\alpha| = \prod_{i=1}^{q} (\gamma_i+\alpha).$$

Thus (2.18) becomes

$$p(\sigma_e^2,\sigma_u^2,\lambda|\mathbf{y}) \propto (\sigma_e^2)^{-(n+2-p-q)/2}(\sigma_u^2)^{-(q+2)/2}$$

$$.\exp[-1/2(\mathbf{x'x}-\hat{\theta}'\mathbf{W'x})/\sigma_e^2].J_\lambda^{(n-p)/n}.[\prod_{i=1}^{q} (\gamma_i+\alpha)]^{-1/2} . \qquad (2.19)$$

2.4.3 Posterior Distribution of Functions of the Variance Ratio and of λ

In some animal breeding applications, interest may be in making inferences on the variance ratio or functions thereof, rather than on the variance components themselves. The variance component σ_u^2 is related to additive genetic variance (V_A) as $\sigma_u^2=rV_A$, where r is the additive relationship among members of a family. For example, if $\mathbf{u}$ is a vector of sire transmitting abilities, then $\sigma_u^2=1/4V_A$ and $\sigma_e^2=V_P-1/4V_A$, where V_P is phenotypic variance. Hence, $\alpha= 4h^{-2}-1$, with $h^2=V_A/V_P$ being heritability in the narrow sense. In general, $\alpha=r^{-1}h^{-2}-1$.

We consider the marginal posterior distribution of α and λ first. In order to integrate σ_e^2 out of (2.19) we make the change of variables $\sigma_e^2/\sigma_u^2=\alpha$, with σ_e^2 and λ remaining as before. The determinant of the Jacobian of this transformation is σ_e^2/α^2. The joint posterior distribution of σ_e^2, α and λ is then:

$$p(\sigma_e^2,\alpha,\lambda|\mathbf{y}) \propto (\sigma_e^2)^{-(n+2-p)/2}.\exp[-1/2\,(\mathbf{x'x}-\hat{\theta}'\mathbf{W'x})/\sigma_e^2]$$

$$.J^{(n-p)/n}.\alpha^{(q-2)/2}.[\prod_{i=1}^{q} (\gamma_i+\alpha)]^{-1/2}. \qquad (2.20)$$

Using formula A.2.1.6 in Box and Tiao (1973)

$$\int (\sigma_e^2)^{-(n-p+2)/2}\exp(-S_{\alpha,\lambda}/2\sigma_e^2)d\sigma_e^2 = S_{\alpha,\lambda}^{-(n-p)/2}\Gamma[(n-p)/2],$$

where $S_{\alpha,\lambda}$ is the mixed model residual sum of squares, which is clearly a function of α and of λ, and $\Gamma(.)$ is the gamma function. Using this, the posterior distribution of α and λ is:

$$p(\alpha,\lambda|y) \propto \alpha^{(q-2)/2}[\prod_{i=1}^{q}(\gamma_i+\alpha)]^{-1/2}.[J_\lambda{}^{2/n}/S_{\alpha,\lambda}]^{(n-p)/2} \tag{2.21}$$

$$(1-r)/r<\alpha<\infty$$

$$-\infty<\lambda<\infty.$$

In a "sire" model, the range of α is between 3 and ∞; in applications outside genetics, $0<\alpha<\infty$. When λ is known, (2.21) gives the exact posterior density (apart from a constant) of the variance ratio for any mixed model with two variance components. This density can be simplified letting $\mathbf{z}=\mathbf{x}/(J_\lambda)^{1/n}$, in which case:

$$S_{\alpha,\lambda} = J_\lambda^{2/n}(\mathbf{z}'\mathbf{z}-\hat{\theta}_z'\mathbf{W}'\mathbf{z}) = J_\lambda^{2/n}S^*_{\alpha,\lambda},$$

where

$$\hat{\theta}_z = (\mathbf{W}'\mathbf{W}+\Sigma)^{-1}\mathbf{W}'\mathbf{z}.$$

Thus,

$$p(\alpha,\lambda|y) \propto \alpha^{(q-2)/2}[\prod_{i=1}^{q}(\gamma_i+\alpha)]^{-1/2}S^{*-(n-p)/2}_{\alpha,\lambda}. \tag{2.22}$$

The posterior distribution of heritability and of λ can be obtained making the change of variables $(\lambda,\alpha)\rightarrow(\lambda,h^2)$ with $\alpha=r^{-1}h^{-2}-1$; the determinant of the Jacobian of the transformation is $r^{-1}h^{-4}$. Then:

$$p(h^2,\lambda|y) \propto \frac{(1-rh^2)^{(q-2)/2}.S^{*2-(n-p)/2}_{h,\lambda}}{(h^2)^{(q+2)/2}\{\prod[\gamma_i+(1-rh^2)/rh^2]\}^{1/2}} \tag{2.23}$$

$$0<h^2<1$$

$$-\infty<\lambda<\infty.$$

2.4.4 Posterior Distribution of λ

The exact posterior distribution of λ is obtained by integrating α out of (2.22) so

$$p(\lambda|y) \propto \int \frac{\alpha^{(q-2)/2}}{[\prod_i^q (\gamma_i+\alpha)]^{1/2} S^{*\,(n-p)/2}_{\alpha,\lambda}} d\alpha . \tag{2.24}$$

The range of integration is indicated in (2.21). We have not been able to solve this integral analytically but it can be evaluated using numerical integration. Computations are easier using a parameterization in α^{-1} because its range is between 0 and 3. The main difficulties in the computation of the integral arise in calculating the eigenvalues of the matrix **P**, or of **Z'MZ** when the elements of **u** are not correlated. This calculation is at least as complex as the inversion of a qxq matrix so when q is large, finding these eigenvalues may be difficult or impossible. Also, repeated calculation of $S_{\lambda,\alpha}$ at different values of α needs to be done in the course of integration; fortunately, this sum of squares can be calculated iteratively, without recourse to matrix inversion techniques. Finally, note that if α were known the posterior distribution of λ would reduce to:

$$p(\lambda|\alpha,y) \propto [S^{*}_{\lambda,\alpha}]^{-(n-p)/2}. \tag{2.25}$$

This state of knowledge corresponds to the rather unrealistic situation where the investigator knows α but not λ!

2.5 Estimation of the Transformation

2.5.1 From the Marginal Distribution of λ

The distributions (2.24) and (2.25) describe the plausibility of the values of λ for the situations where α is unknown or known, respectively. Consider the more general case where α is not known. The density (2.24) can be plotted against λ and this would give a full description of the uncertainty the investigator has with respect to λ, posterior to data. In practice, however, interest may be in one or just a few values of λ that have a high posterior density, e.g., the mode of (2.24). Finding this mode must be done numerically as the equations that need to be solved are not tractable analytically and do not seem to shed further light into the problem. Because of this, we consider in subsequent sections estimation of λ from joint distributions.

When α is known, (2.25) is clearly maximum when $S^*_{\lambda,\alpha}$ is minimum. Equivalently, this involves finding the value of λ that minimizes the residual mixed model sum of squares

$$S^*_{\lambda,\alpha} = \mathbf{z}'\mathbf{z}-\mathbf{z}'\mathbf{W}\hat{\theta}_Z.$$

Because $\hat{\theta}_Z$ can be found iteratively, this series of computations may be feasible, although tedious. Alternatively, an approximation can be obtained expanding $S^*_{\lambda,\alpha}$ with a Taylor series about an arbitrary λ^*, and retaining the constant and linear and quadratic terms on λ. This leads to an expression of the form $a+b\lambda+c\lambda^2$ which can be dealt with analytically. The goodness of the approximation depends, of course, on the closeness between λ^* and the value which minimizes $S^*_{\lambda,\alpha}$. In general, a situation with α known and λ unknown is unlikely to be encountered.

2.5.2 From the Joint Distribution of λ and α

We consider finding the α and λ components of the mode of (2.22). The logarithm of this expression is

$$\ell n\ p(\alpha,\lambda|\mathbf{y}) = \text{Cnst.}+1/2(q-2)\ell n\ \alpha-1/2 \sum_{i=1}^{q} \ell n(\gamma_i+\alpha)-1/2(n-p)\ \ell n\ S^*_{\alpha,\lambda}.$$

Given α, this expression is maximum with respect to λ when $S^*_{\alpha,\lambda}$ is minimum. Also

$$\frac{\delta}{\delta\alpha}\ell n\ p(\alpha,\lambda|\mathbf{y}) = 1/2(q-2)/\alpha-1/2 \sum_{i=1}^{q} (\gamma_i+\alpha)^{-1}-1/2(n-p)\frac{\delta}{\delta\alpha}S^*_{\alpha,\lambda}/S^*_{\alpha,\lambda}. \qquad (2.26)$$

Now $S^*_{\alpha,\lambda}=\mathbf{z}'\mathbf{z}-\mathbf{z}'\mathbf{W}(\mathbf{W}'\mathbf{W}+\Sigma)^{-1}\mathbf{W}'\mathbf{z}$ so

$$\frac{\delta}{\delta\alpha}S^*_{\alpha,\lambda} = -\mathbf{z}'\mathbf{W}\left[\frac{\delta}{\delta\alpha}(\mathbf{W}'\mathbf{W}+\Sigma)^{-1}\right]\mathbf{W}'\mathbf{z}.$$

Using results for differentiation of inverses (Searle 1979) and letting $C_{\beta u}$, and C_{uu} be appropriate partitions of $(\mathbf{W}'\mathbf{W}+\Sigma)^{-1}$, one can write

$$\frac{\delta}{\delta\alpha} S^*_{\alpha,\lambda} = z'(XC_{\beta u}+ZC_{uu})A^{-1}\hat{u}_z = v(\alpha,\lambda),$$

where $\hat{u}_z$ is the **u**-component of $\hat{\theta}_z$. Using this in (2.26) and setting to 0 gives the functional iteration:

$$\alpha^{[k+1,m]} = \{[\sum_{i=1}^{q} (\gamma_i+\alpha)^{-1}+(n-p)v(\alpha,\lambda)/S^*_{\alpha,\lambda}]^{[k,m]}\}(q-2). \tag{2.27}$$

The scheme starts using a trial value of α (α^o) and finding the value of λ that minimizes $S^*_{\alpha,\lambda}$, say λ^o. Given α^o and λ^o, iteration proceeds with (2.27) to obtain a new value of α, and so on. The calculations stop when changes in α and λ are negligible. We have not experimented with this procedure yet. However, it is possible to develop alternative algorithms, e.g., Newton-Raphson, if the above scheme proves to be unsatisfactory. The main computational difficulty is the evaluation of $v(\alpha,\lambda)$, which requires matrix inversion; partitioning and absorption techniques, as well as diagonalization may be useful here. Also, the proposed algorithm is based on first differentials only so it is probably slow to converge. If convergence is to a global maximum, the values of α and λ so obtained give the mode of (2.22).

2.5.3 From the Joint Distribution of σ^2_e,σ^2_u and λ

We now consider maximization of (2.19) jointly with respect to the variance components and the transformation parameter λ. Gianola et al. (1986) and Foulley et al. (1987) introduced a very useful result which is reproduced here because it will be employed in later developments. Let θ_1 and θ_2 be parameter vectors and suppose we wish to find the mode of the marginal posterior density $f(\theta_1|y)$ or, equivalently, of $\ell n\, f(\theta_1|y)$. Then

$$\frac{\delta}{\delta\theta_1} \ell n\, f(\theta_1|y) = [f(\theta_1|y)]^{-1}.\frac{\delta}{\delta\theta_1}\int f(\theta_1,\theta_2|y)\, d\theta_2$$

$$= [f(\theta_1|y)]^{-1}.\int\frac{\delta}{\delta\theta_1} f(\theta_1,\theta_2|y)\, d\theta_2$$

$$= [f(\theta_1|y)]^{-1}.\int[\frac{\delta}{\delta\theta_1}\ell n\, f(\theta_1,\theta_2|y)].f(\theta_1,\theta_2|y)\, d\theta_2$$

$$= \int [\frac{\delta}{\delta\theta_1}\ell n\, f(\theta_1,\theta_2|y)].f(\theta_1|\theta_2,y)\, d\theta_2$$

$$= E[\frac{\delta}{\delta\theta_1} \ell n\ f(\theta_1,\theta_2|\mathbf{y})], \tag{2.28}$$

where the expectation is taken with respect to the distribution $\theta_2|\theta_1,\mathbf{y}$.

Put $\theta_1'=[\sigma_u^2,\sigma_e^2,\lambda]$ and $\theta_2'=[\boldsymbol{\beta}',\mathbf{u}']$. The joint posterior density of $\boldsymbol{\beta},\mathbf{u},\sigma_u^2,\sigma_e^2$ and λ is given in (2.15) and (2.17). The distribution $\boldsymbol{\beta},\mathbf{u}|\sigma_u^2,\sigma_e^2,\lambda$ can be obtained from (2.17) regarding σ_u^2,σ_e^2, and λ as constants. This is a normal distribution with mean $\hat{\theta}$, and covariance matrix $(\mathbf{W}'\mathbf{W}+\Sigma)^{-1}\sigma_e^2$, as stated earlier. Differentiating the log of (2.15) with respect to σ_u^2 and σ_e^2 yields

$$\frac{\delta}{\delta\sigma_u^2} \ell n\ p(\boldsymbol{\beta},\mathbf{u},\sigma_u^2,\sigma_e^2,\lambda|\mathbf{y}) = -1/2(q+2)/\sigma_u^2+1/2\mathbf{u}'\mathbf{A}^{-1}\mathbf{u}/\sigma_u^4 \tag{2.29}$$

and,

$$\frac{\delta}{\delta\sigma_e^2} \ell n\ p(\boldsymbol{\beta},\mathbf{u},\sigma_u^2,\sigma_e^2,\lambda|\mathbf{y}) = -1/2(n+2)/\sigma_e^2+1/2(\mathbf{x}-\mathbf{X}\boldsymbol{\beta}-\mathbf{Zu})'(\mathbf{x}-\mathbf{X}\boldsymbol{\beta}-\mathbf{Zu})/\sigma_e^4. \tag{2.30}$$

Likewise, differentiation with respect to λ gives

$$\frac{\delta}{\delta\lambda} \ell n\ p(\boldsymbol{\beta},\mathbf{u},\sigma_u^2,\sigma_e^2,\lambda|\mathbf{y}) = \frac{(n-p)}{n}\frac{\delta}{\delta\lambda} \ell n\ J_\lambda - \frac{(\delta\mathbf{x})'}{\delta\lambda}(\mathbf{x}-\mathbf{X}\boldsymbol{\beta}-\mathbf{Zu})/\sigma_e^2.$$

Now

$$\frac{\delta}{\delta\lambda} \ell n\ J_\lambda = \frac{\delta}{\delta\lambda}(\lambda-1)\sum_{i=1}^{n} \ell n\ y_i = \sum_{i=1}^{n} \ell n\ y_i$$

and

$$\frac{\delta\mathbf{x}}{\delta\lambda} = \{\frac{\delta}{\delta\lambda}[(y_i^\lambda-1)/\lambda]\} = \{[1+y_i^\lambda(\lambda \ell n\ y_i-1)]/\lambda^2\} = \mathbf{x}^*.$$

Hence

$$\frac{\delta}{\delta\lambda} \ell n\ p(\boldsymbol{\beta},\mathbf{u},\sigma_u^2,\sigma_e^2,\lambda|\mathbf{y}) = (n-p)\ \overline{\ell n\ y} - \mathbf{x}^{*\prime}(\mathbf{x}-\mathbf{X}\boldsymbol{\beta}-\mathbf{Zu})/\sigma_e^2. \tag{2.31}$$

Taking expectations of (2.29), (2.30) and (2.31) as indicated in (2.28) and setting to 0 gives the iterative scheme:

$$\sigma_u^{2[k+1,m+1]} = [\hat{\mathbf{u}}'\mathbf{A}^{-1}\hat{\mathbf{u}}+\mathrm{tr}\mathbf{A}^{-1}\mathbf{C}_{uu}\sigma_e^2]^{[k,m+1]}/(q+2) \tag{2.32}$$

$$\sigma_e^{2[k+1,m+1]} = [\hat{\mathbf{e}}'\hat{\mathbf{e}}+(p+q-\mathrm{tr}\mathbf{A}^{-1}\mathbf{C}_{uu})\sigma_e^2]^{[k,m+1]}/(n+2), \tag{2.33}$$

where $\hat{\mathbf{e}}=\mathbf{x}-\mathbf{W}\hat{\theta}$ and

$$(n-p)\,\overline{\ell n\ y}-[\mathbf{x}^{*\prime}(\mathbf{x}-\mathbf{X}\boldsymbol{\beta}-\mathbf{Z}\hat{\mathbf{u}})]^{[k,m]} = 0. \tag{2.34}$$

Iteration starts using guessed values of the variance components and of λ, and solving (2.34) numerically for λ. Then (2.32) and (2.33) are iterated upon using the "new" λ and the current value of the variance components. If convergence is to a global maximum, the values so obtained are the components of the mode of (2.19). It should be noted that (2.32) and (2.33) are expressions arising in the expectation-maximization algorithm of Dempster et al. (1977), assuming λ known. Also, advantage should be taken of the fact that

$$\mathbf{C}_{uu} = (\mathbf{Z}'\mathbf{M}\mathbf{Z}+\mathbf{A}^{-1}\sigma_e^2/\sigma_u^2)^{-1} = \mathbf{L}\mathbf{U}^{-1}\,(\mathbf{D}+\mathbf{I}\sigma_e^2/\sigma_u^2)^{-1}\mathbf{U}\mathbf{L}',$$

where the matrices **L**, U, and **D** are as defined earlier. Note that the only part of the inverse that depends on α is $(\mathbf{D}+\mathbf{I}\sigma_e^2/\sigma_u^2)$, which is diagonal.

The expectation-maximization algorithm is notoriously slow to converge (Thompson 1979), especially when applied to non-linear models (Hoschele et al. 1987). Because second-order iterative schemes can accelerate convergence considerably, we describe a Newton-Raphson procedure for maximizing (2.19). From the expression preceding (2.28), one can obtain by differentiation

$$\frac{\delta^2}{\delta\theta_1\delta\theta_1'}\,\ell n\ f(\theta_1|\mathbf{y}) = E[\frac{\delta^2}{\theta_1\delta\theta_1'}\,\ell n\ p(\boldsymbol{\beta},\mathbf{u},\sigma_u^2,\sigma_e^2,\lambda|\mathbf{y})]$$

$$+E[\frac{\delta}{\delta\theta_1}\,\ell n\ p(\boldsymbol{\beta},\mathbf{u},\sigma_u^2,\sigma_e^2,\lambda|\mathbf{y})$$

$$\cdot\frac{\delta}{\delta\theta_1'}\,\ell n\ p(\boldsymbol{\beta},\mathbf{u}|\sigma_u^2,\sigma_e^2,\lambda,\mathbf{y})]. \tag{2.35}$$

Evaluating these derivatives and their expectations is relatively easy under normality assumptions. With the first derivatives given in (2.29), (2.30) and (2.31), the second derivatives can be calculated readily and their expectations evaluated. Taking the expectation of the second term in (2.35) is more involved because results for the variance of quadratic forms under normality need to be invoked. With all derivatives and expectations evaluated, the Newton-Raphson algorithm (Dahlquist and Bjorck 1974) can be constructed.

2.6 Analysis of the Effects After Transformation

We have addressed so far the situation where finding the transformation (marginally or in conjunction with all or some of the dispersion parameters) is the focal point of the analysis. However, in animal breeding, the main interest is in making inferences or decisions about the breeding values **u**, or about linear functions of $\boldsymbol{\beta}$ and **u**. In this case, the dispersion parameters and λ should be viewed as "nuisances". The Bayesian approach (Zellner 1971; Box and Tiao 1973; Gianola and Fernando 1986) provides an automatic mechanism to eliminate nuisance parameters via integration from the appropriate posterior density. If the integration is done exactly, the degree of uncertainty about these nuisances posterior to the data is taken into account. If the integration is approximate, the uncertainty is accommodated only partially; the goodness of the approximation depends, of course, on the amount of information available. We now consider two approaches to inference about $\theta'=[\boldsymbol{\beta},\mathbf{u}']$.

2.6.1 Analysis Conditional on α and λ

We wish to obtain the marginal posterior distribution of θ. This is given by

$$p(\theta|\mathbf{y}) = \int_0^\infty \int_{R_\alpha} \int_{R_\lambda} p(\boldsymbol{\beta},\mathbf{u},\sigma_e^2,\alpha,\lambda|\mathbf{y})\, d\lambda\, d\alpha\, d\sigma_e^2$$

$$= \int_0^\infty \Big[\int_{R_\alpha} \int_{R_\lambda} p(\boldsymbol{\beta},\mathbf{u},\sigma_e^2|\alpha,\lambda,\mathbf{y}).p(\alpha,\lambda|\mathbf{y})\, d\lambda\, d\alpha \Big]\, d\sigma_e^2. \tag{2.36}$$

Now, the term inside brackets is the expected value of the conditional density $p(\boldsymbol{\beta},\mathbf{u},\sigma_e^2|\alpha,\lambda,\mathbf{y})$ taken with respect to the distribution of α and λ given **y**; this distribution and how to calculate its modal value were given in previous sections. Now, we can expand the conditional density at these modal values $(\hat{\alpha}, \hat{\lambda})$:

$$p(\boldsymbol{\beta},\mathbf{u},\sigma_e^2|\alpha,\lambda,\mathbf{y}) \simeq p(\boldsymbol{\beta},\mathbf{u},\sigma_e^2|\hat{\alpha},\hat{\lambda},\mathbf{y})+\nabla'[\alpha-\hat{\alpha},\lambda-\hat{\lambda}]'+ \ldots,$$

where ∇ is the vector of first derivatives of the conditional density with respect to α and λ, evaluated at $\hat{\alpha}$ and $\hat{\lambda}$. Taking expectations with respect to $\alpha,\lambda|\mathbf{y}$, we see that the linear term of the expansion vanishes to an approximation when $E(\alpha)\simeq\hat{\alpha}$ and $E(\lambda)\simeq\hat{\lambda}$. This occurs when the density $p(\alpha,\lambda|\mathbf{y})$ is symmetric or approximately so, e.g., posterior distributions tend towards normality as sample size increases (Zellner 1971). Using this approximation in (2.36)

$$p(\theta|\mathbf{y}) \simeq \int_0^\infty p(\boldsymbol{\beta},\mathbf{u},\sigma_e^2|\alpha=\hat{\alpha},\lambda=\hat{\lambda},\mathbf{y})\, d\sigma_e^2. \tag{2.37}$$

Now the joint density $p(\boldsymbol{\beta},\mathbf{u},\sigma_e^2|\alpha,\lambda,\mathbf{y})$ can be obtained from (2.17) making the change of variables $U=\sigma_e^2$, $V=\alpha$ and then regarding the resulting expression as a function of $\boldsymbol{\beta},\mathbf{u}$, and σ_e^2, i.e., with α and λ viewed as constants. With the determinant of the Jacobian of the transformation being σ_e^2/α^2, one obtains:

$$p(\boldsymbol{\beta},\mathbf{u},\sigma_e^2|\alpha,\lambda,\mathbf{y}) \propto (\sigma_e^2)^{-(n+q+2)/2}\exp\{-1/2[(\mathbf{x}'\mathbf{x}-\hat{\theta}'\mathbf{W}'\mathbf{x})$$

$$+(\theta-\hat{\theta})'(\mathbf{W}'\mathbf{W}+\Sigma)(\theta-\hat{\theta})]/\sigma_e^2\}. \tag{2.38}$$

Integrating σ_e^2 out of (2.38) as indicated in (2.37):

$$p(\boldsymbol{\beta},\mathbf{u}|\alpha,\lambda,\mathbf{y}) \propto [\mathbf{x}'\mathbf{x}-\hat{\theta}'\mathbf{W}'\mathbf{x}+(\theta-\hat{\theta})'(\mathbf{W}'\mathbf{W}+\Sigma(\theta-\hat{\theta})]^{-(n+q)/2}. \tag{2.39}$$

Now, the mixed model residual sum of squares depends only on α and on the data, so it can be factored out of (2.39) to yield

$$p(\boldsymbol{\beta},\mathbf{u}|\alpha,\lambda,\mathbf{y}) \propto \left[1+\frac{(\theta-\hat{\theta})'(\mathbf{W}'\mathbf{W}+\Sigma)(\theta-\hat{\theta})}{(n-p)\hat{\sigma}_e^2}\right]^{-(n-p+p+q)/2}, \tag{2.40}$$

where $\hat{\sigma}_e^2$ is the mixed model residual sum of squares divided by n-p. The expression in (2.40) is the kernel of a multivariate t-distribution with mean vector $\hat{\theta}$, covariance matrix

$$(\mathbf{W}'\mathbf{W}+\Sigma)^{-1}\sigma_e^2(n-p)/(n-p-2)$$

and n-p degrees of freedom (Box and Tiao 1973). Thus, any linear combination $\mathbf{K}'\hat{\boldsymbol{\beta}}+\mathbf{M}'\hat{\mathbf{u}}$ has a t-posterior distribution with mean vector $\mathbf{K}'\hat{\boldsymbol{\beta}}+\mathbf{M}'\hat{\mathbf{u}}'$ and covariance matrix:

$$[\mathbf{K}'\mathbf{M}]\begin{bmatrix}\mathbf{X}'\mathbf{X} & \mathbf{X}'\mathbf{Z} \\ \mathbf{Z}'\mathbf{X} & \mathbf{Z}'\mathbf{Z}+\mathbf{A}^{-1}\alpha\end{bmatrix}^{-1}\begin{bmatrix}\mathbf{K} \\ \mathbf{M}\end{bmatrix}\hat{\sigma}_e^2\,(n\text{-}p)/(n\text{-}p\text{-}2).$$

In summary, the analysis would proceed as follows:

1. estimate α and λ as $\hat{\alpha}$ and $\hat{\lambda}$, the components of the mode of the joint posterior density $p(\alpha,\lambda|\mathbf{y})$;
2. given $\hat{\alpha}$ and $\hat{\lambda}$, conduct a mixed model analysis on $\mathbf{y}^{\hat{\lambda}}$.

   Note that this differs from

   $$(y^{\hat{\lambda}}-1)/\hat{\lambda}$$

   only in the scale factor $\hat{\lambda}$ and in a linear shift $1/\hat{\lambda}$;
3. "estimate" σ_e^2 as indicated above, and complete inferences using the t-distribution in (2.40).

2.6.2 Analysis Conditional on α

Suppose now that α is known or that it has been estimated as above. Then

$$p(\boldsymbol{\beta},\mathbf{u}|\alpha,\mathbf{y}) \propto \int_{-\infty}^{\infty} p(\boldsymbol{\beta},\mathbf{u},\lambda|\alpha,\mathbf{y})\, d\lambda.$$

Using an argument similar to the one leading to (2.38) we obtain the joint posterior density

$$p(\boldsymbol{\beta},\mathbf{u},\lambda|\alpha,\mathbf{y}) \propto J_{\lambda}^{(n-p)/n}.[\mathbf{x}'\mathbf{x}-\hat{\theta}'\mathbf{W}'\mathbf{x}+(\theta-\hat{\theta})'(\mathbf{W}'\mathbf{W}+\Sigma)(\theta-\hat{\theta})]^{-(n+q)/2}. \quad (2.41)$$

It was seen previously that the mixed model residual sum of squares $S_{\alpha,\lambda}=J_{\lambda}^{2/n}.S_{\alpha,\lambda}^{*}$, and using this in (2.41) yields

$$p(\boldsymbol{\beta},\mathbf{u},\lambda|\alpha,\mathbf{y}) \propto J_{\lambda}^{-(p+q)/n}.[S_{\alpha,\lambda}^{*}+J_{\lambda}^{-2/n}(\theta-\theta)'(\mathbf{W}'\mathbf{W}+\Sigma)(\theta-\hat{\theta})]^{-(n+q)/2}. \quad (2.42)$$

The marginal density of $\boldsymbol{\beta}$ and $\mathbf{u}$ given y is obtained integrating λ out of the above expression. This is difficult so we proceed by expanding the expression in brackets about $\hat{\lambda}$, the λ-component of the density $p(\alpha,\lambda|y)$ which can be calculated as described earlier. Let

$$Q(\theta,\alpha,\lambda) = J_{\hat{\lambda}}^{-2/n}(\theta-\hat{\theta})'(\mathbf{W}'\mathbf{W}+\Sigma)(\theta-\hat{\theta}).$$

Retaining the linear and quadratic terms of an expansion of $S^*_{\alpha,\lambda}$ and the constant term for the expansion of $Q(\theta,\alpha,\lambda)$, one can write:

$$S^*_{\alpha,\lambda}+Q(\theta,\alpha,\lambda) \simeq S^*_{\alpha,\hat{\lambda}} + \left[\frac{\delta}{\delta\lambda} S^*_{\alpha,\lambda}\right]_{\lambda=\hat{\lambda}}(\lambda-\hat{\lambda})$$

$$+\left[\ 1/2\ \frac{\delta^2}{(\delta\lambda)^2} S^*_{\alpha,\lambda}\right]_{\lambda=\hat{\lambda}}(\lambda-\hat{\lambda})^2$$

$$+\ Q(\theta,\alpha,\hat{\lambda}). \qquad (2.43)$$

The linear term on λ vanishes because the first derivative of the sum of squares $S^*_{\alpha,\lambda}$ is null at $\lambda=\hat{\lambda}$. Using P to denote the coefficient of the term $(\lambda-\hat{\lambda})^2$, putting (2.43) in (2.42) and integrating this expression with respect to λ gives

$$p(\boldsymbol{\beta},\mathbf{u}|\alpha,y) \propto \int_{R_\lambda} J_{\hat{\lambda}}^{-(p+q)/n}\left[S^*_{\alpha,\hat{\lambda}}+Q(\theta,\alpha,\hat{\lambda})+P(\lambda-\hat{\lambda})^2\right]^{-(n+q)/2} d\lambda$$

$$\propto \int_{R_\lambda} J_{\hat{\lambda}}^{-(p+q)/n}\left[S^*_{\alpha,\hat{\lambda}}+Q(\theta,\alpha,\hat{\lambda})\right]^{-(n+q)/2}$$

$$\cdot\left[\ 1+\frac{P(\lambda-\hat{\lambda})^2}{S^*_{\alpha,\hat{\lambda}}+Q(\theta,\alpha,\hat{\lambda})}\right]^{-(n+q)/2} d\lambda \ .$$

The term in square brackets may be taken outside the integral sign with a modification in the exponent to take into account the "loss in degrees of freedom" due to estimating λ as $\hat{\lambda}$. Thus

$$p(\boldsymbol{\beta},\mathbf{u}|\alpha,\mathbf{y}) \propto [S^*_{\alpha,\hat{\lambda}}+Q(\theta,\alpha,\hat{\lambda})]^{-(n+q-1)/2}$$

$$\cdot \int_{R_\lambda} J_\lambda^{-(p+q)/n} p^*(\lambda|\alpha,\mathbf{y})d\lambda, \tag{2.44}$$

where

$$p^*(\lambda|\alpha,\mathbf{y}) = \left[1+\frac{P(\lambda-\hat{\lambda})^2}{S^*_{\alpha,\hat{\lambda}}+Q(\theta,\alpha,\hat{\lambda})}\right]^{-(n+q)/2}.$$

Now, the integral in (2.44) can be viewed as the expectation of $J_\lambda^{-(p+q)/n}$ with respect to the above distribution, whose mode is also $\hat{\lambda}$. Then approximately:

$$E[J_\lambda^{-(p+q)/2}] \simeq J_{\hat{\lambda}}^{-(p+q)/2},$$

which does not depend on θ. Then

$$p(\boldsymbol{\beta},\mathbf{u}|\alpha,\mathbf{y}) \propto [S^*_{a,\hat{\lambda}}+Q(\theta,\alpha,\hat{\lambda})]^{-(n+q-1)/2}$$

$$\propto \left[1+\frac{Q(\theta,\alpha,\hat{\lambda})}{S^*_{\alpha,\hat{\lambda}}}\right]^{-(n+q-1)/2}$$

$$\propto \left[1+\frac{(\theta-\hat{\theta})'(\mathbf{W}'\mathbf{W}+\Sigma)(\theta-\hat{\theta})}{(n-p)\hat{\sigma}_e^2}\right]^{-(n-p-1+p+q)}, \tag{2.45}$$

which is a multivariate t-distribution with mean vector $\hat{\theta}$, covariance matrix $(\mathbf{W}'\mathbf{W}+\Sigma)^{-1}\hat{\sigma}_e^2$ (n-p)/(n-p-2), and n-p-1 degrees of freedom. This is essentially the result given in the preceding section, the modification being in the loss of one degree of freedom due to the integration of λ or, equivalently, estimation of the transformation parameter. Hence, the analysis proceeds as described previously.

2.7 Extensions and Conclusions

The basic theory of the Box-Cox transformation has been extended in several directions, including the estimation of a vector of transformation parameters (Box and Tiao 1973), autocorrelated errors (Savin and White 1978), and heterogeneous variance in regression models (Gaudry and Dagenais 1979). Spitzer (1982) and Seaks and Layson (1983) wrote expository papers on the subject, Box and Tidwell (1962) studied transformation of covariates, and Poirier (1978) discussed the Box-Cox transformation when the response variate is truncated-normal. All these authors considered fixed models and, in principle, it would be possible to extend their results to mixed models using the developments presented in this paper. Zellner (1971) gives a detailed account of the Bayesian approach using a diffuse prior for θ.

Solomon (1985) was apparently the first to study the transformation in balanced random effects models; estimation of the variance and covariance components and of the transformation by maximum likelihood in a one-way layout was considered. The techniques described in our paper illustrate that it is possible to carry this analysis further, e.g., to a general mixed model with unequal subclass numbers. In particular, the estimation of the transformation by maximization of the joint posterior density of the variance components and λ is intimately related to the method of restricted maximum likelihood (Patterson and Thompson 1971). Harville (1974) showed that restricted maximum likelihood estimates correspond to the elements of the mode of the marginal posterior density of the variance parameters taking a flat prior for these and for the fixed effects. In this paper, non-informative priors of the Jeffreys type (Zellner 1971; Box and Tiao 1973) were considered instead, an approach that Harville (1977) qualified as "intriguing". Solomon (1985) stated that in a variance component analysis by restricted maximum likelihood, the necessary factorization of the likelihood into a part which is location invariant cannot be carried out because of a dependence on λ. Our results indicate that restricted maximum likelihood estimates of variance components can be obtained simultaneously with a modal estimate of the transformation parameter λ.

Records available for analysis in animal breeding invariably arise from populations that have been subjected to some degree of selection or assortative mating. This topic is discussed by other contributors to this volume so we deal with it briefly. Henderson (1975) found that if the data are from one cycle of selection upon a normal distribution, and if the selection was based on location invariant functions of the records, then an analysis conducted with all records used for selection will be free of biases due to such selection; he assumed that the dispersion parameters were known. Gianola and Fernando (1986) gave a more general result which indicates that posterior distributions, joint or marginal, are unaffected by selections based on subsets of the records used to construct such distributions; this holds for any distribution and is not restricted to selections based on location invariant functions. Thus, inferences from such distributions apply to the parameters as if selection had not occurred. However, the distributions have to be specified correctly, both in functional form and in parameterization. In this respect, if the distribution of two traits is not normal and if, say, bivariate normality is assumed, then the analysis will not be free of the effect of selection or assortative mating. Hence, an analysis of transformations which renders a multivariate distribution approximately normal, would be useful to alleviate selection biases. This would require an extension of the procedures described here.

The analysis assumed that the transformation achieved normality and constancy of variance at the same time. In practice, it is important to answer whether this is accomplished, which can be done in probabilistic terms. Box and Cox (1964) and Box and Tiao (1973) describe a procedure based on imposing successive constraints to the model, e.g., normality, homoscedasticity, lack of interaction. The plausibility of each constraint is assessed by examining whether, given a particular constraint, a parameter value falls inside or outside of a high posterior density region. The procedure can be applied to the mixed model situation using the results described in this paper.

There is a question on the metric in which the results should be reported. Clearly, a sire evaluation based on an analysis following a logarithmic transformation of milk yield will be difficult to interpret for most people. This makes a reasonable case for reporting results retransformed to the raw scale. However, because of general lack of invariance, it is dangerous to carry selection decisions on the back transformed functions of the records. Also, this would contradict the objective of an analysis of transformations. Because the assumptions apply to the transformed scale, the procedures can be claimed optimal at that level provided that the transformation was successful and that the assumptions hold.

In the study of biometric relations between offspring and parent, Wright (1921) emphasized the importance of finding scales of measurement in which the effects of factors are always combined by addition. Later (Wright 1968) he wrote extensively about transformations that normalize the form of the distribution and remove interaction, and suggested the desirability of systematic methods for finding "normalizing functions". With mixed model methodology being probably the most powerful tool of statistical analysis available in quantitative genetics, it is hoped that results presented in this paper make a contribution in the direction indicated by Wright.

Acknowledgements. The support of the Illinois Agriculture Experiment Station and of Grant US-805-84 from BARD - The United States-Israel Binational Agricultural Research and Development Fund for conducting the research reported here is acknowledged.

References

Box GEP, Cox DR (1964) An analysis of transformations. J R Stat Soc Series B 26:211-252

Box GEP, Tiao GC (1973) Bayesian inference in statistical analysis. Addison-Wesley, Reading

Box GEP, Tidwell PW (1962) Transformation of the independent variables. Technometrics 4:531-550

Brotherstone S, Hill WG (1986) Heterogeneity of variance amongst herds for milk production. Anim Prod 42:297-303

Bulmer MG (1980) The mathematical theory of quantitative genetics. Clarendon Press, Oxford

Dahlquist G, Bjorck A (1974) Numerical methods. Prentice Hall, Englewood Cliffs

Dempster AP, Laird NM, Rubin DB (1977) Maximum likelihood from incomplete data via the EM algorithm. J R Stat Soc Series B 39:1-38

Everett RW, Keown JF, Taylor JF (1982) The problem of heterogeneous within-herd error variances when identifying elite cows. J Dairy Sci 65 suppl 1 (Abstr):100

Fernando RL, Gianola D (1986) Optimal properties of the conditional mean as a selection criterion. Theor Appl Genet 72:822-825

Foulley JL, Im S, Gianola D, Hoschele Ina (1987) Empirical Bayes estimation of parameters for n polygenic binary traits. Genet Sel Evol 19:197-224

Gaudry MJI, Dagenais MG (1979) Heteroscedasticity and the use of the Box-Cox transformations. Econ Letters 2:225-229

Gianola D (1986) On selection criteria and estimation of parameters when the variance is heterogeneous. Theor Appl Genet 72:671-677

Gianola D, Fernando RL (1986) Bayesian methods in animal breeding theory. J Anim Sci 63:217-244

Gianola D, Foulley JL, Fernando RL (1986) Prediction of breeding values when variances are not known. In: Dickerson GE, Johnson RK (eds) Proc 3rd World Congr Genet Appl Livest Prod. Agric Commun, Univ Nebraska, Lincoln, Nebraska XII:356-370

Goffinet B (1983) Selection on selected records. Genet Sel Evol 15:91-97

Haldane JBS (1930) A mathematical theory of natural and artificial selection. III. Selection intensity as a function of a mortality rate. Proc Camb Phil Soc Biol Sci 27:131-136

Harville DA (1974) Bayesian inference for variance components using only error contrasts. Biometrika 61:383-385

Harville DA (1977) Maximum likelihood approaches to variance component estimation and related problems. J Am Stat Assoc 72:320-338

Harville DA (1985) Decomposition of prediction error. J Am Stat Assoc 80:132-138

Henderson CR (1973) Sire evaluation and genetic trends. In: Proc Anim Breed Genet Symp in Honor of Dr. J.L. Lush. ASAS and ADSA, Champaign, Illinois, pp 10-41

Henderson CR (1975) Best linear unbiased estimation and prediction under a selection model. Biometrics 31:423-447

Henderson CR (1984) Applications of linear models in animal breeding. Univ Guelph, Guelph

Henderson CR, Quaas RL (1976) Multiple trait evaluation using relatives' records. J Anim Sci 43:1188-1197

Hill WG (1984) On selection among groups with heterogeneous variance. Anim Prod 39:473-477

Hill WG, Edwards MR, Ahmed MKA, Thompson R (1983) Heritability of milk yield and composition at different levels and variability of production. Anim Prod 36:59-68

Hoschele I, Gianola D, Foulley JL (1987) Estimation of variance components with quasi-continuous data using Bayesian methods. Z Tierz Züchtungsbiol 104:334-349

Judge GC, Griffiths WE, Hill RC, Lutkepol H, Lee TC (1985) The theory and practice of econometrics. Wiley. New York

Kackar RN, Harville DA (1981) Unbiasedness of two-stage estimation and prediction procedures for mixed linear models. Comm Stat Theor Meth A10:1249-1261

Lofgren DL, Vinson WE, Pearson RE, Cassell BG (1984) Effect of herd mean and standard deviation on cow index. J Dairy Sci suppl 1 (Abstr) 67:185-186

Mirande SL, Van Vleck LD (1985) Trends in genetic and phenotypic variances for milk production. J Dairy Sci 68:2278-2286

Patterson HD, Thompson R (1971) Recovery of interblock information when block sizes are unequal. Biometrika 58:545-554

Pericchi LR (1981) A Bayesian approach to transformations to normality. Biometrika 68:35-43

Poirier DL (1978) The use of the Box-Cox transformation in limited dependent variable models. J Am Stat Assoc 73:284-287

Powell RL, Norman HD, Weinland BT (1983) Cow evaluation at different milk yield of herds. J Dairy Sci 66:148-154

Rook AJ (1982) The effects of different environmental correction methods on genetic parameter estimates of traits in on-farm, individually fed, pig performance tests. PhD Thesis, Wye College, Univ London

Savin NE, White KJ (1978) Estimation and testing for functional form and autocorrelation: a simultaneous approach. J Econometrics 8:1-12

Seaks TG, Layson SK (1983) Box-Cox estimation with standard econometric problems. Rev Econ Stat 65:160-164

Searle SR (1974) Prediction, mixed models and variance components. In: Proschan F and Serfling RJ (eds) Proc Conf Reliabilty and Biometry. SIAM, Philadelphia

Searle SR (1979) Notes on variance component estimation: a detailed account of maximum likelihood and kindred methodology. Paper BU-673-M, Biometrics Unit, Cornell Univ, Ithaca.

Solomon PJ (1985) Transformations for components of variance and covariance. Biometrika 72:233-239

Spitzer JJ (1982) A primer on Box-Cox estimation. Rev Econ Stat 64:307-313

Thompson R (1979) Sire evaluation. Biometrics 35:339-353

Wright S (1921) Systems of mating. I. The biometric relations between parent and offspring. Genetics 6:111-123

Wright S (1968) Evolution and the genetics of populations. Vol 1 Univ Chicago Press, Chicago

Zellner A (1971) An introduction to Bayesian inference in econometrics. Wiley, New York

3 Models for Discrimination Between Alternative Modes of Inheritance

R.C. Elston[1]

Two extreme types of data may be available for discriminating between alternative modes of inheritance: data on two inbred lines, their F_1 and the backcrosses, or data on a sample of relatives drawn from a random mating population. For the former kind of data a ten-parameter model is proposed that subsumes, as special cases, one-locus, two-locus, polygenic, or mixed major locus/polygenic inheritance for a quantitative trait. For the latter kind of data three basic models are compared for detecting major gene segregation: the generalized major gene (transmission probability) model, the mixed major gene/polygenic inheritance model and the flexible regressive models. In all cases it is assumed that the environmental variation is normally distributed. The principle of maximizing expected entropy can be used to choose the best-fitting specific genetic hypothesis and, using the general model as a baseline, the likelihood ratio criterion can be used to test for departure from any specific hypothesis. It is recognized that these segregation analysis models can only suggest, never prove, the existence of monogenic segregation.

3.1 Introduction

In any study in which we hope to discriminate among modes of inheritance we are very much limited by the kind of data it is feasible to collect. At one extreme it may be possible to produce inbred (isogenic) lines and mate them in a well-planned experiment designed to answer specific questions about the inheritance of a trait of interest. At the other extreme no experimentation may be possible, and it may be necessary to draw inferences from no more than measures of the trait that have been observed on a sample of relatives drawn from a random mating population. Most genetic studies for the improvement of livestock probably lie somewhere between these two extremes. I shall therefore review some of the models I and my associates have developed for these two extreme situations, in the hope that this will stimulate others to modify and apply them to the specific problems of interest to animal breeders.

[1] Department of Biometry and Genetics, Louisiana State University Medical Center, New Orleans, Louisiana, USA

Throughout this paper I shall restrict myself to the case in which a single univariate trait has been measured on each animal, and the most important aim is to detect and estimate the parameters for monogenic segregation if it exists. For the most part I shall restrict myself to models for quantitative traits, making only passing reference to dichotomous traits. After describing the models and corresponding likelihoods, I shall discuss the general problem of making inferences about the mode of inheritance from a sample of data.

3.2 Data on Inbred Lines, Their F_1 and Backcrosses

When two or more inbred lines are available, it is a relatively simple matter to obtain data on the F_1 and backcross generations as well. Although only further breeding tests can unequivocally distinguish between the various modes of inheritance (Wright 1934), it is possible to use data from these generations alone to determine whether they are compatible with many of the simpler modes of inheritance that are of interest. I shall assume the trait is quantitative, and such that, conditional on genotype, it is normally distributed. As did Elston and Stewart (1973), I shall assume that all environmental effects are random and independent, with constant variance. These assumptions were relaxed by Elston (1984), who also indicated the appropriate log likelihood terms for F_2 data.

Assume that each sample observation is on an individual from one of five classes: the two homozygous parental lines (denoted by the subscripts 1 and 3), the F_1 (denoted by the subscript 2) and the two backcrosses (denoted by the double subscripts 12 and 32, respectively). Let n_i be the number of individuals in the i-th class. Thus there are measurements on n_1 individuals from one parental line and on n_3 from the other, the measurements on the j-th such individuals being respectively z_{1j} and z_{3j}; there are measurements on n_{12} individuals from the backcross to the first parental line and on n_{32} from the backcross to the other, the measurements on the j-th such individuals being respectively z_{12j} and z_{32j}; and there are measurements on n_2 individuals from the F_1, the measurement on the j-th such individual being z_{2j}. Denote the total sample size as $N=n_1+n_2+n_3+n_{12}+n_{32}$.

For convenience, define

$$f_{ij}(\mu_u) = -(z_{ij}-\mu_u)^2/2\sigma^2,$$

where i=1, 2, 3, 12 or 32; μ_u is the mean of the distribution for the particular genotype u, and σ^2 is the common environmental variance. With this notation the log likelihoods appropriate for various modes of inheritance will now be given.

3.2.1 One Locus

If the two parental lines differ at one locus only, and this is the only locus segregating in the backcross progeny, the log likelihood of the data is

$$\ell n\mathrm{L} = \mathrm{K}\text{-}\mathrm{N}\ell n\sigma + \sum_{i=1}^{3}\sum_{j=1}^{n_i} f_{ij}(\mu_i) + \sum_{j=1}^{n_{12}} \ell n\left[1/2e^{f_{12j}(\mu_1)} + 1/2e^{f_{12j}(\mu_2)}\right]$$

$$+\sum_{j=1}^{n_{32}} \ell n\left[1/2e^{f_{32j}(\mu_3)} + 1/2e^{f_{32j}(\mu_2)}\right], \qquad (3.1)$$

where the constant $\mathrm{K}=[-\mathrm{N}\ell n(2\pi)^{1/2}]$ will be kept the same throughout this paper. The first summation in this expression is over the parental and F_1 lines; the last two are over the backcross progeny to the two parental lines, respectively. There is a total of four parameters: the two parental means (μ_1 and μ_3), the F_1 mean (μ_2) and the common environmental variance (σ^2). These are reduced to three parameters in the cases of additive genes [$\mu_2=(\mu_1+\mu_3)/2$], parent 1 dominant ($\mu_1=\mu_2$) or parent 3 dominant ($\mu_3=\mu_2$).

3.2.2 Polygenic Inheritance

If the trait difference between the two parental lines is due to a large number of additive loci with equal effects, then, in the limit as the number of loci tends to infinity, the log likelihood is

$$\ell n\mathrm{L} = \mathrm{K} - \mathrm{N}\ell n\sigma + \sum_{i=1}^{3}\sum_{j=1}^{n_i} f_{ij}(\mu_i) + \sum_{j=1}^{n_{12}} f_{12j}(\mu_{1\infty}) + \sum_{j=1}^{n_{32}} f_{32j}(\mu_{3\infty}), \qquad (3.2)$$

where $\mu_{1\infty}=(\mu_1+\mu_2)/2$ and $\mu_{3\infty}=(\mu_3+\mu_2)/2$. In this case also there are only four independent parameters, and these are reduced to three if at every locus there is intralocus additivity [$\mu_2=(\mu_1+\mu_3)/2$] or dominance ($\mu_1=\mu_2$ or $\mu_3=\mu_2$).

If there is a large number of additive loci but we are not in the limiting situation in which the variance is the same in the backcross and parental generations, it is more appropriate to allow for this fact by defining, in the log likelihood (3.2),

$$f_{12j}(\mu_{1\infty}) = -(z_{12j}-\mu_{1\infty})^2/2\sigma^2_\infty,$$

$$f_{32j}(\mu_{3\infty}) = -(z_{32j}-\mu_{3\infty})^2/2\sigma^2_\infty, \tag{3.3}$$

i.e., by introducing another parameter, σ^2_∞, to allow for genetic variation in the backcross generations.

3.2.3 Mixed Major Locus and Polygenic Inheritance

Segregation due to a polygenic background can be easily introduced into the log likelihood (3.1) for one-locus segregation by introducing four new parameters, μ_{12}, μ_{21}, μ_{32}, and μ_{23}, as follows:

$$\ell n L = K - N \ell n \sigma + \sum_{i=1}^{3} \sum_{j=1}^{n_i} f_{ij}(\mu_i) + \sum_{j=1}^{n_{12}} \ell n \left[1/2 e^{f_{12j}(\mu_{12})} + 1/2 e^{f_{12j}(\mu_{21})} \right]$$

$$+ \sum_{j=1}^{n_{12}} \ell n \left[1/2 e^{f_{32j}(\mu_{32})} + 1/2 e^{f_{32j}(\mu_{23})} \right] . \tag{3.4}$$

The only difference between this and (3.1) is in the backcross generations, in each of which the two means are allowed to be shifted from the parental and F_1 values by the polygenic component. Here also, however, it is often more appropriate to allow for extra genetic variation in the backcross generations by defining, analogous to (3.3),

$$f_{12j}(\mu_{12}) = -(z_{12j}-\mu_{12})^2/2\sigma^2_\infty$$

$$f_{12j}(\mu_{21}) = -(z_{12j}-\mu_{21})^2/2\sigma^2_\infty$$

$$f_{32j}(\mu_{32}) = -(z_{32j}-\mu_{32})^2/2\sigma^2_\infty$$

$$f_{32j}(\mu_{23}) = -(z_{32j}-\mu_{23})^2/2\sigma^2_\infty. \tag{3.5}$$

Additivity at the major locus corresponds to $\mu_2=(\mu_1+\mu_3)/2$. Sometimes it is found useful to constrain the parameter estimates using one of the following sets of symmetry conditions:

$$\text{Symmetry A: } (\mu_{12}-\mu_{21})/(\mu_1-\mu_2) = (\mu_{32}-\mu_{23})/(\mu_3-\mu_2)$$

or

$$\text{Symmetry B: } (\mu_{12}-\mu_{21})^2-(\mu_1-\mu_2)^2 = (\mu_{32}-\mu_{23})^2-(\mu_3-\mu_2)^2.$$

Either of these symmetry conditions, which are discussed in further detail in the context of the two-locus hypothesis, may be used with or without the assumption of major locus additivity; they put a mild restriction on the extra parameters μ_{12}, μ_{21}, μ_{32} and μ_{23}, without imposing the stronger condition of lack of polygenic dominance. Complete additivity at both the major locus and the polygenic loci corresponds to the restrictions

$$\mu_2 = (\mu_1+\mu_3)/2$$

$$\mu_{12}+\mu_{21} = \mu_1+\mu_2$$

and

$$\mu_{32}+\mu_{23} = \mu_3+\mu_2 \,. \qquad (3.6)$$

3.2.4 Two Loci

Provided the number of loci involved is no greater than two, it is not difficult to allow for linkage between the loci. Let the recombination fraction be θ. Then a backcross observation has a probability $(1-\theta)/2$ of coming from a parental distribution, a probability $(1-\theta)/2$ of coming from the F_1 distribution, and a probability $\theta/2$ of coming from each of the two relevant recombination distributions. Let μ_{12} and μ_{21} now be the means of the two recombinant distributions when the backcross is to parent 1, and μ_{32} and μ_{23} be the means when the backcross is to parent 3. The log likelihood of the whole sample is then, analogous to (3.1),

$$\ell n L = K - N \ell n \sigma + \sum_{i=1}^{3} \sum_{j=1}^{n_i} f_{ij}(\mu_i)$$

$$+\sum_{j=1}^{n_{12}} \ell n \left[1/2(1-\theta) \left[e^{f_{12j}(\mu_1)} + e^{f_{12j}(\mu_2)} \right] + 1/2\theta \left[e^{f_{12j}(\mu_{12j})} + e^{f_{12j}(\mu_{21})} \right] \right]$$

$$+\sum_{j=1}^{n_{32}} \ell n \left[1/2(1-\theta) \left[e^{f_{32j}(\mu_3)} + e^{f_{32j}(\mu_2)} \right] + 1/2\theta \left[e^{f_{32j}(\mu_{32j})} + e^{f_{32j}(\mu_{23})} \right] \right] . \quad (3.7)$$

Additivity between the two loci, without assuming equal effects or intralocus additivity, is equivalent to the conditions

$$\mu_{12}+\mu_{21} = \mu_1+\mu_2 \text{ and } \mu_{32}+\mu_{23} = \mu_3+\mu_2 . \quad (3.8)$$

Less restrictive are the symmetry conditions; in combination with (3.8), symmetry A implies the two loci have the same dominance ratio if the loci are in coupling in the parental lines, whereas symmetry B has the same implication if the loci are in repulsion in the parental lines. If the loci are equal and additive in effect, the restrictions (3.6) apply. Finally, of course, $\theta=1/2$ implies that the two loci are unlinked.

A superficial comparison of expressions (3.7) and (3.4) reveals that they are identical when $\theta=1$, though it is to be understood that one would not use the definitions (3.5) in the log likelihood for two linked loci (3.7). Using definitions (3.5) in log likelihood (3.7), however, results in the log likelihood of a general model with ten unknown parameters (μ_1, μ_2, μ_3, μ_{12}, μ_{21}, μ_{32}, μ_{23}, σ^2, σ^2_∞, θ), of which mixed major locus/polygenic inheritance ($\theta=1$) and two-linked-locus inheritance ($\sigma^2_\infty=\sigma^2$, $0\leq\theta<1/2$) are special cases. This fact can be used in hypothesis testing, as indicated below.

3.3 Pedigree Data from a Random Mating Population

When it is not possible to set up breeding experiments, it is hardly feasible to consider estimating the parameters of the two-locus models described above. There is, however, no difficulty in formulating the likelihood of pedigree data in a very general way that encompasses all of the above models (Elston and Stewart 1971; Lange and Elston 1975; Cannings et al. 1978; Ott 1979; Elston 1981a), especially if random mating is assumed. Here, I shall first give the likelihoods appropriate for a random pedigree under the monogenic, polygenic and mixed major gene/polygenic models, and then briefly describe the regressive models recently introduced by Bonney (1984). In this section z_i will denote the phenotype of the i-th member of a pedigree of n individuals. I shall assume that the pedigree contains either no parents or both parents of every member in it; and it must contain the parents of every sibship that contains more than one sib. These conditions can always be satisfied by including in the pedigree, if necessary, individuals with missing values of z. Those individuals who have no parents in the pedigree form the set M, the "marry-ins", while those with both parents in the pedigree form the set I, those "in" the pedigree.

3.3.1 One Locus

Let the possible genotypes that the i-th individual can have at a major locus be indexed by u_i. In the case of two alleles A and a, for example, u_i=AA, Aa or aa. Denote the density of z conditional on u $g_u(z)$, which we shall assume, as before, is a normal density with mean μ_u and variance σ^2. If the value of z is missing, $g_u(z)=1$. Let ψ_u be the probability of genotype u in the population, and let $p_{u_F u_M u}$ be the probability that an offspring has genotype u given parents that have genotypes u_F and u_M. Then the pedigree likelihood for a one-locus model can be written

$$L = \sum_{u_1} \dots \sum_{u_n} \prod_{i=1}^{n} p(u_i) g_i(z_i) \qquad (3.9)$$

$$\text{where } p(u_i) = \begin{cases} \psi_{u_i} & \text{if } i \in M \\ p_{u_{F_i} u_{M_i} u_i} & \text{if } i \in I. \end{cases}$$

In the case of two alleles at one autosomal locus, the values of $p_{u_F u_M u}$ are as given in Table 3.1, and the likelihood contains seven unknown parameters: ψ_{AA}, ψ_{Aa}, ψ_{aa}, μ_{AA}, μ_{Aa}, μ_{aa} and σ^2. However, $\psi_{AA}+\psi_{Aa}+\psi_{aa}=1$ and under panmictic conditions $\psi_{Aa}=2(\psi_{AA}\,\psi_{aa})^{1/2}$, leaving only five independent unknown parameters. As before, we can place an additivity $[\mu_{Aa}=(\mu_{AA}+\mu_{aa})/2]$ or dominance ($\mu_{Aa}=\mu_{AA}$ or $\mu_{Aa}=\mu_{aa}$) restriction on the means. Elston and Stewart (1971) noted that it would be impractical to consider the probabilities $p_{u_F u_M u}$ as unknown parameters, since this would involve a further 12 independent parameters to be estimated. However, we can express the $p_{u_F u_M u}$ for a two-allele autosomal locus in terms of the three transmission probabilities

τ_{AA} = P(AA parent transmits A)

τ_{Aa} = P(Aa parent transmits A)

τ_{aa} = P(aa parent transmits A), (3.10)

as follows:

$$p_{u_F u_M u} = \begin{cases} \tau_{u_F} \tau_{u_M} & \text{if } u = AA \\ \tau_{u_F}(1-\tau_{u_M})+\tau_{u_M}(1-\tau_{u_F}) & \text{if } u = Aa \\ (1-\tau_{u_M})(1-\tau_{u_F}) & \text{if } u = aa. \end{cases}$$

The utility of this parameterization will be discussed later.

Table 3.1 Genetic matrix for two alleles at one autosomal locus

	u_M								
u_F	AA			Aa			aa		
AA	(1	0	0)[a]	(1/2	1/2	0)	(0	1	0)
Aa	(1/2	1/2	0)	(1/4	1/2	1/4)	(0	1/2	1/2)
aa	(0	1	0)	(0	1/2	1/2)	(0	0	1)

[a] Each entry is ($p_{u_F u_M AA}$ $p_{u_F u_M Aa}$ $p_{u_F u_M aa}$); from Elston and Stewart (1971).

3.3.2 Polygenic Inheritance

The likelihood for the limiting case of polygenic inheritance, as the number of loci with equal and additive effects tends to infinity, can be expressed in a very similar manner. Let s_i denote the polygenotypic value ("breeding value") of individual i, and let s_{F_i} and s_{M_i} be those of i's parents. These polygenotypic values are assumed to be normally distributed with mean 0 and variance σ_g^2 over the whole population; denote this density $\phi(s,\sigma_g^2)$. Conditional on the parental values s_F and s_M, on the other hand, s is normally distributed with mean $(s_F+s_M)/2$ and variance $\sigma_g^2/2$, denoted $\phi[s-(s_F+s_M)/2,\sigma_g^2/2]$. Then, using the symbol $\int_s$ to denote integration from $-\infty$ to ∞ with respect to s of everything following, the pedigree likelihood for (infinite) polygenic inheritance is

$$L = \int_{s_1} \ldots \int_{s_n} \prod_{i=1}^{n} p(s_i)\, g_{s_i}(z_i), \tag{3.11}$$

where now

$$p(s_i) = \begin{cases} \phi(s_i,\sigma_g^2) & \text{if } i \in M \\ \phi[s_i-(s_{F_i}+s_{M_i})/2,\sigma_g^2/2] & \text{if } i \in I \end{cases}$$

$$\text{and } g_{s_i}(z_i) = \frac{1}{(2\pi)^{1/2}\sigma} e^{f_i(\mu+s_i)} .$$

There are thus three unknown model parameters, the population mean μ, the additive genetic variance σ_g^2 and the environmental variance σ^2.

3.3.3 Mixed Major Gene and Polygenic Inheritance

Assuming the polygenic effects and the major locus effects are additive, it is easy to combine expressions (3.9) and (3.11) to obtain the following likelihood for mixed major gene/polygenic inheritance:

$$L = \int_{s_1} \ldots \int_{s_n} \sum_{u_1} \ldots \sum_{u_n} \prod_{i=1}^{n} p(s_i)p(u_i)\, g_{s_i u_i}(z_i), \tag{3.12}$$

where $p(s_i)$ and $p(u_i)$ are as before, and

$$g_{s_i u}(z) = \frac{1}{(2\pi)^{1/2}\sigma} e^{f_i(\mu_u + s_i)} .$$

This model contains one more unknown parameter than the one-locus model, namely σ_g^2. It has been extended to allow for a sibling environmental correlation with the addition of a further parameter, so that it is no longer necessary, fixing the major genotypes, for the parent-offspring correlation to equal the sib-sib correlation (Morton and MacLean 1974).

Whereas a fast algorithm is available for calculating the likelihoods (3.9) and (3.11) (Elston and Stewart 1971), this is not the case for the likelihood (3.12), for which numerical integration is necessary (Morton and MacLean 1974; Hasstedt 1982). This motivated Bonney (1984) to develop regressive models, which I shall now describe.

3.3.4 Regressive Models

When human geneticists have used the pedigree models described above to analyze quantitative data, they have typically first adjusted their data for the additive effects of age, sex and/or generation, using a multiple regression analysis that ignores the dependencies among pedigree members. There is, however, no major difficulty in introducing covariates into the above models. To allow for a common regression on the covariate x, for example, we can substitute $\alpha_u + \beta x_i$ for μ_u in (3.9) and (3.12) or $\alpha + \beta x_i$ for μ in (3.11). The same basic computational algorithm can be used, the only difference being the existence of an extra unknown parameter (β) to estimate.

Now let us introduce into the likelihood (3.9), for all i ε I, the parental values z_{F_i} and z_{M_i} as covariates, i.e., for i ε I, let

$$\mu_{u_i} = \alpha_{u_i} + \beta_M z_{M_i} + \beta_F z_{F_i} .$$

Let us also introduce, for one spouse of each spouse pair, the phenotypic value of the other spouse as a covariate. The result is Bonney's **Class A regressive model**. Like the mixed model, it allows for familial correlations over and above that due to major gene segregation. In the special case $\beta_M = \beta_F$, the model is equivalent to one in which these extra correlations - ρ_{PO}: parent-offspring; ρ_{SS}: sib-sib; and ρ_{FM}: father-mother - satisfy the condition

$$\rho_{SS} = 2\rho_{PO}^2/(1+\rho_{FM}).$$

The introduction of other phenotypic values as covariates allows one to model a wide variety of polygenic and/or familial environmental effects. For example, including for each sib, as an additional random covariate, the mean value of z for all older sibs results in Bonney's **Class D regressive model**. If the pedigree structures consist of nuclear families, this model includes the mixed model with sibling environmental correlation as a special case (setting $\beta_M=\beta_F$ and $\rho_{FM}=0$). This flexibility of the regressive models, coupled with the relative ease of likelihood computation, give them great utility in the analysis of traits with appreciable environmental components in their etiology.

3.4 Choice of Genetic Hypothesis

The likelihoods given above, when maximized numerically over unknown parameters, give us a basis for choosing the mode of inheritance, or genetic hypothesis, that best fits the data: the hypothesis that gives rise to the largest likelihood is the best-fitting one. However, if that likelihood is obtained by maximizing the likelihood (or, equivalently, the log likelihood) over a set of unknown parameters, it is clear that any constraints that are put on the parameters, such as (3.6) or (3.8), will automatically lower the maximum likelihood. To discount this effect, and in general to allow for the fact that different hypotheses depend on different numbers of unknown parameters, Akaike (1977) suggested that we choose the hypothesis that maximizes the expected entropy. For this purpose, we choose the hypothesis that leads to the smallest AIC information criterion, defined as follows:

$$\text{AIC} = -\,2\ell n(\text{maximum likelihood})$$
$$+\,2(\text{number of independently estimated parameters}).$$

Akaike (1977) showed that in many situations this criterion, which can be considered a realization of Occam's razor, gives useful solutions to practical problems.

Apart from choosing the "best" hypothesis, however, we should also consider whether the data in some sense fit that hypothesis and whether, in the same sense, other less well-fitting hypotheses can also satisfactorily explain the data. Elston and Stewart (1973) suggested some goodness-of-fit tests for quantitative data derived from inbred lines, and an attempt has been made to adapt these tests to pedigree data (Elston et al. 1978). However, unlike goodness-of-fit tests for dichotomous data (Elston 1981b), this is an area in which further theoretical research is still needed.

Provided we are willing to assume a general model, the likelihood ratio criterion can be used to test any specific genetic hypothesis that is subsumed by that model. For then, under fairly general conditions, we can assume an asymptotic chi-square distribution (if the hypothesis is true) for the test statistic

$$2[\ell n(\text{maximum likelihood under the general model}) - \ell n(\text{maximum likelihood under the hypothesis})]. \quad (3.13)$$

The appropriate number of degrees of freedom is given by the difference between the numbers of independent parameters estimated when the two likelihoods are maximized. As indicated above, all the modes of inheritance discussed for experimental data can be considered to be specific hypotheses under a general model with ten unknown parameters. Thus, using this general model as a baseline, we can use the criterion (3.13) to test whether the data depart significantly from any particular genetic hypothesis.

In the case of pedigree data from an outbred population, two different general models have been suggested to test the specific hypothesis of one-locus segregation. The first of these models, the "generalized major gene model", is based on the likelihood (3.9). Under a model in which the three transmission probabilities can take on arbitrary values in the interval [0, 1], we test the null hypothesis that they are Mendelian, i.e.,

$$H_0 : \begin{cases} \tau_{AA} = 1 \\ \tau_{Aa} = 1/2 \\ \tau_{aa} = 0. \end{cases}$$

The second model, the "mixed model", is based on the likelihood (3.12). Under this model we test whether there is significant departure from the null hypothesis of polygenic inheritance, i.e., we test the null hypothesis

$$H_0 : \begin{cases} \psi_{AA} = 0 \\ \psi_{Aa} = 0 \end{cases} \qquad \text{or } H_0 : \mu_{AA} = \mu_{Aa} = \mu_{aa} .$$

Thus in the generalized major gene model, monogenic segregation corresponds to the null hypothesis, which is accepted by default if it is not rejected. This can happen if the data contain insufficient information about the transmission probabilities, and to guard against this possibility one should check that the data contain sufficient information at least to reject the null hypothesis that there is **no** transmission from one generation to the next, i.e., H_0 : $\tau_{AA}=\tau_{Aa}=\tau_{aa}$. In the mixed model, on the other hand, monogenic segregation is detected by rejecting a null hypothesis.

Since the generalized major gene model does not allow for the possibility of polygenic transmission, the presence of such a component may be interpreted by the model as the presence of monogenic segregation, or it may affect the estimated transmission probabilities to such an extent that a monogenic mechanism, acting in concert with a polygenic mechanism, is not detected. The mixed model does not have this limitation, but it does have the disadvantage that all transmission from one generation to the next is assumed to be either monogenic or polygenic: any environmentally caused transmission that is not confounded with a polygenic mechanism may therefore tend to be interpreted as monogenic segregation. The generalized major gene model, on the other hand, allows for a variety of environmental transmission mechanisms, especially if the transmission probabilities are allowed to be sex-dependent (Demenais and Elston 1981).

MacLean et al. (1975) demonstrated by simulation studies that under the mixed model much more information is available if a quantitative trait is analyzed, rather than if the quantitative trait is first converted to a dichotomous phenotype by dividing the continuous distribution at a threshold value; and this can be expected to be true under the generalized major gene model as well. Unfortunately, however, this extra information is easily misinterpreted, because of the model's assumption that, conditional on the major genotype, the phenotype is normally distributed. An overall non-normal population distribution of the phenotype can only be accommodated under the mixed model by the presence of a monogenic mechanism, so that any such non-normality in the sample data will tend to be interpreted by this model as the presence of monogenic segregation. The generalized major gene model, on the other hand, can accommodate a population distribution that approximates a mixture of three normal distributions without assuming any transmission from one generation to the next. Any polygenic mechanism that gives rise to a non-normal population distribution will tend to be interpreted as a monogenic mechanism by either model; for this reason the scale on which the phenotype is measured, or how it is transformed prior to analysis, can be of crucial importance. This is illustrated, for example, by a recent analysis of serum dopamine-β-hydroxylase levels using the generalized major gene model (Asamoah et al. 1987). Analysis of the untransformed measures indicated significant departure from Mendelian transmission probabilities, but no significant departure from an environmental hypothesis in which all the transmission probabilities are equal. The converse was found to be the case, however, when all the measures were log-transformed prior to analysis. It would therefore appear prudent to assume a more general type of distribution than normal, for the distribution of phenotype conditional on genotype. Use of the Box and Cox (1964) power transformation to induce approximate conditional normality, simultaneously estimating the power parameter, is one possible solution (Wilson et al. 1984).

It is clear from the above discussion that in many respects the generalized major gene model and the mixed model are complementary with respect to power and robustness. A more general model that combines features of both models can thus be expected to be more useful than either. Such a model has been proposed (Boyle and Elston 1979; Lalouel et al. 1983):

the mixed model described above with the probabilities $p_{u_F u_M u}$ expressed as functions of unknown transmission probabilities. Similarly, transmission probabilities can be introduced into regressive models, and we have begun using such models for the analysis of both quantitative and dichotomous traits [in the latter case, assuming $g_u(z)$ is a logistic function; Bonney 1986]. Our initial experience with these models is promising and we believe they will soon be widely used. We should always be cautious, however, not to overinterpret the results of analyses based on any of these models: they can only suggest, never prove, monogenic segregation.

Acknowledgement. This work was supported in part by a Public Health Service Research grant (GM 28356) from the National Institute of General Medical Sciences.

References

Akaike H (1977) On entropy maximization principle. In: Krishnaiah PR (ed) Applications of statistics. North Holland, Amsterdam, pp 27-41

Asamoah A, Wilson AF, Elston RC, Dalferes E Jr, Berenson GS (1987) Segregation and linkage analyses of dopamine-β-hydroylase activity in a six generation pedigree. Am J Med Genet 27:613-621

Bonney GE (1984) On the statistical determination of major gene mechanisms in continuous human traits: regressive models. Am J Med Genet 18:731-749

Bonney GE (1986) Regressive logistic models for familial disease and other binary traits. Biometrics 42:611-625

Boyle CR, Elston RC (1979) Multifactorial genetic models for quantitative traits in humans. Biometrics 35:55-68

Box GEP, Cox DR (1964) An analysis of transformations. J R Stat Soc B 26:211-252

Cannings C, Thompson EA, Skolnick MH (1978) Probability functions on complex pedigrees. Adv Appl Prob 10:26-61

Demenais FM, Elston RC (1981) A general transmission probability model for pedigree data. Hum Hered 31:93-99

Elston RC (1981a) Segregation analysis. Adv Hum Genet 11:63-120

Elston RC (1981b) Testing one- and two-locus hypotheses for the genetic differences of a qualitative trait between two homozygous lines. In: Gershon ES, Matthysse S, Breakefield XO, Ciaranello RD (eds) Genetic research strategies for psychobiology and psychiatry. The Boxwood Press, Pacific Grove, pp 283-293

Elston RC (1984) The genetic analysis of quantitative trait differences between two homozygous lines. Genetics 108:733-744

Elston RC, Stewart J (1971) A general model for the genetic analysis of pedigree data. Hum Hered 21:523-542

Elston RC, Stewart J (1973) The analysis of quantitative traits for simple genetic models from parental, F_1, and backcross data. Genetics 73:695-711

Elston RC, Namboodiri KK, Kaplan E (1978) Resolution of major loci for quantitative traits. In: Morton NE, Chung CS (eds) Genetic epidemiology. Academic Press, New York, pp 223-246

Hasstedt SJ (1982) A mixed model approximation for large pedigrees. Comput Biomed Res 15:295-307

Lalouel JM, Rao DC, Morton NE, Elston RC (1983) A unified model for complex segregation analysis. Am J Hum Genet 35:816-826

Lange K, Elston RC (1975) Extensions to pedigree analysis. I. Likelihood calculations for simple and complex pedigrees. Hum Hered 25:95-106

MacLean CJ, Morton NE, Lew R (1975) Analysis of family resemblance. IV. Operational characteristics of segregation analysis. Am J Hum Genet 27:365-384

Morton NE, MacLean CJ (1974) Analysis of resemblance III. Complex segregation of quantitative traits. Am J Hum Genet 26:489-503

Ott J (1979) Maximum likelihood estimation by counting methods under polygenic and mixed models in human genetics. Am J Hum Genet 31:161-175

Wilson AF, Elston RC, Siervogel RM, Weinshilbourm R, Ward LJ (1984) Linkage relationships between a major gene for catechol-o-methyltransferase activity and 25 polymorphic marker systems. Am J Med Genet 19:525-532

Wright (1934) The results of crosses between inbred strains of guinea pigs differing in numbers of digits. Genetics 19:537-551

Discussion Summary
PART I: GENERAL

R.D. Anderson[1]

1 Statistical Methods in Animal Improvement: Historical Overview

The use of unbiased as opposed to biased predictors dominated the discussion. Addressing the issue in general terms, the author advanced the viewpoint that given the "fairness" of unbiased predictors (an attribute of dubious relevance in the mind of at least one member of the audience), together with the fact that breeders are accustomed to using such predictors, the criterion of unbiasedness should continue to receive emphasis in the future. Addressing the topic in a specific context, namely, bias arising through association between sire and herd values, Dr. Henderson suggested that perhaps herd (herd-year-season) subclasses having more than ten observations should be treated as fixed rather than random effects, to eliminate such biases.

Another point emphasized in the discussion was that contrary to the popular belief among many animal breeders, BLUP was first outlined in a manuscript written by Dr. Henderson in 1949, rather than at the time of the J.L. Lush Symposium in 1972.

2 Mixed Model Methodology and the Box-Cox Theory of Transformations: A Bayesian Approach

The following comments emerged from the discussion of the paper delivered by Dr. Gianola:

1. The transformation may be applied not only to the observed data, but also to the expected values. It was claimed by a member of the audience that such an approach would lead to maximization of genetic gain, and thereby resolve the problem of deciding whether or not the use of a transformation can be justified on anything other than theoretical grounds.
2. A non-parametric model based on ranks rather than a parametric model may be a useful alternative.

[1] Department of Animal Science, Massey University, Palmerston North, New Zealand

3. Different transformations may be considered for different components in the model. For example, one transformation for the residual effects and another for the remaining random effects in the model.
4. Family structures can be accommodated, and parameters pertaining to other genetic effects, such as dominance, can be included in the model.
5. The posterior distribution of heritability is transformation-dependent. The likelihood for the transformation parameter is often flat, but integrating it out analytically is seemingly impossible.
6. Extension of the theory to the multivariate case promises to be a challenging problem because of the difficulty in specifying prior distributions for the variance-covariance parameters.

3 Models for Discrimination Between Alternative Modes of Inheritance

In the discussion of Dr. Elston's paper, it was suggested that there may be a strong connection between the formation of the inverse of the numerator relationship matrix using Dr. Henderson's method and the rapid calculation of the likelihood of pedigrees outlined by Dr. Elston. It was noted that although none of the likelihoods given in the paper took account of the probability of any particular family pedigree being included in the population, the problem can be readily handled. On the question of the application of the techniques described in the paper to animal breeding problems, notably, the detection of major genes from the analysis of field-collected data, Dr. Elston stated there would be major difficulties, one of which is the need to acquire a very large number of observations.

Part II: Design of Experiments and Breeding Programs

4 Considerations in the Design of Animal Breeding Experiments

W.G. Hill[1]

Criteria for design of experiments to estimate within-population genetic parameters such as heritabilities and correlations are reviewed. The design and relative efficiency of sib covariance, offspring-parent regression and combinations of these estimators are described. Problems of objectives and models for design and analysis of selection experiments are discussed. The design of field experiments, for example to estimate covariances between indicator and production traits, is reviewed in terms of convenience and efficiency.

4.1 Introduction

A substantial amount of information in animal breeding comes from analysis of data collected for other purposes. Dairy cattle records are obtained primarily as a management aid to individual producers, are regulated nationally to make them a credible basis for purchase of breeding stock, and then are analyzed for breeding value estimation. Almost all our estimates of the genetic parameters for milk production come from analysis of such data. In other species data are collected solely to enable breeding value estimation and selection decisions, for example in central performance tests of pigs. These data are also used for estimation of parameters such as heritabilities and correlations, which can be used both to make the breeding program more efficient and to be a source for other studies.

Such sets of data from the field are not designed to give information about, for example, genetic variances and population or breed differences *per se*, and whether they do or do not give useful data depends on the questions to be asked and the data structure. In dairy cattle, for example, it is easy to get estimates of heritability of first lactation production which have low sampling variance and are close to unbiased. It is harder to estimate genetic correlations between successive lactations, because account has to be taken of culling and this can be done formally only if all the data on which culling decisions are made are included, which is unlikely with such a complex process. If estimates of correlations are required between yield and variates such as hormone levels which are regarded as putative indicators of

[1] Institute of Animal Genetics,University of Edinburgh, Edinburgh, Scotland

genetic merit for production, then new data have to be gathered and analyzed. In these cases the problem is to design the relevant trial, which might combine field records of production and field or experimental station data on the potential covariates. In addition to this half-way house of combined field and test data, fully designed experiments are required; for example, selection experiments with farm or laboratory animals intended to elucidate information on covariances with production or on more fundamental biological problems.

A basic issue in designing any experiment is the specification of the population or populations on which inferences are to be made. We are familiar with the ideas of fixed and random effects, which in animal breeding are exemplified by breeds and sires within breeds, respectively, but there are some grey areas. Thus, Taylor (1976a) in his multibreed designs considers a sample of breeds and wishes to make inferences about the population of breeds. From experience we know that heritabilities and correlations of traits are rather similar in different breeds or populations, so when analyzing data on one breed we rather hope its application is more general. Similarly, we hope that inferences can be drawn from experiments with laboratory animals to farm livestock. Again we have to rely on both experience and a general appeal to biological principles, but must not forget that the argument is tenuous. Thus, while we would like to make the inferential pathway a direct one, time and facilities may not always permit this, and a standard error of a biased parameter estimate may not be an adequate description of its utility. An example where incorrect standard errors have frequently been applied has been in the analysis of selection experiments for which a standard error of realized heritability or line means computed using the usual independent homoscedastic error assumptions does not describe the error among conceptual replicates of such experiments (Hill 1972). Analysis by mixed model technology using the relationship matrix can in principle, however, give appropriate errors.

In addition to the issues of inference, the experimental design should be influenced by the precision required of an estimate which, in turn, depends on how the estimate is to be used. For example, if parameter estimates are required for selection index construction it was shown some years ago that for weighting individual and family information a rough estimate of intra-class correlation is sufficient, whereas before incorporating a variable, which is not itself of economic importance, a good estimate of genetic correlation is required (Sales and Hill 1976 a,b).

In this paper I shall concentrate on estimates of parameters such as heritabilities and genetic correlations within populations. Designs for population comparisons are usually quite different but straightforward. The usual trap (into which no geneticist would surely now fall!) is to base a breed comparison on only a few families, and to analyze the data as if observations were independent. Once the proper hierarchical structure of family data is appreciated such problems can be avoided, but only at the expense of using large numbers of families unless the same reference sires, for example, can be used in both populations with the assumption of no sire x breed cross-interaction. In designs to estimate parameters within populations we are mainly interested in variables which are not recorded as part of a regular recording or breeding program. Thus, we are not primarily concerned with the heritability of milk production but with the heritability of efficiency of conversion of feed to milk and its correlation with production, and with the correlation or covariances with yield of potential indicator variables, such as hormone or metabolite levels.

The paper will first comprise a review of the basic structures used in estimating genetic parameters in animals: the correlation of collateral relatives, e.g. sibs, the regression of progeny on parent and selection experiments, and will then turn to field experiments. None of the divisions is clear in practice, for in a selection experiment it is possible to estimate sib covariances and single generation progeny on parent regressions. Further, since early papers on design were written it has become possible by modern analytical techniques to extract further information, for example from sib covariances and from offspring of selected parents. One insurmountable problem is that the optimal design for estimating some parameter is a function of the parameter value, so it is necessary to use *a priori* (guessed) values and, therefore, desirable to evaluate the robustness of the design.

Many papers on design have concentrated on optimal structure for estimating heritability rather than, for example, the genetic variance. In principle, the design should be optimized to estimate the relevant parameters in a selection index, for example, or to optimize the accuracy of the index *per se*. For selection on individual and family performance this is simply the intra-class correlation of sibs, but for indices involving several traits more complex functions are relevant (Sales and Hill 1976a,b). The problem of which parameters or combination of parameters should be estimated will not be addressed except in passing, but note that the invariance property of maximum likelihood (ML) estimators means that *a priori* decisions as to what is to be estimated do not have to be made in ML analysis.

4.2 Formal Designs

4.2.1 Intra-Class Correlation of Sibs

Heritability. The design of experiments to estimate heritability from the intra-class correlation of sibs was discussed by Robertson (1959a). It is useful to review this classical work and other approaches to the problem as a way of illustrating some of the principles. It is assumed that no selection is practiced, for unless data on the parental and, if relevant, earlier generations are available, biases cannot be avoided.

It is clear that family sizes should be equal, for without prior knowledge of their performance most information will be gained by adding a record to the smallest family. Robertson assumed that the total number of animals recorded was T, comprising s families of size n, and gave an expression for $V(\hat{\tau})$, where τ is the intra-class correlation, in terms of T, n and τ. Differentiating gives the optimal family size as $n_0=1+1/\tau$, i.e., approximately $1/\tau$. As an illustration of the robustness of the design, if $\tau=0.1$ then $n_0=11$, but if family sizes of 2, 4,

8, 10, 12, 16 or 32 are used, it can be shown that $V(\hat{\tau})$ is increased by 202, 41, 3.2, 0.3, 0.2, 4.2 and 36%, respectively. Thus, a substantial amount of leeway about the optimum can be allowed with little loss of efficiency, except that, as Robertson (1959a) pointed out, small families should be avoided.

A minimax approach to the optimization was suggested by Taylor (1976a) who showed that, for a given family size, $V(\hat{\tau})$ is at a maximum when $\tau=(n-2)/[2(n-1)]$ and that for this value of τ, $V(\hat{\tau})$ is minimized when $n=4$. Although this small group size may be appropriate for the application of interest to Taylor, namely a multibreed experiment in which the groups comprise samples of, perhaps, widely divergent breeds, it is likely to be inefficient for most within-breed analyses where intra-class correlations of sibs are unlikely to greatly exceed 0.1.

These calculations assume that costs are proportional to numbers of offspring recorded. If there is an additional cost per family of C times the cost of an individual record, the optimal family size is increased to $2C+1/\tau$, approximately (Hill 1980b).

In a hierarchical design with full- and half-sib families, Robertson (1959a) showed that for estimating the half-sib correlation it is best to have $1/\tau$ dams per sire and one progeny per dam; similarly, for estimating the dam-family correlation it is best to have a single sire and $1/\tau$ progeny per dam, where τ is the appropriate correlation. If equal information on both correlations is required, it is best to take 3 or 4 dams per sire and $1/(2\tau)$ progeny per dam, or in a state of complete ignorance, 10 progeny per dam. Taylor (1976b) suggested minimax designs for the hierarchical case.

Genetic Correlations. The estimation of genetic correlations from data on sibs in a one-way classification was discussed by Robertson (1959b) for the case of traits with equal heritabilities and, more generally, by Tallis (1959). If both traits are recorded on each animal, their heritabilities are approximately equal and the genetic and environmental correlations are approximately equal, then the optimal design for estimating the genetic correlation is the same as for estimating the heritability, i.e., family sizes of $1/\tau$. If the heritabilities differ, but the genetic and phenotypic correlations are equal, the family sizes should lie between $1/\tau_1$ and $1/\tau_2$, where τ_1 and τ_2 are the intra-class correlations for the two traits. If the genetic and environmental correlations differ, family sizes should be increased. The case where the traits are recorded on different animals, for example milk production of heifers and growth rate of steers, has been discussed by Guiard and Herrendorfer (1977). Unless the genetic correlation is very high, the optimal family size for trait i is close to $1/\tau_i$.

The intra-class correlation structure can also be used to estimate genotype x environment interaction, or its converse the genetic correlation (r_A) between environments, by splitting families into several environments. In this case a reasonable assumption is that heritabilities are the same in each, and Robertson (1959b) showed that the optimal family size in each environment is $n_o=(1+r_A^2)^{1/2}/[\tau(1-r_A^2)]$, i.e., for $r_A=0.5$, 0.7, 0.8 and 0.9, $n_o=1.5/\tau$, $2.4/\tau$, $3.6/\tau$ and $7.1/\tau$, respectively. It can be seen that when the genetic correlation is high the optimal family sizes for detecting interaction are much larger than for estimating the heritability.

Taylor and Hnizdo (1987) have discussed designs for the estimation of correlations among breeds in the multibreed design. As for estimation of variances between breeds, optimal numbers of animals per breed are small.

4.2.2 Offspring-Parent Regression

No Selection. In this simple case it is assumed that data are available only on parents, with records either on one parent or on mid-parent mean, and on progeny, which may be in full- or half-sib families. It is not necessary that the same trait be recorded on parents and offspring. Let σ_X^2 and σ_Y^2 be the phenotypic variances in parents and progeny, respectively, τ the intra-class correlation of family members in the progeny, ρ the correlation of parent and progeny, and β the regression coefficient. With equal progeny family sizes (n), M families (assuming M is sufficiently large that terms of order 1/M can be ignored relative to unity), and no selection of parents,

$$V(\hat{\beta}) = [\tau-\rho^2+(1-\tau)/n]\sigma_Y^2/(M\sigma_X^2). \tag{4.1}$$

Latter and Robertson (1960) discuss the optimal choice of family size. Assuming the cost of recording parents is k per family (k=1 or 2 unless recording costs of parents have an additional cost element over that of progeny) and the cost of rearing a family is C relative to recording one offspring, the total cost is proportional to T=(k+C+n)M. Replacing M in (4.1) and differentiating gives the optimal family size, $n_o=[(k+C)(1-\tau)/(\tau-\rho^2)]^{1/2}$. For example, with son on sire regression and no additional costs per family (i.e., k=1 and C=0), if the heritability is 0.4 and there are half-sib families with no environmental correlation among sibs, then $\tau=0.1$ and $\beta=h^2/2=0.2$, so $n_o=15^{1/2}\doteq 4$. As Latter and Robertson (1960) pointed out, optimal family sizes are generally smaller for the offspring-parent regression than sib covariance estimators of heritability, in this example 4 and 11, respectively.

Selection and Assortative Mating of Parents. Provided the regression coefficient is linear (and in many applications, such as predicting selection response, it is assumed to be linear), the sampling variance of the regression coefficient can be reduced by rearing and recording progeny only from the more extreme parents and, where both parents are recorded, by positive assortative mating of the parents. Perfect assortative mating doubles the variance of mid-parent value compared to random mating and, therefore, halves the variance of the regression coefficient (Hill 1970). If parents are selected high and low, the benefit of assorting within the selected groups is much smaller, but not zero, so it is assumed throughout the following that assorting is practiced if regression on mid-parent is being computed.

In some situations, for example experimental populations of *Drosophila*, there may be no constraints on the ratio (R) of total numbers of parents and progeny scored, whereas in other cases R may be fixed, usually by constraints of field or laboratory facilities. If numbers are sufficiently large that infinite population normal assumptions can be used, and the highest and lowest proportions p of individuals are selected, making 2Mp families in all, the sum of squares of the independent variate is 2Mp(1+ix), where i is the selection intensity and x the abscissa on the standard normal curve. Simulation indicates this value is a good approximation, even when the numbers are small. Thus, the total cost is proportional to T=M[k+2p(C+n)] and

$$V(\hat{\beta}) = [\tau-\rho^2+(1-\tau)/n]\sigma_Y^2/[2Mp(1+ix)\sigma_X^2]$$

$$= [\tau-\rho^2+(1-\tau)/n][k+2p(C+n)]\sigma_Y^2/[2Tp(1+ix)\sigma_X^2]. \qquad (4.2)$$

If the ratio R=2np/k is fixed, n can be replaced by Rk/2p in (4.2) and the optimum found by differentiating solely with respect to p. If R is not fixed, partial differentiation with respect to n and p is necessary.

The solution for R fixed and C=0 is given by

$$x^2/[p(1+ix-x^2)] = 2(1-\tau)/[kR(\tau-\rho^2)], \qquad (4.3)$$

and a graph of $W(p)=x^2/[p(1+ix-x^2)]$ against p is given by Hill and Thompson (1977). For C≠0, the solution is more involved. If R can vary, the simultaneous equations for p and n are, respectively,

$$n = [(k/(2p)+C)(1-\tau)/(\tau-\rho^2)]^{1/2}$$

$$2(n+C)/k = W(p) \qquad (4.4)$$

as derived from Hill (1970). The optimal selection intensity is high, typically less than 20% selected, and optimal family sizes large, even when the costs per family are small. Sampling variances of regression or heritability estimates are typically halved relative to practicing no selection (see examples in Table 4.1). The designs are robust to poor prior estimates of the parameters (Hill 1970; Hill and Thompson 1977).

Unequal Family Sizes. Unequal family sizes are a consequence of high-low selection, but in the previous calculations the size of each selected family has been assumed to be the same. It is clear that, at least in principle, it is better to breed different numbers from each

Table 4.1 Optimal half-sib family sizes (n) and percentage selected p for estimation of heritability of a trait having observations on male parents and progeny, with one progeny per dam. The variance of the estimate, $V(\hat{h}^2)$, is expressed relative to the total number of observations, T, i.e., as VT

Heritability														
0.05			0.1			0.2			0.4			0.6		
p	n	VT	p	n	VT	p	n	VT	p	n	VT	p	n	VT
Intra-class sib correlation, no selection, no records on parents[a]:														
-	81	1.6	-	41	3.1	-	21	5.8	-	11	10.5	-	7	14.0
Progeny-parent regression, no selection[b]:														
-	9	4.9	-	7	5.2	-	5	5.6	-	4	5.8	-	4	5.5
Maximum likelihood, no selection[c]:														
-	70	1.5	-	30	2.8	-	10	4.4	-	4	5.5	-	4	5.2
Progeny-parent regression with selection[d]:														
2.7	39	1.6	3.9	23	1.9	5.2	15	2.4	6.4	11	2.7	6.7	10	2.6
Maximum likelihood with selection[e]:														
3.2	65	1.0	4.1	32	1.5	5.4	17	2.1	6.4	11	2.7	6.7	10	2.5

[a] Robertson (1959a), slightly modified.
[b] Latter and Robertson (1960).
[c] Hill and Nicholas (1974); Thompson (1976a), with model 2a(i) slightly modified to put onto common base.
[d] Hill (1970).
[e] Thompson (1976a), model 2a(ii) slightly modified as above.

family, according to the parental score, with the extreme animals having most offspring, the next most extreme rather less, and those near the middle none at all. Although it might be feasible to derive general rules for optimization, because the number of potential parents in any application is finite, it would seem most appropriate to compute the optimal sizes for the animals actually available. An example was simulated by sampling 30 random normal deviates and allocating the same number of offspring to the animals ranking 1 and 30, 2 and 29 and so on, and taking 120 progeny in all (R=4 assuming son on sire regression, with h^2=0.4 as in previous examples). It was found that with equal family sizes the top and bottom three individuals should be chosen and each have 20 progeny, whereas with unequal family sizes the optimum was to take 30, 12, 11, 5 and 2 offspring from the first, second, ..., fifth ranking top and bottom. This extra inequality, however, reduced $V(\hat{h}^2)$ by less than 10% for this specific example.

In many situations practical problems interfere: family sizes may be constrained by reproductive rate and infertility, it may not be possible to rear all progeny in one location so they have to be allocated to different groups with concurrent loss of efficiency, and so on. Nevertheless, selection greatly reduces the sampling variance of heritability estimates, both below those possible for offspring-parent regression without selection and below those possible using intra-class correlation (Table 4.1). The finer points of unequal family sizes of selected individuals are likely to add little more, for the real benefit derives from excluding as parents those near the mean that give no information about the slope of the linear regression.

4.2.3 Joint Sib and Offspring-Parent Analyses

Parameters such as heritability can be estimated from different kinds of relationships provided they can be assumed to have the same expectation; for example, from both the half-sib and offspring-parent covariances if there are no environmental covariances. These estimates are usually correlated (Hill and Nicholas 1974), but weighting procedures or, preferably, ML methods can be used to combine the data and make allowance for selection (Thompson 1976a). These authors give optimal designs for the case where records are taken on parents and offspring and examples of the efficiencies of the estimates using large sample theory. It is not possible here to review the calculations but, as illustration, solutions and comparisons with regression or sib correlation alone are given in Table 4.1 for the case of son on sire regression with one progeny per dam and with no restriction on the relative numbers (R) of parents and offspring scored.

Gianola (1979) discusses estimation of genetic variances, for which the offspring-parent and sib estimators are very highly correlated, rather than of heritabilities for the case of no selection. He considers family sizes at which the methods are equally efficient for specified heritability. For half-sib families the iso-efficiency family size ranges from 10 to 20 for heritability in the range 0.05 to 0.5, and then increases markedly for higher values. For estimation of genetic variances, it can be shown that optimal family sizes are very similar to those for estimation of heritability for the two specific cases of half-sib structure or offspring-parent without selection.

In the absence of selection the half-sib intra-class correlation estimator of heritability is more efficient than the offspring-parent regression at low heritabilities, and is less efficient at high heritabilities (Robertson 1959a). Combining both pieces of information by ML improves efficiency at intermediate heritabilities (h^2=0.2), but adds little at more extreme values. The optimal family sizes for ML estimation are similar to those for the more efficient method, i.e., they are very large at low heritabilities. Selection of parents increases the efficiency of the regression estimator considerably, and this is always more efficient than the half-sib estimator over all heritability values. Including ML with selection improves the efficiency at low heritability levels, but makes little difference at high levels. Hill and Thompson (1977) show that for the case of selection with R fixed, designs for use with regression alone are close to optimal for ML analysis so the simple formulae given previously can be used for design. For variable R the optimal designs are similar to those given for estimation by regression alone (Hill 1970) when heritability is moderate or high, but for low heritabilities optimal family sizes are larger (Thompson 1976a), as illustrated in Table 4.1.

Genetic Maternal Effects. If more than simple sib or offspring-parent correlations are available, other parameters can be estimated, for example, those of genetic maternal effects. Designs and analysis are discussed by Eisen (1967) and Thompson (1976b).

4.2.4 Genetic Correlations

The formulae for optimal design with offspring-parent regression are appropriate whether heritability or, more generally, the genetic regression of trait Y on trait X is to be estimated for evaluating schemes for indirect selection. If regressions of several traits on X are required, some compromise selection intensity in designing the offspring-parent regression analysis is needed. Because the designs are usually robust to poor prior values of the parameters it should be easy to find some compromise structure.

For two or more traits, the genetic covariances and correlations among them are often required in addition to their genetic variances and heritabilities. As Thompson (1976a) pointed out, it is not obvious what should be the criterion of optimality. Robertson (1959b) discussed the sampling variance of the genetic correlation coefficient (r_A), using results of Reeve (1955) for offspring-parent regression in the absence of selection. Robertson argued that for the case where the genetic and phenotypic correlations are equal, $V(\hat{r}_A)$ is approximately proportional to $[V(\hat{h}^2_X)\ V(\hat{h}^2_Y)]^{1/2}$, where $V(\hat{h}^2_X)$ and $V(\hat{h}^2_Y)$ are the sampling variances of the estimates of heritabilities in the same analysis. Therefore, designs which are efficient for heritability estimation are also efficient for estimation of the genetic correlation. For example, with half-sib families and regression on sire, family sizes of around 6 are likely to be near optimal for a wide range of traits.

Selection of Parents. Although biases in estimates can be avoided by ML techniques (Thompson 1976a) it is not obvious how, if at all, selection should be practiced if correlations among traits are required; for example, the sampling variance of the estimate of genetic correlation between Y and Z may be increased by selection on X. With two traits, an option is to allocate the parents into two groups and select within one group high and low for trait X; in the other, high and low selection for Y would be carried out. As the number of traits increases, this becomes less feasible. For two traits, Cameron and Thompson (1986) showed that it was better to select lines on orthogonal indices of the traits if these were correlated, i.e., HH, HL, LH and LL where, for example, HL denotes selecting high for X and low for Y. However, they also gave a more efficient method and suggested a useful general criterion for evaluating efficiency. This is illustrated here for two traits where there are three parameters to be estimated: h_X^2, h_Y^2 and $r_A h_X h_Y$ after standardization.

Cameron and Thompson (1986) suggest the D-optimality criterion for choice of design, namely minimizing the determinant of the sampling covariance matrix of the estimates of the three parameters. This also gives the optimum for any full-rank transformation of h_X^2, h_Y^2 and $r_A h_X h_Y$, for example, into standardized genetic regressions such as $r_A h_Y/h_X$. Rather than splitting the population into subgroups selected on different traits or indices, Cameron and Thompson (1986) show that it is better to select all extreme individuals using an elliptical selection criterion, in which all those lying outside a contour ellipse for the bivariate normal density are selected. In order to find the optimal design, the choice of the parameters for constructing the ellipse and the selection intensity depends on *a priori* assumptions; however, the estimators obtained are robust to poor prior values. For bivariate data, the variance of the estimates can be reduced by about 40% relative to designs in which half the individuals are selected on one trait and half on another; as the genetic correlation deviates from zero, the variances are reduced further.

4.3 Selection Experiments

4.3.1 Single Generation Experiments

Although the use of single generation selection experiments has been advocated (Bohren 1975), there is no important distinction between these and offspring-parent regression analyses for estimating heritability, and expectations of the estimators differ only if the regression is non-linear. In practice "realized" heritability is estimated from the selection experiment as the ratio of response to selection differential, for example the difference in mean of progeny of high and low parents as a ratio of the selection differential. Information on variation among the selected individuals is, therefore, lost compared with the regression estimate. Although this loss is small when selection is intense (Hill 1971), there seems little point in incurring it. A sampling error of the realized heritability can be estimated from the variance among replicates; however, it is likely to be estimated with more precision in a regression analysis that uses all the information, not just that among replicates.

The single generation experiment can be used to evaluate directly alternative selection methods, for example mass selection compared to selection on family means, but this is only necessary if the genetic assumptions are in question. Otherwise, comparisons can be made from estimates of the relevant parameters such as heritability and intra-class correlation of sibs. It might be efficient to compare selection schemes by selecting on the differences in criteria.

4.3.2 Multiple Generation Experiments

In the foregoing discussion of estimators involving data on only one or two generations, it is clear that inferences are to be made about the present population. In a selection experiment spanning several generations the objective may be to make inferences to the base population, to make inferences about potential breeding programs spanning a similar number of generations, or to find out how parameters change as a consequence of selection. In many cases, the objectives are all of these and more! For any rational, and certainly quantitative analysis of design, it is obviously necessary to be clear as to what is required (see also Hill 1980a).

Genetic parameters such as heritability must change as a consequence of selection for two conceptually distinct reasons: the reduction in variance (assuming truncation) among the selected individuals, and the change in gene frequency due to the selection. The former arises even if the trait is determined by infinitely many additive genes of infinitesimal effect at unlinked loci. The effects of known selection for this infinitesimal model are predictable from a knowledge of the base population parameters (Bulmer 1980). Further, the parameters of the base can be estimated by mixed model methods if the infinitesimal model holds: just as selection effects can, in principle, be removed by BLUP if heritability is known, so can the base population heritability in a multigeneration experiment be estimated by REML without bias by conditioning results on the selection applied (Thompson 1982). Individual animal models being developed permit such an analysis, with the necessary assumption that selection be based only on data included in the analysis.

Changes in variances due to changes in gene frequency are not predictable *a priori* without making assumptions about the distribution of gene effects and frequency and, therefore, cannot be removed in an analysis intended to estimate base population parameters (unless data on all generations but the first are ignored!). In view of the common observation that responses remain close to linear over several generations in many selection experiments (Falconer 1981), and the fact that the informational content of a selection experiment rises with generation number, we are often satisfied to analyze the experiment as if no variance change occurred. Simplistic analyses (e.g., Hill 1971) have ignored the predictable changes due to selection with the infinitesimal model but with better analytical techniques becoming available, such simplifications now seem unjustified.

In earlier discussions of the design of selection experiments (e.g., Hill 1971, 1980a) the mean performance of the selection line was regarded as a Markov process or random walk. The genetic drift variance is assumed to accumulate linearly with generations, at a rate approximately proportional to V_A/N_e, where V_A is the additive genetic variance and N_e the effective population size. The variance of observed generation means is assumed to equal the genetic drift variance plus a term V_P/M, approximately, due to measurement of mean genotype from mean phenotype on a finite number M of individuals, with V_P being the phenotypic variance. The design, in terms of selection intensity, numbers of individuals and generations can then be optimized within the constraints imposed by time and facilities. An extensive discussion is given by Hill (1980a) and need not be repeated here; for example, Nicholas (1980) has considered the relevant scale of experiments for specific objectives.

Analysis. Analysis of an experiment to estimate heritability from the regression of response on selection differential utilizes only the information from differences in generation means in relation to differences between the mean of selected parents and the rest. There is clearly further information available on genetic parameters such as heritability from within generation variability, and it is argued above that there is no justification in ignoring this information when analyzing single generation experiments. In the context of mixed model analysis, the drift variances and covariances among generation means appear as average relationships or coancestry among individuals in the same or different generations (Sorensen and Kennedy 1983). Put simply, the mean covariance of performance of collateral individuals is $2FV_A \sim tV_A/N_e$, where F is the inbreeding coefficient and t is generation number. Thus, standard mixed model procedures such as REML which do not depend on prior estimates of parameters can, in principle, be used. Although all the relevant information can be extracted in the multiple generation case, given sufficient computing power, the extra information will not be large despite the vast number of relationships available. This is so because they involve many distant relationships and are within a selected group, whereas generation means differ as a result of several cycles of selection. This latter point, although not quantified here, is relevant to design, for it has been argued that selection experiments do not require concurrent control or similar populations for elimination of environmental trend (Blair and Pollak 1984); however, estimates of trend then depend on prior values (Thompson 1986). Such an argument would hold only if a small part of the information came from the between-generation comparisons. Thus, I do not believe that developments in methods of analysis are likely to have radical influences on design but we do require more formal calculations of optima for the multigeneration experiments, such as has been done for the offspring-parent case by Thompson (1976a).

For the infinitesimal model where inferences are made to the base population, estimates of the variance in response can be made without replication using the estimates of genetic variance obtained from the analysis of generation means. Estimates of variance of response of correlated traits depend on estimates of genetic variance for these traits. As discussed previously (Hill 1980a), variances of correlated responses cannot, therefore, be obtained from an analysis using only responses and selection differentials. Either *a priori* variance estimates are required, or they have to be obtained from a mixed model analysis using all relationships but with relatively higher sampling error, because less information than for the selection trait will come from the analysis of generation means.

If the selection experiment is being used to predict or describe responses over several generations in the expectation that these may change, and without the sole intention of describing the base population, there seems to be no real alternative to replication of the selection lines. However, as anyone who has designed selection experiments knows, it is difficult to reach any satisfactory compromise between degrees of freedom among replicates and size of replicates. It may be best to use as many replicates as is compatible with the inbreeding levels which can be tolerated over the anticipated duration of the experiment. Another problem which arises is whether selection should be practiced in only the one direction of practical interest, for example, for high growth rate, with a control, or whether the more efficient divergent selection can be used (Hill 1980a,b). Under the infinitesimal model, divergent selection is clearly appropriate. If the experiment is short, asymmetry is unlikely unless there are genes of very large effect. Only if long-term results are required and parameter estimation is not the objective, is unidirectional selection likely to be necessary.

Efficiency. Results have been given elsewhere to illustrate the efficiency of selection experiments. For example, Table 3 of Hill (1971) shows that for full-sib families a selection experiment of five-generation duration, with a total of T recorded overall, gives sampling variances of the heritability estimate of 0.36/T and 0.85/T for h^2=0.1 and 0.4, respectively, whereas one round of offspring-parent regression with selection gives variances of 0.82/T and 1.21/T, respectively. These figures relate to selection in both sexes; with selection in only one sex the corresponding figures are 0.94/T and 2.30/T for the selection experiment and 2.52/T and 3.94/T for the regression analysis. The long experiment is relatively more efficient when heritability is low, essentially because the measurement error drops faster than the genetic drift error. Some discussions of efficiency (e.g., Nicholas 1980) have been given solely in terms of the drift variance, appropriate only if the experiments are long-term.

Overlapping Generations. For farm animal populations in which generations overlap some of these discussions of design and analysis are oversimplistic. The main problems are that animals cannot be attributed to specific generations, that the variance-covariance structure of mean performances each year cannot be expressed as a simple function of effective population size, particularly in early years, and that the number of years to keep breeding animals is an additional variable to be taken into account when optimizing design. There have been no substantial studies of design with overlapping generations but methods of analysis are now being clarified. The variance-covariance structure of yearly means was derived by Johnson (1977). Pitfalls in the calculation of cumulative selection differentials were pointed out by James (1986) and Thompson (1986), together with ways of computing them. Atkins and Thompson (1986) used these results to compute realized heritabilities from yearly means by weighted least-squares. A full analysis would involve using REML and here the conceptual problems are no greater for overlapping than for non-overlapping generations, because information is put together through the relationship matrix.

4.4 Field Experiments

Let us now consider what information can be obtained from adequately designed field experiments, particularly those using data collected for other purposes. For dairy cattle, for example, heritability estimates for production traits come from existing data but if some hormone, metabolite or other variable is thought to be a useful predictor of breeding value for milk, specific data have to be collected. In many such cases the additive genetic variance or heritability of the indicator trait is not, *per se*, of relevance, but only its accuracy as a predictor of the production trait. This has the important consequence that if we want to know whether the indicator trait (X) is a good predictor of yield (Y), we can do the experiment the other way round and select on Y. For example, the regression of X recorded on progeny on Y recorded on mid-parents equals $Cov_A/\sigma^2_{AY}=(r_A h_X/h_Y)(\sigma_{PX}/\sigma_{PY})$. It is usually possible to estimate the phenotypic standard deviation of the indicator variable without substantial error, and h_Y and σ_{PY} on the yield trait are likely to be known. Hence the accuracy, $r_A h_X$, of selection on X, or the genetic covariance between X and Y can be estimated. Further, in the same experiment with selection on Y, all other potential indicators X_2, X_3 can be examined, whereas if parents are selected on X, for example, the efficacy of X_2 cannot be well tested, as discussed further by Hill (1985). Such experiments also yield estimates of repeatability of the indicator variables. They do not give estimates of their genetic variance or of heritability, but these are only required when constructing selection indices to combine information on relatives for the indicator variables, which is less critical than the estimate of accuracy of prediction of the correlated trait.

As an example, assume that metabolite X, and also X_2, X_3, ... are to be recorded in bulls as potential indicators of yield in dairy cattle. Alternative procedures for estimating $Cov_A(X,Y)$ include: taking a sample of young bulls and correlating X with their half sister's yield; regressing daughters' yield on father's value of X in the normal progeny testing scheme; selecting young bulls of high and low predicted breeding value for yield and regressing X on Y. The first of these is inefficient (Table 4.1) and correction needs to be made for selection. In the second scheme it is necessary to wait a long time and, again, correction is necessary for selection. Also, if the regression calculated is that of X on $\hat{A}_Y$ from progeny test, precision is lost because the bulls are selected to have high values of $\hat{A}_Y$. It is, however, necessary to work out the sampling errors of the estimates. We assume $E((\hat{A}_Y)= E(A_Y)$, i.e., BLUP or a similar procedure is used, and $V(A_Y|\hat{A}_Y)=(1-\rho^2)V_{AY}$. It follows that:

$$E(X|\hat{A}_Y) = \beta_{AX.AY} = r_A\sigma_{AX}/\sigma_{AY}$$

and

$$V(X|\hat{A}_Y) = (1-\rho^2_{X\hat{A}Y})V_{PX} = (1-\rho^2 r_A^2 h_X^2)V_{PX}.$$

Note that unless both the heritability of X and the genetic correlation (r_A) between X and Y are high, the accuracy of assessment of the breeding value of Y has little influence on the variance about regression. It, therefore, follows that

$$V(\hat{\beta}_{AX,\hat{A}Y}) = V(X|\hat{A}_Y)/\Sigma(\hat{A}_Y-\bar{\hat{A}}_Y)^2$$

is mostly influenced by the sum of squares of $\hat{A}_Y$ values. The latter is affected by the selection practiced, being increased by the use of young bulls of high and low predicted breeding value, rather than of only high progeny tested bulls, but reduced in proportion to the accuracy of selection, because $V(\hat{A}_Y)=\rho^2 V_{AY}$.

One practical design might involve taking a group of young bulls of high predicted breeding value which are to be used for progeny testing, and a contemporary group of bulls of low predicted breeding value, perhaps from the same herd to avoid environmental influences. The indicator traits would be recorded on each of the two groups, and the latter would be disregarded, without progeny testing them. The alternative would be to use a group of high and low ranking progeny tested bulls. Although this has the benefit that $V(\hat{A}_Y)$ is increased because selection is more accurate, particular animals would have to be bred.

A related type of design is where selected lines have been maintained for a variety of commercial traits, as advocated by Land (1981). Provided control or divergent selected lines are maintained, the lines can be used to predict the relevant genetic regressions. For example, populations selected for growth rate, feed efficiency, fatness and reproductive rate in pigs would serve such a function, as would populations selected on a range of indices, HH, HL, LH and LL for growth and fatness, where HH denotes an index of selection for high growth and high fatness, and the indices are chosen to maximize information (Cameron and Thompson 1986). Such designs generate differences in the population, whereas random mating populations maintained without selection lead to less efficient estimates of correlations and regression.

4.5 Concluding Remarks

This paper has left a lot of loose ends and there remains much further work to be done on design and analysis of experiments to estimate genetic parameters. Provided objectives are clear, this can be done in a rational way. Particular areas which have been identified are: the optimization of selection experiments to be analyzed by mixed model methods using all relationships; the design of selection experiments with overlapping generations; and the sensitivity of responses to departures from the infinitesimal model, and their implications to inferences on parameters in the base population.

Acknowledgements. I am grateful to Robin Thompson for many constructive points, and to him, Neil Cameron and Charles Smith for comments on the manuscript.

References

Atkins KD, Thompson R (1986) Predicted and realised responses to selection for an index of bone length and body weight in Scottish Blackface sheep. 1. Response in the index and component traits. Anim Prod 43:421-435

Blair HT, Pollak EJ (1984) Estimation of genetic trend in a selected population with and without the use of a control population. J Anim Sci 58:878-886

Bohren BB (1975) Designing artificial selection experiments for specific objectives. Genetics 80:205-220

Bulmer MG (1980) The mathematical theory of quantitative genetics. Clarendon Press, Oxford

Cameron ND, Thompson R (1986) Design of multivariate selection experiments to estimate genetic parameters. Theor Appl Genet 72:466-476

Eisen EJ (1967) Mating designs for estimating direct and maternal genetic variances and direct maternal covariances. Canad J Genet and Cytology 9:13-22

Falconer DS (1981) Introduction to quantitative genetics. 2nd edn Longman, London

Gianola D (1979) Estimation of genetic covariance from joint offspring-parent and sib-sib statistics. Genetics 93:1039-1049

Guiard V, Herrendorfer G (1977) Estimation of the genetic correlation coefficient by half-sib analysis if the characters are measured on different offsprings. Biometrical J 19:31-36

Hill WG (1970) Design of experiments to estimate heritability by regression of offspring on selected parents. Biometrics 26:566-571

Hill WG (1971) Design and efficiency of selection experiments for estimating genetic parameters. Biometrics 27:293-311

Hill WG (1972) Estimation of realised heritabilities from selection experiments. I. Divergent selection. Biometrics 28:747-765

Hill WG (1980a) Design of quantitative genetic selection experiments. In: Robertson A (ed) Selection experiments in laboratory and domestic animals. Common Agric Bur, Slough, pp 1-13

Hill WG (1980b) Experimental design in quantitative genetics and animal breeding. Unpublished notes for course in Göttingen, West Germany

Hill WG (1985) Detection and genetic assessment of physiological criteria of merit. In: Land RB, Robinson DW (eds) Genetics of reproduction in sheep. Butterworths, London, pp 319-331

Hill WG, Nicholas FW (1974) Estimation of heritability by both regression of offspring on parent and intra-class correlation of sibs in one experiment. Biometrics 30:447-468

Hill WG, Thompson R (1977) Design of experiments to estimate offspring-parent regression using selected parents. Anim Prod 24:163-168

James JW (1986) Cumulative selection differentials and realized heritabilties with overlapping generations. Anim Prod 42:411-415

Johnson DL (1977) Variance-covariance structure of group means with overlapping generations. In: Pollak E, Kempthorne O, Bailey TB Jr (eds) Proceedings of the International Conference on Quantitative Genetics. Iowa State Univ Press, Ames, pp 851-858

Land RB (1981) An alternative philosophy for animal breeding. Livest Prod Sci 8:95-99

Latter BDH, Robertson A (1960) Experimental design in the estimation of heritability by regression methods. Biometrics 16:348-353

Nicholas FW (1980) Size of population required for artificial selection. Genet Res 35:85-105

Reeve ECR (1955) The variance of the genetic correlation coefficient. Biometrics 11:357-374

Robertson A (1959a) Experimental design in the evaluation of genetic parameters. Biometrics 15:219-226

Robertson A (1959b) The sampling variance of the genetic correlation coefficient. Biometrics 15:469-485

Sales J, Hill WG (1976a) Effect of sampling errors on efficiency of selection indices. 1. Use of information from relatives for single trait improvement. Anim Prod 22:1-17

Sales J, Hill WG (1976b) Effect of sampling errors on efficiency of selection indices. 2. Use of information on associated traits for improvement of a single important trait. Anim Prod 23:1-14

Sorensen DA, Kennedy BW (1983) The use of the relationship matrix to account for genetic drift variance in the analysis of genetic experiments. Theor Appl Genet 66:217-220

Tallis GM (1959) Sampling errors of genetic correlation coefficients calculated from analyses of variance and covariance. Aust J Stat 1:35-43

Taylor S CS (1976a) Multibreed designs. 1. Variation between breeds. Anim Prod 23:133-144

Taylor S CS (1976b) Multibreed designs. 2. Genetic variation within and between breeds. Anim Prod 23:145-154

Taylor S CS, Hnizdo E (1987) Multibreed designs. 3. Inter-breed relationships. Anim Prod 44:39-53

Thompson R (1976a) Design of experiments to estimate heritability when observations are available on parents and offspring. Biometrics 32:283-304

Thompson R (1976b) The estimation of maternal genetic variances. Biometrics 32:903-917

Thompson R (1982) Methods of estimation of genetic parameters. In: Proc 2nd World Congr Genet Appl Livest Prod. Garsi, Madrid 5:95-103

Thompson R (1986) Estimation of realized heritability in a selected population using mixed model methods. Genet Sel Evol 18:475-483

5 Use of Mixed Model Methodology in Analysis of Designed Experiments

B.W. Kennedy[1]

Applications of mixed model methods in the analysis of designed experiments are illustrated and discussed for purposes of: (1) increasing rate of selection response to create genetically diverse lines rapidly or to demonstrate feasibility of selection, (2) estimation of genetic parameters free of bias from selection and inbreeding, (3) estimation of response to selection with or without controls, and (4) verification of experimental design prior to the experiment. For traits controlled by a large number of additive loci, use of the numerator relationship matrix in the mixed model equations accounts for changes in additive genetic variance due to inbreeding, assortative mating and gametic disequilibrium resulting from selection. If the number of loci is small, non-normality of the genotypic distribution and changes in variance due to gene frequency changes (including fixation) are not accounted for but these seem to be of small consequence, at least for short-term selection. Use of mixed model methods do not require prior knowledge of base population heritability which can be estimated from the data unaltered by selection. If dominance effects are important, properties of the dominance relationship matrix and use of mixed model methods are not yet well understood in inbred and selected populations. Simulation results indicate that use of mixed model methods can be effective in randomly mated populations, even if the number of loci is small. There is evidence of bias in selected populations, however, particularly when gene frequencies are extreme. Properties of mixed model methods under dominance and other non-additive genetic models need more study.

5.1 Introduction

Mixed model methods have been applied widely in animal breeding to data derived from field populations of livestock. Such data are usually highly unbalanced and generally involve overlapping generations and confounding between genetic and environmental factors. Principal applications of mixed model methods have been for genetic evaluation, estimation of genetic parameters, estimation of genetic trends and identification and estimation of important

[1] Centre for Genetic Improvement of Livestock, University of Guelph, Guelph, Canada.

environmental influences. Mixed model methods have received relatively little use in analysis of experimental data, although some applications have been made recently (e.g., Blair and Pollak 1984; Hough and Benyshek 1986). Sorensen and Kennedy (1986) examined some of the properties of mixed model methods in these types of application.

This paper examines further the use of mixed model methods in the analysis of data from selection experiments, not only for estimation of genetic parameters and response to selection, but also to increase selection response. Some consideration to applications of mixed model methods to the design of experiments is given also. The focus is on selection experiments involving livestock, as opposed to laboratory species, where control over environmental conditions is usually less precise, and frequently, because of cost considerations or the necessity to maintain population size, breeding females are retained for several reproductive cycles. The thrust of the paper is on use, properties and interpretation rather than on detailed descriptions of mixed model methods. These are provided elsewhere (e.g., Henderson 1984; Sorensen and Kennedy 1986).

5.2 Mixed Model Methods

Formal development of the general theory of mixed model methods applied to animal breeding was first presented by Henderson (1963), although applications had been made earlier (e.g., Henderson 1949). Mixed model methods can provide simultaneously estimates of fixed effects and predictions of random variables with defined statistical properties, given the assumptions of the model. A full treatment of mixed linear model theory and its applications to animal breeding is in Henderson (1984).

Let the model be

$$\mathbf{y} = \mathbf{Xb}+\mathbf{Zu}+\mathbf{e}, \tag{5.1}$$

where $\mathbf{y}$ is a vector of observations, $\mathbf{b}$ is a vector of fixed effects, $\mathbf{u}\sim(\mathbf{0},\mathbf{G})$ is a vector of random variables, $\mathbf{e}\sim(\mathbf{0},\mathbf{R})$ is a vector of random residuals and $\mathbf{X}$ and $\mathbf{Z}$ are incidence matrices. Also, it is generally assumed that var($\mathbf{y}$)=$\mathbf{ZGZ'}+\mathbf{R}$, i.e., cov($\mathbf{u}$,$\mathbf{e}'$)= $\mathbf{0}$. The model is perfectly general in that it can represent observations on any number of traits, and $\mathbf{b}$ and $\mathbf{u}$ can represent any number of fixed effects and random variables, respectively.

Solutions for $\mathbf{b}$ and $\mathbf{u}$ are obtained from

$$\begin{bmatrix} \mathbf{X'R^{-1}X} & \mathbf{X'R^{-1}Z} \\ \mathbf{Z'R^{-1}X} & \mathbf{Z'R^{-1}Z+G^{-1}} \end{bmatrix} \begin{bmatrix} \tilde{\mathbf{b}} \\ \hat{\mathbf{u}} \end{bmatrix} = \begin{bmatrix} \mathbf{X'R^{-1}y} \\ \mathbf{Z'R^{-1}y} \end{bmatrix}, \tag{5.2}$$

assuming **G** and **R** are non-singular. If **G** and **R** are known, or at least known except for an unknown constant σ_e^2, then $\tilde{\mathbf{b}}$ is a generalized least-squares solution to **b**, $\mathbf{K}'\tilde{\mathbf{b}}$ is the best linear unbiased estimator (BLUE) of **K'b**, provided **K'b** is estimable, and $\hat{\mathbf{u}}$ is the best linear unbiased predictor (BLUP) of **u** (Henderson 1973). Selection can alter the variance-covariance structures of **y**, **u** and **e**, but if these vectors are jointly multivariate normal prior to selection and selection has operated on a linear function of **y**, say **L'y**, then solution of (5.2) still leads to BLUE of **K'b** and BLUP of **u** if **L'X=0** (Henderson 1975).

As mentioned, (5.1) is a very general model and **u** can comprise several random factors. Genetic effects, additive and non-additive, as well as non-genetic effects such as environmental effects common to animals reared together (e.g., litter effects) or those common to repeated records on the same animal (permanent environmental effects) can be represented in **u**. A brief outline of some specific cases follows.

First consider a simple additive genetic model for the situation where single records on a single trait are available. In this case

$$\mathbf{y} = \mathbf{Xb}+\mathbf{Za}+\mathbf{e}, \tag{5.3}$$

where $\mathbf{a}\sim(\mathbf{0},\mathbf{A}\sigma_a^2)$ is the vector of additive genetic values of animals in the population, $\mathbf{e}\sim(\mathbf{0},\mathbf{I}\sigma_e^2)$ is the vector of random environmental effects, and (5.2) reduces to

$$\begin{bmatrix} \mathbf{X'X} & \mathbf{X'Z} \\ \mathbf{Z'X} & \mathbf{Z'Z}+\mathbf{A}^{-1}\lambda \end{bmatrix} \begin{bmatrix} \tilde{\mathbf{b}} \\ \hat{\mathbf{a}} \end{bmatrix} = \begin{bmatrix} \mathbf{X'y} \\ \mathbf{Z'y} \end{bmatrix}, \tag{5.4}$$

where $\lambda=\sigma_e^2/\sigma_a^2=(1-h^2)/h^2$. If the trait is recorded on all animals, **Z** is an identity matrix (**I**). If not, **Z** is similar to **I** except it will be augmented by columns of zeros corresponding to animals without records. **A** is the matrix of additive relationships and is comprised of off-diagonal elements a_{ij} equal to the numerator of Wright's (1922) coefficient of relationship between the ith and j^{th} animals, and diagonal elements a_{ii} equal to $1+F_i$, where F_i is the coefficient of inbreeding of the ith animal.

Although the statistical properties of the estimates derived from (5.4) are well-defined, their genetic properties are less well understood. In (5.3), E(**a**)=**0** and σ_a^2 (or alternatively h^2) is a constant, even though the data may span a number of generations and selection may have operated. It is well recognized, however, that both the genetic mean and variance can and do change in animal populations. Some of these changes can occur as a consequence of chance sampling in a population of finite size and other changes are directed by selection. Fortunately, inclusion of $\mathbf{A}^{-1}$ in (5.4) accommodates many of these changes.

For example, it is well established that in closed lines or populations of finite size, inbreeding accumulates with each generation as does genetic drift (e.g., Falconer 1981). Within-line additive genetic variance is reduced and between-line additive genetic variance is increased. For any line, the effect of drift on the genetic mean cannot be predicted but the genetic variance between lines due to drift can be predicted from knowledge of the population structure (Hill 1971). Further, the effect of drift on the genetic means of control populations can be estimated, in retrospect, and the variances of both least-squares and mixed model estimators of genetic means of generations can be expressed as a function of the relationship matrix, which can be interpreted in terms of drift variance (Sorensen and Kennedy 1983, 1986). The relationship matrix, because it accounts for the correlated structure among observations due to common additive genetic effects, accounts for drift variance.

Genetic means and variances change also as a result of selection. Felsenstein (1965) has shown that directional selection causes negative covariances between the frequencies of genes at different loci, that is negative linkage or gametic disequilibrium. Bulmer (1971) showed that with many additive loci, the change in variance is almost exclusively due to the generation of gametic disequilibrium and not to changes in gene frequency. Bulmer (1971) has derived formulae to predict changes in genetic variance each generation from selection under an infinitesimal additive model. It can be demonstrated that the selection model of Henderson (1975), which is statistical in nature and makes use of a result of Pearson (1903), leads to the same result as Bulmer (1971), if selection is examined at the gametic level and recombination is allowed for. Other genetic mechanisms, such as assortative mating, also alter genetic variability (Bulmer 1980; Falconer 1981).

Henderson's (1975) selection model assumes normality and implicit in this is the assumption of a large, strictly speaking infinite, number of loci. Selection leads to departures from normality, but such departures are small as long as the number of loci is large (Bulmer 1980). Additionally, Bulmer (1971, 1980) has shown, also assuming normality, that the Mendelian sampling component of an animal's breeding value is unaffected by selection in previous generations. Sorensen and Kennedy (1984a) made use of this result, in conjunction with matrix results of Henderson (1976) and Thompson (1977) on the structure of **A**, to argue that if **A** is complete for a group of animals its use accommodates changes in variance due to gametic disequilibrium which result from selection.

Simulation work has verified that genetic evaluations from use of (5.4) are not detectably biased by selection and that changes in genetic variance due to gametic disequilibrium and inbreeding (Sorensen and Kennedy 1984a) as well as assortative mating (Kemp 1985) are accommodated. However, if the trait is controlled by relatively few loci, changes in genetic variance due to gene frequency changes, including fixation, are not accounted for by (5.4) and biased evaluations can result (Maki-Tanila and Kennedy 1986).

Essentially, the assumptions in (5.3) that $E(\mathbf{a})=\mathbf{0}$ and σ_a^2 or h^2 is a constant refer to the base population of animals prior to the imposition of selection and the accumulation of inbreeding. Indeed, one of the properties of $\hat{\mathbf{a}}$ is that if generations do not overlap the elements of $\hat{\mathbf{a}}$ for the base population animals, which are assumed unrelated, sum to zero. This is consistent with having zero expectation; however, the solutions for animals in other generations do not. In effect, the mixed model equations require assumed knowledge of heritability prior to selection but not subsequent to it. The relationship matrix accommodates subsequent changes, given that the number of loci involved is large.

In practice, heritability in the base population is never known exactly, but can only be estimated. However, Kackar and Harville (1981) have shown that, in the absence of selection, use of estimates of variances obtained by any of the commonly used variance component estimation procedures leads to unbiased (but not best) predictors of genetic values. Gianola et al (1986) have argued from a Bayesian viewpoint that use of restricted maximum likelihood (REML) estimates of variances, obtained from the data, in the evaluation process yields the closest possible approximation to the expected breeding value given the data, assuming normality and the absence of good prior information on the variances. Sorensen and Kennedy (1986), through simulation, showed that in selected populations, estimation of heritability by minimum variance quadratic unbiased estimation (MIVQUE) from the data, and use of that estimate in the evaluation process, led to seemingly unbiased predictors of additive genetic value, over repeated sampling. However, prediction error was greater than when the true value of heritability was used.

Extension of (5.3) to more complicated genetic models and models involving non-genetic random factors is straightforward. For example, dominance effects can be added to the model such that

$$\mathbf{y} = \mathbf{Xb}+\mathbf{Za}+\mathbf{Zd}+\mathbf{e}, \tag{5.5}$$

where $\mathbf{d}\sim(\mathbf{0},\mathbf{D}\sigma_d^2)$ is the vector of dominance genetic effects of animals in the sample, and (5.4) is expanded to

$$\begin{bmatrix} \mathbf{X'X} & \mathbf{X'Z} & \mathbf{X'Z} \\ \mathbf{Z'X} & \mathbf{Z'Z}+\mathbf{A}^{-1}\lambda & \mathbf{Z'Z} \\ \mathbf{Z'X} & \mathbf{Z'Z} & \mathbf{Z'Z}+\mathbf{D}^{-1}\tau \end{bmatrix} \begin{bmatrix} \tilde{\mathbf{b}} \\ \hat{\mathbf{a}} \\ \hat{\mathbf{d}} \end{bmatrix} = \begin{bmatrix} \mathbf{X'y} \\ \mathbf{Z'y} \\ \mathbf{Z'y} \end{bmatrix}, \tag{5.6}$$

where $\tau=\sigma_e^2/\sigma_d^2$ and other elements are as defined previously. Evaluation for total genetic merit is then $\hat{\mathbf{g}}=\hat{\mathbf{a}}+\hat{\mathbf{d}}$.

However, unlike **A**, the structure and properties of **D** are generally not known for **D** of selected populations or where inbreeding is involved. For example, inbreeding coefficients of animals are an integral part of **A**, but construction of **D** to allow for inbreeding is not as well understood. As an approximation, one might simply ignore these complications and compute **D** as one would for unselected and non-inbred populations, i.e., $\mathbf{D}=\{d_{ij}\}$, where d_{ij}, the dominance relationship between animals i and j, is a function of the additive relationships between sires (p) and dams (m) of i and j, and is equal to 0.25 $(a_{p_ip_j}a_{m_im_j}+a_{p_im_j}a_{m_ip_j})$. How well or poorly this approximation might behave is not known. Ignoring inbreeding contributions to **A** in evaluation for additive genetic merit can lead to bias.

Smith (1984) has proposed a probabilistic method for computing **D** for inbred populations based on a genomic table for calculating relationships (Smith and Allaire 1985). Maki-Tanila and Kennedy (1986) applied this algorithm to simulated data under dominance genetic models with a finite number of loci and did not detect significant bias in estimates of genetic merit $(\hat{\mathbf{a}}+\hat{\mathbf{d}})$ under random mating with inbreeding, although theoretically some bias would be expected. Genetic merit declined over generations, because of inbreeding depression, but estimates of genetic merit tended to reflect this more when **D**, computed according to Smith (1984), was included in the evaluation. If **D** was ignored and analysis was (incorrectly) according to an additive model only, genetic merit was overestimated. With selected populations, Maki-Tanila and Kennedy (1986) showed that including **D** in the evaluation process improved prediction of total genetic value, but evaluations showed significant bias, particularly when gene frequencies were extreme. More work is required to understand better the properties of **D** and resulting estimates of **d** from (5.6) when selection and inbreeding have operated.

Epistatic effects can also be evaluated. For example, (5.3) can be expanded to include an additive x additive effect according to

$$\mathbf{y} = \mathbf{Xb}+\mathbf{Za}+\mathbf{Zi}+\mathbf{e}, \tag{5.7}$$

where **i** is a vector of additive x additive effects, $\text{var}(\mathbf{i})=\mathbf{H}\sigma_i^2$ and other elements are as defined previously. Expansion of (5.7) to include other forms of epistatic effects, for example, additive x dominance effects, is possible. Henderson (1984, 1985a), based on results of Cockerham (1954), has expressed **H** as the Hadamard product (see Styan 1973) of **A** such that $\mathbf{H}=\mathbf{A}*\mathbf{A}=\{a_{ij}^2\}$ for non-inbred populations. If σ_i^2 is known, then **i** is predicted from

$$\begin{bmatrix} \mathbf{X'X} & \mathbf{X'Z} & \mathbf{X'Z} \\ \mathbf{Z'X} & \mathbf{Z'Z}+\mathbf{A}^{-1}\lambda & \mathbf{Z'Z} \\ \mathbf{Z'X} & \mathbf{Z'Z} & \mathbf{Z'Z}+\mathbf{H}^{-1}\theta \end{bmatrix} \begin{bmatrix} \tilde{\mathbf{b}} \\ \hat{\mathbf{a}} \\ \hat{\mathbf{i}} \end{bmatrix} = \begin{bmatrix} \mathbf{X'y} \\ \mathbf{Z'y} \\ \mathbf{Z'y} \end{bmatrix},$$

where $\mathbf{H}^{-1}\theta=(\mathbf{A}*\mathbf{A})^{-1}\sigma_e^2/\sigma_i^2$. The form of **H** in the presence of inbreeding and selection is not well defined.

More detail on applications of mixed model methods in animal breeding and illustrations of their computation are in Henderson (1984), and supplementary discussion and illustration on some aspects are in Kennedy and Schaeffer, Chapter 23, this Volume. The balance of this paper is devoted to assessing the usefulness of mixed model methods in analysis of experimental data derived from both control and selected populations.

5.3 Selection of Breeding Animals

In field populations, animals are selected routinely on the basis of additive genetic value as estimated through use of mixed model methods. Further, genetic evaluation systems are becoming increasingly sophisticated and complex as a consequence of invoking models that are biologically more accurate and realistic (Kennedy and Schaeffer, Chap. 23, this Vol.). In contrast, little use, if any, has been made of mixed model methods in the conduct of selection experiments, that is, in choosing the parents of the next generation. Most selection methods used are quite simple, at least from a statistical viewpoint.

The appropriateness of use of mixed model methods for selection of breeding animals in an experiment depends greatly on the intent of the experiment. Some experiments are designed to test genetic theory and because properties of mixed model methods rest on a number of theoretical assumptions, their use (if at all) in such situations, must be careful and judicious. Such experiments must be carefully planned and designed to obviate the need for mixed model methods to disentangle confounding between genetic and environmental influences in the selection of animals. This is not always simple to do nor is it always done in experiments with large animals.

The primary intent of many selection experiments with large animals is to test or demonstrate that selection response is feasible for the trait or traits of concern. In these experiments, the experimental herd or flock is functioning as a model for a commercial unit, and the test is to see whether commercial production can be improved by carrying out a selection program similar to that of the experiment. In this context, it makes little sense to use selection methods that are likely to be less efficient than those used routinely in the field. The use of mixed model methods for selection in livestock populations is increasing greatly. By the time any newly initiated selection experiment with large animals is completed, selection of dairy and beef cattle, swine and possibly sheep according to mixed model methods under an animal model such as (5.3), or more sophisticated ones, is likely to be the norm at least in commercial populations that maintain a pedigreed breeding structure. In such applications, the experiment should be conducted using selection aids that are at least as advanced, and preferably more so, than those used by industry.

Also, the intent of some selection experiments is to create genetically divergent lines rapidly to allow for subsequent detailed study of biological differences between animals of widely different genetic types. Efficiency of these experiments is likely to be increased if selection is through use of mixed model methods.

It is surprising, therefore, that mixed model methods have not found wider use in animal selection experiments. For example, several experiments have been carried out to increase litter size at birth in pigs, none of which has been successful (Bichard and David 1985). Indeed, the most successful approach to date seems to have been the screening of large field populations for "hyperprolific" sows, and selecting among those (Legault 1985), although this approach is not likely to be optimum in the long term (Avalos and Smith 1987). Bichard and David (1985) concluded that the most likely explanation for lack of demonstrated response to selection for increased litter size is that past experiments have been inadequate. Litter size is difficult to select for because the trait is lowly heritable and sex-limited, although it is quite variable. Bichard and David (1985) correctly pointed out that use of information from relatives would likely improve response to selection, and Avalos and Smith (1987) have estimated that gains of up to half a pig per year are theoretically possible from selection on a family index. It seems odd that attempts to select for a trait like litter size have not been based on selection criteria which make maximum use of the data.

There is no general body of theory that allows prediction of response to selection on genetic evaluations from mixed model methods that are based on information from all relatives, obtained over repeated cycles of selection with overlapping generations, even when relatively simple assumptions are made about the inheritance of the trait (e.g., infinitesimal additive model). As an alternative, Belonsky and Kennedy (1984, 1988) used computer simulation to compare expected response to selection on mixed model genetic evaluations with the more traditional practice of selection on the individual's phenotypic performance. Their experimental unit was a 100 sow and 4 boar breeding herd which was selected over a 10-year period for a single trait with heritability of 0.10, 0.30 or 0.60. Additionally, two forms of culling existing breeding animals were considered. In the first, an animal was removed from the breeding herd only for considerations of age or reproductive failure, that is, there was no culling on merit for the selected trait. In the second, removal of a breeding animal also occurred any time a potential replacement with a better estimated genetic value was available. Genetic evaluation was on phenotypic performance deviated from contemporary average, or by BLUP under an animal model. A summary of results is given in Table 5.1.

Differences in estimated expected response were considerable, particularly at low heritability (h^2=0.10) where BLUP selection gave 80 and 55% greater expected response than phenotypic selection, with and without "genetic" culling, respectively. In Belonsky and Kennedy's (1984, 1988) simulation, the trait was measured in both sexes, but for traits measured only in females, such as litter size, differences in response between selection on phenotypic performance and BLUP of additive genetic merit would be expected to be even greater.

Of interest also was the impact of culling of existing breeding animals on estimated genetic merit when a potentially better replacement was available. The effect was, of course, to shorten the generation interval, and expected rates of response increased by 34 to 57% depending upon heritability and type of selection. However, such culling is avoided in the conduct of many selection experiments because of confounding between genetic and some environmental effects related to time and age. For example, in selection experiments for increased weaning weight in beef cattle, adjustments are usually made for the age of dam of the calf. If traditional methods of adjustment such as least-squares, are employed using the data, these are biased if cows are culled on their own or on their progeny's weaning weights. Their use can lead to incorrect selection decisions and retarded selection response. It is well established, however, that mixed model methods can eliminate this and other forms of culling bias (Henderson et al. 1959). Additionally, they can allow for unbiased genetic comparisons of animals raised in different time periods, which is important if breeding animals are to be culled when better replacements are available. If the intent of the experiment is to create genetically divergent lines quickly, or to demonstrate that selection response is possible and practical, mixed model methods for selection of animals may allow additional culling and, hence, greater selection response than might be possible otherwise.

Table 5.1 Average additive genetic value of progeny[a,b]

	Selection method			
	No "genetic" culling		"Genetic" culling	
Heritability	Phenotypic	BLUP	Phenotypic	BLUP
0.10	0.58±0.03[c]	0.90±0.03	0.78±0.03	1.41±0.05
0.30	1.78±0.05	2.22±0.06	2.40±0.06	3.14±0.07
0.60	3.59±0.09	3.95±0.10	5.10±0.08	5.31±0.07

[a] Abridged from Belonsky and Kennedy (1988).
[b] Response to 10 yr of selection in standard deviation units.
[c] Standard error of average of 30 replicates.

The simulations of Belonsky and Kennedy (1984, 1988) assumed an infinitesimal additive genetic model as well as prior knowledge of heritability in the base population, that is, they met the theoretical assumptions implicit with the method. This, of course, represents an idealized situation which is seldom realized in practice, although it may represent a very good approximation in many situations. Some of the ramifications of violation of these assumptions are discussed in the two subsequent sections on estimation of genetic variances and response to selection.

5.4 Estimation of Genetic Variances

Reliable estimates of genetic variances and covariances are needed to formulate breeding plans, predict response to selection and estimate genetic merits of animals. Usually, in the analysis of data from selection experiments, relatively simple methods based on parent-offspring regressions or intra-class correlations between sibs are used to estimate genetic parameters. Little use is made of more complex methods of variance component estimation that invoke mixed models such as (5.3) or (5.5). Indeed, if the intent of the experiment is only to estimate genetic parameters, simple methods of estimation can be applied appropriately to experiments designed optimally for this purpose (Hill 1970; Hill and Nicholas 1974; Thompson 1976; Thompson and Cameron 1986). Such experiments are usually of short term and involve only parent and offspring generations.

In many experiments with large animals, however, estimation of genetic variances and covariances is only one of several concerns. These experiments are usually of longer term and they involve several generations, which may overlap; also, records may be affected by fixed effects such as sex, age of dam, time and location. Frequently, selection has been practiced. In such cases, a mixed model approach, with the model correctly specified, is likely to offer considerable advantages over simpler traditional methods. Applications of mixed linear models to genetic parameter estimation have been outlined in detail previously, along with illustrative examples (Henderson 1984, 1985b; Sorensen and Kennedy 1984b, 1986), so only a bare sketch will be given here.

First, consider the simplest case of estimating heritability under the additive genetic model (5.3). Traditional estimators based on analysis of variance, such as Henderson's (1953) methods, implicitly assume random mating, a defined breeding structure and a large population size, such that inbreeding and random genetic drift are not important. In most animal experiments, population size is not large and drift must be allowed for. Sorensen and Kennedy (1982) proposed a simple modification to Henderson's (1953) method 3 for the usual sire model which makes use of **A** in the estimation process. Reductions in sums of squares due to sires are taken in the usual manner but their expectations are taken under the true model (5.3), which allows for inbreeding and relationships other than half-sibs. Further description of the method and numerical examples are in Sorensen and Kennedy (1982, 1984b, 1986). The method is useful for analysis of data from control populations because it accounts for the effects of inbreeding on additive genetic variance and, as a result, provides an unbiased estimate of the additive genetic variance in the base population.

However, the modified method 3 estimator does not use all the information available. For example, in an experiment with non-overlapping generations, only the part of the relationship matrix comprising diagonal blocks of relationships within a generation is used. The omission of relationships across generations has important implications in the analysis of data generated by selection. It is well recognized that estimates of heritability obtained from the usual analysis of variance methods are biased if selection has been practiced (Robertson 1977), and modification of these methods to incorporate collateral relationships still leads to estimators which are also biased by selection, although not by drift (Sorensen and Kennedy 1984b). Regressions of offspring on parent records can be free of selection bias. However, with many fixed effects and data spanning several generations, simple regression estimators are of limited use because, for example, the regression model can generally only correct for fixed effects on the offspring and not on the parent's records. Thompson (1977), however, proposed a maximum likelihood estimator for a random model that combined all genetic relationships among animals, and that approach can be extended to mixed models.

Henderson (1984, 1985b) presented generalizations of Rao's (1971) MIVQUE and Patterson and Thompson's (1971) REML based on his mixed model equations. MIVQUE applied to the additive genetic model (5.3) leads to the following equations

$$\begin{bmatrix} \hat{\sigma}_a^2 \\ \hat{\sigma}_e^2 \end{bmatrix} = \begin{bmatrix} n-2\lambda \mathrm{tr}(\mathbf{A}^{-1}\mathbf{C})+\lambda^2 \mathrm{tr}(\mathbf{A}^{-1}\mathbf{C})^2 & \mathrm{tr}(\mathbf{A}^{-1}\mathbf{C})-\lambda \mathrm{tr}(\mathbf{A}^{-1}\mathbf{C})^2 \\ n\lambda-\lambda^2 \mathrm{tr}(\mathbf{A}^{-1}\mathbf{C}) & m-r(\mathbf{X})-n+\lambda \mathrm{tr}(\mathbf{A}^{-1}\mathbf{C}) \end{bmatrix}^{-1} \begin{bmatrix} \hat{\mathbf{a}}'\mathbf{A}^{-1}\hat{\mathbf{a}} \\ \mathrm{SSE} \end{bmatrix}$$

where $\mathrm{SSE}=\mathbf{y}'\mathbf{y}-\tilde{\mathbf{b}}'\mathbf{X}'\mathbf{y}-\hat{\mathbf{a}}'\mathbf{Z}'\mathbf{y}$ from (5.4), m is the number of records, n is the number of animals in **a** and **C** is the part of the inverse of the coefficient matrix of (5.4) corresponding to the **a** equations. Iteration on the MIVQUE equations can lead to REML estimates, but the $\mathrm{tr}(\mathbf{A}^{-1}\mathbf{C})^2$ term can be avoided with some REML computing algorithms. With MIVQUE and REML, note that the computation of $\hat{\mathbf{a}}$, the quadratic in $\hat{\mathbf{a}}$ and its expectation all involve $\mathbf{A}^{-1}$. This seems to be a requirement for unbiased estimation in selected populations.

Sorensen and Kennedy (1984b) used computer simulation to examine the properties of the MIVQUE estimator of additive genetic variance under an infinitesimal additive model in selected populations. MIVQUE seemingly provided unbiased estimates of additive genetic variance in the base population, i.e., before selection operated, when "good" prior values of heritability were used in the estimation process. Use of REML, which is iterative, can obviate the need for good prior values. As pointed out previously, heritability declines with inbreeding and selection, but for most purposes, our interest is in the value of heritability prior to these changes, that is, in the unselected base population.

When one is interested in an estimate of heritability in a later generation (t), after selection has operated in previous generations, Sorensen and Kennedy (1984b) proposed the simple expedient of pretending that the individuals of generation t were unrelated and not inbred ($\mathbf{A}_t=\mathbf{I}$), and creating relationships from there. Relationships in later generations are essentially traced back to generation t and only the data from generation t on are used. Simulation indicated that MIVQUE under these conditions gave unbiased (within the limits of sampling error) estimates of additive genetic variance in generation t. The fact that individuals in generation t were in gametic disequilibrium did not seem to bias the MIVQUE estimator.

Sorensen and Kennedy (1984b, 1986) assumed an infinitesimal additive model in their simulations, and the genetic assumptions implicit in the use of MIVQUE and REML for estimating heritability are similar to those already discussed for predicting $\hat{\mathbf{a}}$. Essentially, a large number of additive loci is assumed. If this assumption is realistic, changes in genetic variance due to inbreeding and gametic disequilibrium are "recognized" and accommodated in variance component estimation by MIVQUE and REML. If the trait is controlled by a small number of loci, changes in genetic variance due to gene frequency changes would not be recognized, similar to the case when predicting $\hat{\mathbf{a}}$. Further examination of the consequences of this would be useful.

It should be mentioned that because MIVQUE and REML require calculation of **C**, the procedures can be very demanding computationally if the number of animals is large. This might render use of these methods prohibitive at present for large field populations, but the procedures are computationally feasible for many experimental situations. Further, the order of the equations can be reduced through extension of a gametic or reduced animal model (Quaas and Pollak 1980) to MIVQUE estimation of genetic variances (Sorensen and Kennedy 1986). Computationally less demanding approximations to MIVQUE and REML have been proposed. Henderson's approximate MIVQUE (Henderson 1984), which involves approximations to both $\hat{\mathbf{a}}$ and **C**, is subject to selection bias (Belonsky and Kennedy, unpublished data). Schaeffer (1986) has proposed a pseudo-expectation approach which obtains $\hat{\mathbf{a}}$ according to (5.4) but uses $\hat{\mathbf{a}}'\mathbf{Z}'\mathbf{P}\mathbf{y}$ and $\mathrm{tr}(\mathbf{Z}'\mathbf{P}\mathbf{Z}\mathbf{A})$ where $\mathbf{P}=\mathbf{I}-\mathbf{X}(\mathbf{X}'\mathbf{X})^{-}\mathbf{X}'$, to estimate σ_a^2; however, it is also biased by selection (Simianer and Schaeffer, unpublished data). This section has addressed heritability estimation only, and joint heritability and genetic correlation estimation is still more demanding computationally. Extensions of MIVQUE and REML to variance-covariance estimation for multiple trait models are in Henderson (1984).

Little attention has been paid to non-additive genetic effects in large animal experiments, except in the context of experiments involving crosses between breeds or inbred lines. Scant use has been made of mixed model methods in such experiments, although procedures have been proposed (e.g., Henderson 1977). Most selection experiments have focused on additive genetic effects, but dominance and other non-additive effects may be important, particularly for traits known to show inbreeding depression. Henderson (1984, 1985a,b) has outlined MIVQUE and REML procedures for estimating dominance and additive x additive genetic variances from data that may involve several generations of mating. If (5.5) and (5.7) are combined in a single analysis, REML of σ_e^2, σ_a^2, σ_d^2 and σ_i^2 can be obtained by iterating on

$$\hat{\sigma}_e^2 = (\mathbf{y}'\mathbf{y}-\hat{\mathbf{b}}'\mathbf{X}'\mathbf{y}-\hat{\mathbf{a}}'\mathbf{Z}'\mathbf{y}-\hat{\mathbf{d}}'\mathbf{Z}'\mathbf{y}-\hat{\mathbf{i}}'\mathbf{Z}'\mathbf{y})/[m-r(\mathbf{X})]$$

$$\hat{\sigma}_a^2 = [\hat{\mathbf{a}}'\mathbf{A}^{-1}\hat{\mathbf{a}}+\hat{\sigma}_e^2 tr(\mathbf{A}^{-1}\mathbf{C}_{aa})]/n$$

$$\hat{\sigma}_d^2 = [\hat{\mathbf{d}}'\mathbf{D}^{-1}\hat{\mathbf{d}}+\hat{\sigma}_e^2 tr(\mathbf{D}^{-1}\mathbf{C}_{dd})]/n$$

$$\hat{\sigma}_i^2 = [\hat{\mathbf{i}}'\mathbf{H}^{-1}\hat{\mathbf{i}}+\hat{\sigma}_e^2 tr(\mathbf{H}^{-1}\mathbf{C}_{ii})]/n.$$

$\mathbf{C}_{aa}$, $\mathbf{C}_{dd}$ and $\mathbf{C}_{ii}$ represent those portions of the inverse of the coefficient matrix corresponding to the blocks of additive, dominance and additive x additive effect equations, respectively. In this application, the full rank order of the mixed model equations is 3n+r(**X**), and solution may be computationally difficult. Henderson (1984, 1985a) has given a procedure for compressing all sources of genetic variance into a single set of equations $\mathbf{Z}'\mathbf{Z}+\mathbf{M}^{-1}\sigma_e^2$, for $\mathbf{M}=\mathbf{A}\sigma_a^2+\mathbf{D}\sigma_d^2+\mathbf{H}\sigma_i^2$, which has order n, and this may have reduced computational demands.

Properties of these estimators are not known for inbred and selected populations. Some of the difficulties involve determining **D** and **H** and accounting for inbreeding depression under these conditions. The probabilistic procedure of Smith (1984) for computing **D** in inbred populations may be a useful approximation and a helpful start to resolution of part of this problem. The implications of using these procedures when dominance is important and the number of loci is small can only be speculated, as discussed in the next section. There certainly is a need for more analytical and simulation work on the properties of MIVQUE and REML estimates of dominance and epistatic variance components obtained from data that involve inbreeding and selection. These methods are potentially valuable for the analysis of experimental data and their properties and the consequences of their use need better definition.

5.5 Estimation of Selection Response

The use of mixed model methods for estimation of selection response in animal experiments has been proposed by Blair and Pollak (1984) and Sorensen and Kennedy (1984a, 1986). Blair and Pollak (1984) examined the behavior of different estimators of selection response in sheep data and Sorensen and Kennedy (1984a, 1986) studied properties of mixed model estimators of selection response through computer simulation. Both studies made comparisons with traditional methods based on (least-squares) estimates of phenotypic generation means, and concluded that mixed model methods can offer advantages over least-squares techniques. Although use of mixed model methods to estimate genetic changes in field populations has been widely practiced and accepted (Henderson 1973), their proposed use for selection experiments has been controversial.

Under the additive genetic model (5.3), the mixed model estimator of the genetic mean of generation t is simply the average estimated genetic value of all individuals in generation t, i.e., $\hat{g}_t=\mathbf{k}_t'\hat{\mathbf{a}}$, where $\hat{\mathbf{a}}$ is from (5.4). If the model (5.3) is correct and h^2 in the base population is known, then $\hat{g}_t$ is an unbiased estimate of the genetic mean. The variance of $\hat{g}_t$ is

$$\mathrm{var}(\mathbf{k}_t'\hat{\mathbf{a}}) = [\bar{a}_t h^2 - \bar{c}_t(1-h^2)]\sigma^2, \tag{5.8}$$

where $\bar{a}_t$ is the average additive genetic relationship among the individuals of generation t, $\bar{c}_t$ is the average element of the prediction error variance-covariance matrix of estimated genetic values of generation t, and h^2 and σ^2 are heritability and the phenotypic variance in the base population (Sorensen and Kennedy 1986). The least-squares estimate of the generation mean ($\bar{y}_t$), which does not depend upon an assumed h^2, is also unbiased (if properly corrected for fixed effects) with:

$$\mathrm{var}(\bar{y}_t) = [\bar{a}_t h^2 + (1-h^2)/n]\sigma^2. \tag{5.9}$$

If n, the number of individuals in t, is large, (5.9) approaches the usual expression for drift variance, i.e., $\bar{a}_t h^2\sigma^2=2\bar{F}_{t+1}h^2\sigma^2$. Because $\bar{c}_t\geq 0$, the variance of the mixed model estimator is less than that of the least-squares estimator for a given set of data. Selection was ignored in the derivation of (5.8) and (5.9), but simulation has shown that they approximate empirical results well when selection has operated (Sorensen and Kennedy 1986).

These properties of mixed model estimators of response rest on assumptions that h^2 is known and the model is described correctly. Questions then arise as to the consequences of errors in these assumptions. Related to this is the issue of whether $\hat{g}_t$ is a predictor or estimator of response. This is more than a semantic quibble, and has been discussed by Sorensen and Kennedy (1986). The distinction will be illustrated here through a simple example.

Consider four individuals (generation 0) 1, 2, 3 and 4 with respective phenotypic records of 10, 9, 8 and 7. Individuals 1 and 2 are selected to have progeny on the basis of their phenotypic records and, assuming h^2=0.5 and (5.3) is correct, the predicted response is 0.5 with a standard error of 0.25σ, assuming a population size of 4 is maintained. This can be computed in the usual way as $\tilde{g}_1=h^2(\bar{y}_s-\bar{y}_0)=0.5(9.5-8.5)$, where $\bar{y}_s$ is the phenotypic mean of records on the selected parents and $\bar{y}_0$ is the phenotypic mean of records of all individuals in generation 0, or from (5.4) and (5.8) with four offspring (without records) of 1 and 2 added to **a** and **Z**= diag{1 1 1 1 0 0 0 0}. Note that there are no data on the progeny yet and $\tilde{g}_1$ is predicted simply from h^2 and the records of individuals in the initial generation.

Suppose the four offspring 5, 6, 7 and 8 in the next generation now have records of 10, 9, 8 and 7, respectively. The data suggest no response to selection, that is $\bar{y}_1-\bar{y}_0=0$. However, applying (5.4) we obtain $\hat{a}'=1/6(4\ 1\ -1\ -4\ 5\ 3\ 1\ -1)$ which leads to $\hat{g}_1=1/3$ with a standard error of 0.32σ. The mixed model estimate of response falls between the predicted response and the observed phenotypic response. But is $\bar{y}_1-\bar{y}_0$ correct? It is an unbiased estimate of response, assuming no environmental trend, but with a standard error of 0.66σ, about twice that of the mixed model estimate. If the model (5.3) is indeed correct and the data were affected by drift or sampling error, then $\hat{g}_1$ is a better estimator of response than $\bar{y}_1-\bar{y}_0$. Nonetheless, one obviously would not design a selection experiment to minimize the variance of the mixed model estimator of response because it contains a predictive element which increases with increasing unreliability of the data. Also, if the model is wrong, either because h^2 is incorrect or the genetic model assumed is inappropriate, $\hat{g}_1$ may be biased, although it may still have a smaller mean squared error than $\bar{y}_1-\bar{y}_0$. For example, if true h^2 is 0.25 and not 0.5 as assumed, but the rest of the model is still appropriate, then the correct $\hat{g}_1$, given the data, is 0.21, not greatly different from the 1/3 obtained under the incorrect assumption that $h^2=0.5$.

Also, prior knowledge of heritability in the base population does not seem to be required for unbiased estimation of selection response. Application of unbiased estimates of σ_a^2 and σ_e^2 in (5.4), as obtained, for example, from use of MIVQUE or REML procedures outlined in the previous section, seem to lead to unbiased estimates of $\hat{g}_t$ over repeated sampling (Sorensen and Kennedy 1986). The variance of $\hat{g}_t$ is, of course, increased over (5.8), but simulation results indicate that it is still less than that of the least-squares estimator.

What if the genetic model is inappropriate? For example, implicit with the use of mixed model methods under (5.3) is the assumption of a large number of loci. Maki-Tanila and Kennedy (1986) examined properties of mixed model estimators of genetic change for traits controlled by relatively few additive loci. Under random mating, they found the methods gave good predictors of generation genetic means, with smaller sampling errors than least-squares, even for models with as few as two loci and where initial gene frequencies were extreme. With selection, there was no significant bias, but there was the suggestion of a tendency of the mixed model estimator to slightly overestimate response when frequencies of the favorable alleles were high, and to slightly underestimate response when frequencies were low. Again, sampling variances were smaller than with least-squares. Maki-Tanila and Kennedy's (1986) simulations were for three generations of selection only. Mixed model methods do not account for the effect of gene frequency changes on genetic variance, and problems associated with this could be more serious over the longer term.

The forces of natural selection can counter artificial selection. If so, potential selection response is reduced and mixed model estimates of response can be biased if "fitness" is genetically correlated with the trait under artificial selection. In the simulated pig selection experiment described earlier, Belonsky (1984) examined the effect of natural selection on reproductive ability in conjunction with artificial selection for a performance trait. A threshold model (Falconer 1981) was assumed for reproductive ability, with underlying heritabilities of 0.10, 0.30 or 0.60. Similar heritabilities were assumed for the performance trait under artificial selection. Genetic correlations between the two traits of 0 and -0.5 were examined. In all situations, mixed model estimates of generation genetic means of the performance trait reflected the reduction in genetic improvement as a result of the selection on reproductive ability, and no significant bias was detected even when the reproductive trait was negatively correlated genetically with the performance trait under selection, although theoretically there is a bias. This form of natural selection, even when severe, had no noticeable effect on the estimation of selection response.

The model used, e.g., (5.3), may also be wrong with respect to the type of gene action assumed. For example, dominance effects may be important. Maki-Tanila and Kennedy (1986) examined this question also for traits controlled by relatively few loci. With dominance, there was inbreeding depression. In both random mated and selected populations, the effects of inbreeding depression were underestimated if analysis was according to the additive genetic model (5.3) only, and dominance was ignored. Of course, the degree of inbreeding depression was gene frequency-dependent. When dominance was included in the model as in (5.5) and genetic evaluation ($\hat{\mathbf{g}}=\hat{\mathbf{a}}+\hat{\mathbf{d}}$) was according to (5.6), then estimates of generation genetic means were improved, although there was evidence of bias particularly with selection. The least-squares estimator was unbiased throughout but had greater sampling errors. Further investigation of the properties of mixed model estimators of generation means under dominance models and selection is needed. Also, the whole question of accounting for epistatic genetic effects in experimental data needs investigation.

Lastly, there has been controversy over the use of mixed model methods to estimate realized heritability. Blair and Pollak (1984) regressed estimated genetic means on average cumulative selection differentials to estimate realized heritability and this has been criticized (Thompson and Cameron 1986) because it is more a measure of assumed heritability than of true heritability, although Blair and Pollak (1984) did devote considerable attention in their study to the effect of assumed heritability on the estimates. The debate may be somewhat academic. Given the Bulmer (1971) effect, realized heritability has an interpretation only in the context of specific selection applied and duration of the experiment. As Falconer (1981) points out, it is primarily a description of response and may not provide a valid estimate of heritability in the base population. The key concern is estimated selection response, rather than realized heritability. With mixed model methods, that too can depend upon the assumed value of base population heritability. As we have seen, however, we can obtain unbiased estimates of base population heritability from selected data and apply these to mixed model methods to obtain unbiased estimates of selection response, over repeated sampling, if the genetic model is correct. Thus the estimated selection response need not depend upon some prior notion of heritability; this parameter can be estimated from the data generated by the selection experiment. Of more critical importance is the appropriateness of the genetic model assumed with respect to number of loci and importance of dominance and other non-additive effects. The model must be realistic and estimates of genetic value from it must be relatively insensitive to violations of the assumptions. We have seen that for many, but not all situations, this can be the case.

5.6 Design

The important area of experimental design has been covered by Hill, Chapter 4, this Volume, and only a brief reference to two points relating to mixed model applications will be given here. Firstly, experiments with large animals are very expensive and mistakes from inadequate design can be extremely costly. Computing is relatively inexpensive, and it would seem prudent to simulate and analyze data for the planned experimental design under hypothesized genetic models prior to initiating the experiment. Expected selection response and its sampling variance can be measured as can sensitivity to the genetic model. This is important if the actual selection of animals is to be based on mixed model evaluations, as there is not a good body of theory in this case from which to predict selection response.

The second matter relates to the need for controls. The theory of control populations has been well outlined by Hill (1972). Controls can deviate from base population values, because of drift and other factors, and an ideal experiment would have a number of replicated controls. However, most experimenters are faced with constraints of fixed resources, and the inclusion of control lines can reduce the size of the selection lines which can reduce selection response and increase the variances of estimated selection response. If the assumed model is correct or nearly so, controls are not necessary and mixed model methods can be used to estimate selection response unbiasedly and with relatively small sampling error (although inclusion of a control may contribute to smaller sampling errors of estimates of base population genetic variances). However, what is gained in reduced sampling error and drift variance might well be lost through assumption and use of an incorrect genetic model. Indeed, one could argue that if the model is known, there is no need for the experiment. This is the dilemma. Perhaps selection experiments should not be conducted without adequate controls, but many are because of limited facilities. We have seen that mixed model methods are usually not highly sensitive to violations of assumptions, but can be in some situations, such as when dominance effects are important and gene frequencies are extreme. In cattle experiments, at least, a workable compromise between the need for a control as a check on genetic theory and the need for estimates of selection response with small variance might be to freeze a sample of semen and embryos from the base population at the time the experiment is initiated for subsequent use at the end of the experiment, as suggested by Smith (1977), and use mixed model methods throughout the rest of the experiment.

5.7 Conclusions

Applications of mixed model methods to designed experiments can be useful for selection of breeding animals, estimation of genetic parameters, estimation of selection response and to check the design of the experiment itself. The dependency of mixed model methods on an assumed value of heritability in the base population is not as critical as generally imagined; it can be estimated from the data of the experiment free of prior notions about its value. Of more critical concern are assumptions about the genetic model used in the mixed model analysis.

Acknowledgements. I am grateful to Daniel Sorensen and Asko Maki-Tanila, not only for their contributions to much of the material presented in this paper, but also for their contributions to my understanding of the subject. Support was provided by the Natural Sciences and Engineering Research Council of Canada, the Ontario Ministry of Agriculture and Food, and IBM Canada.

References

Avalos E, Smith C (1987) Genetic improvement of litter size in pigs. Anim Prod 44:153-164

Belonsky GM (1984) Genetic evaluation of a swine herd under selection. MSc Thesis, Univ Guelph, Guelph

Belonsky GM, Kennedy BW (1984) Within-herd genetic evaluation of swine. J Anim Sci 59 suppl (Abstr) 1:164

Belonsky GM, Kennedy BW (1988) Selection on individual phenotype and best linear unbiased prediction of breeding value in a closed swine herd. J Anim Sci 66:1124-1131

Bichard M, David PJ (1985) Effectiveness of genetic selection for prolificacy in pigs. J Reprod Fert suppl 33:127-138

Blair HT, Pollak EJ (1984) Estimation of genetic trend in a selected population with and without the use of a control population. J Anim Sci 58:878-886

Bulmer MG (1971) The effect of selection on genetic variability. Am Nat 105:201-211

Bulmer MG (1980) The mathematical theory of quantitative genetics. Clarendon Press, Oxford

Cockerham CC (1954) An extension of the concept of partitioning hereditary variance for analysis of covariances among relatives when epistasis is present. Genetics 39:859-882

Falconer DS (1981) Introduction to quantitative genetics. 2nd edn Longman, London

Felsenstein J (1965) The effect of linkage on directional selection. Genetics 42:349-363

Gianola D, Foulley JL, Fernando RL (1986) Prediction of breeding values when variances are not known. In: Dickerson GE, Johnson RK (eds) Proc 3rd World Congr Genet Appl Livest Prod. Agric Commun, Univ Nebraska, Lincoln, Nebraska XII:356-370

Henderson CR (1949) Estimation of changes in herd environment. J Dairy Sci (Abstr) 32:706

Henderson CR (1953) Estimation of variance and covariance components. Biometrics 9:226-252

Henderson CR (1963) Selection index and expected genetic advance. In: Statistical genetics and plant breeding. NAS-NRC 982, Washington DC pp 141-163

Henderson CR (1973) Sire evaluation and genetic trends. In: Proc Anim Breed Genet Symp in Honor of Dr. J.L. Lush. ASAS and ADSA, Champaign, Illinois pp 10-41

Henderson CR (1975) Best linear unbiased estimation and prediction under a selection model. Biometrics 31:423-447

Henderson CR (1976) A simple method for computing the inverse of a numerator relationship matrix used in prediction of breeding values. Biometrics 32:69-83

Henderson CR (1977) Prediction of merits of a single cross. Theor Appl Genet 49:273-282

Henderson CR (1984) Applications of linear models in animal breeding. Univ Guelph Press, Guelph

Henderson CR (1985a) Best linear unbiased prediction of nonadditive genetic merits in noninbred populations. J Anim Sci 60:111-117

Henderson CR (1985b) MIVQUE and REML estimation of additive and nonadditive genetic variances. J Anim Sci 61:113-121

Henderson CR, Kempthorne O, Searle SR, von Krosigk CM (1959) The estimation of environmental and genetic trends from records subject to culling. Biometrics 15:192-218

Hill WG (1970) Design of experiments to estimate heritability by the regression of offspring on selected parents. Biometrics 26:566-571

Hill WG (1971) Design and efficiency of selection experiments for estimating genetic parameters. Biometrics 27:293-311

Hill WG (1972) Estimation of genetic change. I. General theory and design of control populations. Anim Breed Abstr 40:1-15

Hill WG, Nicholas FW (1974) Estimation of heritability by both regression of offspring on parent and intra-class correlation of sibs in one experiment. Biometrics 30:447-468

Hough JD, Benyshek LL (1986) Estimates of genetic trend using the reduced animal model versus least-squares using a control population. J Anim Sci 63 suppl (Abstr) 1:178

Kackar RN, Harville DA (1981) Unbiasedness of two-stage estimation and prediction procedures for mixed linear models. Commun Stat Theor Meth A10:1249-1261

Kemp RA (1985) The effects of positive assortative mating and preferential treatment of progeny on the estimation of breeding values. PhD Thesis, Univ Guelph, Guelph

Legault C (1985) Selection of breeds, strains and individual pigs for prolificacy. J Reprod Fert suppl 33:151-166

Maki-Tanila A, Kennedy BW (1986) Mixed model methodology under genetic models with a small number of additive and non-additive loci. In: Dickerson GE, Johnson RK (eds) Proc 3rd World Congr Genet Appl Livest Prod. Agric Commun, Univ Nebraska, Lincoln, Nebraska XII:443-448

Patterson HD, Thompson R (1971) Recovery of inter-block information when block sizes are unequal. Biometrika 58:545-554

Pearson K (1903) Mathematical contributions to the theory of evolution. XI. On the influence of natural selection on the variability and correlation of organs. Phil Trans R Soc London A 200:1-66

Quaas RL, Pollak EJ (1980) Mixed model methodology for farm and ranch beef cattle testing programs. J Anim Sci 51:1277-1287

Rao CR (1971) Minimum variance quadratic unbiased estimation of variance components. J Mult Anal 1:445-456

Robertson A (1977) The effect of selection on the estimation of genetic parameters. Z Tierz Züchtungsbiol 94:131-135

Schaeffer LR (1986) Pseudo-expectation approach to variance component estimation. J Dairy Sci 69:2884-2889

Smith C (1977) Use of stored frozen semen and embryos to measure genetic trends in farm livestock. Z Tierz Züchtungsbiol 94:119-127

Smith SP (1984) Dominance relationship matrix and inverse for an inbred population. Unpublished mimeo, Department of Dairy Science, Ohio State Univ, Columbus

Smith SP, Allaire FR (1985) Efficient selection rules to increase non-linear merit: application in mate selection. Genet Sel Evol 17:387-406

Sorensen DA, Kennedy BW (1982) Estimation of genetic variances in control and selected populations. In: Proc 2nd World Congr Genet Appl Livest Prod. Garsi, Madrid VII:220-225

Sorensen DA, Kennedy BW (1983) The use of the relationship matrix to account for genetic drift variance in the analysis of genetic experiments. Theor Appl Genet 66:217-220

Sorensen DA, Kennedy BW (1984a) Estimation of response to selection using least-squares and mixed model methodology. J Anim Sci 58:1097-1106

Sorensen DA, Kennedy BW (1984b) Estimation of genetic variances from unselected and selected populations. J Anim Sci 59:1213-1223

Sorensen DA, Kennedy BW (1986) Analysis of selection experiments using mixed model methodology. J Anim Sci 68:245-258

Styan GPH (1973) Hadamard products and multivariate statistical analysis. Linear Algebra and Its Applications 6:217-240

Thompson R (1976) Designs of experiments to estimate heritability when observations are available on parents and offspring. Biometrics 32:283-304

Thompson R (1977) The estimation of heritability with unbalanced data. II. Data available on more than two generations. Biometrics 33:497-504

Thompson R, Cameron ND (1986) Estimation of genetic parameters. In: Dickerson GE, Johnson RK (eds) Proc 3rd World Congr Genet Appl Livest Prod. Agric Commun, Univ Nebraska, Lincoln, Nebraska XII:371-381

Wright S (1922) Coefficients of inbreeding and relationship. Am Nat 56:330-338

6 Statistical Aspects of Design of Animal Breeding Programs: A Comparison Among Various Selection Strategies

L. Dempfle[1]

The goal of animal breeding is to change populations in such a way that overall economic merit is increased as fast as possible over many generations. Available genetic variation and any possible genetic variation arising from mutations need to be effectively utilized. Because of the conflict between maximizing short- and long-term responses compromises need to be found. To compare different strategies, a full-sib structure was simulated and five strategies (family selection, restricted and non-restricted within family selection, mass selection and index selection) were investigated. Comparisons were made for the first generation, short-term and long-term responses. In the first generation, the correlation between estimated and true breeding value and the expected value of the standardized selection differential have to be considered. In the short term the reduction in variance due to linkage disequilibrium also has to be taken into account and, in the long term, the reduction in genetic variance due to drift must be considered. This last factor depends on both the systematic selection pressure and effective population size.

Effective population size itself is largely influenced by the correlation between selection criteria of family members and depends on whether there are additional restrictions on the selection process.

All procedures studied produced predictable results for both precision of estimation and effective population size. However, in applied breeding these two aspects need not be completely connected. One can try to estimate breeding values as precisely as possible but some restrictions are required to maintain a reasonable effective population size. When this is done the additional efficiency in estimating, e.g., the family component, may not be great.

6.1 Introduction

The ultimate goal of animal breeding is to change populations in such a way that their overall economic merit is increased at the fastest rate. Both short- and long-term improvement must

[1] Institut für Tierwissenschaften, der Technischen Universität München, D-8050 Freising-Weihenstephan, FRG.

be considered. The latter could be thought of as conserving the genetic variation. Related to this conservation is the possible utilization of new genetic variation. In exceptional cases the breeding goal may be simply to maintain current performance levels, e.g., where the population is at a limit or for traits showing overdominance.

In designing a breeding scheme all these aspects have to be considered. Because of the well-known conflict between short- and long-term genetic gains, compromises appropriate to the given resources must be made.

In the following sections some aspects of design are considered. Emphasis is on simple and hopefully robust designs. Simplicity and robustness can often be achieved by purposely restricting selection to within certain subpopulations, where these can be full sib or half-sib families, or animals within test stations. Here, I mainly use full sibs to illustrate the principles.

6.2 Full-Sib Structures

6.2.1 First Generation

The following situation is assumed: in each generation we test N candidates of each sex, belonging to f full-sib groups of size k (fk=N). The linear model for an observation is

$$y_{ij} = \mu+b_i+w_{ij}$$

$$= \mu+Ab_i+Eb_i+Aw_{ij}+Ew_{ij}$$

with

$$\sigma_b^2 = \sigma_{Ab}^2+\sigma_{Eb}^2$$

$$\sigma_w^2 = \sigma_{Aw}^2+\sigma_{Ew}^2$$

$$\sigma_P^2 = \sigma_{Ab}^2+\sigma_{Eb}^2+\sigma_{Aw}^2+\sigma_{Ew}^2$$

$$\sigma_A^2 = \sigma_{Ab}^2+\sigma_{Aw}^2;\ \sigma_E^2 = \sigma_{Eb}^2+\sigma_{Ew}^2$$

$$h^2 = \sigma_A^2/\sigma_P^2;\ t = \sigma_b^2/\sigma_P^2;\ r = \sigma_{Ab}^2/\sigma_A^2$$

where

σ_P^2 = phenotypic variance

σ^2_A = additive genetic variance

σ^2_{Ew} = environmental variance within families

σ^2_{Eb} = environmental variance between families

σ^2_{Aw} = additive genetic variance within families

σ^2_{Ab} = additive genetic variance between families

h^2, t and r are the heritability, intra-class correlation and genetic correlation of members of the same family, respectively.

We have $0 \leq h^2 \leq t/r \leq 1$ and $h^2 \leq (1-t)/(1-r)$. If μ is known or accurately estimated we can use a selection criterion of the kind:

$$I_{ij} = b_1(\bar{y}_{i.}-\mu)+b_2(y_{ij}-\bar{y}_{i.}).$$

The basic variance covariance matrix is

$$\mathrm{Var}\begin{bmatrix} A_{ij} \\ \bar{y}_i-\mu \\ y_{ij} \\ y_{ik} \end{bmatrix} = \begin{bmatrix} \sigma^2_A & \sigma^2_{Ab}+\sigma^2_{Aw}/k & \sigma^2_A & \sigma^2_{Ab} \\ \sigma^2_{Ab}+\sigma^2_{Aw}/k & \sigma^2_b+\sigma^2_w/k & \sigma^2_b+\sigma^2_w/k & \sigma^2_b+\sigma^2_w/k \\ \sigma^2_A & \sigma^2_b+\sigma^2_w/k & \sigma^2_P & \sigma^2_b \\ \sigma^2_{Ab} & \sigma^2_b+\sigma^2_w/k & \sigma^2_b & \sigma^2_P \end{bmatrix}$$

The variance covariance matrix needed for the selection criterion is

$$\mathrm{Var}\begin{bmatrix} A_{ij} \\ \bar{y}_i-\mu \\ y_{ij}-\bar{y}_{i.} \\ y_{ik}-\bar{y}_{i.} \end{bmatrix} = \begin{bmatrix} h^2 & h^2[r+(1-r)/k] & h^2(1-r)(k-1)/k & -h^2(1-r)/k \\ h^2[r+(1-r)/k] & t+(1-t)/k & 0 & 0 \\ h^2(1-r)(k-1)/k & 0 & (1-t)(k-1)/k & -(1-t)/k \\ -h^2(1-r)/k & 0 & -(1-t)/k & (1-t)(k-1)/k \end{bmatrix} \sigma^2_P$$

Four special cases can be considered:

Case 1: $b_1=1$; $b_2=0$ between-family selection

Case 2a: $b_1=0$; $b_2=1$ within-family selection (unrestricted)

Case 2b: restricted within-family selection

Case 3: $b_1=1$; $b_2=1$ phenotypic selection

Case 4: $b_1 = \frac{(1-t)\,[1+(k-1)r]}{(1-r)\,[1+(k-1)t]}$; $b_2=1$ "optimum" selection.

It should be noted that Case 4 can produce the other three cases by taking appropriate values of t; for t=r it becomes Case 3, for t=1 it gives Case 2, and for t=0 and k large it tends to Case 1. The variance of the selection criterion is

$$\mathrm{Var}(I_{ij}) = \{b_1^2[t+(1-t)/k]+b_2^2(1-t)(k-1)/k\}\sigma_P^2.$$

The covariance between the breeding value and the criterion is

$$\mathrm{cov}(A_{ij},I_{ij}) = \{b_1h^2[r+(1-r)/k]+b_2h^2(1-r)(k-1)/k\}\sigma_P^2.$$

The covariance between index values of two members of the same family is

$$\mathrm{Cov}(I_{ij},I_{ik}) = \{b_1^2[t+(1-t)/k]-b_2^2(1-t)/k\}\sigma_P^2.$$

The expected genetic gain in any path is given by

$$\delta G = \mathrm{Cov}(A_{ij},I_{ij})\mathrm{Var}(I_{ij})^{-1}[\mathrm{Var}(I_{ij})^{1/2}E(I_s)],$$

where $E(I_s)$ is the expected superiority of selected animals expressed in standard units. Thus

$$\delta G = E(I_s)\frac{\mathrm{Cov}(A_{ij},I_{ij})}{\sqrt{\mathrm{Var}(I_{ij})}}$$

$$= E(I_s) \frac{h^2\{b_1[r+(1-r)/k]+b_2(1-r)(k-1)/k\}}{\sqrt{\{b_1^2[t+(1-t)/k]+b_2^2(1-t)(k-1)/k\}}} \sigma_P$$

$$= E(I_s) f_{AI} \sigma_A .$$

For the first generation, the additive genetic standard deviation depends on the population, and for the correlation we have

$$f_{AI} = \frac{h\{b_1[r+(1-r)/k]+b_2(1-r)(k-1)/k\}}{\sqrt{\{b_1^2[t+(1-t)/k]+b_2^2(1-t)(k-1)/k\}}} .$$

The ratio of $f_{AI}\sigma_A$ for any case relative to that of the optimal index is independent of h^2 and depends only on t and k. Values of this ratio are given in Table 6.1.

If we calculate breeding values of animals (i.e., all animals of a family) and then take one individual at random from each family, this produces a set of unrelated estimated breeding values. Selecting among these animals, the expected superiority could be determined using order statistics from independent variables (in the above case, we would select f out of kf individuals). However, in practical terms we have k members of a family, whose selection criteria are correlated, and we select from these partly correlated, partly uncorrelated variates.

In the following calculation, we assume that tests are separate for the two sexes (or that the genetic correlation between sexes is zero), the population size remains constant and 40 animals of each sex are tested in each generation. For example, with family sizes of 8 males (5 families) 5 males must be selected from 40 tested with or without regard to family structure.

For each of the four cases, the correlation between the selection criteria of animals within a family is:

Case 1: = 1, regardless of t and k

Case 2: = -1, -0.333, -0.143, 0 for k equal to 2, 4, 8 and ∞

Case 3: = t, regardless of k

Case 4:

t	k = 2	4	8	∞
0.01	0.796	0.852	0.902	0.990
0.10	0.761	0.803	0.840	0.900
0.50	0.500	0.500	0.500	0.500
0.90	-0.357	-0.088	0.014	0.100
0.99	-0.914	-0.306	-0.126	0.010

Table 6.1 The correlation between estimated and true breeding values relative to that for the optimal index (t: intra-class correlation; f: number of full-sib groups; k: size of group)

Case[a]	t	f = 20 k = 2	10 4	5 8	- ∞
1	0.01	0.948	0.943	0.956	0.995
2		0.319	0.333	0.292	0.100
3		0.898	0.766	0.622	0.199
1	0.1	0.938	0.923	0.927	0.949
2		0.346	0.384	0.375	0.316
3		0.928	0.842	0.760	0.600
1	0.5	0.866	0.791	0.750	0.707
2		0.500	0.612	0.661	0.707
3		1.000	1.000	1.000	1.000
1	0.9	0.567	0.429	0.370	0.316
2		0.824	0.903	0.929	0.949
3		0.737	0.660	0.628	0.600
1	0.99	0.208	0.143	0.120	0.100
2		0.978	0.990	0.993	0.995
3		0.277	0.229	0.212	0.199

[a] 1: Between-family selection; 2: unrestricted within-family selection; 3: phenotypic selection.

For the expected value of I_s we have the following factors:

1. **Infinite versus finite population size.** The well-known difference between selection from finite and infinite populations is indicated below, as ratios of responses (finite/infinite) for different fractions selected, or for a given fraction selected from populations differing in size [R=100. $E(I_s)/E(I_\infty)$].

Selection of one out of n:

n -	2	3	4	5	10	20	50	100	400
R -	70.68	77.54	80.96	83.07	87.69	90.50	92.90	94.11	95.43

Selection of 50% out of n:

n -	2	4	6	10	50	100	400
R -	70.68	83.08	88.22	92.61	98.50	99.25	99.75

2. **Correlated selection criteria.** The selection differential is the mean of the f highest order statistics from fk variables. The fk variables have a correlation structure of

$$\begin{bmatrix} \mathbf{V}_1 & & \\ & \mathbf{V}_2 & \\ & & \mathbf{V}_3 \end{bmatrix}$$

where variables within a family have a variance-covariance matrix $\mathbf{V}_i$ and the variables are equicorrelated in a balanced design

$$V_i = \begin{bmatrix} 1 & f & f \\ f & 1 & f \\ f & f & 1 \end{bmatrix} \sigma_i^2 \text{ where } \frac{-1}{k-1} < f < 1.$$

Moments for the order statistics from an equicorrelated set are given, e.g., by Owen and Steck (1962), as

$$E(Z_{(i)}) = (1-f)^{1/2}E(X_{(i)})$$

$$Var(Z_{(i)}) = f + (1-f)Var(X_{(i)}),$$

where the X_i are independent variables and the $X_{(i)}$ order statistics. When compared with independent variables, the expected values for the correlated variables are smaller if $f > 0$, and larger if $f < 0$.

Values for the expected means of the f highest order statistics from fk unequally correlated variables are given by Rawlings (1976) and Hill (1976) for a very restricted set of values of f, k and $f > 0$.

No values were found for $f < 0$, but from the equicorrelated case it can be concluded that these expected values will be higher than those for $f=0$ where, in the present case, the differences increase with decreasing k.

The standardized selection differentials for various values of k and $f(I_{ij},I_{ik})$ are listed in Table 6.2.

Table 6.2 Expected selection differentials for a range of selection strategies

$f(I_{ij},I_{ik})$	f = 20 k = 2	10 4	5 8
0.0	0.798[a]	1.271	1.647
0.0	0.783	1.242	1.596
0.5	0.777[b]	1.215	1.497
0.9	0.771[b]	1.180	1.326
1.0	0.768[b]	1.147	1.163
0.0	0.564[c]	1.029	1.424
-1/(k-1)	0.798[d]	1.188	1.522

[a] Expected selection differential in an infinite population when the fraction selected is 0.5, 0.25, 0.125.

[b] Expected selection differential in a finite population for the given family structure and correlation (Hill 1976; Rawlings 1976).

[c] Expected selection differential when the best individual is selected from the uncorrelated members of families of size k.

[d] Expected selection differential when the best individual is selected from correlated members of families when the correlation is -1/(k-1).

In Case 2, a slightly different procedure can be considered. In the situation given, the f largest of fk deviations (from family means) are selected regardless of family. Hence, one family can provide between zero and min(f,k) animals. The alternative is to select the best individual from each family (Case 2b: restricted within-family selection). In this case, we can work with y_{ij} directly and the conditional distribution is

$$(y_{ij}|\text{family } i) \sim N(\mu+b_i,\sigma^2_w) \sim N(\mu+b_i,\sigma^2_{Aw}+\sigma^2_{Ew}).$$

As all variables are conditionally independent, the expected genetic gain is

$$\delta G = i_{1|k}\frac{\sigma^2_{Aw}}{\sqrt{(\sigma^2_{Aw}+\sigma^2_{Ew})}} = i_{1|k}\frac{(1-r)\sigma^2_A}{\sqrt{[(1-t)\sigma^2_P]}} = i_{1|k}h^2\frac{(1-r)}{\sqrt{(1-t)}}\sigma_P,$$

where $i_{1|k}$ is the highest of k order statistics. Instead of working directly with y_{ij}, $y_{ij}-\bar{y}_i$ can be used. Within a family, we then have

$$\text{Var}\begin{bmatrix} y_{ij}-\bar{y}_i \\ y_{ik}-\bar{y}_{i.} \end{bmatrix} = \begin{bmatrix} (1-t)(k-1)/k & -(1-t)/k \\ -(1-t)/k & (1-t)(k-1)/k \end{bmatrix}\sigma^2_P$$

with correlation

$$\frac{-(1-t)k}{k(1-t)(k-1)} = \frac{-1}{k-1}\ .$$

The covariance between A_{ij} and $y_{ij}-\bar{y}_{i.}$ is $[h^2(1-r)(k-1)/k]\sigma^2_P$, so the genetic progress is

$$\delta G = E(I_s)\frac{h^2(1-r)(k-1)/k}{\sqrt{(1-t)(k-1)/k}}\sigma_P\ .$$

From the formula of standardized superiority we obtain

$$\delta G = (1-f)^{1/2}(EX_{(i)})\frac{h^2(1-r)(k-1)/k}{\sqrt{[(1-t)(k-1)/k]}}\sigma_P\ .$$

Combining correlations and expected values we obtain

$$= i_{1|k}(1+\frac{1}{k-1})^{1/2}\frac{h^2(1-r)(k-1)/k}{\sqrt{[(1-t)(k-1)/k]}}\sigma_P$$

$$= i_{1|k}\left(\frac{k}{k-1}\right)^{1/2} \frac{h^2(1-r)(k-1)/k}{\sqrt{[(1-t)(k-1)/k]}}\sigma_P = i_{1|k}h^2\frac{(1-r)}{\sqrt{(1-t)}}\sigma_P.$$

A comparison with the general formula for δG shows the equivalence to unrestricted within-family selection, except for $E(I_s)$. However, for larger k, even the two values of $E(I_s)$ tend to the same value.

The expected genetic gains for various selection strategies are given in Table 6.3.

Table 6.3 Expected genetic gain for various selection strategies ($f\sigma_A$ of index selection is taken as unity)

Case	t	f = 20 k = 2	10 4	5 8
1	0.01	0.728	1.082	1.112
2a		0.250[a]	0.410	0.466
3		0.703	0.951	0.993
4		0.771[b]	1.180	1.326
1	0.5	0.680	0.907	0.872
2a		0.392[a]	0.761	1.055
3		0.777	1.215	1.497
4		0.777	1.215	1.497
1	0.9	0.435	0.492	0.430
2a		0.645[a]	1.122	1.483
3		0.568	0.779	0.833
4		0.783[b]	1.242	1.596
1	0.99	0.160	0.164	0.140
2a		0.766[a]	1.230	1.585
3		0.213	0.263	0.247
4		0.783[b]	1.242	1.596

[a] For Case 2 all values are underestimated, since the i value used applies to independent variables. In reality, these values are negatively correlated, thus the true i values are bigger.

[b] The values for the optimal index are only approximate.

6.2.2 Short- to Medium-Term Results

The genetic parameters change due to selection. This is especially true for the genetic variance due to:

1. Random change in allele frequency due to small effective numbers.
2. Systematic change in allele frequency due to selection.

3. Linkage (negative covariances between loci).

For short-term selection we can ignore the first factor.

The second factor can only be dealt with when allele frequencies and allele effects are known. This factor can be ignored for the infinitesimal model (many loci each with a small effect). Further, this effect should be small if allele frequencies are distributed uniformly between zero and one.

The third factor is a change in variance with little or no change in allele frequency. This factor can be most easily visualized if we think of stabilizing selection. If in a population we select all animals which are clustered closely around the mean, then the variance of the selected animals approaches zero, and if h^2 is very high then the variance of the breeding values also tends to be zero. However, besides additive genetic variance at each locus, there are negative covariances between allelic effects at different loci. With directional selection the effect is very similar.

To quantify this effect of linkage, the following approach is proposed, assuming normality and equicorrelation.

Test fk males and select the top f individuals. For small numbers of fk and f, means and variances of order statistics are easily known. Variance (mean squared deviation within the selected group) can be calculated as

$$\mathrm{Var}(I_s) = E\,\frac{1}{f}\sum_{i=1}^{f}(I_{(i)}-\bar{I}.)^2$$

$$= E\mathbf{I}'_{(i)}(\mathbf{I} - \tfrac{1}{f}\mathbf{1}\mathbf{1}')\mathbf{I}_{(i)} = E\mathbf{I}'_{(i)}\mathbf{B}\mathbf{I}_{(i)}$$

$$= \mathrm{tr}[\mathbf{B}\ \mathrm{Var}(\mathbf{I}_{(i)})] + (E\mathbf{I}_{(i)})\mathbf{B}(E\mathbf{I}_{(i)}).$$

If both f and fk are very large, then censoring becomes truncation selection and

$$\mathrm{Var}(I_s) = [1-i(i-x)]\sigma_I^2.$$

We are, however, interested in the variance of A rather than I and, therefore, expected values and variances of associated variables are needed. These are given by Watterson (1959) and in this case are

$$EA_{[i]} = f\sigma_A EI_{(i)}$$

$$\mathrm{Var}(A_{[i]}) = (1-f^2)\sigma_A^2 + f^2\sigma_A^2\,\sigma_{i|n}^2$$

$$\mathrm{Cov}(A_{[i]}, A_{[j]}) = f^2\sigma_A^2\sigma_{j,iln}.$$

For large f and fk the variance of the breeding values after selection tends to

$$\mathrm{Var}(A_i) = \sigma_A^2[1-f^2 i(i-x)].$$

For small f a finite correction is useful and,

$$\mathrm{Var}(A_i) = \sigma_A^2[1-f^2 i(i-x)]\,\frac{(f-1)}{f}.$$

If the same selection is applied to males and females, we obtain in the next generation,

$$\mathrm{Var}(A_i) = \sigma_s^2+\sigma_d^2+\sigma_{Aw}^2$$

$$= 1/4\sigma_A^2[1-f^2 i(i-x)]+1/4\sigma_A^2[1-f^2 i(i-x)]+1/2\sigma_A^2$$

$$= \sigma_A^2[1-0.5f^2 i(i-x)].$$

The third term (σ_{Aw}^2) is due to Mendelian segregation within loci, and its value is hardly changed by selection (Fisher 1918). Since for most practical cases i(i-x) is between 0.64 and 0.90, "usable" additive genetic variance can be considerably reduced in the first generation following selection. However, to repeat the selection cycle, a different σ_A^2 and σ_P^2 (and thus also h^2 and t) are used. Repetition of this scheme over time leads quickly to a stable value which under mass selection (Dempfle 1975) is given by:

$$\tilde{V}_A = \frac{V_A-V_E+[(V_A+V_E)^2+4V_AV_E i(i-x)]^{0.5}}{2[1+i(i-x)]},$$

where V_A and V_E is additive genetic variance prior to selection and environmental variance, respectively.

If we stop selection, the "usable" additive genetic variance reverts quickly to the old value σ_A^2 (apart from changes due to drift and systematic selection pressure).

As a consequence of this linkage effect, the "usable" genetic variance is distributed differentially between and within families in the selected population. This is illustrated with the following example.

Given $\sigma_P=1$; $h^2=0.5$; t=0.5 and r=0.5, then

$$\sigma_P^2 = \sigma_{Ab}^2+\sigma_{Eb}^2+\sigma_{Aw}^2+\sigma_{Ew}^2$$

$$\sigma_P^2 = 0.25+0.25+0.25+0.25.$$

If 1 of 8 are selected then i(i-x) is 1.647(1.647-1.15)= 0.818, and $\tilde{V}_A$ is 0.37. Thus, σ_P^2=0.87=0.12+0.25+0.25+0.25; $\tilde{h}^2$=0.43; t=0.43; $\tilde{r}$=0.32.

This leads to changes in relative efficiency of index selection and within-family selection. Table 6.4 gives a comparison between index selection and within-family selection. The use of an optimal index was always assumed. Relevant index coefficients change during the first few generations. Properties of index selected population are also given in Table 6.4. The last column gives the ratio between the expected genetic gain from within-family selection relative to that due to optimal index selection. However, the same i values were used in the calculations for both selection schemes. In reality, the effects of finiteness and of the correlation (f_{II}) on the i value are different for the two procedures.

Table 6.4 Equilibrium parameters under index selection, and ratio of genetic progress due to within-family relative to index selection. Starting values are σ_P^2=1, r=0.5, selection 1 of 8, same order statistics

				Optimal index selection				
h^2	t	$\tilde{\sigma}_P^2$	$\tilde{h}^2$	$\tilde{r}$	$\tilde{t}$	$\tilde{f}_{AI}$	$\tilde{f}_{II}$	$\delta G_w/\delta G_I$
0.01	0.01	1.00	0.010	0.490	0.01	0.157	0.896	0.308
	0.10	1.00	0.010	0.493	0.10	0.130	0.833	0.381
	0.50	1.00	0.010	0.496	0.50	0.100	0.492	0.688
	0.90	1.00	0.010	0.490	0.90	0.160	0.004	0.923
	0.99	1.00	0.008	0.390	0.99	0.518	-0.135	1.000
0.10	0.10	0.99	0.090	0.442	0.090	0.378	0.783	0.436
	0.50	0.99	0.094	0.462	0.496	0.306	0.427	0.708
	0.90	0.98	0.082	0.382	0.898	0.538	-0.067	0.968
0.50	0.25	0.86	0.419	0.306	0.128	0.689	0.513	0.654
	0.50	0.87	0.424	0.321	0.424	0.662	0.225	0.825
	0.75	0.81	0.383	0.193	0.691	0.867	-0.074	0.969
0.90	0.50	0.63	0.841	0.152	0.207	0.922	0.103	0.885
0.99	0.50	0.56	0.982	0.098	0.106	0.991	0.091	0.892

If we select from all tested individuals, then we are actually selecting the same number of individuals out of the tested ones. Thus, the effect of finiteness is the same for the procedures (except for restricted within-family selection).

The effect of the correlation between selection criteria is much higher for index selection. In some situations correlations can be as high as 0.9 and, thus, expected values of order statistics are considerably reduced. On the other hand, with restricted within-family selection correlations are always -1/(k-1) and, thus, order statistics are higher than with independent variables. Both factors, therefore, operate in the same direction because the efficiency of within-family selection is underestimated.

With restricted within-family selection, we select the best out of k (not the best f out of kf) and, thus, expected values of the largest order statistics are reduced more compared to selection from an infinite population. This reduction is, as indicated in Table 6.4, much greater with small families. On the other hand, negative correlations (equicorrelation) increase expected values.

An example will serve to illustrate this. Let us assume that $\sigma_P^2=1$, $h^2=0.5$, $t=0.25$, $k=8$, $f=5$. Then, the expected gain due to restricted within-family selection is

$$\delta G_w = i_{1|8} \frac{(1-r)\sigma_A^2}{\sqrt{[(1-t)\sigma_P^2]}} = \frac{1.424 \times 0.5 \times 0.5}{\sqrt{0.75}} = 0.411$$

or

$$E(I_s) f_{AI} \sigma_A = 1.522 \times 0.382 \times 0.707 = 0.411.$$

Expected selection gain with an optimal index is (at equilibrium $\sigma_P^2=0.86$, $h^2=0.42$, $r=0.31$, $t=0.13$, $f_{AI}=0.69$, $f_{II}=0.513$), $\delta G_I=E(I_s)f_{AI}\sigma_A=1.497 \times 0.689 \times 0.6=0.619$. The ratio $\delta Gw/\delta GI$ is thus 0.664, slightly more favorable to within-family selection than shown in Table 6.4. This is especially true because restricted within-family selection must always give a lower response than unrestricted within-family selection in a single cycle of selection.

6.2.3 Long-Term Results

Without knowing allele frequencies and allele effects precise long-term predictions cannot be made. Thus, long-term predictions are not possible in practical situations. However, comparative statements can be made. More precise predictions are possible only for the infinitesimal model.

After a few generations equilibrium is reached and we have

$$\delta G_t = E(I_s) f(t) \sigma_A(t).$$

In each generation the additive genetic variance is expected to be changed by

$$(1 - \frac{1}{2N_e}),$$

where N_e is the effective population size. Thus, the accumulated selection gain is

$$\sum_{j=1}^{t} \delta G(j) = \sum_{j=1}^{t} E(I_s) f(j) \sigma_A(j) \quad \text{with } \sigma_A(0)=\sigma_A.$$

In the general case all quantities change with time. For two special cases of mass selection ($b_1=b_2=1$) this expression simplifies to:

$$= \sum_{j=1}^{\infty} \delta G(j) \frac{\sigma^2_{A(j)}}{\sqrt{(\sigma^2_{A(j)}+\sigma^2_E)}} = \begin{cases} E(I_s) \dfrac{\sigma^2_A}{\sqrt{(\sigma^2_A+\sigma^2_E)}} \ .2Ne & \text{if } \sigma^2_A \ll \sigma^2_E \\ \text{or} \\ E(I_s) \dfrac{\sigma^2_A}{\sqrt{(\sigma^2_A+\sigma^2_E)}} \ .4Ne & \text{if } \sigma^2_A \gg \sigma^2_E \end{cases}$$

and N is large.

The second expression is approximately $i\sigma_A 4N$. For these two special cases, restricted within-family selection ($b_1=0$, $b_2=1$) leads to

$$\sum_{j=1}^{\infty} \delta G(j) = \sum_{j=1}^{\infty} E(I_s) \frac{(1-r)\,\sigma^2_A(j)}{\sqrt{(1-t)}\sqrt{(\sigma^2_A(j)+\sigma^2_E)}} = \sum_{j=1}^{\infty} E(I_s) \frac{0.5\sigma^2_A(j)}{\sqrt{(0.5\sigma^2_A(j)+\sigma^2_E)}}$$

$$= \begin{cases} E(I_s) \dfrac{0.5\sigma^2_A}{\sqrt{(0.5\sigma^2_A+\sigma^2_{Ew})}} \ 2Ne & \text{if } \sigma^2_A \ll \sigma^2_E \\ \text{or} \\ E(I_s) \dfrac{0.5\sigma^2_A}{\sqrt{(0.5\sigma^2_A+\sigma^2_{Ew})}} \ 4Ne & \text{if } \sigma^2_A \gg \sigma^2_E \end{cases}$$

and N is large.

The second expression is approximately $i\sigma_A 4N\sqrt{2}$.

For index selection or for h^2 not near 0 or 1, no simple expression was found. Here, numerical calculation was used.

In all expressions for long-term results, N_e is needed. For random mating and for restricted within-family selection, N_e is known. Robertson (1961) and Burrows (1984) have dealt with effective population size under selection. Unfortunately, neither has given effective size for the use of negative correlations between members of a family (within-family selection or index selection if t is large). Thus, comparisons can only be made between restricted within-family selection, mass selection, and index selection.

For restricted within-family selection, the variance of family size is zero and effective size approaches twice the number of breeding animals.

With random selection from a population of f families each with k male offspring and k female offspring, effective size is slightly greater than the number of breeding animals since sampling is from a hypergeometric distribution.

When selection is based on the phenotypic value or on an index, the effective population size with a fixed number of breeding animals is expected to be different from the size of a random mating population. There are two reasons for this. First, if there is a high correlation between selection criteria of members of the same family (f_{II}), then family size of "successful" offspring has no longer a binomial or Poisson distribution. In the extreme case of family selection (b1=1, b2=0), either all or no animals of a family are selected. The second aspect is that the selection criterion is heritable; thus, individuals from a "successful" family also have more "successful" offspring. The latter aspect was especially considered by Robertson (1961), whereas the first aspect was considered in a very systematic way by Burrows (1984).

Burrows (1984) derived approximate formulae similar to those of Robertson (1961), where reductions in effective size due to both correlation between members of a family and proportion selected (α) is given. He expressed this reduction in terms of

$$N_{e(\alpha)} = R(\alpha, f_{II}) N_{e(random)}.$$

If f=5, k=8 and population size is constant, then α=0.125, and for $R(\alpha,f)$ we have (figures interpolated between α=0.12 and α=0.13)

f_{II}	0.1	0.2	0.3	0.4	0.5	0.6	0.7	0.8	0.9
$R(0.125, f_{II})$	0.775	0.619	0.504	0.418	0.351	0.297	0.252	0.213	0.177

Thus, for various selection schemes with large f and α=0.125 and low h^2, approximate N_e values are given in Table 6.5. Some of these values were checked by Monte Carlo simulation with f=5, k=8 and varied heritability. These results are presented in Table 6.6.

If equilibrium values of f_{II} are used, these results agree fairly well with figures of Burrows (1984). However, the values compared (values of f_{II}<0.2) constitute only a very narrow range and do not cover most of the selection index cases.

Using the $R(\alpha, f_{II})$ values given by Burrows (1984) and the appropriate genetic parameters (equilibrium values), accumulated selection responses were obtained and these are given in Table 6.7.

Table 6.5 Approximate effective population sizes

t	Restricted within -family selection	Random selection	Mass selection	Index selection
0.01	4f	2f	2.00f	0.35f
0.10	4f	2f	1.55f	0.43f
0.50	4f	2f	0.70f	0.70f
0.90	4f	2f	0.35f	2.00f

Table 6.6 Effective population sizes from Monte Carlo simulation

h^2	$f_{II}(0)$	$\tilde{f}_{II}$	F_{30}	± SE	N_e	$\hat{R}(0.125,f)$
Restricted within-family selection			0.547	0.003	18.55	1.66
~0	~0	~0	0.735	0.004	11.18	1.00
0.01	0.005	0.003	0.746	0.006	10.84	0.97
0.10	0.05	0.03	0.796	0.006	9.38	0.84
0.20	0.10	0.06	0.831	0.006	8.41	0.75
0.30	0.15	0.08	0.849	0.006	7.94	0.71
0.40	0.20	0.10	0.855	0.006	7.75	0.69
0.50	0.25	0.12	0.861	0.006	7.60	0.68
0.60	0.30	0.14	0.863	0.005	7.55	0.68
0.70	0.35	0.15	0.864	0.005	7.53	0.67
0.80	0.40	0.15	0.861	0.006	7.61	0.68
0.90	0.45	0.14	0.843	0.006	8.09	0.72
1.00	0.50	0.09	0.827	0.005	8.51	0.76

6.3 Discussion

To predict genetic response in the short and the long term it is not sufficient to only calculate

$$i \cdot f \cdot \sigma_A.$$

It is imperative to consider the effect of finiteness, the effect of correlation between the estimated breeding values, and the effect on the effective population size.

The following five different procedures were investigated for a full-sib family structure: family selection, unrestricted within-family selection, restricted within-family selection, mass selection and index selection with constant and with changing weights.

These procedures differ in the work needed to put them into practice and in the precision with which breeding values are estimated. They also differ in the correlations among estimates and, thus, in expected superiority and effective population size. Each procedure determines in a predictable way the precision of the estimator and the effective population size, and, for that reason, comparisons are useful. The results show that effective population size can be very important.

Table 6.7 Accumulated selection response up to generation j (σ_P=1) (index selection 1, same index coefficients in all generations; index selection 2, optimal index at each generation)

Starting values		Generation	Within -family selection	Mass selection	Index selection	
h^2	t				1	2
0.01	0.005	5	0.03	0.07	0.06	0.06
		10	0.06	0.13	0.08	0.08
		20	0.10	0.21	0.08	0.08
		100	0.23	0.35	0.08	0.08
	0.50	5	0.04	0.06	0.06	0.06
		10	0.08	0.09	0.09	0.09
		20	0.15	0.11	0.11	0.11
		100	0.33	0.12	0.12	0.12
	0.99	5	0.32	0.03	0.36	0.36
		10	0.61	0.03	0.68	0.69
		20	1.10	0.03	1.21	1.22
		100	2.67	0.03	2.67	2.77
0.10	0.05	5	0.32	0.69	0.59	0.59
		10	0.61	1.23	0.77	0.77
		20	1.07	1.99	0.84	0.82
		100	2.40	3.28	0.85	0.83
	0.10	5	0.33	0.68	0.56	0.57
		10	0.62	1.19	0.75	0.75
		20	1.10	1.88	0.82	0.81
		100	2.47	2.86	0.83	0.82
	0.50	5	0.45	0.57	0.57	0.57
		10	0.84	0.86	0.86	0.88
		20	1.49	1.09	1.09	1.14
		100	3.35	1.17	1.17	1.24
	0.90	5	1.01	0.39	1.14	1.16
		10	1.92	0.47	2.07	2.14
		20	3.48	0.49	3.43	3.72
		100	8.44	0.49	5.42	7.36
0.50	0.25	5	1.84	3.00	2.64	2.65
		10	3.48	5.23	4.09	4.02
		20	6.22	8.42	5.48	4.90
		100	14.59	13.81	6.22	4.99

Table 6.7 Continued

Starting values h²	Starting values t	Generation	Within -family selection	Mass selection	Index selection 1	Index selection 2
	0.50	5	2.26	2.69	2.69	2.77
		10	4.30	4.24	4.24	4.64
		20	7.77	5.68	5.68	6.79
		100	18.88	6.26	6.26	8.00
	0.75	5	3.24	2.24	3.54	3.72
		10	6.27	2.94	6.15	6.97
		20	11.74	3.09	9.19	12.64
		100	36.61	3.10	9.96	35.80
0.90	0.45	5	3.91	5.10	4.97	4.97
		10	7.52	9.04	8.65	8.66
		20	13.88	14.96	13.89	13.72
		100	37.66	25.58	21.66	17.12
	0.50	5	4.11	4.98	4.98	5.06
		10	7.93	8.61	8.61	9.00
		20	14.73	13.40	13.40	15.03
		100	41.91	16.96	16.96	23.51
0.99	0.50	5	4.54	5.63	5.63	5.64
		10	8.77	10.09	10.09	10.14
		20	16.41	17.14	17.14	17.37
		100	50.55	29.70	29.70	34.27

In these procedures the increases in effective population size were achieved by sacrificing some precision and/or by restricting selection to exploit only within-family variation. In practical applications, this sacrifice may be necessary only to a small extent. We can try to maximize precision via index or BLUP selection, taking into account relationships, and then apply some restriction on family size, i.e., we must not strictly select on estimated breeding values alone.

It may, however, happen that the increase in precision may not be usable. If we select strictly within families, then the inclusion of collateral relatives does not improve precision, because these contribute only to the between-family component.

Practical breeders have always considered family structure and have avoided selecting from only a few families. This approach is well justified. However, we should be able to give better guidelines for finding an optimal compromise.

References

Burrows PM (1984) Inbreeding under selection from unrelated families. Biometrics 40:357-366

Dempfle L (1975) A note on increasing the limit of selection through selection within families. Genet Res Camb 24:127-135

Fisher RA (1918) The correlation between relatives on the supposition of Mendelian inheritance. Trans R Soc Edinb 52:399-433

Hill WG (1976) Order statistics of correlated variables and implications in genetic selection programmes. Biometrics 32:889-902

Owen DB, Steck GP (1962) Moments of order statistics from the equicorrelated multivariate normal distribution. Ann Math Stat 33:1281-1291

Rawlings JO (1976) Order statistics for a special class of unequally correlated multinormal variates. Biometrics 32:875-887

Robertson A (1961) Inbreeding in artificial selection programmes. Genet Res 2:189-194

Watterson GA (1959) Linear estimation in censored samples from multivariate normal populations. Ann Math Statist 30:814-824

7 Optimal Designs for Sire Evaluation Schemes

R.L. Fernando and D. Gianola[1]

Maximizing genetic progress in a breeding program involves using the best method of evaluation and the available information to make optimum breeding decisions. These are based on the current information ($\mathbf{y}_0$) and may include choice and number of individuals culled, choice and number of new animals to include in the evaluation, allocation of progeny testing resources among animals evaluated, non-random mating, or any combination of the above. Let $\mathbf{u}_i$ be the merit of the candidates available for breeding and $\mathbf{y}_i$ be the additional records obtained, following the ith decision based on $\mathbf{y}_0$, for i=1,...s. Suppose all the information is used to select a constant number of candidates. This paper considers how breeding decisions based on $\mathbf{y}_0$ and selections based on $\mathbf{y}_0$ and $\mathbf{y}_i$ can be made such that expected genetic progress is maximized. It is shown that choosing the alternative corresponding to the largest conditional mean of merit of candidates selected at the final stage (given $\mathbf{y}_0$), and selecting those with the largest conditional means of merit given $\mathbf{y}_0$ and $\mathbf{y}_i$ maximizes expected genetic progress.

7.1 Introduction

Accuracy of evaluation of the genetic value of animals is but one of the factors affecting genetic progress in an animal breeding program. Selection intensity and generation interval also affect the rate of response to selection. These factors interact in a complicated and often conflicting manner (e.g., Dickerson and Hazel 1944; Robertson 1957). An optimum design for sire evaluation needs to consider the interaction of these factors on genetic progress.

Cochran (1951) showed that evaluating animals by the conditional mean of the genetic values given the data has the attractive property of maximizing expected merit of candidates selected by truncation. This holds when the merit variable and the data are identically and independently distributed from candidate to candidate. Bulmer (1980), Goffinet (1983), and Fernando and Gianola (1986) have shown that if a constant proportion of candidates is kept, selection based on the conditional mean maximizes the expected merit of the selected

[1] Department of Animal Sciences, University of Illinois, Urbana, Illinois, USA

candidates, for any distribution of the merit variables and the data. In populations undergoing selection and non-random mating the joint distribution of the merit variables and the data may be very complicated. However, the conditional distribution of the merit variables given the data used for making the breeding decisions remains unaltered (Gianola and Fernando 1986). Thus, animals can be evaluated using the conditional mean ignoring selection (Goffinet 1983; Fernando and Gianola 1984a).

Breeding decisions, often based on prior evaluations, affect the accuracy of future evaluations, selection intensity and generation interval. For example, sires with "good" evaluations are often used with higher frequency, and this results in more accurate evaluations of these sires. Further, new candidates for evaluation are likely to be relatives of those having better evaluations. Maximizing genetic progress in a breeding program involves not only using the best method of evaluation, but also using the available information to make optimum breeding decisions. These decisions may involve the following: (1) choice and number of animals to be culled; (2) choice and number of new animals to include in the evaluation; (3) allocation of progeny testing resources among animals evaluated; (4) non-random mating; or (5) any combination of the above.

In this paper we consider how breeding decisions, based on prior information, can be made such that expected genetic progress is maximized. A rule for choosing one among alternative breeding strategies is presented (see also Fernando 1984; Fernando and Gianola 1984b). This rule is in the spirit of those developed by James (1979) for finding optimum family size in progeny testing, Van Raden et al. (1984) in multiple stage selection, and Smith and Allaire (1985) for mate selection.

7.2 Theory

Let the data available for breeding decisions be denoted by $\mathbf{y}_0$ and let $\mathbf{u}_0$ be the merit variables of the animals evaluated at this stage. The decision process can be described as follows: when $\mathbf{y}_0 \varepsilon R_i^o$, the ith strategy is followed; R_i^o (i=1,...s) are mutually exclusive and exhaustive subsets of the sample space of $\mathbf{y}_0$. This results in more data ($\mathbf{y}_i$) and possibly a new set of animals to be evaluated with merit variables $\mathbf{u}_i$. These animals will typically include some of the animals evaluated using $\mathbf{y}_0$, and some new ones. Now, a subset of individuals with merits in $\mathbf{u}_i$ is to be selected based on $\mathbf{y}_0$ and $\mathbf{y}_i$ for further use. The quantity to be maximized is the expected genetic merit of the individuals eventually kept. Principles discussed here should be applicable to other optimality criteria.

The above process involves two decision rules. The first one, based on $\mathbf{y}_0$, is used to choose among the different breeding strategies. The second rule is based on $\mathbf{y}_0$ and $\mathbf{y}_i$, and is used to select animals for further breeding. Both rules need to be constructed such that expected genetic progress is maximized.

If a constant (over repeated sampling) number of animals is to be selected, regardless of the definition of the first rule, expected genetic progress is maximized with respect to the second rule by selecting animals with the highest conditional means, given $\mathbf{y}_o$ and $\mathbf{y}_i$ (Goffinet 1983; Fernando and Gianola 1984a). Unfortunately, this is not true in general if selection is by truncation (Fernando and Gianola 1986).

Let T_i be the arithmetic mean of the merit variables ($\mathbf{u}_i$) of the animals selected using the second rule. Now, the expected merit of the selected animals can be written (Fernando 1984) as:

$$E(T_i) = \sum_{i=1}^{s} \{\int_{R_i^o} \int_{R^{1i}} E(T_i|\mathbf{y}_o,\mathbf{y}_i)\, f(\mathbf{y}_o,\mathbf{y}_i)\, d\mathbf{y}_i d\mathbf{y}_o\}, \tag{7.1}$$

where $E(T_i|\mathbf{y}_o,\mathbf{y}_i)$ is the conditional mean of T_i given $\mathbf{y}_o$ and $\mathbf{y}_i$, $f(\mathbf{y}_o,\mathbf{y}_i)$ is the joint density of $\mathbf{y}_o$ and $\mathbf{y}_i$, and R^{1i} is the space of $\mathbf{y}_i$. Dividing and multiplying by the marginal distribution of $\mathbf{y}_o$, (7.1) becomes:

$$\sum_{i=1}^{s} \int_{R_i^o} \int_{R^{1i}} [E(T_i|\mathbf{y}_o,\mathbf{y}_i)\, f(\mathbf{y}_o,\mathbf{y}_i)/f(\mathbf{y}_o)]\, d\mathbf{y}_i\, f(\mathbf{y}_o) d\mathbf{y}_o$$

$$= \sum_{i=1}^{s} \{\int_{R_i^o} E[E(T_i|\mathbf{y}_o,\mathbf{y}_i)|\mathbf{y}_o]\, f(\mathbf{y}_o) d\mathbf{y}_o\}. \tag{7.2}$$

Now, in (7.2) let $B_i(\mathbf{y}_o)=E[E(T_i|\mathbf{y}_o,\mathbf{y}_i)|\mathbf{y}_o]$, and write

$$E(T_i) = \int_{R_1^o} B_1(\mathbf{y}_o) f(\mathbf{y}_o) d\mathbf{y}_o + \ldots + \int_{R_s^o} B_s(\mathbf{y}_o) f(\mathbf{y}_o) d\mathbf{y}_o$$

as

$$E(T_i) = \int_{R^o} B_C(\mathbf{y}_o) f(\mathbf{y}_o) d\mathbf{y}_o, \tag{7.3}$$

where

$$C = C(\mathbf{y}_o) = i \text{ when } \mathbf{y}_o \varepsilon R_i^o. \tag{7.4}$$

The value of $C(\mathbf{y}_o)$ in (7.4) corresponds to the alternative chosen at the first stage. Because animals are selected at the second stage by the conditional mean given $\mathbf{y}_o$ and $\mathbf{y}_i$, each $B_i(\mathbf{y}_o)$ is maximized with respect to the second rule, for i=1,...s. The question then is how R^o should be partitioned into s subsets such that selecting the ith strategy when $\mathbf{y}_o \varepsilon R_i^o$ leads to optimality. From (7.3) and (7.4), this is accomplished by defining R_i^o to be the set of values of $\mathbf{y}_o$ for which

$$B_i(\mathbf{y}_o) > B_j(\mathbf{y}_o), \quad (i,j) = 1,...s; \; i \neq j.$$

Thus, selecting the breeding strategy corresponding to the largest $B_i(\mathbf{y}_o)$ (first rule), and selecting the animals with the highest values of $E(\mathbf{u}_i|\mathbf{y}_o,\mathbf{y}_i)$ (second rule) maximizes expected genetic merit with respect to both rules. Note that $B_i(\mathbf{y}_o)$ is the conditional mean, given $\mathbf{y}_o$ and the ith strategy, of the breeding values of the animals selected by the second rule. Formally, this is

$$B_i(\mathbf{y}_o) = \underset{\mathbf{y}_i|\mathbf{y}_o}{E} [E(T_i|\mathbf{y}_o,\mathbf{y}_i)|\mathbf{y}_o].$$

Calculation of $B_i(\mathbf{y}_o)$ involves taking expectations of the ordered elements of the random vector $\hat{\mathbf{u}}_i|\mathbf{y}_o$ where $\hat{\mathbf{u}}_i=E(\mathbf{u}_i|\mathbf{y}_o, \mathbf{y}_i)$ is also known as the best predictor of $\mathbf{u}_i$ (Henderson 1974). In conducting the above calculations it is useful to note that

$$E(\hat{\mathbf{u}}_i|\mathbf{y}_o) = E[E(\mathbf{u}_i|\mathbf{y}_o,\mathbf{y}_i)|\mathbf{y}_o] = E(\mathbf{u}_i|\mathbf{y}_o). \tag{7.5}$$

This can be seen by writing (7.5) as

$$E[E(\mathbf{u}_i|\mathbf{y}_o,\mathbf{y}_i)|\mathbf{y}_o] = \int_{R^{1i}} [\int_{R^u} \mathbf{u}_i f(\mathbf{u}_i|\mathbf{y}_o,\mathbf{y}_i)d\mathbf{u}_i] f(\mathbf{y}_i|\mathbf{y}_o)d\mathbf{y}_i, \tag{7.6}$$

where R^u is the range of $\mathbf{u}_i$. Because

$$f(\mathbf{u}_i|\mathbf{y}_o,\mathbf{y}_i).f(\mathbf{y}_i|\mathbf{y}_o)=f(\mathbf{u}_i,\mathbf{y}_i|\mathbf{y}_o)=f(\mathbf{u}_i|\mathbf{y}_o).f(\mathbf{y}_i|\mathbf{u}_i,\mathbf{y}_o) \tag{7.7}$$

(7.6) can be also written as

$$\int_{R^u} \mathbf{u}_i f(\mathbf{u}_i|\mathbf{y}_o)[\int_{R^{1i}} f(\mathbf{y}_i|\mathbf{u}_i,\mathbf{y}_o)\, d\mathbf{y}_i]d\mathbf{u}_i. \tag{7.8}$$

The inner integral is equal to one for all values of $\mathbf{u}_i$ and $\mathbf{y}_o$; thus, (7.8) is equal to $E(\mathbf{u}_i|\mathbf{y}_o)$.

7.3 Numerical Examples

7.3.1 Allocation of Progeny Testing Resources

A hypothetical single trait progeny test example is used to illustrate how the results from the previous section can be used to choose among alternative sire sampling schemes.

Suppose 5 progeny records are available for each of 10 unrelated sires. The means of the progeny groups are 1.0, 0.9, 0.8, 0.7, 0.6, 0.5, 0.4, 0.3, 0.2, 0.1, and the heritability of the trait is 0.5. The goal is to select two sires such that the expected value of their mean breeding value is maximized. The decision will be based on the 50 progeny records already available, plus 50 additional records to be obtained. The question is how these 50 new records should be distributed among the 10 sires. The best one out of the five alternative sampling schemes given in Table 7.1 is to be chosen.

Table 7.1 Allocation of progeny testing resources: alternative sampling schemes

	Distribution of progeny records									
	Sires									
Sampling Scheme	1	2	3	4	5	6	7	8	9	10
1	5	5	5	5	5	5	5	5	5	5
2	10	10	10	10	10	0	0	0	0	0
3	25	25	0	0	0	0	0	0	0	0
4	15	12	10	8	5	0	0	0	0	0
5	12	10	8	6	4	3	2	2	2	1

Let $\mathbf{y}_o$ represent the progeny records already available, and $\mathbf{y}_i$ (i=1,...,5) represent records to be obtained under the ith sampling scheme. Because the goal is to select the two best sires, we need to calculate:

$$B_i(\mathbf{y}_o) = E[(\hat{u}_{[10]} + \hat{u}_{[9]})/2|\mathbf{y}_o],\ i=1,...5, \tag{7.9}$$

where $\hat{u}_{[10]}$ and $\hat{u}_{[9]}$ are the two largest values of the vector of predicted breeding values, $E(\mathbf{u}_i|\mathbf{y}_o,\mathbf{y}_i)$. The vector $\mathbf{u}$ represents the transmitting abilities of the ten sires.

As shown in the Appendix 7A.1, (7.9) was calculated for i=1, ...5, assuming that the breeding values of the sires and the data have a multivariate normal distribution. The values of $B_i(\mathbf{y}_o)$ for i=1,...5, were 0.476, 0.494, 0.454, 0.493, and 0.488, respectively. Thus, given the realized value of $\mathbf{y}_o$, the second sampling scheme would be expected to give the largest mean breeding value of the two selected sires.

Suppose now that the goal is to select a single sire using the same five sampling schemes of Table 7.1. The sampling scheme chosen would now be the one giving the largest:

$$B_i(\mathbf{y}_o) = E[\hat{u}_{[10]}|\mathbf{y}_o],\ i=1,...,5\ . \qquad (7.10)$$

As shown in the Appendix, $B_i(\mathbf{y}_o)$, for i=1,...5, was calculated to give 0.525, 0.559, 0.536, 0.561, 0.551. Thus, if only one sire is to be selected, the fourth sampling scheme would be optimum.

7.3.2 Sampling New Candidates

Consider a hypothetical situation where data are available on five unrelated sires for a trait with heritability equal to 0.25. The conditional means of their transmitting abilities ($\mathbf{v}_o$), given the realized data ($\mathbf{y}_o$) and subclass numbers are given in Table 7.2. Suppose five new young sires are to be sampled with the objective of keeping one of them after further evaluation based on the already available data ($\mathbf{y}_o$) and five additional offspring records obtained from each of the ten sires. Six different schemes for sampling the new animals among sons of the ten sires are considered (Table 7.2). To find the optimum sampling scheme

$$B_i(\mathbf{y}_o) = E[\hat{v}_{i[5]}|\mathbf{y}_o],\ i=1,...5 \qquad (7.11)$$

needs to be calculated, where $\mathbf{v}_i$ is the vector of transmitting abilities of the five new animals, and $\hat{v}_{i[5]}$ is the largest element of $E(\mathbf{v}|\mathbf{y}_o,\mathbf{y}_i)$.

Multivariate normality and the random model

$$\begin{aligned} \mathbf{y}_o &= \mathbf{Z}_o\mathbf{u}_o+\mathbf{e}_o \\ \mathbf{y}_i &= \mathbf{Z}_i\mathbf{u}_i+\mathbf{e}_i \\ E(\mathbf{u}_i) &= \mathbf{0},\ E(\mathbf{e}) = \mathbf{0} \\ Var(\mathbf{u}_i) &= \mathbf{G}_i,\ Var(\mathbf{e}) = \mathbf{I}\sigma_e^2 \end{aligned} \qquad (7.12)$$

were assumed. Above, $\mathbf{u}_i'=[\mathbf{v}_o',\mathbf{v}_i']$, $\mathbf{Z}_o$ and $\mathbf{Z}_i$ are known matrices relating $\mathbf{u}_i$ to $\mathbf{y}_o$ and $\mathbf{y}_i$, respectively, and $\mathbf{e}'=[\mathbf{e}_o',\mathbf{e}_i']$. With these assumptions, (7.11) was estimated by Monte-Carlo simulation as described below.

Table 7.2 Comparison of alternative strategies for sampling new candidates

	Sires						
	1	2	3	4	5		
Conditional means (stage 1)	0.615	0.615	0.0	-0.059	-0.176		
Number of records (stage 1)	19	1	1	2	2		
Alternative	Sampling scheme (stage 2)					B_i[a]	SE[b]
1	5	0	0	0	0	0.75	0.004
2	0	5	0	0	0	0.76	0.005
3	3	2	0	0	0	0.83	0.004
4	2	3	0	0	0	0.83	0.004
5	2	2	1	0	0	0.80	0.004
6	1	1	1	1	1	0.73	0.003

[a] $B_i(y_o)$: estimated conditional mean of the breeding values of the selected animals given $\mathbf{y}_o$ and the strategy.
[b] SE: standard error of the above estimate.

Using Henderson's mixed model techniques (Henderson 1974), the conditional mean of $\mathbf{u}_i$ given $\mathbf{y}_o$ can be calculated as:

$$E(\mathbf{u}_i|\mathbf{y}_o) = (\mathbf{Z}_o'\mathbf{Z}_o + \mathbf{G}_i^{-1}\sigma_e^2)^{-1}\mathbf{Z}_o'\mathbf{y}_o. \tag{7.13}$$

From the equivalence given in (7.5), (7.13) is also equal to $E[E(\mathbf{u}_i|\mathbf{y}_o,\mathbf{y}_i)|\mathbf{y}_o]$. The conditional mean of $\mathbf{u}_i$ given $\mathbf{y}_o$ and $\mathbf{y}_i$ can also be written as:

$$E(\mathbf{u}_i|\mathbf{y}_o,\mathbf{y}_i) = (\mathbf{Z}'\mathbf{Z}+\mathbf{G}_i^{-1}\sigma_e^2)^{-1}(\mathbf{Z}_o'\mathbf{y}_o+\mathbf{Z}_i'\mathbf{y}_i), \tag{7.14}$$

where $\mathbf{Z}'=[\mathbf{Z}_o',\mathbf{Z}_i']$. Now, the conditional variance of (7.14) given $\mathbf{y}_o$ is:

$$Var[E(\mathbf{u}_i|\mathbf{y}_o,\mathbf{y}_i)|\mathbf{y}_o] = \mathbf{C}Var(\mathbf{Z}_i'\mathbf{y}_i|\mathbf{y}_o)\mathbf{C}' = \mathbf{A}, \tag{7.15}$$

where $\mathbf{C}=(\mathbf{Z}'\mathbf{Z}+\mathbf{G}^{-1}\sigma_e^2)^{-1}$. Then the random vector defined as:

$$\mathbf{s} = E[E(\mathbf{u}_i|\mathbf{y}_o,\mathbf{y}_i)|\mathbf{y}_o]+\mathbf{Hz}, \tag{7.16}$$

where $\mathbf{HH'}=\mathbf{A}$ and $\mathbf{z}$ is a vector of standard normal independent random variables, is identical in distribution to the random vector $[E(\mathbf{u}_i|\mathbf{y}_o,\mathbf{y}_i)|\mathbf{y}_o]$. Thus, (7.11) can be calculated for the different sampling schemes by generating $\mathbf{s}$ using a pseudostandard normal random number generator in (7.16), and taking the mean of the highest ranking values of $\mathbf{s}$ corresponding to transmitting abilities of the new bulls. The results in Table 7.2 are based on 10,000 samples of $\mathbf{s}$.

In sampling schemes 1 and 2 all the new bulls sampled are sons of sires 1 or 2, respectively (Table 7.2). Sires 1 and 2 had identical conditional means, given $\mathbf{y}_o$. However, the evaluation for sire 1 was based on 19 records while that of sire 2 was based on 1 record. Thus, the diagonals of $\mathbf{A}$ were larger in the second sampling scheme (0.292) than in the first (0.206). This might lead one to expect scheme 2 to be better than 1 because $\mathbf{s}$ in (7.16) has a larger variance. However, the two schemes were practically the same (Table 7.2). This is so because $\mathbf{A}$ in scheme 2 had larger off-diagonals (0.142) than in scheme 1 (0.056). This results in a lower selection intensity with the second sampling scheme (Hill 1976). Thus, it seems that the expected advantage due to the larger diagonals in $\mathbf{A}$ with the second scheme were lost due to the larger off-diagonals. In the third and fourth schemes (Table 7.2) the new bulls were sampled from the two better sires. These two schemes were the best.

7.3.3 Two Stage Selection

Suppose data are available on ten unrelated sires for a trait with heritability equal to 0.25. The objective is to use this information to cull five of the sires, and to select the best one from the remaining five sires after obtaining five more records on each of these.

Assume that each of the first five sires has a conditional mean of 0.05 based on ten records each; the five remaining sires have a conditional mean of 0 based on one record each. Two strategies for culling are considered. The first one is to keep the first five and cull the remaining animals. In the second one, the first five animals are culled and the remaining sires are kept. In order to choose the better strategy one needs to calculate:

$$B_i(\mathbf{y}_o) = E(\hat{u}_{i[5]}|\mathbf{y}_o),\ i=1,2, \tag{7.17}$$

where $\mathbf{u}_i$ is the vector of transmitting abilities of the five sires selected for further evaluation, and $\hat{u}_{i[5]}$ is the largest element of $E(\mathbf{u}_i|\mathbf{y}_o,\mathbf{y}_i)$. Assuming multivariate normality and a random model, (7.17) was computed for each of the two strategies using tables of order statistics (David 1970). The values for the two strategies were 0.418 and 0.549. The diagonal elements of the conditional variance-covariance matrix ($\mathbf{A}$) were 0.1 and 0.223 for the two strategies. The off-diagonal elements were null in both cases. Thus, even though the first five sires had slightly higher conditional means given $\mathbf{y}_o$, selecting the sires with the lower predictions and lower accuracy was better. This agrees with Van Raden et al. (1984).

7.4 Discussion

The procedure described here for choosing an optimum breeding strategy from a finite set of alternatives requires calculating $B_i(\mathbf{y}_o)$ for each strategy. This calculation is impossible for most real problems by analytical methods. However, $B_i(\mathbf{y}_o)$ may be estimated by Monte-Carlo simulation although this may not be feasible in many real situations. In these cases, small representative examples may be studied to get some insight into the problem of finding optimum breeding strategies. Other approximations to $B_i(\mathbf{y}_o)$ may be considered as well. For example, the rules described by James (1979) and Van Raden et al. (1984) may be used, where applicable.

Even for a simple problem, the number of possible alternative breeding strategies can be very large. Thus, an efficient method of searching for the optimum strategy without calculating $B_i(\mathbf{y}_o)$ for every possible strategy is required. This problem has been addressed by Jansen and Wilton (1984) and by Smith and Allaire (1985) in the context of mate selection. These authors discuss how integer linear programming can be used to search for an optimum mating scheme.

In principle, the approach presented here can also be used to maximize expected genetic progress over many stages of decision making. However, the required calculations may not be feasible. For example, let $\mathbf{y}_o$ be the data initially available. Suppose that based on $\mathbf{y}_o$, one of s_1 alternatives needs to be chosen. Let $\mathbf{y}_{1i}$ be the data that accumulates following the first stage. Then, based on $\mathbf{y}_o$ and $\mathbf{y}_{1i}$ one of s_2 strategies must be chosen. More data become available following this second decision stage, say $\mathbf{y}_{2ij}$. Finally, a fixed number of animals is to be selected based on $\mathbf{y}_o$, $\mathbf{y}_{1i}$, and $\mathbf{y}_{2ij}$. Three rules are then needed to maximize the expected breeding value of the animals selected at the final stage.

Regardless of the decisions made at the first and second stages, selecting animals based on the conditional mean given all the data will maximize the expected breeding value of the selected animals, with respect to the third rule. The optimum rule for the second stage would be the one discussed here. For each of the s_2 strategies,

$$B_{2ij}(\mathbf{y}_o,\mathbf{y}_{1i}) = E[E(T_{ij}|\mathbf{y}_o,\mathbf{y}_{1i},\mathbf{y}_{2ij})|\mathbf{y}_o,\mathbf{y}_{1i}],\ j=1...s_2, \tag{7.18}$$

where T_{ij} is the arithmetic mean of the breeding values of the selected candidates, needs to be calculated. This gives the expected breeding value of the selected candidates, given $\mathbf{y}_o$ and $\mathbf{y}_{1i}$, for each of the s_2 alternatives that could be adopted at the second stage. Selecting the strategy corresponding to the largest value of (7.18) maximizes expected genetic progress, as defined here, with respect to the second rule, regardless of the decision made at the first stage.

To select the optimum strategy at the first stage,

$$B_{1i}(\mathbf{y}_o) = E[E(T_{ij}|\mathbf{y}_o,\mathbf{y}_{1i},\mathbf{y}_{2ij})|\mathbf{y}_o],\ i=1,...s_1 \tag{7.19}$$

needs to be calculated. This is very difficult in most situations because $\mathbf{y}_{1i}$ is not yet realized at this stage. The rule adopted at the second stage will depend on the realized value of $\mathbf{y}_{1i}$; i.e., j in (7.19) is a random variable. Thus, calculating (7.19) involves taking expectations over the rules adopted at the second stage, given $\mathbf{y}_0$. Monte-Carlo simulation could be used to estimate (7.19) but in most problems this would require tedious computation.

Comparison of strategies 1 and 2 in the second example showed that the optimum strategy not only depends on the conditional means (7.13) and conditional variances (diagonals of **A**) of the final predictor, given $\mathbf{y}_0$, but also on the conditional covariances among the predictors (off-diagonals of **A**). Thus, methods based only on the conditional means and conditional variances may not be suitable when there are large conditional covariances associated with some of the alternatives.

References

Bulmer MG (1980) The mathematical theory of quantitative genetics. Clarendon Press, Oxford

Cochran WG (1951) Improvement by means of selection. Proc 2nd Berkeley Symp Math Stat Prob, pp 449-470

David HA (1970) Order statistics. Wiley, New York

Dickerson GE, Hazel LN (1944) Effectiveness of selection on progeny performance as a supplement to earlier culling in livestock. J Agric Res 69:459-476

Fernando RL (1984) Selection and assortative mating. PhD Thesis, Univ Illinois, Urbana

Fernando RL, Gianola D (1984a) Optimal properties of the conditional mean as a selection criterion. J Anim Sci 59 suppl (Abstr) 1:177

Fernando RL, Gianola D (1984b) An optimal two stage selection procedure. J Anim Sci 59 suppl (Abstr) 1:177

Fernando RL, Gianola D (1986) Optimal properties of the conditional mean as a selection criterion. Theor Appl Genet 72:822-825

Gianola D, Fernando RL (1986) Bayesian methods in animal breeding theory. J Anim Sci 63:217-244

Goffinet B (1983) Selection on selected records. Genet Sel Evol 15:91-98

Henderson CR (1974) General flexibility of linear model techniques for sire evaluation. J Dairy Sci 57:963-972

Hill WG (1976) Order statistics of correlated variables and implications in genetic selection programmes. Biometrics 32:889-902

James JW (1979) Optimum family size in progeny testing when prior information is available. Z Tierz Züchtungsbiol 95:194-203

Jansen GB, Wilton JW (1984) Selecting mating pairs with linear programming techniques. J Dairy Sci 67 suppl (Abstr) 1:246

Robertson A (1957) Optimum group size in progeny testing and family selection. Biometrics 13:442-450

Smith SP, Allaire FR (1985) Efficient selection rules to increase non-linear merit: application in mate selection. Genet Sel Evol 17:387-406

Van Raden PM, Freeman AE, Rothschild MF (1984) Maximizing genetic gain under multiple-stage selection. J Dairy Sci 67:1761-1766

APPENDIX 7A.1

Computation of the conditional mean of ordered predicted breeding values

Calculation of (7.9), for i=1...,5, is shown here under multivariate normality. The model for a record on the kth progeny of the jth sire is

$$y_{jk} = u_j + e_{jk} \qquad j = 1,\ldots,10 \quad k = 1,\ldots,n_j \tag{7A.1}$$

where $u_j \sim N.I.D.(0,\sigma_u^2)$, $e_{jk} \sim N.I.D.(0,\sigma_e^2)$ and $Cov(u_j,e_{jk})=0$ for all j's and k's. Let

$\bar{y}_{(o)j}$ = mean of progeny records from the jth sire in $\mathbf{y}_o$;

$\bar{y}_{(i)j}$ = mean of progeny records from the jth sire in $\mathbf{y}_i$;

$n_{(i)j}$ = number of progeny to be sampled from the jth sire in the ith sampling scheme;

$\sigma_o^2 = Var[\bar{y}_{(o)j}] = \sigma_u^2+\sigma_e^2/5;$

$\sigma_i^2 = Var[\bar{y}_{(i)j}] = \sigma_u^2+\sigma_e^2/n_{(i)j}.$

The variance of transmitting abilities is $\sigma_u^2=h^2(\sigma_u^2+\sigma_e^2)/4$, with h^2 (heritability) equal to 0.50 in the example. Then

$$\hat{u}_{ij} = E(u_j|\bar{y}_{(o)j},\bar{y}_{(i)j})$$

$$= \frac{\sigma_u^2[(\sigma_i^2-\sigma_u^2)\bar{y}_{(o)j}+(\sigma_o^2-\sigma_u^2)\bar{y}_{(i)j}]}{\sigma_o^2\sigma_i^2-\sigma_u^4} \tag{7A.2}$$

if $n_{(i)j} \neq 0$, and

$$\hat{u}_{ij} = \bar{y}_{(o)j}\sigma_u^2/\sigma_o^2 \tag{7A.3}$$

if $n_{(i)j} = 0$.

Further, from (7.5),

$$E(\hat{u}_{ij}|y_o) = (7A.3)\ . \tag{7A.4}$$

Also,

$$\mathrm{Var}(\hat{u}_{ij}|y_o) = \left[\frac{\sigma_u^2(\sigma_o^2-\sigma_u^2)}{\sigma_o^2\ \sigma_i^2-\sigma_u^4}\right]^2 \sigma_i^2 \left[1 - \frac{\sigma_u^4}{\sigma_i^2\ \sigma_o^2}\right] \tag{7A.5}$$

if $n_{(i)j} \neq 0$, and

$$\mathrm{Var}\ (\hat{u}_{ij}|\mathbf{y}_o) = 0 \tag{7A.6}$$

if $n_{(i)j} = 0$.

In order to calculate (7.9), let z_{ij} be a random variable distributed as $(\hat{u}_{ij}|\mathbf{y}_o)$. Then,

$$E(z_{i[10]}) = E(u_{i[10]}|\mathbf{y}_o],$$

and so on. Using theory of order statistics (David 1970),

$$E(z_{i[10]}) = \sum_{\substack{j=1\\ n_{(i)j}\neq 0}}^{10} \left\{ \int_{-\infty}^{\infty} \left[\prod_{\substack{k=1\\ k\neq j}}^{10} P(z_{ik}<t)\right] tf(t)dt \right\} +$$

$$\sum_{\substack{j=1 \\ n_{(i)j}=0}}^{10} \left\{ \prod_{\substack{k=1 \\ k\neq j}}^{10} P(z_{ik}<\mu_{ij})\mu_{ij} \right\} \tag{7A.7}$$

where f(t) is the density function of z_{ij} and $\mu_{ij}=E(z_{ij})$. Taking $v=(t-\mu_{ij})/\sigma_{ij}$, where $\sigma_{ij}^2=Var(z_{ij})$, the integral in (7A.7) can be written as

$$\int_{-\infty}^{\infty} \left\{ \prod_{\substack{k=1 \\ k\neq j}}^{10} P[z_{ik}<(v\sigma_{ij}+\mu_{ij})] \right\} (v\sigma_{ij}+\mu_{ij})\, \phi(v)dv,$$

where v~N(0,1), $\phi(v)$ is the standard normal density function, and

$$P[z_{ik}<(v\sigma_{ij}+\mu_{ij})] = \Phi[(v\sigma_{ij}+\mu_{ij}-\mu_{ik})/\sigma_{ik}]$$

if $n_{(i)k}\neq 0$. If $n_{(i)k}=0$,

$$P[z_{ik}<(v\sigma_{ij}+\mu_{ij})] = 1 \text{ if } \mu_{ik}<(v\sigma_{ij}+\mu_{ij})$$

and

$$P[z_{ik}<(v\sigma_{ij}+\mu_{ij})] = 0 \text{ if } \mu_{ik}>(v\sigma_{ij}+\mu_{ij}).$$

Now, (7A.7) can be written as a single integral as

$$E(z_{i[10]}) = \int_{-\infty}^{\infty} \sum_{\substack{j=1 \\ n_{(i)j}\neq 0}}^{10} \left[\left\{ \prod_{\substack{k=1 \\ k\neq j}}^{10} P(z_{ik}<v\sigma_{ij}+\mu_{ij}) \right\} (v\sigma_{ij}+\mu_{ij}) \right] \phi(v)dv$$

$$+ \sum_{\substack{j=1 \\ n_{(i)j}=0}}^{10} \left\{ \prod_{\substack{k=1 \\ k\neq j}}^{10} P(z_{ik}<\mu_{ij})\mu_{ij} \right\}. \tag{7A.8}$$

Finally,

$$E(z_{i[9]}) = \sum_{\substack{j=1 \\ n_{(i)j}\neq 0}}^{10} \left\{ \sum_{\substack{k=1 \\ k\neq j}}^{10} \int_{-\infty}^{\infty} P(z_{ik}>t) \prod_{\substack{l=1 \\ l\neq k \\ l\neq j}}^{10} P(z_{il}<t)tf(t)dt \right\}$$

$$+ \sum_{\substack{j=1 \\ n_{(i)j}=0}}^{10} \left\{ \sum_{\substack{k=1 \\ k\neq j}}^{10} P(z_{ik}>\mu_{ij}) \prod_{\substack{l=1 \\ l\neq k \\ l\neq j}}^{10} P(z_{il}<\mu_{ij})\mu_j \right\} . \qquad (7A.9)$$

Discussion Summary
PART II: DESIGN OF EXPERIMENTS AND BREEDING PROGRAMS

J.W. James[1]

4 Considerations in the Design of Animal Breeding Experiments

It was pointed out that single generation selection experiments allowed heritability estimates to be obtained without use of individual pedigrees, although at the expense of difficulty in obtaining a standard error.

In using an elliptical selection criterion for bivariate parameter estimation, only the phenotypic variances and correlation need to be known well, and these normally are. The method is quite robust.

The use of selection for production traits should enable one to obtain good estimates of genetic covariances with indicator traits, although genetic variances of such traits would not be well estimated. However, errors in the covariances are more important for application purposes.

Using standard errors as a measure of accuracy assumes approximate normality of estimates, but this may not be the case. In the very large experiments needed for accurate estimates, the approximation should be good. In smaller experiments, some checks have been made using exact measures, and the conclusions were similar to those found from standard error calculations.

In estimating parameters from multigeneration data, it is important to be clear as to what the reference population is, and this will usually be the base population.

5 Use of Mixed Model Methodology in Analysis of Designed Experiments

The elements of the dominance relationship matrix are calculated from the coefficients of identity as defined by Gillois and Harris. Because the inclusion of dominance effects in models is desirable, a search for efficient computing methods comparable to those devised by Henderson for the additive relationship matrix would be worthwhile.

[1] Department of Wool Science, University of New South Wales, Kensington, Australia

Although estimates of genetic change can be obtained from selection experiments without controls, such estimates do not make use of between-generation data. This is a considerable fraction of the total information in a selection experiment and controls are, therefore, desirable. For a fixed total number of animals, inclusion of controls will give greatest accuracy.

When mixed model methods are used to increase selection response, one is also increasing the rate of inbreeding.

6 Statistical Aspects of Design of Animal Breeding Programs: A Comparison Among Various Selection Strategies

In the calculations it was assumed that parameters were known. However, if parameters are estimated, the results would be only slightly different, with the index being a little less efficient.

An alternative way of getting more long-term response would be to select less intensely. There is a general trade-off because any procedure which increases short-term response is expected to reduce long-term response. The problem is to find the best balance between the two.

For complex breeding programs, calculation of effective population size will be difficult. For systems such as MOET (multiple ovulation and embryo transfer) in dairy cattle programs, the best answer may be to simulate.

If different strategies are compared at the same effective population size and some strategies are better at all sizes, then these will be best no matter what time scale is considered. However, this approach will not be useful if the strategies require substantially different facilities.

7 Optimal Designs for Sire Evaluation Schemes

Apart from computational difficulties, perhaps the major problem is to identify the strategies which should be tested, because the number of possible alternatives is infinitely large. One possibility for identifying strategies would be to consider the effect of adding a single extra observation. Once the best way to add the extra observation has been found, further addition can be considered. While this may not provide the best possible answer, it ought to lead to a reasonable solution. This method is probably analogous to the exchange algorithms used in experimental design studies.

In the examples presented, the best strategy was defined as that which maximized expected genetic gain. Since varying amounts of information were available, this is equivalent to a linear utility function. Other objectives, such as maximizing the probability of a correct ordering of breeding values, would not give the same ranking of strategies. Another possible objective would be to maximize the probability that the breeding value of a selected animal is above a specified minimum. However, there is not necessarily a universally acceptable objective.

In principle, the method can be easily extended to selection over several generations, but it would then become computationally unfeasible.

Part III: Estimation of Genetic Parameters

8 Computational Aspects of Likelihood-Based Inference for Variance Components

D.A. Harville[1] and T.P. Callanan[2]

In this paper, iterative algorithms for computing restricted maximum likelihood (REML) estimates of variance components are discussed in the context of a possibly unbalanced, mixed linear model that contains a single set of m random effects. The coverage includes the Newton-Raphson algorithm, the method of scoring, the EM algorithm, and the method of successive approximations.

Before applying these various algorithms, it may be advisable to reparameterize, to "linearize" the likelihood equations, or to eliminate one of the likelihood equations (by absorption). The computations required to implement the algorithms can sometimes be reduced by diagonalizing or tridiagonalizing a certain mxm matrix. The computations required to obtain likelihood-based confidence intervals are similar in nature to those encountered in REML estimation, but are more extensive.

8.1 Introduction

Statistical procedures derived on the basis of mixed-effects linear models are used extensively in the agricultural and biological sciences. In particular, animal breeders often estimate the heritability of a trait by first expressing the heritability in terms of the variance components of a mixed-effects linear model and by then making use of a statistical procedure for estimating variance components. The estimate of heritability may be of direct interest, or it may be used indirectly in making inferences about the genetic merit of an animal or population of animals.

Likelihood-based methods for estimating variance components are rapidly gaining favor among animal breeders and other practitioners. Of these methods, restricted maximum likelihood (REML) estimation, in which the likelihood function is taken to be that associated with a set of error contrasts, is becoming the method of choice. This method was proposed, and its use advocated, by Thompson (1962) and Patterson and Thompson (1971) and has been reviewed by, for example, Harville (1977).

1 Department of Statistics, Iowa State University, Ames, Iowa, USA
2 Applied Statistics, Management Services Division, Eastman Kodak Company, Rochester, New York, USA

Unfortunately, the computations required to implement REML estimation are very extensive, prohibitively so, in many cases. In general, closed-form expressions for the REML estimates do not exist, and the estimates must be computed by using an iterative numerical method to locate the maximum of the likelihood function. At least four categories of iterative algorithms have been proposed: Newton-Raphson, scoring, EM, and successive approximations. The computational burden imposed by these algorithms is determined by the number of iterations required to attain convergence and also by the extensiveness of the computations required to carry out each iteration.

For purposes of making inferences about variance components or functions of variance components, the use of likelihood-based methods seems to have been limited primarily to point estimation. Their use in obtaining confidence intervals or in making significance tests does not seem to have been widely investigated.

In the present paper, we discuss the computational aspects of likelihood-based inferences about variance components or functions of variance components. In particular, we review, in a common framework and using common notation, various iterative algorithms for computing REML estimates, and introduce some possible improvements.

The Newton-Raphson and method-of-scoring algorithms depend on the choice of parameterization; the versions of these algorithms corresponding to the two most common parameterizations are considered in detail in Sections 8.7 and 8.12. In Sections 8.8 and 8.9, we discuss how the performance of the Newton-Raphson approach might be improved by applying it to a "concentrated" log likelihood function, instead of the original log likelihood function, and by "linearizing" the likelihood equations. In Section 8.13, we discuss two iterative algorithms obtained via the EM approach; one of these was developed by Laird (1982); the other is new. Use of the method of successive approximations is considered in Sections 8.13 and 8.14.

In Sections 8.10 and 8.11, we discuss how the computational burden imposed by the various algorithms can be kept to a minimum by using: (1) diagonalization (Dempster et al. 1984) or tridiagonalization (Smith and Graser 1986) and (2) mixed-model equations (Henderson 1963). In Section 8.15, we indicate how the likelihood function associated with a set of error contrasts can be used to obtain an approximate confidence interval for a function of the variance components. In Section 8.16, we apply the likelihood-based estimation and confidence-interval procedures to an example analyzed previously by Harville and Fenech (1985) (via a different approach), and evaluate the performance of the various iterative algorithms in the context of this example.

We have chosen, as the basic framework for our presentation, a mixed-effects linear model (introduced in Sect. 8.2) that contains a single set of uncorrelated and homoscedastic random effects. By limiting our discussion to models of this type, we require a less elaborate notation (Sects. 8.3 and 8.4) and are able to derive an appealing expression for the log likelihood function associated with the error contrasts (Sect. 8.4). This expression is itself useful computationally, leads directly to computationally useful expressions for the derivatives, and provides considerable insight into the nature of the likelihood function.

Mixed-effects linear models that contain a single set of uncorrelated and homoscedastic random effects are often employed in practice, and results for such models can sometimes be applied directly to models that contain more than one set of random effects (Harville and Fenech 1985, Sect. 8.1). In any case, extensions to more complex mixed-effects linear models are relatively straightforward and are summarized in Section 8.17.

8.2 Model

Let **y** represent an $n \times 1$ observable random vector. Suppose that **y** follows the mixed-effects linear model

$$\mathbf{y} = \mathbf{X}\boldsymbol{\beta}+\mathbf{Zs}+\mathbf{e}, \tag{8.1}$$

where $\boldsymbol{\beta}$ is a $p \times 1$ vector of unknown parameters and where **s** and **e** are unobservable random vectors of dimensions $m \times 1$ and $n \times 1$ that are distributed independently as $\text{MVN}(\mathbf{0},\sigma_s^2\mathbf{I})$ (multivariate normal with mean vector **0** and variance-covariance matrix $\sigma_s^2\mathbf{I}$) and $\text{MVN}(\mathbf{0},\sigma_e^2\mathbf{I})$, respectively. Thus, the elements of **s** consist of a single set of "random effects", while the elements of **e** are random "errors".

The quantities **X** and **Z** are given matrices. Define $p^*=\text{rank}(\mathbf{X})$, $r=\text{rank}(\mathbf{X},\mathbf{Z})-\text{rank}(\mathbf{X})$, and $f=n-\text{rank}(\mathbf{X},\mathbf{Z})$. It is assumed that $r>0$ and $f>0$.

Let $\gamma=\sigma_s^2/\sigma_e^2$ represent the variance ratio. Then, $\text{var}(\mathbf{y})=\sigma_e^2\mathbf{I}+ \sigma_s^2\mathbf{ZZ}'=\sigma_e^2\mathbf{V}$, with $\mathbf{V}=\mathbf{I}+\gamma\mathbf{ZZ}'$. The variance components σ_s^2 and σ_e^2 are taken to be unknown parameters satisfying $\sigma_s^2\geq 0$ and $\sigma_e^2>0$. Consequently, $\gamma\geq 0$.

8.3 Analysis of Variance (ANOVA) and ANOVA-Related Notation

Consider the linear system

$$\begin{bmatrix} \mathbf{X}'\mathbf{X} & \mathbf{X}'\mathbf{Z} \\ \mathbf{Z}'\mathbf{X} & \mathbf{Z}'\mathbf{Z} \end{bmatrix} \begin{bmatrix} \tilde{\boldsymbol{\beta}} \\ \tilde{\mathbf{s}} \end{bmatrix} = \begin{bmatrix} \mathbf{X}'\mathbf{y} \\ \mathbf{Z}'\mathbf{y} \end{bmatrix}, \tag{8.2}$$

which consists of the normal equations (NE) obtained by treating **s**, like **β**, as a vector of unknown parameters and applying ordinary least-squares. If we absorb the first set of NE (8.2) into the second set, we obtain the reduced NE

$$\mathbf{C}\tilde{\mathbf{s}} = \mathbf{q}, \tag{8.3}$$

where $\mathbf{C}=\mathbf{Z}'(\mathbf{I}-\mathbf{P_X})\mathbf{Z}$ and $\mathbf{q}=\mathbf{Z}'(\mathbf{I}-\mathbf{P_X})\mathbf{y}$ with $\mathbf{P_X}=\mathbf{X}(\mathbf{X}'\mathbf{X})^-\mathbf{X}'$. [$(\mathbf{X}'\mathbf{X})^-$represents an arbitrary generalized inverse of **X'X**.]

Subsequently, we let $\tilde{\boldsymbol{\beta}}$ and $\tilde{\mathbf{s}}$ represent the two parts of an arbitrary solution to NE (8.2). Note that defining $\tilde{\mathbf{s}}$ in this way is equivalent to defining it to be an arbitrary solution to reduced NE (8.3).

A significance test of the null hypothesis that γ=0 (or, equivalently, that σ_s^2=0) can be obtained from the following ANOVA table:

Source	df	SS	MS	EMS	Test statistic
β	p*	S_β			
s after **β**	r	S_s	S_s/r	$\sigma_e^2 + \kappa\sigma_s^2$	$F = (f/r)(S_s/S_e)$
Residual	f	S_e	S_e/f	σ_e^2	

Here,

$$S_\beta = \mathbf{y}'\mathbf{P_X}\mathbf{y},\ S_s = \tilde{\mathbf{s}}'\mathbf{q},$$

$$S_e = \mathbf{y}'\mathbf{y}-\tilde{\boldsymbol{\beta}}'\mathbf{X}'\mathbf{y}-\tilde{\mathbf{s}}'\mathbf{Z}'\mathbf{y} = \mathbf{y}'\mathbf{y}-S_\beta-S_s, \tag{8.4}$$

and $\kappa=(1/r)\mathrm{tr}(\mathbf{C})$.

The traditional method for the point estimation of σ_s^2 and σ_e^2, which is known as the method of fitting constants or as Henderson's Method 3 (Searle 1971, Sect. 10.4), consists of equating the two mean squares S_s/r and S_e/f to their expectations and then solving for σ_s^2 and σ_e^2. The estimators obtained by this approach are:

$$\tilde{\sigma}_e^2 = S_e/f \text{ and } \tilde{\sigma}_s^2 = (1/\kappa)[(S_s/r)-\tilde{\sigma}_e^2].$$

The corresponding estimator of γ is

$$\tilde{\gamma} = \tilde{\sigma}_s^2/\tilde{\sigma}_e^2 = (1/\kappa)(F-1).$$

8.4 Likelihood Function

The REML estimates of σ_s^2 and σ_e^2 (or of functions of σ_s^2 and σ_e^2) are obtained by maximizing the likelihood function associated with a set of n-p* linearly independent error contrasts, whereas the conventional maximum likelihood estimates are obtained by maximizing the likelihood function associated with the complete data vector **y**. By error contrast, we mean a linear unbiased estimator of zero, that is, a linear form $\boldsymbol{\ell}'\mathbf{y}$ such that $E(\mathbf{P}'\mathbf{y})\equiv 0$ or, equivalently, such that $\mathbf{X}'\boldsymbol{\ell}=\mathbf{0}$, where $\boldsymbol{\ell}$ is a non-random vector that is functionally independent of the parameters.

The likelihood function associated with the set of n-p* error contrasts is invariant to the choice of error contrasts, in the sense that the likelihood function associated with any one set differs from that associated with any other set by no more than a multiplicative constant (e.g., Harville 1974).

We now derive the likelihood function associated with a particular set of n-p* linearly independent error contrasts.

It is known that rank(**C**)=r (e.g., Marsaglia and Styan 1974, p.441). Let $\Delta_1,...,\Delta_r$ represent the non-zero (and hence positive) characteristic values of **C**, and let $\mathbf{D}=\text{diag}(\Delta_1,...,\Delta_r)$. Take **R** to be a matrix of dimensions $m x r$ whose columns are orthonormal characteristic vectors of **C** corresponding to the values $\Delta_1,...,\Delta_r$, respectively. That is, **R** is an mxr matrix such that $\mathbf{R}'\mathbf{R}=\mathbf{I}$ and $\mathbf{CR}=\mathbf{RD}$.

If **s**, like $\boldsymbol{\beta}$, were treated as a vector of unknown parameters, then (1) a necessary and sufficient condition for a linear function $\lambda'\mathbf{s}$ to be estimable would be the existence of a vector $\boldsymbol{\ell}$ such that $\lambda'=\boldsymbol{\ell}'\mathbf{C}$ and (2) there would exist an $rx1$ vector of linearly independent estimable functions of **s** (e.g., Thompson 1955, Sect. 2.3). In fact, one such vector is the vector

$$\mathbf{t} = (t_1,...,t_r)' = \mathbf{D}^{1/2}\mathbf{R}'\mathbf{s} = \mathbf{D}^{-1/2}(\mathbf{RD})'\mathbf{s} = \mathbf{D}^{-1/2}\mathbf{R}'\mathbf{Cs}.$$

Let

$$\tilde{\mathbf{t}} = (\tilde{t}_1,...,\tilde{t}_r)' = \mathbf{D}^{1/2}\mathbf{R}'\tilde{\mathbf{s}} = \mathbf{D}^{-1/2}\mathbf{R}'\mathbf{C}\tilde{\mathbf{s}} = \mathbf{D}^{-1/2}\mathbf{R}'\mathbf{q},$$

so that, if **s** were a vector of unknown parameters, $\tilde{t}_1,...,\tilde{t}_r$ would be the best unbiased estimators of $t_1,...,t_r$, respectively.

Let

$$\tilde{\mathbf{e}} = \mathbf{y} - \mathbf{X}\tilde{\boldsymbol{\beta}} - \mathbf{Z}\tilde{\mathbf{s}} = \mathbf{y} - \mathbf{X}(\mathbf{X}'\mathbf{X})^-\mathbf{X}'(\mathbf{y}-\mathbf{Z}\tilde{\mathbf{s}}) - \mathbf{Z}\tilde{\mathbf{s}}$$

$$= (\mathbf{I}-\mathbf{P_X})\mathbf{y} - (\mathbf{I}-\mathbf{P_X})\mathbf{Z}\tilde{\mathbf{s}} = \mathbf{My}$$

with $\mathbf{M}=\mathbf{I}-\mathbf{P_X}-(\mathbf{I}-\mathbf{P_X})\mathbf{ZC}^-\mathbf{Z}'(\mathbf{I}-\mathbf{P_X})$. Recall that

$$\mathbf{P_X X} = \mathbf{X} \text{ and } \mathbf{P_X} = \mathbf{P'_X} = \mathbf{P^2_X}$$

(e.g., Searle 1971, Sect. 1.5), and observe that

$$\mathbf{Z'(I-P_X)M} = \mathbf{Z'M} = \mathbf{0},\ \ \mathbf{M'} = \mathbf{M} = \mathbf{M}^2. \tag{8.5}$$

The matrix $\mathbf{M}$ is symmetric idempotent of rank f, and hence there exists an orthogonal matrix $\mathbf{W}=(\mathbf{W}_1, \mathbf{W}_2)$ such that

$$\begin{bmatrix} \mathbf{W}_1'\mathbf{M}\mathbf{W}_1 & \mathbf{W}_1'\mathbf{M}\mathbf{W}_2 \\ \mathbf{W}_2'\mathbf{M}\mathbf{W}_1 & \mathbf{W}_2'\mathbf{M}\mathbf{W}_2 \end{bmatrix} = \mathbf{W'MW} = \begin{bmatrix} \mathbf{I} & \mathbf{0} \\ \mathbf{0} & \mathbf{0} \end{bmatrix}, \tag{8.6}$$

where the dimensions of $\mathbf{W}_1$ are $n x f$. Note that (8.5) and (8.6) imply that $\mathbf{MW}_2=\mathbf{0}$. Define $\mathbf{z}=(z_1,\ldots,z_f)'=\mathbf{W}_1'\tilde{\mathbf{e}}=\mathbf{W}_1'\mathbf{My}$.

Observing that

$$E(\mathbf{q}) = \mathbf{0} \text{ and } \mathrm{var}(\mathbf{q}) = \sigma_e^2\mathbf{Z'(I-P_X)V(I-P_X)Z} = \sigma_e^2(\mathbf{C}+\gamma\mathbf{C}^2)$$

and making use of (8.5) and (8.6), we find that

$$E(\tilde{\mathbf{t}}) = \mathbf{0},$$

$$\mathrm{var}(\tilde{\mathbf{t}}) = \sigma_e^2\mathbf{D}^{-1/2}\mathbf{R}'(\mathbf{C}+\gamma\mathbf{C}^2)\mathbf{R}\mathbf{D}^{-1/2} = \sigma_e^2(\mathbf{I}+\gamma\mathbf{D})$$

$$= \sigma_e^2\mathrm{diag}(1+\gamma\Delta_1,\ldots,1+\gamma\Delta_r),$$

$$\mathrm{cov}(\tilde{\mathbf{t}},\mathbf{z}) = \sigma_e^2\mathbf{D}^{-1/2}\mathbf{R'Z'(I-P_X)(I+\gamma ZZ')MW}_1 = \mathbf{0},$$

$$E(\mathbf{z}) = \mathbf{0}, \qquad \mathrm{var}(\mathbf{z}) = \sigma_e^2\mathbf{I},$$

and also that

$$\Sigma_i z_i^2 = \mathbf{z'z} = \mathbf{y'MW}_1\mathbf{W}_1'\mathbf{My} = \mathbf{y'M(I-W}_2\mathbf{W}_2')\mathbf{My}$$

$$= \mathbf{y'M}^2\mathbf{y} = \mathbf{y'My} = \mathbf{y'(I-P_X)y}-\tilde{\mathbf{s}}'\mathbf{q} = S_e.$$

We conclude that $\tilde{t}_1,\ldots,\tilde{t}_r,z_1,\ldots,z_f$ are r+f (=n-p*) linearly independent error contrasts whose joint probability density function is

$$h(\tilde{\mathbf{t}},\mathbf{z};\gamma,\sigma_e^2) = (2\pi\sigma_e^2)^{-(r+f)/2}\Pi_i(1+\gamma\Delta_i)^{-1/2}$$

$$\cdot\exp\{-[1/(2\sigma_e^2)][S_e+\Sigma_i\tilde{t}_i^2/(1+\gamma\Delta_i)]\}.$$

The log likelihood function associated with these error contrasts is

$$\ell(\gamma,\sigma_e^2;\mathbf{y}) = -(1/2)\{(r+f)\ell n(2\pi)+(r+f)\ell n(\sigma_e^2)$$

$$+\Sigma_i\ell n(1+\gamma\Delta_i)+(1/\sigma_e^2)[S_e+\Sigma_i\tilde{t}_i^2/(1+\gamma\Delta_i)]\}.$$

8.5 Extended Parameter Space

Suppose that, instead of defining the distribution of **y** in terms of the mixed-effects linear model (8.1), we were simply to start with the assumption that the distribution of **y** is MVN[$\mathbf{X}\boldsymbol{\beta}$, $\sigma_e^2(\mathbf{I}+\gamma\mathbf{ZZ}')$] and (as in the mixed-effects linear model) to regard σ_e^2, γ, and the elements of $\boldsymbol{\beta}$ as unknown parameters.

If the parameter space for γ and σ_e^2 were taken to be

$$\Omega = \{\gamma, \sigma_e^2 : \gamma \geq 0, \sigma_e^2 > 0\},$$

this approach (to defining the distribution of **y**) would be equivalent to that based on model (8.1). Note, however, that the matrix $\sigma_e^2(\mathbf{I}+\gamma\mathbf{ZZ}')$ is positive-definite for all values of γ and σ_e^2 belonging to the slightly larger space

$$\Omega^* = \{\gamma, \sigma_e^2 : \gamma > -1/\lambda^*, \sigma_e^2 > 0\},$$

where λ^* is the largest characteristic value of **ZZ**' or, equivalently, of **Z'Z**. Thus, we could extend the parameter space from Ω to Ω^*, though, as a consequence, the parameter $\sigma_s^2=\gamma\sigma_e^2$ would no longer be interpretable as a variance and γ would no longer be interpretable as a variance ratio.

With regard to the distribution of error contrasts, an even further extension of the parameter space for γ and σ_e^2 is possible. Suppose that, instead of deriving the joint distribution of **z** and $\tilde{\mathbf{t}}$ from the distribution of **y**, we simply started with the assumption that **z** and $\tilde{\mathbf{t}}$ are distributed independently as MVN($\mathbf{0}$,$\sigma_e^2\mathbf{I}$) and MVN[$\mathbf{0}$,$\sigma_e^2(\mathbf{I}+\gamma\mathbf{D})$], respectively. Note that the matrices $\sigma_e^2\mathbf{I}$ and $\sigma_e^2(\mathbf{I}+\gamma\mathbf{D})$ are both positive-definite for all values of γ and σ_e^2 belonging to the space

$$\Omega^+ = \{\gamma, \sigma_e^2 : \sigma_e^2 > 0, \gamma > -1/\Delta^*\},$$

where $\Delta^*=\max(\Delta_1,...,\Delta_r)$. Thus, we could take the parameter space for the joint distribution of the error contrasts to be Ω^+ rather than Ω or Ω^*. It can be shown that $\Delta^*\leq\lambda^*$ and hence that $\Omega^*\subset\Omega^+$ (Harville and Fenech 1985, p.141).

Subsequently, unless otherwise indicated, we act as though the parameter space for the joint distribution of the error contrasts is the set Ω^+. Accordingly, we take the domain of the log likelihood function $\ell(\gamma,\sigma_e^2;\mathbf{y})$ to be Ω^+.

8.6 REML Estimation

By definition, REML estimates of γ and σ_e^2 are values $\hat{\gamma}$ and $\hat{\sigma}_e^2$ at which the log likelihood function ℓ attains a maximum value over the parameter space for γ and σ_e^2. Further, if $\hat{\gamma}$ and $\hat{\sigma}_e^2$ are REML estimates of γ and σ_e^2, then $f(\hat{\gamma},\hat{\sigma}_e^2)$ is a REML estimate of a function $f(\gamma,\sigma_e^2)$ of γ and σ_e^2. In particular, $\hat{\sigma}_s^2=\hat{\gamma}\hat{\sigma}_e^2$ is a REML estimate of $\sigma_s^2=\gamma\sigma_e^2$.

The first-order partial derivatives of ℓ are

$$\frac{\partial\ell}{\partial\gamma} = -(1/2)\left[\sum_i \frac{\Delta_i}{1+\gamma\Delta_i} - \frac{1}{\sigma_e^2}\sum_i \frac{\Delta_i\tilde{t}_i^2}{(1+\gamma\Delta_i)^2}\right], \tag{8.7}$$

$$\frac{\partial\ell}{\partial\sigma_e^2} = -(1/2)(r+f)[\sigma_e^2-\hat{\sigma}_e^2(\gamma)]/(\sigma_e^2)^2, \tag{8.8}$$

where

$$\hat{\sigma}_e^2(\gamma) = \frac{S_e+\sum_i \tilde{t}_i^2/(1+\gamma\Delta_i)}{r+f}.$$

It follows from result (8.8) that, for any fixed value of γ, ℓ attains its maximum, over the interval $0<\sigma_e^2<\infty$, uniquely at the value $\hat{\sigma}_e^2(\gamma)$. (Actually, in the event that $\tilde{t}_1=...=\tilde{t}_r=S_e=0$, ℓ is strictly decreasing over this interval and does not attain a maximum; however, to avoid trivialities, we do not attempt to distinguish between results that are valid for all values of **y** and those that are valid "only" for a set of values having probability one.)

Define

$$\ell_c(\gamma;y) = \ell[\gamma;\hat{\sigma}_e^2(\gamma);y]$$

$$= -(1/2)\{(r+f)[1+\ell n(2\pi)]+\Sigma_i \ell n(1+\gamma\Delta_i)+(r+f)\ell n[\hat{\sigma}_e^2(\gamma)]\}.$$

Adopting terminology employed by Bard (1974, p.65), we refer to the function ℓ_c as the concentrated log likelihood function. Clearly, a value $\hat{\gamma}$ is a REML estimate of γ if and only if ℓ_c attains a maximum value at $\hat{\gamma}$, in which case $\hat{\sigma}_e^2(\hat{\gamma})$ is a REML estimate of σ_e^2.

The quantity $\ell_c(\gamma;y)$ is a continuous function of γ over the interval $-1/\Delta^*<\gamma<\infty$ with

$$\lim_{\gamma\to\infty} \ell_c(\gamma;y) = \lim_{\gamma\to -1/\Delta^*} \ell_c(\gamma;y) = -\infty.$$

It follows that $\ell_c(\gamma;y)$ attains a maximum value over the interval $-1/\Delta^*<\gamma<\infty$.

The first-order derivative of ℓ_c is

$$\frac{d\ell_c}{d\gamma} = -(1/2)\left[\sum_i \frac{\Delta_i}{1+\gamma\Delta_i} - \frac{1}{\hat{\sigma}_e^2(\gamma)} \sum_i \frac{\Delta_i \tilde{t}_i^2}{(1+\gamma\Delta_i)^2}\right]. \tag{8.9}$$

Observing that

$$\bar{\Delta} = \kappa, \quad \sum_i \tilde{t}_i^2 = S_s \tag{8.10}$$

(e.g., Harville and Fenech 1985), we find that, in the special case $\Delta_1=...=\Delta_r$,

$$\hat{\sigma}_e^2(\gamma) = \frac{S_e+(1+\gamma\bar{\Delta})^{-1}\Sigma_i \tilde{t}_i^2}{r+f} = \frac{S_e+(1+\kappa\gamma)^{-1}S_s}{r+f}, \tag{8.11}$$

$$\frac{d\ell_c}{d\gamma} = -(1/2)\left(\frac{r\bar{\Delta}}{1+\gamma\bar{\Delta}}\right)\left[\frac{1+\gamma\bar{\Delta}-(f/r)\Sigma_i \tilde{t}_i^2/S_e}{1+\gamma\bar{\Delta}+\sum_i \tilde{t}_i^2/S_e}\right]$$

$$= -(1/2)\left(\frac{r\kappa}{1+\kappa\gamma}\right)\left[\frac{1+\kappa\gamma-F}{1+\kappa\gamma+(r/f)F}\right]. \tag{8.12}$$

It follows from result (8.12) that, in this special case, ℓ_c attains its maximum uniquely at the method-of-fitting-constants estimate $\tilde{\gamma}=(1/\kappa)(F-1)$. Moreover, using expression (8.11), we find that, when $\Delta_1=...=\Delta_r$, $\hat{\sigma}_e^2(\tilde{\gamma})=S_e/f=\tilde{\sigma}_e^2$. Thus, in the special case $\Delta_1=...=\Delta_r$, the REML estimates of γ, σ_e^2, and σ_s^2 are the same as the method-of-fitting-constants estimates $\tilde{\gamma}$, $\tilde{\sigma}_e^2$, and $\tilde{\sigma}_s^2$.

If the parameter space for γ and σ_e^2 were taken to be Ω, rather than Ω^+, then, in the special case $\Delta_1=...=\Delta_r$, the REML estimates of γ, σ_e^2, and σ_s^2 would be $\hat{\gamma}=\max(0,\tilde{\gamma})$, $\hat{\sigma}_e^2=\min[\tilde{\sigma}_e^2,(S_s+S_e)/(r+f)]$, and $\hat{\sigma}_s^2=\max(0,\tilde{\sigma}_s^2)$, respectively.

Except for the special case $\Delta_1=...=\Delta_r$, there do not exist closed-form expressions for the value(s) of γ that maximize ℓ_c nor, as a consequence, for REML estimates of γ, σ_e^2, and σ_s^2. To obtain these estimates, we must, in general, resort to iterative numerical methods.

8.7 Newton-Raphson Algorithms

The Newton-Raphson method is a well-known iterative procedure for maximizing or minimizing a function. (It can be more accurately described as a method for finding stationary points of the function.)

One approach to the computation of REML estimates of γ and σ_e^2 consists of applying the Newton-Raphson method to the function ℓ. Let $\theta=(\gamma,\sigma_e^2)'$. Define

$$\mathbf{b}(\theta)=\begin{bmatrix}\partial\ell/\partial\gamma \\ \partial\ell/\partial\sigma_e^2\end{bmatrix},\quad \mathbf{B}(\theta)=\begin{bmatrix}\partial^2\ell/\partial\gamma^2 & \partial^2\ell/\partial\gamma\partial\sigma_e^2 \\ \partial^2\ell/\partial\gamma\partial\sigma_e^2 & \partial^2\ell/\partial(\sigma_e^2)^2\end{bmatrix}.$$

As applied to the function ℓ, the (j+1)st iterate of the Newton-Raphson method is

$$\hat{\theta}^{(j+1)}=\hat{\theta}^{(j)}-\mathbf{B}^{-1}(\hat{\theta}^{(j)})\mathbf{b}(\hat{\theta}^{(j)}). \tag{8.13}$$

Expressions for the elements of $\mathbf{B}(\theta)$, analogous to expressions (8.7) and (8.8) for the elements of $\mathbf{b}(\theta)$, are

$$\frac{\partial^2\ell}{\partial\gamma^2}=(1/2)\left[\sum_i\frac{\Delta_i^2}{(1+\gamma\Delta_i)^2}-(2/\sigma_e^2)\sum_i\frac{\Delta_i^2\tilde{t}_i^2}{(1+\gamma\Delta_i)^3}\right], \tag{8.14}$$

$$\frac{\partial^2 \ell}{\partial\gamma\partial\sigma_e^2} = -\,[2(\sigma_e^2)^2]^{-1}\sum_i \Delta_i \tilde{t}_i^2/(1+\gamma\Delta_i)^2\,, \tag{8.15}$$

$$\frac{\partial^2 \ell}{\partial(\sigma_e^2)^2} = (1/2)(r+f)[\sigma_e^2 - 2\hat{\sigma}_e^2(\gamma)]/(\sigma_e^2)^3. \tag{8.16}$$

When the model is parameterized in terms of σ_s^2 and σ_e^2, the log likelihood function becomes

$$\ell_v(\sigma_s^2,\sigma_e^2;\mathbf{y}) = \ell(\sigma_s^2/\sigma_e^2,\sigma_e^2;\mathbf{y}).$$

Let $\sigma=(\sigma_s^2,\sigma_e^2)'$. Define

$$\mathbf{g}(\sigma) = \begin{bmatrix} \partial\ell_v/\partial\sigma_s^2 \\ \partial\ell_v/\partial\sigma_e^2 \end{bmatrix},\ \mathbf{G}(\sigma) = \begin{bmatrix} \partial^2\ell_v/\partial(\sigma_s^2)^2 & \partial^2\ell_v/\partial\sigma_s^2\partial\sigma_e^2 \\ \partial^2\ell_v/\partial\sigma_s^2\partial\sigma_e^2 & \partial^2\ell_v/\partial(\sigma_e^2)^2 \end{bmatrix}.$$

In computing REML estimates, the Newton-Raphson method is sometimes applied to the function ℓ_v, instead of the function ℓ. As applied to ℓ_v, the (j+1)st iterate of the Newton-Raphson method is

$$\hat{\sigma}^{(j+1)} = \hat{\sigma}^{(j)} - \mathbf{G}^{-1}(\hat{\sigma}^{(j)})\mathbf{g}(\hat{\sigma}^{(j)})\,. \tag{8.17}$$

In applying the Newton-Raphson method to a log likelihood function, different parameterizations can be expected to produce non-equivalent sequences of iterates. Thus, for example, even if the starting value $\hat{\sigma}^{(0)}$ employed when the Newton-Raphson method is applied to ℓ_v is equivalent to the starting value $\hat{\theta}^{(0)}=(\hat{\gamma}^{(0)}, \hat{\sigma}_e^{2(0)})'$ employed in applying the method to ℓ, that is, even if $\hat{\sigma}^{(0)}=(\hat{\gamma}^{(0)}\hat{\sigma}_e^{2(0)}, \hat{\sigma}_e^{2(0)})'$, it need not be the case that $\hat{\sigma}^{(1)}$ is equivalent to $\hat{\theta}^{(1)}=(\hat{\gamma}^{(1)},\hat{\sigma}_e^{2(1)})'$, that is, it need not be the case that $\hat{\sigma}^{(1)}=(\hat{\gamma}^{(1)}\hat{\sigma}_e^{2(1)},\hat{\sigma}_e^{2(1)})'$.

Expressions for the elements of $\mathbf{g}(\sigma)$ and $\mathbf{G}(\sigma)$ can be derived by direct differentiation or, alternatively, by making use of the chain rule and by taking advantage of expressions (8.7), (8.8), and (8.14)-(8.16) for the first-order and second-order partial derivatives of ℓ. Expressions for the first-order partial derivatives of ℓ_v are

$$\partial\ell_v/\partial\sigma_s^2 = (\sigma_e^2)^{-1}(\partial\ell/\partial\gamma),$$

$$\partial \ell_v/\partial \sigma_e^2 = -(1/2)(\sigma_e^2)^{-2}\{\sigma_e^2[r+f-\gamma\sum_i \Delta_i/(1+\gamma\Delta_i)]-[(r+f)\hat{\sigma}_e^2(\gamma)-\gamma\sum_i \Delta_i \tilde{t}_i^2/(1+\gamma\Delta_i)^2]\}.$$

Expressions for the second-order partial derivatives of ℓ_v are given in the Appendix.

In applying the Newton-Raphson method to the log likelihood function associated with the error contrasts, other parameterizations, besides the two discussed here, can be considered. Refer, for example, to Hemmerle and Hartley (1973) and to Dempster et al. (1984).

8.8 Concentrated Log Likelihood Function

Instead of applying the Newton-Raphson method to the log likelihood function ℓ (or ℓ_v), we could apply the method to the concentrated log likelihood function ℓ_c. The intuitive appeal of this alternative approach (to the computation of REML estimates of γ and σ_e^2) is that it exploits our knowledge that ℓ assumes its maximum with respect to σ_e^2 (for any fixed value of γ) at $\hat{\sigma}_e^2(\gamma)$, thereby reducing a two-dimensional maximization problem to a one-dimensional problem.

Define

$$k(\gamma) = d\ell_c/d\gamma \quad \text{and} \quad K(\gamma) = d^2\ell_c/d\gamma^2.$$

As applied to the function ℓ_c, the (j+1)st iterate of the Newton-Raphson algorithm is

$$\hat{\gamma}^{(j+1)} = \hat{\gamma}^{(j)}-k(\hat{\gamma}^{(j)})/K(\hat{\gamma}^{(j)}). \tag{8.18}$$

An expression for $K(\gamma)$, analogous to expression (8.9) for $k(\gamma)$, is

$$\frac{d^2\ell_c}{d\gamma^2} = (1/2)\sum_i \Delta_i^2/(1+\gamma\Delta_i)^2-[\hat{\sigma}_e^2(\gamma)]^{-1}\sum_i \Delta_i^2\tilde{t}_i^2/(1+\gamma\Delta_i)^3$$

$$+(1/2)(r+f)^{-1}[\hat{\sigma}_e^2(\gamma)]^{-2}[\sum_i \Delta_i\tilde{t}_i^2/(1+\gamma\Delta_i)^2]^2.$$

The function ℓ can be reparameterized in terms of the non-linear parametric function $\psi=(1+\gamma\bar{\Delta})^{-1}=(1+\kappa\gamma)^{-1}$. Note that ψ equals the ratio of the two expected mean squares σ_e^2 and $\sigma_e^2+\kappa\sigma_s^2$. When parameterized in terms of ψ, the concentrated log likelihood function becomes

$$\ell_R(\psi;\mathbf{y}) = \ell_c[(\psi^{-1}-1)/\kappa;\mathbf{y}].$$

Let

$$m(\psi) = d\ell_R/d\psi = -\kappa^{-1}\psi^{-2}(d\ell_c/d\gamma), \tag{8.19}$$

$$M(\psi) = d^2\ell_R/d\psi^2 = \kappa^{-1}\psi^{-3}[2(d\ell_c/d\gamma)+\kappa^{-1}\psi^{-1}(d^2\ell_c/d\gamma^2)].$$

As applied to the function ℓ_R, the (j+1)st iterate of the Newton-Raphson algorithm is

$$\hat{\psi}^{(j+1)} = \hat{\psi}^{(j)}-m(\hat{\psi}^{(j)})/M(\hat{\psi}^{(j)}). \tag{8.20}$$

In general, the sequence of iterates obtained by applying the Newton-Raphson method to the function ℓ_R, or to any other non-linear reparameterization of the function ℓ_c, will not be equivalent to that obtained by applying the method to ℓ_c.

Corresponding to the (j+1)st iterate $\hat{\gamma}^{(j+1)}$ or $\hat{\psi}^{(j+1)}$ is the value $\hat{\sigma}_e^2(\hat{\gamma}^{(j+1)})$ or $\hat{\sigma}_e^2\{[(1/\hat{\psi}^{(j+1)})-1]/\kappa\}$ of σ_e^2 at which $\ell(\hat{\gamma}^{(j+1)}, \sigma_e^2;\mathbf{y})$ or $\ell\{[(1/\hat{\psi}^{(j+1)})-1]/\kappa,\sigma_e^2;\mathbf{y}\}$ attains its maximum. This value can be regarded as the σ_e^2-component of the (j+1)st iterate.

8.9 Linearization

Newton's method is a well-known iterative method for solving a system of one or more non-linear equations. In effect, the Newton-Raphson method for maximizing or minimizing a function consists of applying Newton's method to the system of equations formed by equating the first-order partial derivatives of the function to zero.

Success with Newton's method is most likely when it is applied to a system consisting of equations that are approximately linear. From results (8.12) and (8.19), we see that, even if $\Delta_1=\ldots=\Delta_r$ (in which case the problem of maximizing ℓ_c or ℓ_R can be easily solved by analytical means), the equation $k(\gamma)=0$ or $m(\psi)=0$, formed by equating the derivative of ℓ_c or ℓ_R to 0, is non-linear (in γ or ψ). Likewise, each of the systems $\mathbf{b}(\theta)=\mathbf{0}$ and $\mathbf{g}(\sigma)=\mathbf{0}$ consists of equations that are non-linear even in the special case $\Delta_1=\ldots=\Delta_r$.

Define

$$k^*(\gamma) = (1+\kappa\gamma)[1+\kappa\gamma+(r/f)F]k(\gamma),$$

$$m^*(\psi) = \psi[1+\psi(r/f)F]m(\psi).$$

Instead of applying Newton's method to the equation $k(\gamma)=0$ or $m(\psi)= 0$, we can apply it to the equivalent equation $k^*(\gamma)=0$ or $m^*(\psi)=0$ obtained by multiplying both sides of $k(\gamma)=0$ or $m(\psi)=0$ by the same strictly positive function.

The appeal of this approach is that, in the special case $\Delta_1=...=\Delta_r$,

$$k^*(\gamma) = -(1/2)r\kappa(1+\kappa\gamma-F),$$

$$m^*(\psi) = (1/2)r(1-\psi F),$$

so that, in this special case, the equations $k^*(\gamma)=0$ and $m^*(\psi)=0$ are linear. More generally, the hope is that, regardless of the values of $\Delta_1,...,\Delta_r$, the equation $k^*(\gamma)=0$ or $m^*(\psi)=0$ will be closer to linearity than the equation $k(\gamma)=0$ or $m(\psi)=0$ and that, as a consequence, the sequence of iterates obtained by applying Newton's method to $k^*(\gamma)=0$ or $m^*(\psi)=0$ will converge more rapidly than those obtained by applying the method to $k(\gamma)=0$ or $m(\psi)=0$.

Define

$$K^*(\gamma) = dk^*/d\gamma = [2\kappa(1+\kappa\gamma)+\kappa(r/f)F]k(\gamma)+(1+\kappa\gamma)[1+\kappa\gamma+(r/f)F]K(\gamma),$$

$$M^*(\psi) = dm^*/d\psi = [1+2\psi(r/f)F]m(\psi)+\psi[1+\psi(r/f)F]M(\psi).$$

As applied to the equation $k^*(\gamma)=0$, the (j+1)st iterate of Newton's method is

$$\hat{\gamma}^{(j+1)} = \hat{\gamma}^{(j)}-k^*(\hat{\gamma}^{(j)})/K^*(\hat{\gamma}^{(j)}), \tag{8.21}$$

and, as applied to the equation $m^*(\psi)=0$, the (j+1)st iterate is

$$\hat{\psi}^{(j+1)} = \hat{\psi}^{(j)}-m^*(\hat{\psi}^{(j)})/M^*(\hat{\psi}^{(j)}). \tag{8.22}$$

It can be shown that algorithms (8.21) and (8.22) are essentially the same, that is, produce equivalent sequences of iterates. Thus, if the starting values $\hat{\gamma}^{(0)}$ and $\hat{\psi}^{(0)}$ are equivalent, that is, if $\hat{\psi}^{(0)}=(1+\kappa\hat{\gamma}^{(0)})^{-1}$, then the jth iterates $\hat{\gamma}^{(j)}$ and $\hat{\psi}^{(j)}$ of algorithms (8.21) and (8.22) are equivalent, that is $\hat{\psi}^{(j)}=(1+\kappa\hat{\gamma}^{(j)})^{-1}$ (for j=1,2,...).

We now consider yet another system of equations to which we can apply Newton's method for purposes of computing REML estimates. Suppose that model (8.1) were parameterized in terms of the expected mean squares $\varepsilon_s=\sigma_e^2+\kappa\sigma_s^2=\sigma_e^2+\bar{\Delta}\sigma_s^2$ and $\varepsilon_e=\sigma_e^2$, in which case the log likelihood function becomes

$$\ell_\varepsilon(\varepsilon_s,\varepsilon_e;y) = \ell_v[(\varepsilon_s-\varepsilon_e)/\bar{\Delta},\varepsilon_e;y].$$

Let $\varepsilon=(\varepsilon_s,\varepsilon_e)'$. Define $\mathbf{q}(\varepsilon)=[q_1(\varepsilon),q_2(\varepsilon)]'$, with $q_1(\varepsilon)=\partial\ell_\varepsilon/\partial\varepsilon_s$ and $q_2(\varepsilon)=\partial\ell_\varepsilon/\partial\varepsilon_e$. Applying the Newton-Raphson method to the function ℓ_ε is equivalent to applying Newton's method to the system $\mathbf{q}(\varepsilon)=\mathbf{0}$.

Define $\mathbf{q}^*(\varepsilon)=[q_1^*(\varepsilon),q_2^*(\varepsilon)]'$, with

$$q_1^*(\varepsilon) = \varepsilon_s^2 q_1(\varepsilon) \text{ and } q_2^*(\varepsilon) = \varepsilon_e^2 q_2(\varepsilon).$$

[Expressions for $q_1^*(\varepsilon)$ and $q_2^*(\varepsilon)$ are given in the Appendix.] Instead of applying Newton's method to the system $\mathbf{q}(\varepsilon)=\mathbf{0}$, we can apply it to the transformed system $\mathbf{q}^*(\varepsilon)=\mathbf{0}$. It is easy to show [by simplifying expressions (8A.1) and (8A.2) or by simplifying ℓ_ε and then differentiating] that, in the special case $\Delta_1=\ldots=\Delta_r$,

$$q_1^*(\varepsilon) = -(1/2)(r\varepsilon_s-S_s),\ q_2^*(\varepsilon) = -(1/2)(f\varepsilon_e-S_e),$$

so that, in this special case, the system $\mathbf{q}^*(\varepsilon)=\mathbf{0}$ is linear.

Define

$$\mathbf{Q}^*(\varepsilon) = \begin{bmatrix} \partial q_1^*/\partial\varepsilon_s & \partial q_1^*/\partial\varepsilon_e \\ \partial q_2^*/\partial\varepsilon_s & \partial q_2^*/\partial\varepsilon_e \end{bmatrix}.$$

[Expressions for the elements of $\mathbf{Q}^*(\varepsilon)$ are given in the Appendix.] As applied to the system $\mathbf{q}^*(\varepsilon)=\mathbf{0}$, the (k+1)st iterate of Newton's method is

$$\hat{\varepsilon}^{(j+1)} = \hat{\varepsilon}^{(j)} - \mathbf{Q}^{*-1}(\hat{\varepsilon}^{(j)})\mathbf{q}^*(\hat{\varepsilon}^{(j)}). \tag{8.23}$$

Note that, since ε_s and ε_e are linear functions of σ_s^2 and σ_e^2, applying the Newton-Raphson method to the function ℓ_ε is equivalent to applying it to the function ℓ_v. Thus, algorithm (8.23) can be viewed as an attempted improvement upon algorithm (8.17).

The technique of "linearizing" an equation before applying Newton's method could prove useful in various other problems. In particular, Harville and Fenech (1985) discuss a confidence interval for γ, each of whose end points is the solution to an equation (in γ) of the form

$$(f/r)S_e^{-1}\sum_i \tilde{t}_i^2/(1+\gamma\Delta_i)-c = 0, \tag{8.24}$$

where c represents an appropriately chosen positive constant. Instead of applying Newton's method to Eq. (8.24), we could apply it to the equation

$$(f/r)S_e^{-1}(1+\gamma\bar{\Delta})\sum_i \tilde{t}_i^2/(1+\gamma\Delta_i)-c(1+\gamma\bar{\Delta}) = 0, \tag{8.25}$$

obtained by multiplying both sides of Eq. (8.24) by $1+\gamma\bar{\Delta}$.

8.10 Computation of Iterates

In Sections 8.7-8.9, we presented six distinct algorithms for computing REML estimates, each of which is based on Newton's method for solving systems of non-linear equations. We now consider the computations required in using the representations given in Sections 8.3-8.9 (and the Appendix) to implement these algorithms.

All of the relevant information about the model and the data is included in the following "statistics": r, r+f, $\Delta_1,...,\Delta_r$, $\tilde{t}_1,...,\tilde{t}_r$, and $\mathbf{y}'\mathbf{y}-S_\beta$. Following Dempster et al. (1984), Harville and Fenech (1985, Sect. 4) outlined a procedure for computing these statistics. This procedure operates on the elements of the symmetric array

$$\begin{bmatrix} \mathbf{X}'\mathbf{X} & \mathbf{X}'\mathbf{Z} & \mathbf{X}'\mathbf{y} \\ \mathbf{Z}'\mathbf{X} & \mathbf{Z}'\mathbf{Z} & \mathbf{Z}'\mathbf{y} \\ \mathbf{y}'\mathbf{X} & \mathbf{y}'\mathbf{Z} & \mathbf{y}'\mathbf{y} \end{bmatrix}$$

and consists essentially of the following four steps:

1. Find p*=rank(**X**), compute a solution to the linear system

   $$\mathbf{X}'\mathbf{X}(\mathbf{H}, \mathbf{a}) = (\mathbf{X}'\mathbf{Z}, \mathbf{X}'\mathbf{y})$$

   [for example, by computing a generalized inverse $(\mathbf{X}'\mathbf{X})^-$ and forming the matrix $\mathbf{H}=(\mathbf{X}'\mathbf{X})^-\mathbf{X}'\mathbf{Z}$ and the vector $\mathbf{a}=(\mathbf{X}'\mathbf{X})^-\mathbf{X}'\mathbf{y}$], and compute $\mathbf{y}'\mathbf{y}-S_\beta=\mathbf{y}'\mathbf{y}-\mathbf{a}'\mathbf{X}'\mathbf{y}$.
2. Compute **C** and **q** from

   $$\mathbf{C} = (\mathbf{Z}'\mathbf{Z})-(\mathbf{Z}'\mathbf{X})\mathbf{H} \text{ and } \mathbf{q} = (\mathbf{Z}'\mathbf{y})-(\mathbf{Z}'\mathbf{X})\mathbf{a}.$$

3. Find r, and compute the characteristic values $\Delta_1,...,\Delta_r$ and the columns of $\mathbf{R}\mathbf{D}^{-1/2}$, that is, corresponding orthogonal characteristic vectors of lengths $1/\sqrt{\Delta_1},...,1/\sqrt{\Delta_r}$, respectively.
4. Compute r+f=n-p* and $\tilde{\mathbf{t}}=(\mathbf{R}\mathbf{D}^{-1/2})'\mathbf{q}$.

Also, to avoid repetition, we should, prior to initiating the iterative process, compute various of the following quantities: f=(r+f)-r, $\kappa=\bar{\Delta}$, S_s [using result (8.10)], S_e [from expression (8.4)], and F.

On each iteration of Newton's method, we shall need to recompute each of the following quantities, taking the value of γ to be the estimate from the previous iteration:

$$\Sigma_i\Delta_i/(1+\gamma\Delta_i),\ \Sigma_i[\Delta_i/(1+\gamma\Delta_i)]^2,\ \Sigma_i\tilde{t}_i^2/(1+\gamma\Delta_i),\ \Sigma_i\Delta_i\tilde{t}_i^2/(1+\gamma\Delta_i)^2,$$

$$\text{and } \Sigma_i\Delta_i^2\tilde{t}_i^2/(1+\gamma\Delta_i)^3.$$

8.11 An Alternative Approach to the Computation of Iterates

The computational approach discussed in Section 8.10 makes direct use of the representations given in Sections 8.3-8.9 (and the Appendix). The primary difficulty with this approach is the heavy burden imposed by the computation of the characteristic values and vectors of **C**. We now investigate the extent to which this difficulty can be circumvented.

Consider, for $-1/\Delta^*<\gamma<\infty$, the linear system

$$(\mathbf{I}+\gamma\mathbf{C})\mathbf{u} = \mathbf{q} \tag{8.26}$$

in **u**. The matrix $\mathbf{I}+\gamma\mathbf{C}$ is (for $-1/\Delta^*<\gamma<\infty$) positive-definite and hence non-singular, implying that linear system (8.26) is consistent and in fact has a unique solution (Harville and Fenech 1985, Appendix). Subsequently, we use **u** to represent the solution to linear system (8.26), that is, we take $\mathbf{u}=(\mathbf{I}+\gamma\mathbf{C})^{-1}\mathbf{q}$. Note that **u** is functionally dependent on γ.

All of the representations given in Sections 8.3-8.9 (and the Appendix) can be re-expressed in terms of quantities associated with linear systems (8.3) and (8.26). Harville and Fenech (1985, Appendix) showed that

$$\sum_i\Delta_i/(1+\gamma\Delta_i)=\mathrm{tr}[(\mathbf{I}+\gamma\mathbf{C})^{-1}\mathbf{C}],\ \sum_i\Delta_i^2/(1+\gamma\Delta_i)^2=\mathrm{tr}\{[(\mathbf{I}+\gamma\mathbf{C})^{-1}\mathbf{C}]^2\}, \tag{8.27}$$

$$\sum_i\tilde{t}_i^2/(1+\gamma\Delta_i) = \mathbf{u}'\tilde{\mathbf{s}} = \mathbf{u}'(\mathbf{C}^-+\gamma\mathbf{I})\mathbf{u}, \tag{8.28}$$

$$\sum_i\Delta_i\tilde{t}_i^2\,/(1+\gamma\Delta_i)^2 = \mathbf{u}'\mathbf{u},\ \sum_i\Delta_i^2\tilde{t}_i^2/(1+\gamma\Delta_i)^3 = \mathbf{u}'(\mathbf{I}+\gamma\mathbf{C})^{-1}\mathbf{C}\mathbf{u}. \tag{8.29}$$

They also showed that $\Pi_i(1+\gamma\Delta_i)=\det(\mathbf{I}+\gamma\mathbf{C})$, which implies that

$$\sum_i \ell n(1+\gamma\Delta_i) = \ell n[\det(\mathbf{I}+\gamma\mathbf{C})].$$

Observe that $\mathbf{I}=(\mathbf{I}+\gamma\mathbf{C})^{-1}+\gamma(\mathbf{I}+\gamma\mathbf{C})^{-1}\mathbf{C}$ and hence that

$$\gamma(\mathbf{I}+\gamma\mathbf{C})^{-1}\mathbf{C} = \mathbf{I}-(\mathbf{I}+\gamma\mathbf{C})^{-1}.$$

Using this result in conjunction with results (8.27) and (8.29), we find that:

$$\gamma\sum_i \Delta_i /(1+\gamma\Delta_i) = m-\mathrm{tr}[(\mathbf{I}+\gamma\mathbf{C})^{-1}], \tag{8.30}$$

$$\gamma^2\sum_i \Delta_i^2/(1+\gamma\Delta_i)^2 = m-2\mathrm{tr}[(\mathbf{I}+\gamma\mathbf{C})^{-1}] + \mathrm{tr}[(\mathbf{I}+\gamma\mathbf{C})^{-2}], \tag{8.31}$$

$$\gamma\sum_i \Delta_i^2\tilde{t}_i^2/(1+\gamma\Delta_i)^3 = \mathbf{u}'\mathbf{u}-\mathbf{u}'(\mathbf{I}+\gamma\mathbf{C})^{-1}\mathbf{u}.$$

For $\gamma\neq 0$, these identities yield expressions for $\Sigma_i\Delta_i/(1+\gamma\Delta_i)$, $\Sigma_i\Delta_i^2/(1+\gamma\Delta_i)^2$, and $\Sigma_i\Delta_i^2\tilde{t}_i^2/(1+\gamma\Delta_i)^3$ alternative to those provided by results (8.27) and (8.29).

By using the various results of (8.27)-(8.31), we can, in applying Newton's method, avoid the computation of characteristic values and vectors. Note, however, that to do so, we must, on each iteration, re-invert the $m\times m$ matrix $\mathbf{I}+\gamma\mathbf{C}$, taking the value of γ to be the estimate from the previous iteration. Thus, the use of results (8.27)-(8.31) would be advantageous when convergence is achieved in a sufficiently small number of iterations, but not otherwise.

In applications where $\mathbf{Z}'\mathbf{Z}$ is diagonal (or, more generally, $\mathbf{Z}'\mathbf{Z}$ and $\mathbf{I}+\gamma\mathbf{Z}'\mathbf{Z}$ are easy to "invert") and p is small relative to m (as in the one-way random model), the computations expended in solving linear systems (8.3) and (8.26) can be greatly reduced. This reduction can be achieved, for example, by recognizing that the matrix

$$(\mathbf{Z}'\mathbf{Z})^-+(\mathbf{Z}'\mathbf{Z})^-\mathbf{Z}'\mathbf{X}[\mathbf{X}'\mathbf{X}-\mathbf{X}'\mathbf{Z}(\mathbf{Z}'\mathbf{Z})^-\mathbf{Z}'\mathbf{X}]^-\mathbf{X}'\mathbf{Z}(\mathbf{Z}'\mathbf{Z})^-$$

is a generalized inverse of $\mathbf{C}$ and by making use of the formula

$$(\mathbf{I}+\gamma\mathbf{C})^{-1} = (\mathbf{I}+\gamma\mathbf{Z}'\mathbf{Z})^{-1}+\gamma(\mathbf{I}+\gamma\mathbf{Z}'\mathbf{Z})^{-1}\mathbf{Z}'\mathbf{X}$$
$$\cdot[\mathbf{X}'\mathbf{X}-\gamma\mathbf{X}'\mathbf{Z}(\mathbf{I}+\gamma\mathbf{Z}'\mathbf{Z})^{-1}\mathbf{Z}'\mathbf{X}]^-\mathbf{X}'\mathbf{Z}(\mathbf{I}+\gamma\mathbf{Z}'\mathbf{Z})^{-1}$$

(Henderson and Searle 1981, Sect. 4).

As discussed by, for example, Harville (1977, Sect. 3), the linear system (8.26) is equivalent to the linear system

$$\begin{bmatrix} \mathbf{X'X} & \gamma\mathbf{X'Z} \\ \mathbf{Z'X} & \mathbf{I}+\gamma\mathbf{Z'Z} \end{bmatrix} \begin{bmatrix} \bar{\boldsymbol{\beta}} \\ \bar{\mathbf{u}} \end{bmatrix} = \begin{bmatrix} \mathbf{X'y} \\ \mathbf{Z'y} \end{bmatrix}, \tag{8.32}$$

in the sense that the lower $m\times1$ subvector of any solution to linear system (8.32) equals **u** and the lower right $m\times m$ submatrix of any generalized inverse of the coefficient matrix of linear system (8.32) equals $(\mathbf{I}+\gamma\mathbf{C})^{-1}$. Thus, our discussion of the use of formulae (8.27)-(8.31) could be recast in terms of linear system (8.32). If linear system (8.32) is re-expressed as a linear system in $\bar{\boldsymbol{\beta}}$ and $\gamma\bar{\mathbf{u}}$ (which can be done for $\gamma\neq0$), it is identical to Henderson's (1963) mixed-model equations.

We complete this section by describing a computational approach proposed by Smith and Graser (1986). Their approach can be regarded as a compromise between the approach discussed in Section 8.10, which is based on the representations given in Sections 8.3-8.9 (and the Appendix), and the alternative approach discussed heretofore in this section, which is based on representations (8.27)-(8.31).

Let $\mathbf{O}$ represent any $m\times m$ orthogonal matrix such that the matrix $\mathbf{O'CO}$ is tridiagonal (e.g., Stewart 1973, Sect. 7.1), and define $\mathbf{Z}^{\#}=\mathbf{ZO}$. Then,

$$\sigma_e^2(\mathbf{I}+\gamma\mathbf{ZZ'}) = \sigma_e^2(\mathbf{I}+\gamma\mathbf{Z}^{\#}\mathbf{Z}^{\#\prime}),$$

that is, var($\mathbf{y}$) can be re-expressed as $\sigma_e^2(\mathbf{I}+\gamma\mathbf{Z}^{\#}\mathbf{Z}^{\#\prime})$. Thus, for purposes of computing REML estimates, $\mathbf{Z}^{\#}$ can be substituted for $\mathbf{Z}$.

With regard to representations (8.27)-(8.31), the effect of substituting $\mathbf{Z}^{\#}$ for $\mathbf{Z}$ is to replace $\mathbf{C}$, $\tilde{\mathbf{s}}$, and $\mathbf{u}$ with $\mathbf{C}^{\#}$, $\tilde{\mathbf{s}}^{\#}$, and $\mathbf{u}^{\#}$, respectively, where $\mathbf{C}^{\#}=\mathbf{O'CO}$ and, letting $\mathbf{q}^{\#}=\mathbf{O'q}$, $\tilde{\mathbf{s}}^{\#}$ and $\mathbf{u}^{\#}$ are the solutions to the linear systems

$$\mathbf{C}^{\#}\tilde{\mathbf{s}}^{\#} = \mathbf{q}^{\#},\ (\mathbf{I}+\gamma\mathbf{C}^{\#})\mathbf{u}^{\#} = \mathbf{q}^{\#}. \tag{8.33}$$

Note that the coefficient matrices $\mathbf{C}^{\#}$ and $\mathbf{I}+\gamma\mathbf{C}^{\#}$ of linear systems (8.33) are both tridiagonal.

Consider the effect of basing the REML computations on the modified versions of representations (8.27)-(8.31) (in which $\mathbf{C}^{\#}$, $\tilde{\mathbf{s}}^{\#}$, and $\mathbf{u}^{\#}$ have been substituted for $\mathbf{C}$, $\tilde{\mathbf{s}}$, and $\mathbf{u}$). In general, fewer computations are required to tridiagonalize $\mathbf{C}$ than to obtain the characteristic values and vectors of $\mathbf{C}$ and fewer computations are required to solve the tridiagonal linear systems (8.33) than linear systems (8.3) and (8.26) (Stewart 1973, Chaps. 7 and 3). Thus, relative to the computational approach based on the (unmodified versions of) representations (8.27)-(8.31), the approach based on the modified versions of these representations requires more preliminary computations but requires fewer computations per iteration. And, relative to the approach based on the representations given in Sections 8.3-8.9, it requires fewer preliminary computations but requires more subsequent computations per iteration. We conclude that the use of the representations given in Sections 8.3-8.9 is advantageous when a sufficiently large number of iterations is required to achieve convergence, the use of representations (8.27)-(8.31) is advantageous when a sufficiently small number of iterations is required, and the use of the modified versions of representations (8.27)-(8.31) is advantageous when an intermediate number of iterations is required.

8.12 Method of Scoring

Algorithms (8.13) and (8.17) were derived by applying the Newton-Raphson method to the functions ℓ and ℓ_V, respectively. Alternative algorithms are obtained if instead we apply the method of scoring to these functions.

The method of scoring is the same as the Newton-Raphson method except that second-order partial derivatives are replaced by their expected values. Thus, as applied to the functions ℓ and ℓ_V, the method of scoring is the same as the Newton-Raphson method except that the matrices $\mathbf{B}(\theta)$ and $\mathbf{G}(\sigma)$ are replaced by the matrices $\mathbf{B}_e(\theta)= E[\mathbf{B}(\theta)]$ and $\mathbf{G}_e(\sigma)=E[\mathbf{G}(\sigma)]$, respectively. More specifically, the (j+1)st iterates of the algorithms obtained by applying the method of scoring to ℓ and ℓ_V are

$$\tilde{\theta}^{(j+1)} = \tilde{\theta}^{(j)}-\mathbf{B}_e^{-1}(\tilde{\theta}^{(j)})\mathbf{b}(\tilde{\theta}^{(j)}), \tag{8.34}$$

$$\tilde{\sigma}^{(j+1)} = \tilde{\sigma}^{(j)}-\mathbf{G}_e^{-1}(\tilde{\sigma}^{(j)})\mathbf{g}(\tilde{\sigma}^{(j)}), \tag{8.35}$$

respectively.

Expressions for the distinct elements of $\mathbf{B}_e(\theta)$ are

$$E(\partial^2\ell/\partial\gamma^2) = -(1/2)\sum_i \Delta_i^2/(1+\gamma\Delta_i)^2,$$

$$E(\partial^2\ell/\partial\gamma\partial\sigma_e^2) = -(1/2)(\sigma_e^2)^{-1}\sum_i \Delta_i/(1+\gamma\Delta_i),$$

$$E[\partial^2\ell/\partial(\sigma_e^2)^2] = -(1/2)(r+f)/(\sigma_e^2)^2.$$

By making use of result (8.27), or alternatively (for the case $\gamma\neq 0$) results (8.30) and (8.31), we can, if we wish, re-express the elements of $\mathbf{B}_e(\theta)$ in terms of quantities associated with linear systems (8.3) and (8.26).

Expressions for the elements of $\mathbf{G}_e(\sigma)$ can be obtained from those for $\mathbf{B}_e(\theta)$ via the relationship

$$\mathbf{G}_e(\sigma) = [\mathbf{H}(\theta)]'\mathbf{B}_e(\theta)\mathbf{H}(\theta),$$

where

$$\mathbf{H}(\theta) = \begin{bmatrix} 1/\sigma_e^2 & -\gamma/\sigma_e^2 \\ 0 & 1 \end{bmatrix}$$

(e.g., Cox and Hinkley 1974, p. 130).

Note that

$$\mathbf{G}_e^{-1}(\sigma)\mathbf{g}(\sigma) = \mathbf{H}^{-1}(\theta)\mathbf{B}_e^{-1}(\theta)[\mathbf{H}^{-1}(\theta)]'[\mathbf{H}(\theta)]'\mathbf{b}(\theta)$$

$$= \mathbf{H}^{-1}(\theta)\mathbf{B}_e^{-1}(\theta)\mathbf{b}(\theta),$$

$$\mathbf{H}^{-1}(\theta) = \begin{bmatrix} \sigma_e^2 & \gamma \\ 0 & 1 \end{bmatrix}.$$

It follows that, even if the starting value $\bar{\sigma}^{(0)}$ for algorithm (8.35) is equivalent to the starting value $\bar{\theta}^{(0)}=(\bar{\gamma}^{(0)}, \bar{\sigma}_e^{2(0)})'$ for algorithm (8.34), that is, even if $\bar{\sigma}^{(0)}=(\bar{\gamma}^{(0)}\bar{\sigma}_e^{2(0)},\bar{\sigma}_e^{2(0)})'$, the vector $\bar{\sigma}^{(j)}$ need not be equivalent to the vector $\bar{\theta}^{(j)}=(\bar{\gamma}^{(j)},\bar{\sigma}_e^{2(j)})'$ (for j=1,2,...). (The second component of $\bar{\sigma}^{(1)}$ equals $\bar{\sigma}_e^{2(1)}$ but, in general, its first component does not equal $\bar{\gamma}^{(1)}\bar{\sigma}_e^{2(1)}$).

It is known (and is a straightforward exercise to verify) that, in the special case $\Delta_1=...=\Delta_r$, algorithm (8.35) converges in a single iteration (from any starting point) to the REML estimate of σ (e.g., Harville 1977, p. 334).

8.13 EM Algorithm and the Method of Successive Approximations

The EM algorithm (e.g., Dempster et al. 1977) can be used to compute REML estimates of σ_s^2 and σ_e^2 (or, equivalently, of γ and σ_e^2). We now describe two distinct ways in which the EM algorithm can be applied to this problem, the first of which is equivalent to the implementation described by Laird (1982) and Dempster et al. (1984). Our approach is relatively simple and direct and leads almost immediately to useful representations.

Let

$$\varepsilon_1(\sigma) = \gamma^2\sum_i \Delta_i \tilde{t}_i^2/(1+\gamma\Delta_i)^2+\sigma_s^2[m-\gamma\sum_i \Delta_i/(1+\gamma\Delta_i)],$$

$$\varepsilon_2(\sigma) = S_e + \sum_i \tilde{t}_i^2/(1+\gamma\Delta_i)^2 + \sigma_e^2[p^* + r - \sum_i (1+\gamma\Delta_i)^{-1}]$$

$$= (n-p^*)\hat{\sigma}_e^2(\gamma) - \gamma\sum_i \Delta_i \tilde{t}_i^2/(1+\gamma\Delta_i)^2 + \sigma_e^2[p^* + \gamma\sum_i \Delta_i/(1+\gamma\Delta_i)].$$

Under model (8.1), we find (using various results from Sect. 8.4) that

$$\tilde{\mathbf{t}} = \mathbf{D}^{1/2}\mathbf{R}'\mathbf{s} + \mathbf{D}^{-1/2}\mathbf{R}'\mathbf{Z}'(\mathbf{I}-\mathbf{P_X})\mathbf{e}, \tag{8.36}$$

$$\mathbf{z} = \mathbf{W}_1'\mathbf{M}\mathbf{e},$$

and hence that

$$E(\mathbf{s}|\tilde{\mathbf{t}},\mathbf{z}) = \gamma\mathbf{R}\mathbf{D}^{1/2}(\mathbf{I}+\gamma\mathbf{D})^{-1}\tilde{\mathbf{t}},$$

$$\text{var}(\mathbf{s}|\tilde{\mathbf{t}},\mathbf{z}) = \sigma_s^2[\mathbf{I} - \gamma\mathbf{R}\mathbf{D}^{1/2}(\mathbf{I}+\gamma\mathbf{D})^{-1}\mathbf{D}^{1/2}\mathbf{R}'],$$

$$E(\mathbf{e}|\tilde{\mathbf{t}},\mathbf{z}) = (\mathbf{I}-\mathbf{P_X})\mathbf{Z}\mathbf{R}\mathbf{D}^{-1/2}(\mathbf{I}+\gamma\mathbf{D})^{-1}\tilde{\mathbf{t}} + \mathbf{M}\mathbf{W}_1\mathbf{z},$$

$$\text{var}(\mathbf{e}|\tilde{\mathbf{t}},\mathbf{z}) = \sigma_e^2[\mathbf{I} - (\mathbf{I}-\mathbf{P_X})\mathbf{Z}\mathbf{R}\mathbf{D}^{-1/2}(\mathbf{I}+\gamma\mathbf{D})^{-1}\mathbf{D}^{-1/2}\mathbf{R}'\mathbf{Z}'(\mathbf{I}-\mathbf{P_X}) - \mathbf{M}\mathbf{W}_1\mathbf{W}_1'\mathrm{M}],$$

implying that

$$E(\mathbf{s}'\mathbf{s}|\tilde{\mathbf{t}},\mathbf{z}) = [E(\mathbf{s}|\tilde{\mathbf{t}},\mathbf{z})]'E(\mathbf{s}|\tilde{\mathbf{t}},\mathbf{z}) + \text{tr}[\text{var}(\mathbf{s}|\tilde{\mathbf{t}},\mathbf{z})] = \varepsilon_1(\sigma),$$

$$E(\mathbf{e}'\mathbf{e}|\tilde{\mathbf{t}},\mathbf{z}) = [E(\mathbf{e}|\tilde{\mathbf{t}},\mathbf{z})]'E(\mathbf{e}|\tilde{\mathbf{t}},\mathbf{z}) + \text{tr}[\text{var}(\mathbf{e}|\tilde{\mathbf{t}},\mathbf{z})] = \varepsilon_2(\sigma).$$

If, in implementing the EM algorithm, **s** and **e** are regarded as the complete data (and $\tilde{\mathbf{t}}$ and **z** the incomplete data), the (j+1)st iterate of the algorithm is

$$\hat{\sigma}^{(j+1)} = \begin{bmatrix} \varepsilon_1(\hat{\sigma}^{(j)})/m \\ \varepsilon_2(\hat{\sigma}^{(j)})/n \end{bmatrix}. \tag{8.37}$$

This representation is essentially the same as that given by Dempster et al. (1984, Sects. 5.2 and 5.3).

Note that expression (8.36) can be rewritten as

$$\tilde{\mathbf{t}} = \mathbf{D}^{1/2}\mathbf{R}'\mathbf{s}+\mathbf{d}$$

with $\mathbf{d}=\mathbf{D}^{-1/2}\mathbf{R}'\mathbf{Z}'(\mathbf{I}-\mathbf{P}_X)\mathbf{e}$. Note also that

$$E(\mathbf{d}) = \mathbf{0},\ \ \mathrm{var}(\mathbf{d}) = \sigma_e^2 I,\ \ \mathrm{cov}(\mathbf{d},\mathbf{z}) = \mathbf{0}.$$

Let

$$\varepsilon_2^*(\sigma) = (n-p^*)\hat{\sigma}_e^2(\gamma)-\gamma\sum_i \Delta_i \tilde{t}_i^2/(1+\gamma\Delta_i)^2+\sigma_s^2\sum_i \Delta_i/(1+\gamma\Delta_i).$$

Under model (8.1), we find that

$$E(\mathbf{d}|\tilde{\mathbf{t}},\mathbf{z}) = (\mathbf{I}+\gamma\mathbf{D})^{-1}\tilde{\mathbf{t}},\ \ \mathrm{var}(\mathbf{d}|\tilde{\mathbf{t}},\mathbf{z}) = \sigma_e^2[\mathbf{I}-(\mathbf{I}+\gamma\mathbf{D})^{-1}],$$

implying that

$$E(\mathbf{z}'\mathbf{z}+\mathbf{d}'\mathbf{d}|\tilde{\mathbf{t}},\mathbf{z}) = \mathbf{z}'\mathbf{z}+[E(\mathbf{d}|\tilde{\mathbf{t}},\mathbf{z})]'E(\mathbf{d}|\tilde{\mathbf{t}},\mathbf{z})+\mathrm{tr}[\mathrm{var}(\mathbf{d}|\tilde{\mathbf{t}},\mathbf{z})] = \varepsilon_2^*(\sigma).$$

If, in implementing the EM algorithm, **s**, **d**, and **z** (rather than **s** and e) are regarded as the complete data, the (j+1)st iterate of the algorithm is

$$\hat{\sigma}^{(j+1)} = \begin{bmatrix} \varepsilon_1(\hat{\sigma}^{(j)})/m \\ \varepsilon_2^*(\sigma^{(j)})/(n-p^*) \end{bmatrix}. \tag{8.38}$$

In general, this iterate differs from the iterate (8.37) obtained from the first implementation.

We now describe an alternative way of arriving at algorithms (8.37) and (8.38). Consider the likelihood equations

$$\partial \ell_v/\partial\sigma_s^2 = 0,\ \ \partial \ell_v/\partial\sigma_e^2 = 0. \tag{8.39}$$

Multiplying both sides of the first of these equations by $-2(\sigma_s^2)^2$ and both sides of the second by $-2(\sigma_e^2)^2$, we obtain the equations

$$\sigma_s^2\gamma\sum_i \Delta_i/(1+\gamma\Delta_i)-\gamma^2\sum_i \Delta_i\tilde{t}_i^2/(1+\gamma\Delta_i)^2 = 0, \tag{8.40}$$

$$\sigma_e^2\,[n-p^*-\gamma\sum_i \Delta_i/(1+\gamma\Delta_i)]-(n-p^*)\hat{\sigma}_e^2(\gamma)+\gamma\sum_i \Delta_i\tilde{t}_i^2/(1+\gamma\Delta_i)^2 = 0, \tag{8.41}$$

respectively. Solutions to the simultaneous equations (8.40) and (8.41) consist of the solutions to the likelihood equations (8.39), along with the further solution $\sigma_s^2=0$, $\sigma_e^2=\hat{\sigma}_e^2(0)$. Subtracting $m\sigma_s^2$ from both sides of Eq. (8.40) and then dividing both sides by -m, we obtain the equivalent equation

$$\sigma_s^2 = \varepsilon_1(\sigma)/m. \tag{8.42}$$

Similarly, subtracting $n\sigma_e^2$ from both sides of Eq. (8.41) and then dividing by -n, we obtain the equivalent equation

$$\sigma_e^2 = \varepsilon_2(\sigma)/n. \tag{8.43}$$

Applying the method of successive approximations (e.g., Harville 1977, Sect. 6.1) to Eqs. (8.42) and (8.43), we obtain algorithm (8.37). Alternatively, if we apply the method of successive approximations to Eq. (8.42) and the equation obtained from Eq. (8.41) by subtracting $(n-p^*)\sigma_e^2$ from both sides and then dividing both sides by $-(n-p^*)$, we obtain algorithm (8.38).

Another possibility is to transform Eqs. (8.40) and (8.41) into the equivalent equations

$$\sigma_s^2 = h_1(\sigma), \quad \sigma_e^2 = h_2(\sigma), \tag{8.44}$$

where

$$h_1(\sigma) = [\gamma\sum_i \Delta_i \tilde{t}_i^2/(1+\gamma\Delta_i)^2]/[\sum_i \Delta_i/(1+\gamma\Delta_i)],$$

$$h_2(\sigma) = [(n-p^*)\hat{\sigma}_e^2(\gamma)-\gamma\sum_i \Delta_i \tilde{t}_i^2/(1+\gamma\Delta_i)^2]/[n-p^*-\gamma\sum_i \Delta_i/(1+\gamma\Delta_i)].$$

Applying the method of successive approximations to the simultaneous Eqs. (8.44), we obtain an iterative algorithm whose (j+1)st iterate is

$$\hat{\sigma}^{(j+1)} = \begin{bmatrix} h_1(\hat{\sigma}^{(j)}) \\ h_2(\hat{\sigma}^{(j)}) \end{bmatrix}. \tag{8.45}$$

Let $h_2^*(\sigma)=\hat{\sigma}_e^2(\gamma)$. Two other iterative algorithms for computing REML estimates of σ_s^2 and σ_e^2 are those whose (j+1)st iterates are

$$\hat{\sigma}^{(j+1)} = \begin{bmatrix} \varepsilon_1(\hat{\sigma}^{(j)})/m \\ h_2^*(\hat{\sigma}^{(j)}) \end{bmatrix}, \tag{8.46}$$

$$\hat{\sigma}^{(j+1)} = \begin{bmatrix} h_1(\hat{\sigma}^{(j)}) \\ h_2^*(\hat{\sigma}^{(j)}) \end{bmatrix}, \tag{8.47}$$

respectively. Each was obtained by Harville (1977, Sect. 6.1) by applying the method of successive approximations to equations derived from the likelihood equations $\partial \ell/\partial\gamma=0$ and $\partial \ell/\partial\sigma_e^2=0$. Algorithm (8.46) can be regarded as an REML version of an algorithm proposed by Henderson (1973, p. 17) and by Hartley and Rao (1967, Sect. 5) for computing maximum likelihood estimates.

The following representations, which can be easily derived by using results (8.28)-(8.30), can be used to re-express the iterates of each of the algorithms (8.37), (8.38) and (8.45)-(8.47) in terms of quantities associated with linear systems (8.3) and (8.26):

$$\varepsilon_1(\sigma) = \gamma^2\mathbf{u}'\mathbf{u}+\sigma_s^2\mathrm{tr}[(\mathbf{I}+\gamma\mathbf{C})^{-1}],$$

$$\varepsilon_2(\sigma) = S_e+\mathbf{u}'(\tilde{\mathbf{s}}-\gamma\mathbf{u})+\sigma_e^2\{p^*+m-\mathrm{tr}[(\mathbf{I}+\gamma\mathbf{C})^{-1}]\},$$

$$\varepsilon_2^*(\sigma) = S_e+\mathbf{u}'(\tilde{\mathbf{s}}-\gamma\mathbf{u})+\sigma_e^2\{m-\mathrm{tr}[(\mathbf{I}+\gamma\mathbf{C})^{-1}]\},$$

$$\begin{aligned} h_1(\sigma) &= \gamma^2\mathbf{u}'\mathbf{u}/\{m-\mathrm{tr}[(\mathbf{I}+\gamma\mathbf{C})^{-1}]\}, \text{ if } \gamma \neq 0, \\ &= 0, \text{ if } \gamma = 0, \end{aligned}$$

$$h_2(\sigma) = \{\mathbf{S}_e+\mathbf{u}'(\tilde{\mathbf{s}}-\gamma\mathbf{u})\}/\{n-p^*-m+\mathrm{tr}[(\mathbf{I}+\gamma\mathbf{C})^{-1}]\},$$

$$h_2^*(\sigma) = [S_e+\mathbf{u}'\tilde{\mathbf{s}}]/(n-p^*).$$

On the (j+1)st iterate of each of the algorithms (8.37), (8.38), and (8.45)-(8.47), the current iterate $\hat{\sigma}^{(j)}=(\hat{\sigma}_s^{2(j)},\hat{\sigma}_e^{2(j)})'$ is transformed into a new iterate $\hat{\sigma}^{(j+1)}=(\hat{\sigma}_s^{2(j+1)}, \hat{\sigma}_e^{2(j+1)})'$. If $\hat{\sigma}_s^{2(j)}\geq 0$ (and $\hat{\sigma}_e^{2(j)}>0$), then (for any of these five algorithms) $\hat{\sigma}_s^{2(j+1)}\geq 0$ (with equality if and only if $\hat{\sigma}_s^{2(j)}=0$) and $\hat{\sigma}_e^{2(j+1)}>0$, as is easily verified. Note that this implies that if both components of the starting value $\hat{\sigma}^{(0)}$ are positive, then both components of every iterate are positive. If $0>\hat{\sigma}_s^{2(j)}>-\hat{\sigma}_e^{2(j)}/\Delta^*$, then, for algorithms (8.45)-(8.47), $\hat{\sigma}_e^{2(j+1)}>0$ (as before) and, for algorithms (8.45) and (8.47), $\hat{\sigma}_s^{2(j+1)}<0$.

8.14 Linearized Method of Successive Approximations

Algorithms (8.21)-(8.23) were derived by applying Newton's method to the linearized equations $k^*(\gamma)=0$, $m^*(\psi)=0$, and $\mathbf{q}^*(\varepsilon)=\mathbf{0}$, respectively. By re-expressing these equations and applying the method of successive approximations, we now derive three additional algorithms.

The equation $k^*(\gamma)=0$ can be rewritten as

$$2r^{-1}\kappa^{-2}k^*(\gamma)+\gamma-\gamma = 0$$

or equivalently as

$$\gamma = \gamma+2r^{-1}\kappa^{-2}k^*(\gamma). \tag{8.48}$$

Applying the method of successive approximations to Eq. (8.48), we obtain the iterative algorithm whose (j+1)st iterate is

$$\hat{\gamma}^{(j+1)}= \hat{\gamma}^{(j)}+2r^{-1}\kappa^{-2}k^*(\hat{\gamma}^{(j)}). \tag{8.49}$$

In the special case $\Delta_1=...=\Delta_r$, algorithm (8.49) converges in one iteration from any starting value $\hat{\gamma}^{(0)}$.

Similarly, defining $\mathbf{q}^+(\varepsilon)=[(2/r)q_1^*(\varepsilon),(2/f)q_2^*(\varepsilon)]'$, the equations $m^*(\psi)=0$ and $\mathbf{q}^*(\varepsilon)=\mathbf{0}$ can be rewritten as

$$2(rF)^{-1}m^*(\psi)+\psi-\psi = 0,$$

$$\mathbf{q}^+(\varepsilon)+\varepsilon-\varepsilon = \mathbf{0},$$

respectively, or, equivalently, as

$$\psi = \psi+2(rF)^{-1}m^*(\psi), \tag{8.50}$$

$$\varepsilon = \varepsilon+\mathbf{q}^+(\varepsilon). \tag{8.51}$$

Applying the method of successive approximations to Eqs. (8.50) and (8.51) (individually), we obtain the iterative algorithms whose (j+1)st iterates are

$$\hat{\psi}^{(j+1)} = \hat{\psi}^{(j)}+2(rF)^{-1}m^*(\hat{\psi}^{(j)}), \tag{8.52}$$

$$\hat{\varepsilon}^{(j+1)} = \hat{\varepsilon}^{(j)}+\mathbf{q}^+(\hat{\varepsilon}^{(j)}). \tag{8.53}$$

In the special case $\Delta_1=...=\Delta_r$, algorithms (8.52) and (8.53), like algorithm (8.49), converge in a single iteration.

Our approach, in deriving algorithms (8.49), (8.52), and (8.53), is somewhat similar to the approach taken by Thompson and Meyer (1986), who proposed two algorithms, each of which depends on a constant (to be specified by the user) labelled k. The first of their algorithms is outlined in the last paragraph of their page 223 [and is given in a special case by their formulae (2.10) and (2.11)], and the second is given by their formulae (2.12) and (2.13). For the choice $k=\bar{\Delta}$, it can be shown that their first algorithm is essentially identical to algorithm (8.53) and that [assuming equivalent starting values and letting $\hat{\varepsilon}_s^{(j)}$ and $\hat{\varepsilon}_e^{(j)}$ represent the ε_s and ε_e components of the jth iterate of algorithm (8.53)] the changes in the ε_s and ε_e components produced on the initial iteration of their second algorithm are $(r/m)(\hat{\varepsilon}_s^{(1)} - \hat{\varepsilon}_s^{(0)})$ and $[f/(r+f)](\hat{\varepsilon}_e^{(1)} - \hat{\varepsilon}_e^{(0)})$, respectively.

Solutions to equations of the general form (8.24) (and hence the end points of a confidence interval for γ) can be computed via an algorithm analogous to algorithm (8.49). Equation (8.25), which was obtained by linearizing Eq. (8.24), can be rewritten as

$$(c\bar{\Delta})^{-1}[(f/r)S_e^{-1}(1+\gamma\bar{\Delta})\sum_i \tilde{\tau}_i^2/(1+\gamma\Delta_i)-c(1+\gamma\bar{\Delta})]+\gamma-\gamma = 0$$

or equivalently as

$$\gamma = \gamma+(c\bar{\Delta})^{-1}[(f/r)S_e^{-1}(1+\gamma\bar{\Delta})\sum_i \tilde{\tau}_i^2/(1+\gamma\Delta_i)-c(1+\gamma\bar{\Delta})]. \tag{8.54}$$

An algorithm for computing the solution to Eq. (8.24) is obtained by applying the method of successive approximations to Eq. (8.54).

8.15 Confidence Intervals and Hypothesis Tests

Suppose that we wish to make inferences about a parametric function $\phi=\phi(\gamma,\sigma_e^2)$ of γ and σ_e^2 (whose domain is the set Ω^+). Let R represent the range of the function ϕ. Denote the REML estimates of γ and σ_e^2 by $\hat{\gamma}$ and $\hat{\sigma}_e^2$, respectively. For each element ϕ^* of R, define $\tilde{\gamma}(\phi^*)$ and $\tilde{\sigma}_e^2(\phi^*)$ to be values of γ and σ_e^2 that maximize $\ell(\gamma,\sigma_e^2;y)$ subject to the constraint $\phi(\gamma,\sigma_e^2)=\phi^*$ (and subject to the constraint that γ and σ_e^2 belong to Ω^+).

By definition, an REML estimate of ϕ is $\hat{\phi}=\phi(\hat{\gamma},\hat{\sigma}_e^2)$. In practice, we may seek a confidence interval for ϕ as well as a point estimate. Standard asymptotic theory suggests that the distribution of the difference $2\{\ell(\hat{\gamma},\hat{\sigma}_e^2;y)-\ell[\tilde{\gamma}(\phi),\tilde{\sigma}_e^2(\phi);y]\}$ may approximate a chi-square distribution with one degree of freedom (e.g., Cox and Hinkley 1974, Sect. 9.3). Thus, the set S(y) of values of ϕ ($\phi\varepsilon R$) that satisfy the inequality

$$\ell(\hat{\gamma},\hat{\sigma}_e^2;\mathbf{y})-\ell[\tilde{\gamma}(\phi),\tilde{\sigma}_e^2(\phi);\mathbf{y}] \le (1/2)\chi^2_{1,\alpha} \tag{8.55}$$

(where $\chi^2_{j,\alpha}$ represents the upper-α point of the chi-square distribution with j degrees of freedom) may constitute an approximate 100(1-α)% confidence set for ϕ.

The confidence set S(**y**) can, of course, be used to carry out an approximate, two-sided, α-level test of the null hypothesis that ϕ equals a specified constant c; if $c\varepsilon S(\mathbf{y})$, the null hypothesis is accepted; otherwise, it is rejected.

In the important special case $\phi=\gamma$, inequality (8.55) is re-expressible as

$$\ell_c(\hat{\gamma};\mathbf{y})-\ell_c(\gamma;\mathbf{y}) \le (1/2)\chi^2_{1,\alpha}. \tag{8.56}$$

If ϕ is a function $\phi=\phi(\gamma)$ of γ alone, then the confidence set for ϕ can be obtained from that for γ by transformation. More specifically, the confidence set for the function $\phi=\phi(\gamma)$ consists of the range of the function when its domain is restricted to the confidence set for γ.

The confidence set for γ defined by inequality (8.56) can be computed graphically by plotting the concentrated log likelihood function $\ell_c(\gamma;\mathbf{y})$ and by then "reading off" the values of γ for which $\ell_c(\gamma;\mathbf{y})$ lies above the horizontal line $\gamma=\ell_c(\hat{\gamma};\mathbf{y})-(1/2)\chi^2_{1,\alpha}$. Alternatively, we can iteratively compute the solutions to the non-linear equation

$$\ell_c(\gamma;\mathbf{y})-\ell_c(\hat{\gamma};\mathbf{y})+(1/2)\chi^2_{1,\alpha} = 0 \tag{8.57}$$

(in γ) by, for example, employing Newton's method. Assuming that the confidence set for γ is an interval (as will be the case except for certain "pathological" data sets), the solutions to equation (8.57) are the end points of the interval.

As applied to Eq. (8.57), the (j+1)st iterate of Newton's method is

$$\tilde{\gamma}^{(j+1)} = \tilde{\gamma}^{(j)}-[\ell_c(\tilde{\gamma}^{(j)};\mathbf{y})-\ell_c(\hat{\gamma};\mathbf{y})+(1/2)\chi^2_{1,\alpha}]/k(\tilde{\gamma}^{(j)}).$$

Prior to initiating the iterative computation of the solutions to Eq. (8.57), it might be helpful (in deciding on the starting values) to plot $\ell_c(\gamma;\mathbf{y})$ for a "few" values of γ. Alternatively, we could take the starting values to be

$$\hat{\gamma}\pm[\chi^2_{1,\alpha}(-b_{11}+b_{12}^2/b_{22})^{-1}]^{1/2}, \tag{8.58}$$

where, with $\hat{\theta}=(\hat{\gamma},\hat{\sigma}_e^2)'$, b_{ij} represents the ijth element of the matrix $\mathbf{B}_e(\hat{\theta})$.

The interval with end points (8.58) can itself be regarded as an approximate 100(1-α)% confidence interval for γ. In fact, the confidence set determined by inequality (8.56) is just one of the several approximate 100(1-α)% confidence sets that can be derived from the likelihood function, refer to, for example, Cox and Hinkley (1974, Sect. 9.3).

For some choices of ϕ, procedures that give exact 100(1-α)% confidence intervals are available. For instance, the interval

$$S_e/\chi^2_{f,\alpha/2} \leq \sigma_e^2 \leq S_e/\chi^2_{f,1-\alpha/2}$$

is an exact 100(1-α)% confidence interval for σ_e^2, and an exact 100(1-α)% confidence interval for γ is described by, for example, Harville and Fenech (1985). Comparisons of the approximate 100(1-α)% confidence interval for γ defined by inequality (8.56) with the exact 100(1-α)% confidence interval (on the basis, for example, of expected length) might be of interest.

8.16 Example

We reconsider the data analyzed by Harville and Fenech (1985). These data, which consist of weights at birth of 62 single-birth male lambs, are reproduced in Table 8.1.

The data come from five distinct population lines. Each lamb was the progeny of 1 of 23 rams, and each lamb had a different dam. Age of dam was recorded as belonging to one of three categories, numbered 1 (1-2 years), 2 (2-3 years) and 3 (over 3 years).

Let y_{ijkd} represent the weight (at birth) of the dth of those lambs that are the offspring of the kth sire in the jth population line and of a dam belonging to the ith age category. Like Harville and Fenech (1985), we assume that y_{ijkd} follows the mixed-effects linear model

$$y_{ijkd} = \mu+\delta_i+\pi_j+s_{jk}+e_{ijkd},$$

where the age effects ($\delta_1,\delta_2,\delta_3$) and the line effects ($\pi_1,...,\pi_5$) are fixed effects, where the sire (within-line) effects ($s_{11},s_{12}, ...,s_{58}$) are random effects that are distributed independently as $N(0,\sigma_s^2)$ (normal with mean 0 and variance σ_s^2), and where the random errors $e_{1111},e_{1121},...,e_{3582}$ are distributed as $N(0,\sigma_e^2)$ independently of each other and of the sire effects. Clearly, this model is expressible as a special case of mixed-effects linear model (8.1).

Define the parametric function $h^2=4\sigma_s^2/(\sigma_s^2+\sigma_e^2)$, which is interpretable as a heritability. We have that $h^2=4\gamma/(1+\gamma)$ and (for $\gamma>0$) $h^2=4/(1+\gamma^{-1})$. Clearly, h^2 is a strictly increasing function of γ over the domain $0\leq\gamma<\infty$. Since a heritability is inherently less than or equal to one, there is an implicit assumption that $0\leq h^2\leq 1$ or, in terms of γ, that $0\leq\gamma\leq 1/3$.

For the lamb-weight data, $\lambda^*=9$, r=18, f=37, S_s=80.296, S_e= 102.235, and F=1.615. Further, $\Delta_1,...,\Delta_r$ and the corresponding observed values of $\tilde{t}_1,...,\tilde{t}_r$ are:

Δ	$\tilde{t}$	Δ	$\tilde{t}$	Δ	$\tilde{t}$
0.8400	2.9062	1.4078	1.2882	2.7482	-2.5893
0.9027	0.4505	1.7077	-2.5006	3.1505	-2.4773
1.0000	-4.8083	1.9329	-2.1365	3.3236	-3.0049
1.0750	-0.7319	2.0000	1.1294	3.5644	1.5521
1.1644	-0.7361	2.0000	-1.4106	4.2340	-1.8835
1.3456	1.2924	2.3293	1.2735	5.0875	0.7676

[The values of some of the $\tilde{t}_i$'s differ from those reported by Harville and Fenech (1985), since our choices for some of the columns of **R** differ (at least in sign) from theirs.] Then, $\kappa=\bar{\Delta}=2.2118$ and $\Delta^*=5.0875$. Note that $-1/\Delta^*=-0.197$, while $-1/\lambda^*=-0.111$. Note also that

Table 8.1 Birth weights (in pounds) of lambs

Sire	Dam age	Wt.	Sire	Dam age	Wt.	Sire	Dam age	Wt.
Line 1			Line 3			Line 5		
1	1	6.2	1	2	9.0	1	1	11.7
2	1	13.0		3	9.5			12.6
3	1	9.5			12.6	2	1	9.0
		10.1	2	1	11.0		3	11.0
		11.4		2	10.1	3	3	9.0
	2	11.8			11.7			12.0
	3	12.9		3	8.5	4	3	9.9
		13.1			8.8	5	2	13.5
4	1	10.4			9.9	6	2	10.9
	2	8.5			10.9		3	5.9
					11.0	7	2	10.0
Line 2					13.9			12.7
1	3	13.5	3	1	11.6		3	13.2
2	2	10.1		3	13.0			13.3
	3	11.0	4	2	12.0	8	1	10.7
		14.0						11.0
		15.5	Line 4					12.5
3	1	12.0	1	1	9.2		3	9.0
4	1	11.5			10.6			10.2
	3	10.8			10.6			
				3	7.7			
					10.0			
					11.2			
			2	1	10.2			
					10.9			
			3	1	11.7			
				3	9.9			

the lamb-weight data are highly unbalanced, as evidenced by the large differences among $\Delta_1,\ldots,\Delta_r$.

The concentrated log likelihood function ℓ_c is depicted in Fig. 8.1. The REML estimates of γ, σ_e^2, and h^2 are $\hat{\gamma}$=0.175, $\hat{\sigma}_e^2(\hat{\gamma})$=2.962, and $4/(1+\hat{\gamma}^{-1})$=0.59, respectively. By way of comparison, the method-of-fitting-constants estimators of γ, σ_e^2, and h^2 are 0.278, 2.763, and 0.87.

Further, $\ell_c(\hat{\gamma};y)$=-110.7416 and $\chi^2_{1,.20}$=1.6424, so that $\ell_c(\hat{\gamma};y)-(1/2)\chi^2_{1,.20}$= -111.5625. Then, as illustrated in Fig. 8.1, the approximate 80% confidence set defined by inequality (8.56) consists of the interval $-0.079\leq\gamma\leq0.729$. By way of comparison, the confidence interval obtained by Harville and Fenech (1985), which is an exact 80% confidence interval, is $-0.008\leq\gamma\leq1.125$. Both of these confidence intervals include as subsets the interval $0\leq\gamma\leq1/3$, corresponding to the entire range $0\leq h^2\leq1$ of heritability values. Thus, as discussed by Harville and Fenech, the lamb-weight data is so limited in extent and so unbalanced that it provides very little information about h^2.

A total of 16 distinct iterative algorithms for computing REML estimates were described in Sections 8.7-8.14: the four Newton-Raphson algorithms (8.13), (8.17), (8.18), and (8.20); the two linearized Newton algorithms (8.22) [or (8.21)] and (8.23); the two method-of-scoring algorithms (8.34) and (8.35); the two EM algorithms (8.37) and (8.38); the three MSA (method of successive approximation) algorithms (8.45)-(8.47), and the three linearized MSA algorithms (8.49), (8.52), and (8.53).

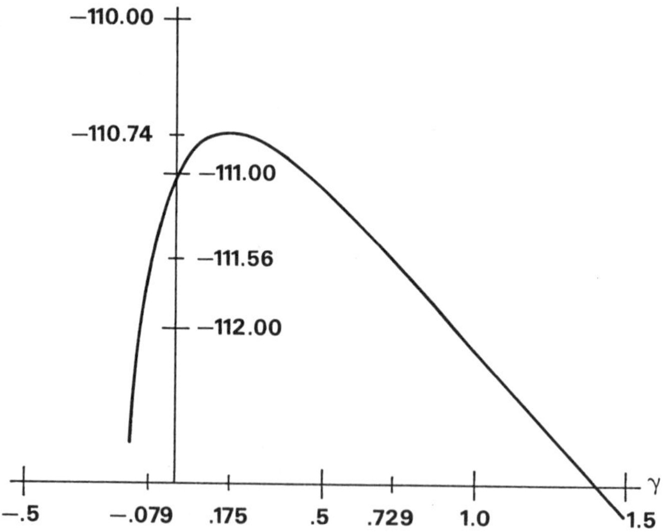

Fig. 8.1 Plot of the concentrated log likelihood function $\ell_c(\gamma;y)$

Table 8.2 The γ-components of the first ten iterates of each algorithm

	Algorithm															
	Newton-Raphson				Linearized Newton		Method of scoring		EM		MSA			Linearized MSA		
Iteration	(8.13) (8.53)	(8.17)	(8.18)	(8.20)	(8.22)	(8.23)	(8.34)	(8.35)	(8.37)	(8.38)	(8.45)	(8.46)	(8.47)	(8.49)	(8.52)	
	Starting values: $\hat{\gamma}^{(0)} = 0.01$, $\hat{\sigma}_e^{2(0)} = 1$															
1	0.099	0.018	0.100	0.225	0.139	0.070	0.326	0.113	0.00348	0.00321	0.012	0.00319	0.012	0.170	0.140	0.155
2	0.191	0.030	0.158	0.178	0.174	0.165	0.217	0.156	0.00321	0.00318	0.014	0.00318	0.014	0.174	0.164	0.170
3	0.252	0.052	0.174	0.175	0.175	0.175	0.186	0.169	0.00319	0.00318	0.017	0.00318	0.017	0.174	0.172	0.174
4	0.236	0.088	0.175				0.178	0.173	0.00319	0.00318	0.020	0.00318	0.020	0.175	0.174	0.174
5	0.185	0.137					0.175	0.174	0.00319	0.00319	0.024	0.00319	0.024		0.174	0.175
6	0.175	0.169						0.174	0.00319	0.00319	0.028	0.00319	0.028		0.175	
7		0.174						0.175	0.00320	0.00319	0.033	0.00319	0.033			
8		0.175							0.00320	0.00320	0.039	0.00320	0.038			
9									0.00320	0.00320	0.045	0.00320	0.044			
10									0.00321	0.00321	0.051	0.00320	0.050			
	Starting values: $\hat{\gamma}^{(0)} = 1$, $\hat{\sigma}_e^{2(0)} = 1$															
1	0.956	0.821	-0.196[a]	0.521	0.210	-0.146	-0.196[a]	0.305	0.585	0.548	0.571	0.523	0.628	0.368	0.326	0.301
2	0.608	0.585	-0.196	0.280	0.174	-0.095	-0.156	0.207	0.474	0.452	0.418	0.453	0.467	0.219	0.218	0.207
3	0.331	0.352	-0.195	0.189	0.175	0.033	-0.091	0.184	0.422	0.409	0.340	0.418	0.380	0.183	0.187	0.182
4	0.224	0.219	-0.194	0.175		0.153	0.023	0.177	0.390	0.381	0.294	0.392	0.326	0.176	0.178	0.176
5	0.181	0.179	-0.193			0.174	0.113	0.175	0.366	0.359	0.264	0.370	0.290	0.175	0.176	0.175
6	0.175	0.175	-0.191			0.175	0.155		0.346	0.341	0.243	0.351	0.264		0.175	
7			-0.188				0.169		0.330	0.325	0.228	0.336	0.246			
8			-0.183				0.173		0.316	0.312	0.217	0.322	0.232			
9			-0.172				0.174		0.304	0.301	0.209	0.311	0.221			
10			-0.145				0.174		0.294	0.291	0.202	0.300	0.212			

Note: No entry is listed for iterations subsequent to that on which convergence to three significant digits was first attained.

[a] The γ-component of the actual iterate, which was smaller than the lower bound $-1/\Delta^* = -0.197$, was truncated at -0.196.

Table 8.3 Number of iterations required for the γ-components of the iterates of each algorithm to attain an accuracy of N significant digits

	Algorithm																
	Newton-Raphson				Linearized Newton		Method of scoring		EM		MSA			Linearized MSA		
N	(8.13)	(8.17)	(8.18)	(8.20)	(8.22)	(8.23)	(8.34)	(8.35)	(8.37)	(8.38)	(8.45)	(8.46)	(8.47)	(8.49)	(8.52)	(8.53)
	Starting values: $\hat{\gamma}^{(0)} = 0.01$, $\hat{\sigma}_e^{2(0)} = 1$																
1	4	6	2	1	2	2	2	2	985	986	25	1001	26	1	2	1
2	6	6	3	3	2	3	6	3	1015	1015	30	1032	32	1	3	2
3	6	8	4	3	3	3	6	7	1140	1140	56	1167	60	4	6	5
4	7	8	4	3	3	4	8	7	1160	1160	60	1189	64	4	7	6
5	7	8	4	4	3	4	11	9	1203	1203	68	1235	74	5	8	7
	Starting values: $\hat{\gamma}^{(0)} = 1$, $\hat{\sigma}_e^{2(0)} = 1$																
1	4	4	14	3	1	4	6	2	17	17	6	18	7	2	2	2
2	6	6	15	4	2	5	7	6	142	141	32	153	36	5	6	5
3	6	6	16	4	3	6	11	7	142	141	32	153	36	5	6	5
4	7	7	16	5	3	6	11	8	194	194	43	210	48	6	8	7
5	7	7	16	5	3	6	13	11	297	296	64	321	71	9	11	9

For purposes of evaluating and comparing the 16 algorithms, each was applied to the lamb-weight data. Two trials were made with each algorithm; one with starting values $\hat{\gamma}^{(0)}=0.01$ and $\hat{\sigma}_e^{2(0)}=1$ and the other with starting values $\hat{\gamma}^{(0)}=1$ and $\hat{\sigma}_e^{2(0)}=1$. In both cases, the starting values were relatively "poor" (note that a γ-value of 1 corresponds to a heritability well outside the permissible range). For each of the 16 algorithms and each choice of starting values, Table 8.2 gives the γ-components of the first ten iterates, and Table 8.3 lists the number of iterations required to achieve an accuracy of N significant digits (N=1,...,5).

All 16 algorithms converged from both choices of starting values though the rate of convergence (as measured by the number of iterations) for the EM and (non-linearized) MSA algorithms was very slow, especially for the EM algorithms and for the MSA algorithm (8.46). The fastest convergence was achieved by the linearized Newton algorithm (8.22). The performance of this algorithm was also found to be superior when the 16 algorithms were applied to the rat-weight data of Dempster et al. (1984); refer to Callanan (1985, Chap. VI).

There are, of course, many ways in which we could attempt to improve on the performance of the 16 algorithms; for instance, if the value of the likelihood function for the (i+1)st iterate is smaller than for the ith iterate, it would seem that the (i+1)st iterate should be adjusted. Also, other types of algorithms, refer, for example, to Gill et al. (1981) or Luenberger (1984), could be considered.

In evaluating the efficiency of the iterative algorithms, the total computing time required to achieve convergence is seemingly a more relevant measure of efficiency than the number of iterations. As discussed in Section 8.11, the computing time depends on whether the preliminary computations (the computations preceding the iteration) include the computation of the characteristic values and vectors of **C**.

Computing times were recorded (in the case of the lamb-weight data) for all algorithms except the linearized MSA algorithms. [Computing times for the linearized MSA algorithms would be similar to those for the EM and (non-linearized) MSA algorithms]. The preliminary computations required only 0.87 s of CPU time (on a National Advanced Systems AS/6 computer) when the computation of the characteristic values and vectors was omitted versus 3.53 s when it was included. When the computation of the characteristic values and vectors was omitted from the preliminary computations, the subsequent computing time for Newton-Raphson, linearized Newton, and method-of-scoring algorithms ranged from 1.44 to 1.50 s per iteration, while that for the EM and MSA algorithms ranged from 0.92 to 0.93. When the computation of the characteristic values and vectors was included in the preliminary computations, the subsequent computing time for Newton-Raphson, linearized Newton, and method-of-scoring algorithms (8.13), (8.17), (8.23), (8.34), and (8.35) ranged from 0.0079 to 0.0090 s per iteration, while that for Newton-Raphson and linearized Newton algorithms (8.18), (8.20), and (8.22) ranged from 0.0048 to 0.0049 and that for the EM and MSA algorithms ranged from 0.0036 to 0.0037. Thus, depending on the choice of algorithm, the total computing time after two or three iterations was less when the computation of the characteristic values and vectors was included in the preliminary computations than when it was omitted.

Differences in efficiency among the iterative algorithms are not especially important in applications of modest size. In such applications, the computing costs for even relatively inefficient algorithms are likely to be small. However, in more extensive applications, the computing costs may be either acceptable or unacceptable depending on the choice of algorithm.

8.17 Extensions

8.17.1 More than One Set of Random Effects

The results given in Sections 8.7-8.15 are for model (8.1), which contains only one set of random effects. Let us consider the extension of those results to the model

$$\mathbf{y} = \mathbf{X}\boldsymbol{\beta} + \mathbf{Z}_1\mathbf{s}_1 + \ldots + \mathbf{Z}_c\mathbf{s}_c + \mathbf{Z}\mathbf{s} + \mathbf{e}. \tag{8.59}$$

Here, $\mathbf{Z}_i$ is a given $n x m_i$ matrix, $\mathbf{s}_i$ is an unobservable random vector whose distribution is MVN($\mathbf{0}, \sigma_i^2\mathbf{I}$) (where σ_i^2 is an unknown, non-negative parameter), and it is assumed that $\mathbf{s}_1, \ldots, \mathbf{s}_c$, $\mathbf{s}$ and $\mathbf{e}$ are statistically independent.

The Newton-Raphson, method-of-scoring, EM, and method-of-successive-approximation algorithms (8.13), (8.17), (8.18), (8.20), (8.34), (8.35), (8.37), (8.38), and (8.45)-(8.47) extend in a straightforward way to model (8.59). Of the three linearized Newton algorithms, algorithms (8.22) and (8.23) can be easily extended to model (8.59), while algorithm (8.21) cannot [as discussed in detail by Callanan (1985, Sect. III.B)]. Similarly, of the three linearized method-of-successive-approximation algorithms, algorithms (8.52) and (8.53) have natural extensions, but algorithm (8.49) does not.

Define $\mathbf{Z}^* = (\mathbf{Z}_1, \ldots, \mathbf{Z}_c)$, and let $\boldsymbol{\Gamma} = \text{diag}(\gamma_1\mathbf{I}, \ldots, \gamma_c\mathbf{I})$, where $\gamma_i = \sigma_i^2/\sigma_e^2$ (i=1,...,c). As discussed by, for example, Callanan (1985), the computations required to carry out any particular iteration of the extended algorithms can be reduced to those associated with the solution (for specified values of $\gamma_1, \ldots, \gamma_c$, and γ) of the linear system

$$\begin{bmatrix} \mathbf{I} + \mathbf{Z}^{*\prime}(\mathbf{I} - \mathbf{P_X})\mathbf{Z}^*\boldsymbol{\Gamma} & \gamma\mathbf{Z}^{*\prime}(\mathbf{I} - \mathbf{P_X})\mathbf{Z} \\ \mathbf{Z}'(\mathbf{I} - \mathbf{P_X})\mathbf{Z}^*\boldsymbol{\Gamma} & \mathbf{I} + \gamma\mathbf{C} \end{bmatrix} \begin{bmatrix} \mathbf{u}^* \\ \mathbf{u} \end{bmatrix} = \begin{bmatrix} \mathbf{Z}^{*\prime}(\mathbf{I} - \mathbf{P_X})\mathbf{y} \\ \mathbf{q} \end{bmatrix}.$$

If prior to commencing the iterative process, the non-zero characteristic values $\Delta_1,...,\Delta_r$ of **C** and the corresponding orthonormal characteristic vectors (the columns of **R**) are computed, then the computations required on each iteration can in effect be further reduced to those associated with the linear system

$$[\mathbf{I}+\mathbf{Z}^{*\prime}(\mathbf{I}-\mathbf{P}_X)\mathbf{Z}^*\Gamma-\gamma\mathbf{Z}^{*\prime}(\mathbf{I}-\mathbf{P}_X)\mathbf{Z}\mathbf{R}(\mathbf{I}+\gamma\mathbf{D})^{-1}\mathbf{R}'\mathbf{Z}'(\mathbf{I}-\mathbf{P}_X)\mathbf{Z}^*\Gamma]\mathbf{u}^*$$

$$= \mathbf{Z}^{*\prime}(\mathbf{I}-\mathbf{P}_X)\mathbf{y}-\gamma\mathbf{Z}^{*\prime}(\mathbf{I}-\mathbf{P}_X)\mathbf{Z}\mathbf{R}(\mathbf{I}+\gamma\mathbf{D})^{-1}\mathbf{R}'\mathbf{q}.$$

The likelihood-based confidence set and test procedures discussed in Section 8.15, in the context of model (8.1), extend in a straightforward way to model (8.59).

8.17.2 Correlated or Heteroscedastic Random Effects

Suppose that $\mathbf{s}\sim\text{MVN}(\mathbf{0},\sigma_s^2\mathbf{A})$, where **A** is a given non-negative definite matrix, rather than, as previously supposed, that $\mathbf{s}\sim\text{MVN}(\mathbf{0},\sigma_s^2\mathbf{I})$. Then, the distribution of **y** is $\text{MVN}[\mathbf{X}\boldsymbol{\beta},\sigma_e^2(\mathbf{I}+\gamma\mathbf{Z}\mathbf{A}\mathbf{Z}')]$.

Take **L** to be any matrix of dimensions $m\times\text{rank}(\mathbf{A})$ such that $\mathbf{A}=\mathbf{L}\mathbf{L}'$. Let $\mathbf{Z}^@=\mathbf{Z}\mathbf{L}$. Then,

$$\sigma_e^2(\mathbf{I}+\gamma\mathbf{Z}\mathbf{A}\mathbf{Z}') = \sigma_e^2(\mathbf{I}+\gamma\mathbf{Z}^@\mathbf{Z}^{@\prime}),$$

that is, var(**y**) can be re-expressed as $\sigma_e^2(\mathbf{I}+\gamma\mathbf{Z}^@\mathbf{Z}^{@\prime})$. Consequently, all of the results of Sections 8.3-8.15, which were derived under the assumption that $\text{var}(\mathbf{y})=\sigma_e^2(\mathbf{I}+\gamma\mathbf{Z}\mathbf{Z}')$, can be extended to the more general case where $\text{var}(\mathbf{y})=\sigma_e^2(\mathbf{I}+\gamma\mathbf{Z}\mathbf{A}\mathbf{Z}')$ simply by substituting $\mathbf{Z}^@$ for **Z**. With this substitution, m, r, **C**, and **q** are replaced by $m^@=\text{rank}(\mathbf{A})$, $r^@=\text{rank}(\mathbf{X},\mathbf{Z}^@)-\text{rank}(\mathbf{X})$, $\mathbf{C}^@=\mathbf{Z}^{@\prime}(\mathbf{I}-\mathbf{P}_X)\mathbf{Z}^@=\mathbf{L}'\mathbf{C}\mathbf{L}$, and $\mathbf{q}^@=\mathbf{Z}^{@\prime}(\mathbf{I}-\mathbf{P}_X)\mathbf{y}=\mathbf{L}'\mathbf{q}$, respectively.

Consider the problem of finding the $r^@$ non-zero characteristic values of $\mathbf{C}^@$, together with a corresponding set of $r^@$ orthonormal characteristic vectors, or, equivalently, the problem of finding a matrix $\mathbf{R}^@$ of dimensions $m^@\times r^@$ such that $\mathbf{R}^{@\prime}\mathbf{R}^@$ equals **I** and $\mathbf{R}^{@\prime}\mathbf{C}^@\mathbf{R}^@$ is diagonal. This problem can in effect be embedded in that of finding **L** by requiring that **L** be such that **L'CL** is diagonal (in addition to satisfying **A=LL'**). Similarly, the problem of finding an orthogonal matrix $\mathbf{0}^@$ such that $\mathbf{0}^{@\prime}\mathbf{C}^@\mathbf{0}^@$ is tridiagonal can be embedded in that of finding **L** by requiring that **L** be such that **L'CL** is tridiagonal. Note that, when rank (**A**)=m, the condition **A=LL'** can be re-expressed as $\mathbf{L}'\mathbf{A}^{-1}\mathbf{L}=\mathbf{I}$.

When $\mathbf{Z}^@$ is substituted for **Z**, we obtain results that involve the matrix **L**. In the Appendix, we give some identities that can be used to re-express these results in terms of **A** or [if rank(**A**)=m] $\mathbf{A}^{-1}$.

Acknowledgements. The authors are indebted to Alan P. Fenech (University of California-Davis), who computed the graph depicted in Fig. 8.1. Harville's work was supported in part by Office of Naval Research Contract N00014-85-K-0418.

References

Bard Y (1974) Nonlinear parameter estimation. Academic Press, New York

Callanan TP (1985) Restricted maximum likelihood estimation of variance components: computational aspects. PhD Diss Dep Stat, Iowa State Univ, Ames

Cox DR, Hinkley, DV (1974) Theoretical statistics. Chapman and Hall, London

Dempster AP, Laird NM, Rubin, DB (1977) Maximum likelihood from incomplete data via the EM algorithm. J R Stat Soc Ser B 39:1-38

Dempster AP, Selwyn MR, Patel CM, Roth AJ (1984) Statistical and computational aspects of mixed model analysis. Appl Stat 33:203-214

Gill PE, Murray W, Wright MH (1981) Practical optimization. Academic Press, New York

Hartley HO, Rao JNK (1967) Maximum-likelihood estimation for the mixed analysis of variance model. Biometrika 54:93-108

Harville DA (1974) Bayesian inference for variance components using only error contrasts. Biometrika 61:383-385

Harville DA (1977) Maximum likelihood approaches to variance component estimation and to related problems. J Am Stat Assoc 72:320-340.

Harville DA, Fenech AP (1985) Confidence intervals for a variance ratio, or for heritability, in an unbalanced mixed linear model. Biometrics 41:137-152

Hemmerle WJ, Hartley HO (1973) Computing maximum likelihood estimates for mixed A.O.V. model using the W transformation. Technometrics 15:819-831

Henderson CR (1963) Selection index and expected genetic advance. In: Hanson WD and Robinson HF (eds) Statistical genetics and plant breeding. NAS-NRC Publ 982, Washington DC pp 141-163

Henderson CR (1973) Sire evaluation and genetic trends. In: Proc Anim Breed Genet Symp in Honor of Dr. J.L. Lush. ASDA and ADSA, Champaign, Illinois pp 10-41

Henderson HV, Searle SR (1981) On deriving the inverse of a sum of matrices. Siam Review 23:53-60

Laird NM (1982) Computation of variance components using the EM algorithm. J Stat Comput Simulat 14:295-303

Luenberger DG (1984) Linear and nonlinear programming. 2nd edn. Addison-Wesley, Reading

Marsaglia G, Styan GPH (1974) Rank conditions for generalized inverses of partitioned matrices. Sankhyā Ser A 36:437-442

Patterson HD, Thompson R (1971) Recovery of interblock information when block sizes are unequal. Biometrika 58:545-554

Searle SR (1971) Linear models. Wiley, New York

Searle SR (1982) Matrix algebra useful for statistics. Wiley, New York

Smith SP, Graser H-U (1986) Estimating variance components in a class of mixed models by restricted maximum likelihood. J Dairy Sci 69:1156-1165

Stewart GW (1973) Introduction to matrix computations. Academic Press, New York

Thompson WA Jr (1955) On the ratio of variances in the mixed incomplete block model. Ann Math Statist 26:721-733

Thompson WA Jr (1962) The problem of negative estimates of variance components. Ann Math Statist 33:273-289

Thompson R, Meyer K (1986) Estimation of variance components: what is missing in the EM algorithm? J Stat Comput Simulat 24:215-230

APPENDIX

8.A.1 Second-Order Partial Derivatives of ℓ_V

The second-order partial derivatives of ℓ_V have the following representations:

$$\frac{\partial^2 \ell_V}{\partial(\sigma_s^2)^2} = (\sigma_e^2)^{-2}(\partial^2 \ell/\partial\gamma^2),$$

$$\frac{\partial^2 \ell_V}{\partial\sigma_s^2\partial\sigma_e^2} = -(\sigma_e^2)^{-2}[\gamma(\partial^2 \ell/\partial\gamma^2)-(1/2)\sum_i \Delta_i/(1+\gamma\Delta_i)+(\sigma_e^2)^{-1}\sum_i \Delta_i \tilde{t}_i^2/(1+\gamma\Delta_i)^2],$$

$$\frac{\partial^2 \ell_V}{\partial(\sigma_e^2)^2} = (1/2)(\sigma_e^2)^{-3}\{\sigma_e^2[r+f-2\gamma\sum_i \Delta_i/(1+\gamma\Delta_i)+\gamma^2\sum_i \Delta_i^2/(1+\gamma\Delta_i)^2]$$

$$-2[(r+f)\hat{\sigma}_e^2(\gamma)-2\gamma\sum_i \Delta_i \tilde{t}_i^2/(1+\gamma\Delta_i)^2$$

$$+\gamma^2\sum_i \Delta_i^2 \tilde{t}_i^2/(1+\gamma\Delta_i)^3]\}.$$

8.A.2 Elements of q*(ε) and Q*(ε)

The elements of the vector **q***(ε) and **Q***(ε)have the following representations:

$$q_1^*(\varepsilon) = (\varepsilon_s^2/\bar{\Delta})(\partial \ell_v/\partial\sigma_s^2), \tag{8A.1}$$

$$q_2^*(\varepsilon) = -(1/2)\{\sigma_e^2[r+f-(\gamma+\bar{\Delta}^{-1})\sum_i \Delta_i/(1+\gamma\Delta_i)]$$

$$-S_e-\sum_i \tilde{t}_i^2/(1+\gamma\Delta_i)+(\gamma+\bar{\Delta}^{-1})\sum_i \Delta_i\tilde{t}_i^2/(1+\gamma\Delta_i)^2\}, \tag{8A.2}$$

$$\frac{\partial q_1^*}{\partial\varepsilon_s} = (\varepsilon_s/\bar{\Delta})\{(\varepsilon_s/\bar{\Delta})[\partial^2\ell_v/\partial(\sigma_s^2)^2]+2(\partial\ell_v/\partial\sigma_s^2)\},$$

$$\frac{\partial q_2^*}{\partial\varepsilon_s} = -(1/\bar{\Delta})[(\gamma+\bar{\Delta}^{-1})(\partial^2\ell/\partial\gamma^2)-(1/2)\sum_i \Delta_i/(1+\gamma\Delta_i)$$

$$+(\sigma_e^2)^{-1}\sum_i \Delta_i\tilde{t}_i^2/(1+\gamma\Delta_i)^2],$$

$$\frac{\partial q_1^*}{\partial\varepsilon_e} = (1+\gamma\bar{\Delta})^2(\partial q_2^*/\partial\varepsilon_s),$$

$$\frac{\partial q_2^*}{\partial\varepsilon_e} = -(1/2)(\sigma_e^2)^{-1}\{\sigma_e^2[r+f-(\gamma+\bar{\Delta}^{-1})^2\sum_i \Delta_i^2/(1+\gamma\Delta_i)^2]$$

$$-2(\gamma+\bar{\Delta}^{-1})[\sum_i \Delta_i\tilde{t}_i^2/(1+\gamma\Delta_i)^2$$

$$-(\gamma+\bar{\Delta}^{-1})\sum_i \Delta_i^2\tilde{t}_i^2/(1+\gamma\Delta_i)^3]\}.$$

8.A.3 Some Identities

With the substitution of $\mathbf{Z}^{@}$ for $\mathbf{Z}$, S_s is replaced by $S^{@}_{s}=\tilde{\mathbf{s}}^{@\prime}\mathbf{q}^{@}$, where $\tilde{\mathbf{s}}^{@}$ is the solution to $\mathbf{C}^{@}\tilde{\mathbf{s}}^{@}=\mathbf{q}^{@}$. Similarly, κ, f, S_e, F, and $\mathbf{u}$ are replaced by $\kappa^{@}=(1/r^{@})\mathrm{tr}(\mathbf{C}^{@})$, $f^{@}=n-\mathrm{rank}(\mathbf{X},\mathbf{Z}^{@})$, $S^{@}_{e}=\mathbf{y}'\mathbf{y}-S_{\beta}-S^{@}_{s}$, $F^{@}=(f^{@}/r^{@})(S^{@}_{s}/S^{@}_{e})$, and $\mathbf{u}^{@}=(\mathbf{I}+\gamma\mathbf{C}^{@})^{-1}\mathbf{q}^{@}$, respectively.

We find that $r^{@}=\mathrm{rank}(\mathbf{X},\mathbf{ZA})-\mathrm{rank}(\mathbf{X})$, $f^{@}=n-\mathrm{rank}(\mathbf{X},\mathbf{ZA})$, $\mathrm{tr}(\mathbf{C}^{@})=\mathrm{tr}(\mathbf{CA})$, and $S^{@}=\tilde{\mathbf{s}}^{@\prime}_{1}\mathbf{Aq}$, where $\tilde{\mathbf{s}}^{@}_{1}$ is any solution to $\mathbf{ACA}\tilde{\mathbf{s}}^{@}_{1}=\mathbf{Aq}$. Moreover,

$$\det(\mathbf{I}+\gamma\mathbf{C}^{@}) = \det(\mathbf{I}+\gamma\mathbf{CA}),$$

$$(\mathbf{I}+\gamma\mathbf{C}^{@})^{-1} = \mathbf{I}-\gamma\mathbf{L}'(\mathbf{I}+\gamma\mathbf{CA})^{-1}\mathbf{CL},$$

$$\gamma(\mathbf{I}+\gamma\mathbf{CA})^{-1}\mathbf{CA} = \mathbf{I}-(\mathbf{I}+\gamma\mathbf{CA})^{-1} = \gamma\mathbf{CA}(\mathbf{I}+\gamma\mathbf{CA})^{-1}$$

(e.g., Searle 1982, Chap. 10), implying in particular that

$$(\mathbf{I}+\gamma\mathbf{C}^{@})^{-1}\mathbf{L}' = \mathbf{L}'[\mathbf{I}-\gamma(\mathbf{I}+\gamma\mathbf{CA})^{-1}\mathbf{CA}] = \mathbf{L}'(\mathbf{I}+\gamma\mathbf{CA})^{-1}$$

and hence that

$$\mathbf{u}^{@} = (\mathbf{I}+\gamma\mathbf{C}^{@})^{-1}\mathbf{L}'\mathbf{q} = \mathbf{L}'\mathbf{u}^{@}_{1},$$

where $\mathbf{u}^{@}_{1}=(\mathbf{I}+\gamma\mathbf{CA})^{-1}\mathbf{q}$. Consequently,

$$\mathrm{tr}[(\mathbf{I}+\gamma\mathbf{C}^{@})^{-1}\mathbf{C}^{@}] = \mathrm{tr}[(\mathbf{I}+\gamma\mathbf{CA})^{-1}\mathbf{CA}],$$

$$\gamma\mathrm{tr}[(\mathbf{I}+\gamma\mathbf{CA})^{-1}\mathbf{CA}] = m-\mathrm{tr}[(\mathbf{I}+\gamma\mathbf{CA})^{-1}],$$

$$\mathrm{tr}\{[(\mathbf{I}+\gamma\mathbf{C}^{@})^{-1}\mathbf{C}^{@}]^{2}\} = \mathrm{tr}\{[(\mathbf{I}+\gamma\mathbf{CA})^{-1}\mathbf{CA}]^{2}\},$$

$$\gamma^{2}\mathrm{tr}\{[(\mathbf{I}+\gamma\mathbf{CA})^{-1}\mathbf{CA}]^{2}\} = m-2\mathrm{tr}[(\mathbf{I}+\gamma\mathbf{CA})^{-1}]+\mathrm{tr}[(\mathbf{I}+\gamma\mathbf{CA})^{-2}],$$

$$\mathbf{u}^{@\prime}\tilde{\mathbf{s}}^{@} = \mathbf{u}^{@\prime}_{1}\mathbf{L}\tilde{\mathbf{s}}^{@} = \mathbf{u}^{@\prime}_{1}\mathbf{AA}^{-}\mathbf{L}\tilde{\mathbf{s}}^{@} = \mathbf{u}^{@\prime}_{1}\mathbf{A}\tilde{\mathbf{s}}^{@}_{1},$$

$$\mathbf{u}^{@\prime}\mathbf{u}^{@} = \mathbf{u}^{@\prime}_{1}\mathbf{A}\mathbf{u}^{@}_{1},$$

$$\mathbf{u}^{@\prime}(\mathbf{I}+\gamma\mathbf{C}^{@})^{-1}\mathbf{C}^{@}\mathbf{u}^{@} = \mathbf{u}^{@\prime}_{1}\mathbf{A}(\mathbf{I}+\gamma\mathbf{CA})^{-1}\mathbf{CA}\mathbf{u}^{@}_{1},$$

$$\gamma\mathbf{u}^{@\prime}_{1}\mathbf{A}(\mathbf{I}+\gamma\mathbf{CA})^{-1}\mathbf{CA}\mathbf{u}^{@}_{1} = \mathbf{u}^{@\prime}_{1}\mathbf{A}\mathbf{u}^{@}_{1}-\mathbf{u}^{@\prime}_{1}\mathbf{A}(\mathbf{I}+\gamma\mathbf{CA})^{-1}\mathbf{u}^{@}_{1}.$$

In the special case where rank(A)=m, we have that $r^@=r$, $f^@=f$, $S^@_s=S_s$,

$$\det(\mathbf{I}+\gamma\mathbf{C}^@) = \det(\mathbf{A})\cdot\det(\mathbf{A}^{-1}+\gamma\mathbf{C}),$$

$$\mathrm{tr}[(\mathbf{I}+\gamma\mathbf{C}^@)^{-1}\mathbf{C}^@] = \mathrm{tr}[(\mathbf{A}^{-1}+\gamma\mathbf{C})^{-1}\mathbf{C}],$$

$$\gamma\mathrm{tr}[(\mathbf{A}^{-1}+\gamma\mathbf{C})^{-1}\mathbf{C}] = m-\mathrm{tr}[\mathbf{A}^{-1}(\mathbf{A}^{-1}+\gamma\mathbf{C})^{-1}],$$

$$\mathrm{tr}\{[(\mathbf{I}+\gamma\mathbf{C}^@)^{-1}\mathbf{C}^@]^2\} = \mathrm{tr}\{[(\mathbf{A}^{-1}+\gamma\mathbf{C})^{-1}\mathbf{C}]^2\},$$

$$\gamma^2\mathrm{tr}\{[(\mathbf{A}^{-1}+\gamma\mathbf{C})^{-1}\mathbf{C}]^2\} = m-2\mathrm{tr}[\mathbf{A}^{-1}(\mathbf{A}^{-1}+\gamma\mathbf{C})^{-1}]+\mathrm{tr}\{[\mathbf{A}^{-1}(\mathbf{A}^{-1}+\gamma\mathbf{C})^{-1}]^2\},$$

Letting $\mathbf{u}^@_2=(\mathbf{A}^{-1}+\gamma\mathbf{C})^{-1}\mathbf{q}$, we also have (in this special case) that

$$\mathbf{u}^{@\prime}\tilde{\mathbf{s}}^@ = \mathbf{u}^@_2\mathbf{A}^{-1}\tilde{\mathbf{s}},\quad \mathbf{u}^{@\prime}\mathbf{u}^@ = \mathbf{u}^{@\prime}_2\mathbf{A}^{-1}\mathbf{u}^@_2,$$

$$\mathbf{u}^{@\prime}(\mathbf{I}+\gamma\mathbf{C}^@)^{-1}\mathbf{C}^@\mathbf{u}^@ = \mathbf{u}^{@\prime}_2\mathbf{A}^{-1}(\mathbf{A}^{-1}+\gamma\mathbf{C})^{-1}\mathbf{C}\mathbf{u}^@_2$$

$$\gamma\mathbf{u}^{@\prime}_2\mathbf{A}^{-1}(\mathbf{A}^{-1}+\gamma\mathbf{C})^{-1}\mathbf{C}\mathbf{u}^@_2 = \mathbf{u}^{@\prime}_2\mathbf{A}^{-1}\mathbf{u}^@_2-\mathbf{u}^{@\prime}_2\mathbf{A}^{-1}(\mathbf{A}^{-1}+\gamma\mathbf{C})^{-1}\mathbf{A}^{-1}\mathbf{u}^@_2.$$

9 Parameter Estimation in Variance Component Models for Binary Response Data

M.W. Knuiman and N.M. Laird[1]

Mixed model analysis of categorical response data is an important but difficult analytical problem. This paper first reviews mixed model analysis from a Bayesian perspective in the linear model setting, and then shows how the general Bayes approach can be extended to categorical data. Some limitations of the Bayes approach in this setting are discussed. Both maximum and quasi-likelihood approaches in the categorical data case are also described.

9.1 Introduction

The analysis of mixed models is fundamental to applied research in many areas. In mixed models, we can distinguish three types of parameters of interest: fixed effects, random effects, and variance components. Inferences about all three types of parameters are generally of interest in animal breeding research. The traditional ANOVA approach to inference in the mixed model is piecemeal, since it requires that distinctions be made about types of parameters. As pointed out in Laird and Ware (1982) and Gianola and Fernando (1986), a Bayes approach offers a unified approach to inference in the mixed model, and in the linear setting with vague priors, leads to similar estimators and confidence statements as the ANOVA approach.

This paper will focus on the estimation in mixed models with categorical response variables. We will first briefly review the linear model case with measured responses, since much of the work on categorical responses draws heavily on the linear model analysis. We then discuss Bayes approaches to the analysis of mixed models with binary responses. Similarities and important differences in the two cases are highlighted. Following this we discuss some non-Bayes approaches to the mixed model analysis for binary response data.

[1] Department of Biostatistics, Harvard School of Public Health, Boston, Massachusetts, USA

9.2 Review of the Linear Case

We assume the Nx1 data vector of responses can be represented as

$$\mathbf{y} = \mathbf{X}\alpha+\mathbf{Z}\mathbf{b}+\mathbf{e} \tag{9.1}$$

where α is a px1 vector of fixed effects, $\mathbf{b}$ is a qx1 vector of random effects, $\mathbf{X}$ and $\mathbf{Z}$ are known design matrices and $\mathbf{b}$ and $\mathbf{e}$ are independent, normally distributed random vectors with zero means and covariance matrices $\mathbf{D}$ and $\mathbf{R}$, respectively. Generally, $\mathbf{D}$ and $\mathbf{R}$ will be known functions of vectors of unknown components, i.e., $\mathbf{D}=\mathbf{D}(\theta_2)$ and $\mathbf{R}=\mathbf{R}(\theta_1)$ where $\theta^T=(\theta_1^T, \theta_2^T)$ are the unknown variance and covariance components.

Classical ANOVA and maximum likelihood approaches to inferences about α, $\mathbf{b}$, and θ have been discussed extensively in the literature; some key references include Henderson (1953), Harville (1976, 1977), Rao (1975), Patterson and Thompson (1971), and Thompson (1980). Maximum likelihood estimates of α and θ are obtained by maximizing the marginal likelihood of the data: $\mathbf{y}\sim \mathbf{N}(\mathbf{X}\alpha,\mathbf{V})$ where

$$\mathbf{V} = \mathbf{Z}\mathbf{D}\mathbf{Z}^T+\mathbf{R}. \tag{9.2}$$

Restricted maximum likelihood (REML) estimates of the variance components are usually preferred, as they adjust for lost degrees of freedom in estimating the fixed effects (Patterson and Thompson 1971; Harville 1977). Estimates of the random effects are obtained by an extension of best linear unbiased estimation (BLUE) to include random effects (Rao 1975; Harville 1976). Formulae for the variances of $\hat{\alpha}$ and $\hat{\mathbf{b}}$ are derived by assuming θ is known, then substituting an estimate for θ. In the linear case, this yields asymptotically valid estimates for var($\hat{\alpha}$), since $\hat{\alpha}$ is asymptotically uncorrelated with $\hat{\theta}$.

The formulae for ML estimates of α, and "extended" BLUE estimates of $\mathbf{b}$, can be conveniently obtained as the solution to:

$$\begin{bmatrix} \mathbf{X}^T\mathbf{R}^{-1}\mathbf{X} & \mathbf{X}^T\mathbf{R}^{-1}\mathbf{Z} \\ \mathbf{Z}^T\mathbf{R}^{-1}\mathbf{X} & \mathbf{Z}^T\mathbf{R}^{-1}\mathbf{Z}+\mathbf{D}^{-1} \end{bmatrix} \begin{bmatrix} \hat{\alpha} \\ \hat{\mathbf{b}} \end{bmatrix} = \begin{bmatrix} \mathbf{X}^T\mathbf{R}^{-1}\mathbf{y} \\ \mathbf{Z}^T\mathbf{R}^{-1}\mathbf{y} \end{bmatrix}. \tag{9.3}$$

The asymptotic variances of $\hat{\alpha}$ and $\hat{\mathbf{b}}$-$\mathbf{b}$ are obtained from the inverse of $\mathbf{M}$, where $\mathbf{M}$ is the matrix on the left-hand side of (9.3) (Harville 1976). In addition, the REML estimates of θ can also be linked to the solution of (9.3), as shown by Thompson (1980) for the case where $\mathbf{R}=\sigma^2\mathbf{I}$ and $\mathbf{D}=\sigma_b^2\mathbf{I}$. The result is easiest to see for a general model by deriving the REML estimates as Bayes estimates; thus, we now turn to inferences in the Bayes setting.

The Bayes approach for the mixed model, as outlined in Lindley and Smith (1972), is to formulate the model in stages. The data model is given by the multivariate normal density $f(\mathbf{y}|\alpha,\mathbf{b},\theta_1)$. The prior for $\mathbf{b}$ is the multivariate normal $f(\mathbf{b}|\theta_2)$; α, θ_1, and θ_2 are assigned independent prior distributions, which we will take as improper [proportional to $(d\alpha,d\theta_1,d\theta_2)$]. Inferences about the separate parameters $(\alpha,\mathbf{b},\theta)$ are preferably based on their marginal posterior distributions: $p(\alpha|\mathbf{y})$, $p(\mathbf{b}|\mathbf{y})$, and $p(\theta|\mathbf{y})$. Even in the normal linear model setting, these marginal posteriors do not have closed forms and, thus, require approximation. The usual compromise is to base inferences about α and $\mathbf{b}$ on $p(\alpha|\theta,\mathbf{y})$ and $p(\mathbf{b}|\theta,\mathbf{y})$, respectively, which are multivariate normal, as is $p(\alpha,\mathbf{b}|\theta,\mathbf{y})$, and inferences about θ on $p(\theta|\mathbf{y})$. However, the posterior mean of $(\alpha,\mathbf{b}|\mathbf{y},\theta)$ is given by the solution to (9.3), and their posterior variance is $\mathbf{M}^{-1}$. Since $p(\alpha,\mathbf{b}|\theta,\mathbf{y})$ is multivariate normal in this case, the joint moments of $(\alpha,\mathbf{b})$ are the same as the marginal moments of $p(\alpha|\theta,\mathbf{y})$ and $p(\mathbf{b}|\theta,\mathbf{y})$. Hence, under the flat prior on α, the Bayes and ANOVA estimates of $(\alpha,\mathbf{b})$ and their estimated variances are identical. As noted in Laird and Ware (1982), $\hat{\mathbf{b}}$ is also empirical Bayes, since it can be expressed as $E(\mathbf{b}|\mathbf{y},\alpha,\theta)$, with α and θ evaluated at their ML estimate.

Since $p(\theta|\mathbf{y})$ has no closed form expression, the mode is used as an approximation for the mean. However, Harville (1974) has shown that the REML likelihood is proportional to $p(\theta|\mathbf{y})$; hence, the Bayes and likelihood approaches to inferences about θ are also equivalent.

To see how the equations for the REML estimates of θ relate to the solution of (9.3), consider the EM approach to estimating variance components outlined in Dempster et al. (1977) and Laird (1978). The EM approach is to regard $(\mathbf{y},\mathbf{b},\mathbf{e})$ as complete data. With complete data, the sufficient statistics for θ will be quadratic forms in $\mathbf{e}$ and $\mathbf{b}$; the exact expressions for these quadratic forms depend upon the parameterization of $\mathbf{R}(\theta_1)$ and $\mathbf{D}(\theta_2)$. Denote these vectors of sufficient statistics for θ_1 and θ_2 as $\mathbf{S}_1$ and $\mathbf{S}_2$, where the dimensions of $\mathbf{S}_1$ and $\mathbf{S}_2$ are equal to the dimensions of θ_1 and θ_2, respectively. Then with complete data we solve:

$$E(\mathbf{S}_1) = \mathbf{S}_1 \text{ and } E(\mathbf{S}_2) = \mathbf{S}_2. \tag{9.4}$$

Assuming the components of $\mathbf{S}_1$ and $\mathbf{S}_2$ are quadratic forms in $\mathbf{b}$ and $\mathbf{e}$, we solve equations of the form

$$\text{tr } \mathbf{R}(\theta_1)\mathbf{R}_k = \mathbf{e}^T\mathbf{R}_k\mathbf{e} \qquad k = 1,\ldots,r \tag{9.5}$$

$$\text{tr } \mathbf{D}(\theta_2)\mathbf{D}_j = \mathbf{b}^T\mathbf{D}_j\mathbf{b}, \qquad j = 1,\ldots,d,$$

where r and d are the dimensions of θ_1 and θ_2, respectively, and $\mathbf{R}_k$ and $\mathbf{D}_j$ define the appropriate quadratic forms. Note that if $\mathbf{R}^{-1}$ and $\mathbf{D}^{-1}$ have linear structure, then

$$\mathbf{R}^{-1}(\theta_1) = \sum_{k=1}^{r} \gamma_k \mathbf{R}_k$$

$$\mathbf{D}^{-1}(\theta_2) = \sum_{j=1}^{d} \psi_j \mathbf{D}_j$$

for appropriate $\mathbf{R}_k$'s and $\mathbf{D}_j$'s, where γ and ψ are one-to-one transformations of θ_1 and θ_2, and the representation in (9.5) holds.

As Laird (1978) shows, to obtain ML estimates of θ, we solve

$$E(\mathbf{S}_1|\theta_1) = E(\mathbf{S}_1|\mathbf{y},\alpha,\theta)$$

$$E(\mathbf{S}_2|\theta_2) = E(\mathbf{S}_2|\mathbf{y},\alpha,\theta),$$

whereas for REML, we solve

$$E(\mathbf{S}_1|\theta_1) = E(\mathbf{S}_1|\mathbf{y},\theta)$$

$$E(\mathbf{S}_2|\theta_2) = E(\mathbf{S}_2|\mathbf{y},\theta). \tag{9.6}$$

However, since $\mathbf{S}_2$ is composed of quadratic forms in $\mathbf{b}$, $\mathbf{b}^T\mathbf{D}_j\mathbf{b}$, this means we set

$$\mathrm{tr}(\hat{\mathbf{D}}\mathbf{D}_j) = \hat{\mathbf{b}}^T\mathbf{D}_j\hat{\mathbf{b}}+\mathrm{tr}(\hat{\Gamma}\mathbf{D}_j), \tag{9.7}$$

where $\hat{\mathbf{b}}$ and $\hat{\Gamma}$ are the posterior mean and variance of $\mathbf{b}$ given $\mathbf{y}$ and θ. But we have already seen that $\hat{\mathbf{b}}$ and $\hat{\Gamma}$ are given by the solution of (9.3), $\hat{\mathbf{b}}$ being the estimate of $\mathbf{b}$, and $\hat{\Gamma}$ being the lower block diagonal of $\mathbf{M}^{-1}$. For $\mathbf{S}_1$, we must solve equations of the form

$$\mathrm{tr}(\hat{\mathbf{R}}\mathbf{R}_k) = \hat{\mathbf{e}}^T\mathbf{R}_k\hat{\mathbf{e}}+\mathrm{tr}(\hat{\Omega}\mathbf{R}_j),$$

where $\hat{\mathbf{e}}$ and $\hat{\Omega}$ are the marginal posterior means and variances of $\mathbf{e}$ given $\mathbf{y}$ and θ. Since

$$\mathbf{y} = \mathbf{X}\alpha+\mathbf{Z}\mathbf{b}+\mathbf{e},$$

it follows that

$$\hat{\mathbf{e}} = \mathbf{y} - \mathbf{X}\hat{\alpha}-\mathbf{Z}\hat{\mathbf{b}},$$

and

$$\mathrm{var}(\hat{\mathbf{e}}|\mathbf{y},\theta) = [\mathbf{X}\ \mathbf{Z}]\mathbf{M}^{-1}\begin{bmatrix}\mathbf{X}^T \\ \\ \mathbf{Z}^T\end{bmatrix}.$$

This shows that the solution of (9.3) can be used to compute REML estimates of θ in the general setting, by equating the conditional and unconditional expectations of the appropriate quadratic forms in **b** and **e**.

9.3 Mixed Model Analysis with Binary Response

In many important applications, the observed responses are categorical. For simplicity, we deal with binary responses. Following the linear mixed model, a general approach is to assume $E(y_i)=p_i$ and

$$\eta = \mathbf{X}\alpha+\mathbf{Z}\mathbf{b},$$

where $\eta_i=h(p_i)$ is some function of p_i (called the link function in generalized linear models terminology) and **X**, **Z**, α, and **b** are as in the preceding section. The link function is usually chosen to be either the logit or probit function. Notice that we no longer need the within-unit variance matrix $\mathbf{R}(\theta_1)$. In this section θ consists only of θ_2, the components of **D**.

In principle the Bayes approach can be extended in a straightforward manner, by assigning vague priors to α and θ, and basing inferences on the posterior distributions of α, **b**, and θ. Technically, this is usually quite infeasible to implement, due to the fact that the posterior distributions do not have closed forms, even in the simplest settings. For basically the same reason, the maximum likelihood approach to estimation of α and θ requires numerical integration, and is often also quite infeasible. Section 9.3.1 discusses an approximate Bayes approach and 9.3.2 discusses likelihood approaches.

9.3.1 Bayes Approach

Due to the technical infeasibility of a true Bayes approach the following approximate Bayes approach is often used (Gianola and Foulley 1983; Stiratelli et al. 1984; Harville and Mee 1984). Inferences about α and **b** are based on their posterior distribution, conditional on θ. Unlike the linear model setting, this posterior is not multivariate normal, nor does it have a mathematically tractable form. The posterior mode is used to approximate the posterior mean, and the inverse second derivative matrix is used to approximate the posterior variance-covariance matrix. Because of non-normality of the joint posterior for $(\alpha,\mathbf{b})$, these are no longer marginal modes. Since these approximations for the posterior moments depend upon θ, an estimate of θ is substituted, as in the normal case. The exact forms for these posterior moments depend upon the specific link function used. However, as anticipated by Thompson (1980), the general forms can be related to a modified version of equations (9.3), as follows.

In the "fixed effects" case, where the **b**'s are treated as fixed parameters, if each y_i is an independent binomial $(1,p_i)$, then Nelder and Wedderburn (1972) show that the solution for the ML estimates of $(\alpha,\mathbf{b})$ is obtained by iteratively solving:

$$\begin{bmatrix} \mathbf{X}^T\mathbf{W}\mathbf{X} & \mathbf{X}^T\mathbf{W}\mathbf{Z} \\ \mathbf{Z}^T\mathbf{W}\mathbf{X} & \mathbf{Z}^T\mathbf{W}\mathbf{Z} \end{bmatrix} \begin{bmatrix} \hat{\alpha} \\ \hat{\mathbf{b}} \end{bmatrix} = \begin{bmatrix} \mathbf{X}^T\mathbf{W}\mathbf{y}^* \\ \mathbf{Z}^T\mathbf{W}\mathbf{y}^* \end{bmatrix}, \tag{9.8}$$

where

$$\mathbf{y}^* = \eta + \mathbf{U}(\mathbf{y}-\mathbf{p}),$$

U is a diagonal matrix with elements

$$u_i = (\partial\eta_i/\partial p_i)$$

and **W** is a diagonal matrix with elements

$$W_i = (\partial p_i/\partial\eta_i)^2/p_i(1-p_i).$$

For the logit link, we have

$$p_i = e^{\eta_i}/(1+e^{\eta_i})$$

hence

$$W_i = p_i(1-p_i)$$

and

$$u_i = 1/W_i,$$

For the probit link

$$p_i = \Phi(\eta_i),$$

$$W_i = \phi(\eta_i)^2/\{\Phi(\eta_i)[1-\Phi(\eta_i)]\}$$

and

$$u_i = 1/\phi(\eta_i),$$

where Φ and ϕ are the distribution and density functions for a standard normal variate.

Gianola and Foulley (1983) have shown that, as in the linear models case, if we modify Eq. (9.8) to:

$$\begin{bmatrix} \mathbf{X}^T\mathbf{W}\mathbf{X} & \mathbf{X}^T\mathbf{W}\mathbf{Z} \\ \mathbf{Z}^T\mathbf{W}\mathbf{Z} & \mathbf{Z}^T\mathbf{W}\mathbf{Z}+\mathbf{D}^{-1} \end{bmatrix} \begin{bmatrix} \hat{\alpha} \\ \hat{\mathbf{b}} \end{bmatrix} = \begin{bmatrix} \mathbf{X}^T\mathbf{W}\mathbf{y}^* \\ \mathbf{Z}^T\mathbf{W}\mathbf{y}^* \end{bmatrix}, \tag{9.9}$$

then solving (9.9) yields the joint posterior mode for $(\alpha,\mathbf{b})$ given $\mathbf{y}$ and θ when the link function is probit. Since $\mathbf{W}$ is a function of $(\alpha,\mathbf{b})$, these equations must be solved iteratively as

$$\mathbf{A}^{(p-1)} \begin{bmatrix} \alpha \\ \mathbf{b} \end{bmatrix}^{(p)} = \mathbf{B}^{(p-1)}$$

for p=1,2,..., until convergence.

It is easily seen that a similar expression holds in the logit model, by noting that for the logit link, the log of the joint posterior density can be written as:

$$L(\alpha,\mathbf{b}) = \sum_i [y_i\,\eta_i - \ell n(1-p_i)] - \mathbf{b}^T\mathbf{D}^{-1}\mathbf{b}/2,$$

hence its first and second derivatives are

$$\frac{\partial L}{\partial(\alpha,\mathbf{b})} = \begin{bmatrix} \mathbf{X}^T(\mathbf{y}-\mathbf{p}) \\ \mathbf{Z}^T(\mathbf{y}-\mathbf{p})-\mathbf{D}^{-1}\mathbf{b} \end{bmatrix}$$

and

$$\frac{\partial^2 L}{\partial^2(\alpha,\mathbf{b})} = -\begin{bmatrix} \mathbf{X}^T\mathbf{W}\mathbf{X} & \mathbf{X}^T\mathbf{W}\mathbf{Z} \\ \mathbf{Z}^T\mathbf{W}\mathbf{X} & \mathbf{Z}^T\mathbf{W}\mathbf{Z}+\mathbf{D}^{-1} \end{bmatrix} = -\mathbf{A},$$

where $\mathbf{W}$ is as defined above for the logit link. Hence using Newton's method for finding the mode yields Eqs. (9.9) at convergence. This is easily generalized for both the probit and logit models to the case where y_i is binomial (n_i,p_i) with $E(y_i)= n_ip_i=\mu_i$. In this case W_i is replaced by $n_ip_i(1-p_i)$ and $(\mathbf{y}-\mathbf{p})$ is replaced by $(\mathbf{y}-\mu)$.

When θ is unknown, an estimate is substituted. Again arguing by analogy to the linear case, the suggestion has been made to estimate θ by its marginal posterior mode, since in the linear models case this gives the preferred REML estimate of θ. To obtain the mode we must maximize

$$L(\theta) = \ell n \int_{R^{p+q}} f(\mathbf{y}|\alpha,\mathbf{b})f(\mathbf{b}|\mathbf{D})d\alpha d\mathbf{b}. \tag{9.10}$$

Unlike the normal case, this integral is not guaranteed to exist for all $\mathbf{y}$. We conjecture that it will whenever the MLE of α exists in the case where $\mathbf{b}=0$. Even though Eq. (9.10) has no closed form, we can utilize the EM approach to develop an expression for the likelihood equations as before. By regarding $(\alpha,\mathbf{b},\mathbf{y})$ as complete data, the likelihood equations for θ are easily seen to reduce to solving exactly the same equations as in the normal case (9.6), since with complete data, θ is estimated from $\mathbf{b}$ alone. If we assume that the posterior of $(\alpha,\mathbf{b})$ is approximately normal with mean and covariance matrix given by the solution of (9.9), then the REML equations for the binary case have exactly the same form as in the normal case (9.7). This approach to obtaining approximate REML estimates was used in Laird (1978), Stiratelli et al. (1984), and Harville and Mee (1984). Drawbacks in the categorical case are:

1. Given a value of θ, Eqs. (9.9) must be solved iteratively because $\mathbf{W}$ is a function of $(\alpha,\mathbf{b})$. Thus this REML implementation requires an iteration within each E-step of the EM algorithm. However, if we use good initial values for $(\hat{\alpha},\hat{\mathbf{b}})$, it may be necessary to take only one Newton's step rather than solving (9.9). Both Laird (1978) and Harville and Mee (1984) report slow convergence, which is a general characteristic of the EM algorithm for variance component problems (Laird et al. 1987).
2. The assumption that the marginal posterior expectation of $\mathbf{b}\mathbf{b}^T$ is well approximated by $\hat{\mathbf{b}}\hat{\mathbf{b}}^T+\hat{\Gamma}$, where $\hat{\mathbf{b}}$ is the solution of (9.9) and $\hat{\Gamma}$ is the lower corner of $\hat{\mathbf{A}}^{-1}$ at the solution, may not be very good. If we have a lot (at least ten) observations on each

component of **b**, the approximation may be appropriate, but in many cases the approximation may be quite poor, as was found in Laird et al. (1984). In such cases, better approximations for the posterior moments will be necessary. For this same reason, the solution of Eqs. (9.9) may not yield very good approximations to the marginal posterior moments of α and **b**. More research is needed before these approximations can safely be adopted.

3. In the normal setting, it is easily seen that the form of the asymptotic variance of $\hat{\alpha}$ is the same whether or not we assume θ is known. This has not been demonstrated to hold for the categorical data setting, thus $\hat{\mathbf{A}}^{-1}$ may not be a valid estimate of the variance-covariance matrix in this case (Zeger et al. 1986).

9.3.2 Likelihood Approaches

Both full and quasi-likelihood approaches to the analysis of mixed models for binary data have been suggested. Gilmour et al. (1985) develop an approximate quasi-likelihood approach for mixed probit models, and Mauritsen (1984) and Anderson and Aitkin (1985) discuss full likelihood approaches for the mixed logit model.

A quasi-likelihood approach is possible if the variance may be expressed as a function of the mean (McCullagh and Nelder 1983). For the mixed probit model the marginal (integrated over the distribution of **b**) mean of y_i (given α and θ) is easily seen to be $\Phi(s_i x_i^T \alpha)$, where $s_i=(1+\mathbf{z}_i^T \mathbf{D} \mathbf{z}_i)^{-1/2}$, $\mathbf{x}_i^T$ is the *ith* row of **X**, and $\mathbf{z}_i^T$ is the ith row of **Z**. The marginal variance has the form

$$\mathrm{var}(\mathbf{y}) = \mathrm{var}(\mathbf{p})+\mathrm{Diag}\ \{E[p_i(1-p_i)]\},$$

where $\mathrm{Diag}(a_i)$ is a diagonal matrix with elements a_i, and the expectation and variance are over the distribution of **b**. Although a convenient closed-form expression for var(**y**) is not available, an approximation based on a linear Taylor expansion for $\Phi(\mathbf{x}_i^T\alpha+\mathbf{z}_i^T\mathbf{b})$ about $s_i\mathbf{x}_i^T\alpha$ has the form $\Delta\mathbf{ZDZ}^T\Delta+\mathbf{V}$, where $\Delta=\mathrm{Diag}[\phi(s_i\mathbf{x}_i^T\alpha)]$ and $\mathbf{V}=\mathrm{Diag}\{\Phi(s_i\mathbf{x}_i^T\alpha)[1-\Phi(s_i\mathbf{x}_i^T\alpha)]-\phi^2(s_i\mathbf{x}_i^T\alpha)\mathbf{z}_i^T\mathbf{D}z_i\}$. Since this variance expression is a function of the mean $\Phi(s_i\mathbf{x}_i^T\alpha)$, a quasi-likelihood in α corresponding to this mean-variance relationship may be constructed. The Newton equation for maximizing this quasi-likelihood has the form

$$\mathbf{X}^T\mathbf{SUSX}\alpha = \mathbf{X}^T\mathbf{SU}\mathbf{y}^*,$$

where

$$\mathbf{U} = \Delta[\Delta\mathbf{ZDZ}^T\Delta+\mathbf{V}]^{-1}\Delta,$$

$$\mathbf{y}^* = \mathbf{SX}\alpha+\Delta^{-1}[\mathbf{y}-\Phi(\mathbf{SX}\alpha)]$$

and $S=\text{Diag}(s_i)$ (McCullagh and Nelder 1983). The solution of these equations may also be obtained from solving a set of equations similar to (9.9), namely

$$\begin{bmatrix} \mathbf{X}^T\mathbf{S}\Delta\mathbf{V}^{-1}\Delta\mathbf{S}\mathbf{X} & \mathbf{X}^T\mathbf{S}\Delta\mathbf{V}^{-1}\Delta\mathbf{Z} \\ \mathbf{Z}^T\Delta\mathbf{V}^{-1}\Delta\mathbf{S}\mathbf{X} & \mathbf{Z}^T\Delta\mathbf{V}^{-1}\Delta\mathbf{Z}+\mathbf{D}^{-1} \end{bmatrix} \begin{bmatrix} \hat{\alpha} \\ \hat{\mathbf{b}} \end{bmatrix} = \begin{bmatrix} \mathbf{X}^T\mathbf{S}\Delta\mathbf{V}^{-1}\Delta\mathbf{y}^* \\ \mathbf{Z}^T\Delta\mathbf{V}^{-1}\Delta\mathbf{y}^* \end{bmatrix}.$$

These equations imply an estimate for $\mathbf{b}$ and $\text{var}(\hat{\mathbf{b}}-\mathbf{b})$ which has no apparent justification.

For the special case where $\mathbf{D}=\sigma^2\mathbf{A}$ where $\mathbf{A}$ is a known matrix, Gilmour et al. (1985) recommend an estimator of σ^2 obtained from equating $\mathbf{b}^T\mathbf{A}^{-1}\mathbf{b}$ to its expectation. The EM approach to variance component estimation outlined in Section 9.2 reveals this to have the form of the REML estimator, and the estimating equation derived by Gilmour et al. (1985) coincides with expression (9.7).

These estimating equations for α have the form of generalized estimating equations (Liang and Zeger 1986); Zeger et al. (1986) show that although an approximation to $\text{var}(\mathbf{y})$ is used, $\sqrt{N}(\hat{\alpha}-\alpha)$ is still asymptotically normal with mean zero. They also point out that estimates for α and θ are not even asymptotically uncorrelated and so one cannot obtain a consistent estimate of $\text{var}(\hat{\alpha})$ simply by substituting $\hat{\theta}$ for θ in the formula for $\text{var}(\hat{\alpha})$ assuming θ is known.

A similar development for the mixed logit model is not strictly possible since in this case the mean of y_i is not $h^{-1}(s_i\,\mathbf{x}_i^T\alpha)$ for some s_i where h(.) is the logit link. However, Zeger et al. (1986) show that such a relationship still holds approximately and the generalized estimating equation approach may be carried through.

A full likelihood approach to the estimation of α and θ requires the maximization of

$$L(\alpha,\theta) = \ell n \int_{R^q} f(\mathbf{y}|\alpha,\mathbf{b})f(\mathbf{b}|\theta)d\mathbf{b}$$

$$= \ell n \int_{R^q} \left\{\prod_i p_i^{y_i}(1-p_i)^{1-y_i}\right\} |\mathbf{D}|^{-1/2}\exp\left\{\frac{1}{2}\mathbf{b}^T\mathbf{D}^{-1}\mathbf{b}\right\} d\mathbf{b},$$

where $p_i=\Phi(\eta_i)$ in the probit case and $p_i=\exp(\eta_i)/[1+\exp(\eta_i)]$ in the logit case. As with (9.10), this integration cannot be carried out analytically. Numerical approaches are required to evaluate $L(\alpha,\theta)$ and its derivatives.

The special case of a mixed logit model with $\mathbf{D}=\mathbf{I}\sigma^2$ and z_i, an indicator function, has been considered by Mauritsen (1984), who extends the work of Pierce and Sands (1975), and by Anderson and Aitkin (1985). In this case the model may be re-expressed as y_{ij} being binomial $(1,p_{ij})$ with

$$\text{logit}(p_{ij}) = \mathbf{x}_{ij}^T\alpha+\sigma b_i \quad i = 1,\ldots,m; \; j = 1,\ldots,n_i$$

for some appropriate $\mathbf{x}_{ij}$, where b_i has a standard normal distribution. In the animal breeding context i might denote sire and j dam. Mauritsen (1984) considers Gauss-Hermite quadrature approximations to $L(\alpha,\theta)$ and its derivatives, whereas Anderson and Aitkin (1985) consider Gaussian quadrature. Anderson and Aitkin (1985) show that the approximate likelihood equations obtained in this way coincide with weighted logistic regression estimating equations for the model which uses $\bar{y}_{ijk}=y_{ij}$ and $\bar{y}_{ijk}$ is binomial $(1,\bar{p}_{ijk})$ with logit $(\bar{p}_{ijk})=\mathbf{x}_{ij}^T\alpha+\sigma S_k$, where $S_1,\ldots,S_K$ are the K quadrature points and the weights depend upon α and σ. Therefore, MLE's for (α,σ) may be obtained via iterative re-weighted logistic regression estimating equations using programs such as GLIM (Baker and Nelder 1978). Note that $\bar{\mathbf{y}}$ has length NK where $N = \sum_i n_i$ so this procedure becomes quite impractical for large K. Also, since the log-likelihood is not quadratic around $\hat{\sigma}$ the standard errors produced in this way might be misleading. The approach of Mauritsen (1984) also becomes impractical as the number of quadrature points becomes large. Ironically, a large number of quadrature points are needed to obtain a good approximation to $L(\alpha,\theta)$ and its derivatives.

Although brute-force numerical approaches to full maximum likelihood analysis may be feasible when the dimension of θ is 1 as in the case considered above, they are quite impractical in general and the Bayes approaches outlined in Section 9.3.1 remain the only viable approach for general models at this time.

Acknowledgements. Dr Laird was supported by grant GM29745 from the National Institutes of Health.

References

Anderson DA, Aitkin M (1985) Variance component models with binary response: interviewer variability. J R Stat Soc Ser B 47:203-210

Baker RJ, Nelder JA (1978) The GLIM system, release 3; generalised linear interactive modelling. R Stat Soc London

Dempster AP, Laird NM, Rubin DB (1977) Maximum likelihood from incomplete data via the EM algorithm. J R Stat Soc Ser B 39:1-38

Gianola D, Fernando RL (1986) Bayesian methods in animal breeding theory. J Anim Sci 63:217-244

Gianola D, Foulley JL (1983) Sire evaluation for ordered categorical data with a threshold model. Genet Sel Evol 15:201-223

Gilmour AR, Anderson RD, Rae AL (1985) The analysis of binomial data by a generalized linear mixed model. Biometrika 72:593-599

Harville DA (1974) Bayesian inference for variance components using only error contrasts. Biometrika 61:383-385

Harville DA (1976) Extension of the Gauss-Markov theorem to include the estimation of random effects. Ann Stat 4:384-395

Harville DA (1977) Maximum likelihood approaches to variance component estimation and to related problems. J Am Stat Assoc 72:320-340

Harville DA, Mee RW (1984) A mixed model procedure for analyzing ordered categorical data. Biometrics 40:393-408

Henderson CR (1953) Estimation of variance and covariance components. Biometrics 9:226-252

Laird NM (1978) Bayesian estimation methods for two-way contingency tables. Biometrika 65:581-590

Laird NM, Ware JH (1982) Random effects models for longitudinal data. Biometrics 38:963-974

Laird NM, Beck GJ, Ware JH (1984) Longitudinal data with binary response. Technical Report, Department of Biostatistics, Harvard School of Public Health, Boston, MA

Laird NM, Lange N, Stram DO (1987) Maximum likelihood computations with repeated measures: application of the EM algorithm. J Am Stat Assoc 82:97-105

Liang KY, Zeger SL (1986) Longitudinal data analysis using generalized linear models. Biometrika 73:13-22

Lindley DV, Smith AFM (1972) Bayes estimates for the linear model. J R Stat Soc, Ser B 34:1-41

Mauritsen RH (1984) Logistic regression with random effects. PhD Thesis, Univ Washington

McCullagh P, Nelder JA (1983) Generalized linear models. Chapman and Hall, London

Nelder JA, Wedderburn RWM (1972) Generalized linear models. J Roy Stat Soc Ser A 135:370-384

Patterson HD, Thompson R (1971) Recovery of inter-block information when block sizes are unequal. Biometrika 58:545-554

Pierce DA, Sands BR (1975) Extra-Bernoulli variation in binary data. Technical Report 46, Department of Statistics, Oregon State Univ, Corvallis, OR

Rao, CR (1975) Simultaneous estimation of parameters in different linear models and applications to biometric problems. Biometrics 31:545-554

Stiratelli R, Laird NM, Ware JH (1984) Random effects models for serial observations with dichotomous response. Biometrics 40:961-972

Thompson R (1980) Maximum likelihood estimation of variance components. Math Operationsforsch Stat Ser Stat 11:1-17

Zeger SL, Liang KY, Albert P (1986) Models for longitudinal data: generalized estimating equation approach. Technical Report. Dep Biostatistics, Johns Hopkins Univ

10 Estimation of Genetic Parameters in Non-Linear Models

S.P. Smith[1]

Various ways of linearizing a non-linear model are described. These procedures allow the estimation of variance components in non-linear models because standard techniques designed for linear models can be adapted. Details are given on how to estimate a sire variance in a paternal half-sib analysis. Results from applying the methods to an actual data set suggest that estimated variances may be insensitive to the way linearization is implemented.

10.1 Introduction

There are many ways genetic parameters or variance components can be estimated in linear models. Some common techniques are Henderson's (1953) methods, the restricted maximum likelihood (REML) of Patterson and Thompson (1971), and the minimum variance quadratic unbiased estimation of Rao (1971).

Random genetic effects can be incorporated in non-linear models as well, and methods for estimating variance components in non-linear models are generally extensions of procedures for linear models. Such methods are derived by linearizing a non-linear model in accordance with large sample theory. By definition then, options for variance component estimation in non-linear models are at least as numerous as for linear models. Because there are several ways to linearize a non-linear model, the number of ways to implement variance component estimation for non-linear models is further increased.

The purpose of this paper is to describe linearization of a general non-linear model, and to briefly consider an efficient numerical strategy to compute REML-type estimates in a particular model. The methods are applied to an actual data set, employing several linearizations.

[1] Animal Genetics and Breeding Unit, University of New England, Armidale, Australia

10.2 Models

Only a special class of non-linear models is considered in detail. This class is represented by the log-likelihood function (conditional on **b**)

$$L(\mathbf{Tb}), \tag{10.1}$$

where **T** is a known matrix and **b** is a vector of parameters. The quantity **Tb** defines a vector where its i-th element is generally associated with the i-th observation. For simplicity, the vector **Tb** is sometimes represented as **u**. The matrix **T** is a design matrix and **b** is a vector of location parameters. For example, (10.1) might represent the linear model

$$\mathbf{y} = \mathbf{Tb}+\mathbf{e}, \tag{10.2}$$

where **y** is an observation vector and **e** is a residual vector. To complete specification of the model, an assumption about the distribution of **e** is required. A typical assumption is that **e** is multivariate normal with $\mathrm{Var}\{\mathbf{e}\}=\sigma^2\mathbf{I}$. In this case the log-likelihood is a function of **b** and σ^2. Thus, (10.1) needs to be written as:

$$L(\mathbf{Tb}, \sigma^2) = \text{const.}-\log|\sigma\mathbf{I}|-(\mathbf{y}-\mathbf{Tb})'(\mathbf{y}-\mathbf{Tb})/(2\sigma^2).$$

A model like (10.2) sometimes applies to non-linear models. However, **y** is frequently assumed to be unobserved. For example, **y** could be a vector of underlying variables associated with binomial responses (Gianola and Foulley 1983). Alternatively, **y** could be a vector of transformed observations where the transformation(s) is unknown (Smith and Hammond 1988). When **y** is a vector of transformations the likelihood is constructed from rank information. For these cases σ^2 cannot be estimated. Indeed, σ^2 can be standardized with no loss in generality and it then suffices to work with (10.1).

Let

$$\mathbf{Tb} = \mathbf{X}\mathbf{b}_f+\mathbf{Z}\mathbf{b}_r\,,$$

where **X** and **Z** are known design matrices, $\mathbf{b}_f$ is a vector of fixed effects and $\mathbf{b}_r$ is a vector of multivariate normal random variables with null mean and covariance matrix **G**. Normality is an appropriate assumption given random genetic effects and polygenic inheritance.

The likelihood (10.1) does not represent all non-linear models. For example, (10.1) is not general enough to represent threshold models for multinomial responses because no provisions have been made for thresholds. Simple threshold models are described by Gianola and Foulley (1983) and Harville and Mee (1984). Complicated threshold systems are described in the methods section of Smith and Allaire (1985). Techniques in this paper and the likelihood (10.1) can be generalized in a straightforward way to threshold models. However, to generalize the machinery in the present paper would only serve to obscure the message.

Multivariate threshold models (e.g., Foulley and Gianola 1984; Hoschele 1986) are important but I have given very little thought to them. For these models, Var{**e**} is less restricted and residual correlations are needed and cannot be ignored. I will only suggest how these correlations can be estimated. Emphasis is on estimating **G** in this paper.

The functional form of (10.1) depends on the model and I will not complicate the discussion by introducing particular examples of (10.1).

10.3 Linearization

By linearization I mean manipulation of quantities related to (10.1) so as to allow the implementation of methods designed for linear models. Linearization techniques are described in this section.

10.3.1 Maximum Likelihood

The value of **b** (say $\mathbf{b}_{ML}$) that maximizes (10.1) is the maximum likelihood estimate. To evaluate $\mathbf{b}_{ML}$, the following system of equations is solved

$$\mathbf{T'f(Tb}_{ML}) = 0, \tag{10.3}$$

where $\mathbf{f(u)}=\partial L(\mathbf{u})/\partial\mathbf{u}$. The left side of (10.3) is the score statistic evaluated at $\mathbf{b}_{ML}$, i.e., the vector of first derivatives of (10.1) with respect to **b**.

Consider expanding the score statistic by a Taylor series around $\mathbf{b}_{ML}$ to give

$$\mathbf{T'f(Tb)} \doteq \mathbf{T'f(Tb}_{ML})\mathbf{-T'W(Tb}_{ML})\mathbf{T(b-b}_{ML}),$$

where $\mathbf{W(u)}=-\partial^2 L(\mathbf{u})/(\partial\mathbf{u}\partial\mathbf{u}')$, and **b** is understood to be the true value. The first term on the right vanishes because of (10.3). Following rearrangement, we obtain

$$\mathbf{T'W(Tb}_{ML})\mathbf{Tb}_{ML} \doteq \mathbf{T'W(Tb}_{ML})\mathbf{Tb+T'f(Tb)}. \tag{10.4}$$

The known statistics on the left approximate expressions on the right that involve **b**. Conditionally on **b**, the mean of the score statistic is **0** and the variance is **I(b)**=E[**T'W(Tb)T**] (Lindgren 1976). **I(b)** is called the information matrix and it can be approximated by $\mathbf{T'W(Tb}_{ML})\mathbf{T}$, the sample information matrix. The matrix **I(b)** can also be approximated by

$\mathbf{I}(\mathbf{b}_{ML})$; however, this option will not be used in the following presentation. Conditionally on $\mathbf{b}$ and provided $\mathbf{T'W}(\mathbf{Tb}_{ML})\mathbf{T}$ behaves as a constant, it follows that the mean of the right side of (10.4) is $\mathbf{T'W}(\mathbf{Tb}_{ML})\mathbf{Tb}$ and the variance is $\mathbf{T'W}(\mathbf{Tb}_{ML})\mathbf{T}$.

The hierarchical model is completed by noting that the mean of $\mathbf{Tb}$ is $\mathbf{Xb}_f$ and the variance is $\mathbf{ZGZ'}$.

When the linear model (10.2) applies, the right side of (10.4) equals $\mathbf{T'Wy}$, where $\mathbf{W}=\sigma^{-2}\mathbf{I}$, and (10.4) holds as an equality. Conditionally on $\mathbf{b}$, the mean of $\mathbf{T'Wy}$ is $\mathbf{T'WTb}$ and the variance is $\mathbf{T'WT}$. These moments are directly analogous to the mean and variance listed above for the general case. Indeed, the right side of (10.4) is $\mathbf{T'W}(\mathbf{Tb}_{ML})\mathbf{y}^+$ where

$$\mathbf{y}^+ = \mathbf{Tb}+\mathbf{W}(\mathbf{Tb}_{ML})^{-1}\mathbf{f}(\mathbf{Tb}),$$

and, hence, the right side of (10.4) is directly analogous to $\mathbf{T'Wy}$. The relationships suggest that $\mathbf{G}$ can be estimated using standard techniques, e.g., Henderson's (1953) methods, or REML applied to the "constructed model"

$$\mathbf{y}^+ = \mathbf{Xb}_f+\mathbf{Zb}_r+\mathbf{e}^+$$

$$\mathrm{Var}(\mathbf{b}_r) = \mathbf{G}$$

$$\mathrm{Var}(\mathbf{e}^+) = \mathbf{W}(\mathbf{Tb}_{ML})^{-1}.$$

We usually do not know $\mathbf{y}^+$. However, we can carry out estimation using the "constructed coefficient matrix"

$$\mathbf{T'W}(\mathbf{Tb}_{ML})\mathbf{T}$$

and the "constructed right-hand side"

$$\mathbf{T'W}(\mathbf{Tb}_{ML})\mathbf{Tb}_{ML}.$$

From (10.4) the constructed right-hand side approximates $\mathbf{T'W}\,(\mathbf{Tb}_{ML})\mathbf{y}^+$.

The procedures suggested here are consistent with the method given in Smith and Allaire (1986). However, these authors derived their variance estimate using asymptotic properties of the maximum likelihood estimates.

Because maximum likelihood can have poor small sample properties, methods of this section have a serious drawback. The matrix $\mathbf{T'W}(\mathbf{Tb}_{ML})\mathbf{T}$ may not behave as a constant. Moreover, some elements of $\mathbf{b}_{ML}$ may be plus or minus infinity. When elements of $\mathbf{b}_{ML}$ are infinite, then estimating $\mathbf{G}$ using the constructed model usually involves an implied edit as the observations associated with infinite estimates do not contribute information. Because the edited observations are usually associated with extreme values, the maximum likelihood procedure implies selective editing and this can bias the estimate of **G**.

Infinite estimates are usually associated with sparse information. For these cases we may wish to use prior information and disallow estimates going to plus or minus infinity. The well behaved estimates can then be used to define a constructed model for use in variance component estimation. This view is consistent with Foulley et al. (1983) who suggested that arbitrary effects can be used to define a constructed model. This approach is developed in the next two sections.

10.3.2 Maximum a Posteriori

Let $p(\mathbf{b}_s)$ be a prior density function for $\mathbf{b}_s$ such that $\mathbf{b}_s$ is multivariate normal with mean $\mathbf{b}_m$ and variance $\mathbf{G}^*$. Elements of $\mathbf{b}_s$ are those from $\mathbf{b}$ that are assigned prior information. In some situations $\mathbf{b}_s=\mathbf{b}_r$, $\mathbf{b}_m=\mathbf{0}$ and $\mathbf{G}^*$ is close to $\mathbf{G}$.

Elements in $\mathbf{b}$ that are not in $\mathbf{b}_s$ are assigned non-informative and independent priors. Thus, the prior for $\mathbf{b}$ [say $c(\mathbf{b})$] is

$$c(\mathbf{b}) \propto p(\mathbf{b}_s)$$

and the log joint distribution,

$$L(\mathbf{Tb})+\log[p(\mathbf{b}_s)] \tag{10.5}$$

is defined. The maximum *a posteriori* estimate of $\mathbf{b}$ (say $\mathbf{b}_{MAP}$) is obtained by maximizing the posterior distribution or equivalently (10.5) (Melsa and Cohn 1978).

Now define the vector $\mathbf{q}$ and the matrix $\mathbf{Q}$ such that $(\mathbf{b}-\mathbf{q})'\mathbf{Q}(\mathbf{b}-\mathbf{q})=(\mathbf{b}_s-\mathbf{b}_m)'\mathbf{G}^{*-1}(\mathbf{b}_s-\mathbf{b}_m)$, i.e., $\mathbf{G}^{*-1}$ is a submatrix of $\mathbf{Q}$ where other elements in $\mathbf{Q}$ are zero, and $\mathbf{b}_m$ is a subvector of $\mathbf{q}$ where other elements in $\mathbf{q}$ are zero. Then, to maximize (10.5), the vector of first derivatives are equated to zero to give

$$\mathbf{T}'\mathbf{f}(\mathbf{Tb}_{MAP})-\mathbf{Q}(\mathbf{b}_{MAP}-\mathbf{q}) = \mathbf{0}. \tag{10.6}$$

Expanding the score statistic by Taylor series around $\mathbf{b}_{MAP}$ and using (10.6) gives

$$\mathbf{T}'\mathbf{f}(\mathbf{Tb}) \doteq \mathbf{Q}(\mathbf{b}_{MAP}-\mathbf{q})-\mathbf{T}'\mathbf{W}(\mathbf{Tb}_{MAP})\mathbf{T}(\mathbf{b}-\mathbf{b}_{MAP}).$$

Following rearrangement we have

$$[\mathbf{T}'\mathbf{W}(\mathbf{Tb}_{MAP})\mathbf{T}+\mathbf{Q}]\mathbf{b}_{MAP}-\mathbf{Qq} \doteq \mathbf{T}'\mathbf{W}(\mathbf{Tb}_{MAP})\mathbf{Tb}+\mathbf{T}'\mathbf{f}(\mathbf{Tb}). \tag{10.7}$$

Now we will take $\mathbf{T'W(Tb_{MAP})T}$ as an approximation of $\mathbf{I(b)}$ that behaves as a constant. Then, conditionally on $\mathbf{b}$, the mean of the right side of (10.7) is $\mathbf{T'W(Tb_{MAP})Tb}$ and the variance is $\mathbf{T'W\,(Tb_{MAP})T}$. The right side of (10.7) is $\mathbf{T'W(Tb_{MAP})y^+}$, where $\mathbf{y^+}=\mathbf{Tb}+\mathbf{W(Tb_{MAP})}^{-1}\,\mathbf{f(Tb)}$ and, consequently, we have constructed a model for $\mathbf{y^+}$.

The matrix $\mathbf{G}$ can be estimated using the constructed coefficient matrix

$$\mathbf{T'W(Tb_{MAP})T}$$

and the constructed right-hand side

$$[\mathbf{T'W(Tb_{MAP})T}+\mathbf{Q}]\mathbf{b_{MAP}}-\mathbf{Qq}.$$

10.3.3 Foulley's Method

It is possible to expand the score statistic by Taylor series around an arbitrary value of $\mathbf{b}$ (say $\mathbf{b}^*$) including cases where $\mathbf{b}^*=\mathbf{b_{ML}}$ or $\mathbf{b_{MAP}}$. After rearrangement, this gives:

$$\mathbf{T'W(Tb^*)Tb^*}+\mathbf{T'f(Tb^*)} \doteq \mathbf{T'W(Tb^*)Tb}+\mathbf{T'f(Tb)}. \tag{10.8}$$

In this case, $\mathbf{T'W(Tb^*)T}$ is taken as an approximation of $\mathbf{I(b)}$. The matrix $\mathbf{G}$ is estimated using the constructed coefficient matrix, $\mathbf{T'W(Tb^*)T}$, and the constructed right-hand side $\mathbf{T'W(Tb^*)y^*}$, where $\mathbf{y^*}$ is

$$\mathbf{Tb^*}+\mathbf{W(Tb^*)}^{-1}\mathbf{f(Tb^*)}.$$

Note that $\mathbf{T'W(Tb^*)y^*}$ equals the left side of (10.8). To evaluate $\mathbf{T'W(Tb^*)y^*}$ it may be better to use the left side of (10.8) so as to avoid matrix inversion.

The above method was suggested by Foulley et al. (1983). For rank regression models, Pettitt (1983) approximated the log-likelihood by a Taylor series around $\mathbf{b}=\mathbf{0}$. Pettitt's approach can be extended to variance component estimation by taking $\mathbf{b}^*=\mathbf{0}$.

10.3.4 The Method of Harville and Mee

Until now, methods for estimating $\mathbf{G}$ involved constructing a linear model by expanding the score statistic by a Taylor series. Harville and Mee (1984) describe a method that appears to have different origins. These authors assign a non-informative prior to $\mathbf{b_f}$ and consider estimating $\mathbf{G}$ by maximum likelihood after $\mathbf{b}$ has been integrated out of the likelihood. Invoking approximations, Harville and Mee develop a strategy that resembles the estimation-maximization (EM) algorithm of Dempster et al. (1977) for REML. Their procedure is equivalent to:

1. Linearizing via the maximum *a posteriori* approach where $\mathbf{b}_s=\mathbf{b}_r$, $\mathbf{b}_m=\mathbf{0}$ and $\mathbf{G}^*$ is set to a prior value;
2. Updating the prior $\mathbf{G}^*$ by one round of the EM algorithm, then going to Step 1 if $\mathbf{G}^*$ has not converged.

For some situations it is better to go several rounds with the EM algorithm before returning to Step 1. This is true when tridiagonalization (Smith and Graser 1986) can be used to simplify the EM algorithm. The importance of continually linearizing for the updated prior $\mathbf{G}^*$ needs to be investigated. It may be that Step 1 is needed only once.

10.3.5 Gilmour's Method

Gilmour et al. (1985) present a procedure for dichotomous data that is similar to the method of Harville and Mee. However, Gilmour's linearization is different to that of Step 1 above and, indeed, is different to methods described previously. Gilmour's method appears to be an iterative extension of the approach of Quaas and Van Vleck (1980). It involves:

1. Approximating the means, variances and covariances of the binomial responses, treating the binomial probabilities as a function of $\mathbf{Tb}$, where $\mathbf{b}_r$ is random. These moments are a function of $\mathbf{b}_f$ and $\mathbf{G}$;
2. Using the moments evaluated in Step 1 to represent a quasi-likelihood for zero-one observations, and with the appropriate link function that connects $\mathbf{b}_f$ to the vector of means, applying the machinery of Nelder and Wedderburn (1972) to develop a system of equations for estimating $\mathbf{b}_f$. This action pretends that the quasi-likelihood is in the exponential family. The equations resemble the normal equations where the weight matrix and the observations are constructed from $\mathbf{b}_f$;
3. Expanding the equations for estimating $\mathbf{b}_f$, to equations for estimating $\mathbf{b}_f$ and $\mathbf{b}_r$. This is directly analogous to substituting equations for generalized least-squares with Henderson's (1973) mixed model equations. The new equations can be constructed directly and are easier to build than the original system. After solving the equations, take the current estimate of $\mathbf{b}_f$ and go to Step 1 if needed.

It turns out that equations created during Step 3 can be manipulated so as to employ the EM algorithm for estimating $\mathbf{G}$. Gilmour's procedure seems *ad hoc* because the quasi-likelihood is unlikely to be in the exponential family. Thus, the appropriateness of iterative reweighted least-squares is uncertain. The quasi-likelihood is not explicitly defined and there is no guarantee that the hypothetical likelihood is fully specified when there is hierarchical sampling. In particular, it may be insufficient to consider only the means, variances and covariances of zero-one observations when randomness is due to sampling $\mathbf{b}_r$ and binomial sampling, i.e., when randomness is not defined by the normal distribution or other distributions in the exponential family. Perhaps it is advantageous to use an estimate of the realized value of $\mathbf{b}_r$ when specifying the first and second moments of zero-one responses.

Despite the above argument, this method is purported to be well suited to cases where $\mathbf{b}_r$ is associated with sparse information. Gilmour's method of estimating variance components had less bias in a simulation study than the procedure of Harville and Mee (Gilmour et al. 1985). However, these results have not been confirmed in a separate analysis (Hoschele 1986).

10.3.6 Remarks

It should be noted that computing $\mathbf{b}_{ML}$ or $\mathbf{b}_{MAP}$ usually involves successive linearization (e.g., Smith, Chap.16, this Vol.). Thus, arrays needed to estimate $\mathbf{G}$ are usually an immediate by-product when estimating or predicting $\mathbf{b}$.

Some methods of estimating $\mathbf{G}$ have been motivated herein by equating the mean and variance of a constructed right-hand side to analogous arrays defined for a linear model. Expectations were evaluated under conditions of repeated sampling, treating $\mathbf{b}$ as fixed. This approach was taken due to convenience, and out of respect for tradition. However, the Bayesian would object to this style of presentation because in most cases only one data set is observed, and repeated sampling is not realistic to assume (e.g., Berger 1980). It is more logical to condition on the observations and consider the stochastic properties of $\mathbf{b}$ and $\mathbf{G}$. This view requires prior distributions for variance components and $\mathbf{b}$. Hoschele (1986) discusses the Bayesian view of estimating dispersion parameters.

Bayesian estimates of $\mathbf{G}$ need not be radically different to those that are obtained using linearization. Indeed, the first step in a Bayesian analysis is to construct a likelihood conditional on $\mathbf{b}$. Moreover, the procedure of Harville and Mee (1984) can be viewed as a Bayesian approach which uses a flat prior for the variance component being estimated. Methods that approximate maximum likelihood are essentially Bayesian. However, to be consistent with the Bayesian philosophy it would be better to use $\mathbf{T}'\mathbf{W}(\mathbf{Tb}^*)\mathbf{T}$ in place of $\mathbf{I}(\mathbf{b}^*)$. The matrix $\mathbf{I}(\mathbf{b}^*)$ involves a hypothetical integration over the sample space. Moreover, for censored observations or when observations are represented by ranks, $\mathbf{I}(\mathbf{b}^*)$ cannot usually be evaluated.

Methods have not been described to estimate residual correlations in multivariate threshold models. It would seem natural to extend the procedures described here; as $\mathbf{W}(\mathbf{Tb}^*)$ is a function of these correlations, $\mathbf{W}(\mathbf{Tb}^*)^{-1}$ can be treated as the dispersion matrix for residuals in the constructed linear model, and the correlations can be estimated accordingly.

This scheme may be dangerous because of the lack of theoretical justification. A more rational approach would be to extend the method of Tallis (1962) to multivariate threshold models. Such an extension would involve maximum likelihood estimation of thresholds, **b** and residual correlations.

10.4 Numerical Methods

Restricted maximum likelihood estimation of variance components has theoretical appeal (Harville 1977). Consequently, REML is becoming the method of choice when its numerical requirements can be met. We have at our disposal powerful techniques to compute REML estimates, e.g., Smith and Graser (1986) and Graser et al. (1987), and these can be applied both to linear and non-linear models. In this section efficient methods are described that allow a paternal half-sib analysis using REML and a non-linear model. It is assumed that a hypothetical underlying variable, **y**, follows the linear model (10.2). Moreover, $\mathbf{b}_r$ is a vector of sire effects and

$$\mathbf{G} = \sigma_S^2 \mathbf{A},$$

where σ_S^2 is the sire variance, and **A** is the matrix of relationships among sires. Methods of estimating σ_S^2 are now described.

10.4.1 Preliminary Absorption

Represent the constructed coefficient matrix as **T'WT** and the constructed right-hand side as **T'Wy**. With these quantities, set up the following array

$$\begin{bmatrix} \mathbf{X'WX} & \mathbf{X'WZ} & \mathbf{X'Wy} \\ \mathbf{Z'WX} & \mathbf{Z'WZ} & \mathbf{Z'Wy} \end{bmatrix}$$

and absorb the rows representing $\mathbf{b}_f$ to give

$$(\mathbf{Z'MZ} \quad \mathbf{Z'My}), \tag{10.9}$$

where $\mathbf{M}=\mathbf{W}-\mathbf{WX}(\mathbf{X'WX})^{-}\mathbf{X'W}$. Absorption can be done without matrix inversion by using the Gaussian elimination strategy described in Smith and Graser (1986). Sometimes a subset of the sire effects are treated as fixed effects. This is done when sires do not come from the population of interest or when the sires do not have a certain minimum number of daughters. For these cases the fixed sires are included in $\mathbf{b}_f$, and the corresponding rows are absorbed when building (10.9).

Absorption or Gaussian elimination need not be time-consuming, or impossible. Full Gaussian elimination of a square, not sparse, matrix requires one-third of the time needed for matrix inversion. For sparse matrices, Gaussian elimination can be extremely fast and much more so than inversion. For example, the model used by Smith (1987) included 2765 effects representing contemporary groups, 2611 fixed sire effects and 539 random sires. Rows representing contemporary groups were absorbed as data were read. After reading all data, the 2611 rows representing fixed sires were absorbed in reasonable time using sparse-matrix Gaussian elimination. The second absorption would have been impossible had matrix inversion been attempted.

Gaussian elimination is absorption one row at a time. The effectiveness of sparse-matrix Gaussian elimination depends on the order in which rows are absorbed; a desirable order is one where absorption does not create a very large number of non-zero off-diagonal elements. The order used for the second absorption implemented by Smith (1987) was determined from the size of the diagonals after the first absorption. Rows corresponding to smaller diagonals were absorbed first. Note that this is the reverse of the order often used for Gaussian elimination when matrices are non-sparse. Typically, rows corresponding to the largest diagonals are absorbed first so as to avoid rounding errors.

10.4.2 Accommodating Relationships

Once (10.9) has been evaluated, it can be further transformed such that relationships are accommodated (Meyer 1986; Smith and Graser 1986). This is done by reparameterizing $\mathbf{b}_r$ as

$$\mathbf{b}_r = \mathbf{L}\mathbf{b}_u,$$

where $\mathbf{L}$ is any matrix such that $\mathbf{LL}'=\mathbf{A}$ and

$$\mathrm{Var}\{\mathbf{b}_u\} = \sigma_s^2\mathbf{I}.$$

To implement the reparameterization, $\mathbf{ZL}$ replaces $\mathbf{Z}$ in (10.9) to give

$$(\mathbf{L}'\mathbf{Z}'\mathbf{MZL} \quad \mathbf{L}'\mathbf{Z}'\mathbf{My}). \tag{10.10}$$

Three methods of evaluating (10.10) are now described.

A Method 1

The first approach is to: (1) determine **A**; (2) evaluate **L** using a Cholesky decomposition of **A**; and (3) apply **L** to the array (10.9). When random sires make up the entire population of known ancestors, arguments of gene flow can simplify the above procedure because **L** can be evaluated directly (Meyer 1986). However, when this short cut can be used, the following more efficient approach is also feasible (Quaas 1982, personal communication).

B Method 2

1. Evaluate **H**, where $\mathbf{H'H}=\mathbf{A}^{-1}$, by adapting the machinery described by Hudson (1986). When sires are ordered correctly, **H** will be lower triangular with no more than three non-zero elements in each row.
2. Solve for $\mathbf{B}_1$, $\mathbf{B}_2$ and $\mathbf{B}_3$ in:

$$\mathbf{H'B}_1 = \mathbf{Z'MZ},\ \ \mathbf{H'B}_2 = \mathbf{B}_1',\ \ \mathbf{H'B}_3 = \mathbf{Z'My}.$$

These equations can be solved using backward substitution. This procedure is quick because **H** is very sparse. When solutions are found $\mathbf{B}_2$=**L'Z'MZL** and $\mathbf{B}_3$= **L'Z'My**.

The above procedure requires the unlikely condition that the sires in the analysis make up the entire population of known ancestors. This implies that all female relatives are ignored. A third method that requires no auxiliary assumptions is now presented.

C Method 3

1. Evaluate the inverse relationship matrix (say $\mathbf{A}_e^{-1}$) for the entire population of known ancestors using the technique of Henderson (1976).
2. Order the rows and columns of $\mathbf{A}_e^{-1}$ such that

$$\mathbf{A}_e^{-1} = \begin{bmatrix} \mathbf{A}^{11} & \mathbf{A}^{12} \\ \mathbf{A}^{21} & \mathbf{A}^{22} \end{bmatrix},$$

where $\mathbf{A}^{22}$ corresponds to sires in the analysis and rows of $\mathbf{A}^{11}$ are ordered to place youngest animals first.

3. Absorb $\mathbf{A}^{11}$ into $\mathbf{A}^{22}$ in row order using sparse-matrix Gaussian elimination to give $\mathbf{A}^{-1} = \mathbf{A}^{22} - \mathbf{A}^{21}(\mathbf{A}^{11})^{-1}\mathbf{A}^{12}$. Note that $\mathbf{A}^{-1}$ is the correct inverse of $\mathbf{A}$ defined for the subpopulation of sires.
4. Evaluate $\mathbf{F'F=A}^{-1}$ and $\mathbf{E'E=Z'MZ}$ using the Cholesky decomposition.
5. Solve for $\mathbf{B}_1$ and $\mathbf{B}_2$ in

$$\mathbf{F'B}_1 = \mathbf{E'} \quad \text{and} \quad \mathbf{F'B}_2 = \mathbf{Z'My},$$

using forward substitution, and evaluate $\mathbf{B}_1\mathbf{B}_1'$. This gives $\mathbf{B}_1\mathbf{B}_1'=\mathbf{L'Z'MZL}$ and $\mathbf{B}_2=\mathbf{L'Z'My}$.

The above procedure has been used in Smith (1987), and it represents another application of sparse-matrix Gaussian elimination. In this case Gaussian elimination will create many off-diagonal elements that are essentially zero. The procedure is very fast if the zero elements are occasionally released from storage. It may seem odd that $\mathbf{Z'MZ}$ is decomposed in Step 4. However, evaluating $\mathbf{E}$, solving for $\mathbf{B}_1$ and multiplying $\mathbf{B}_1\mathbf{B}_1'$ is quicker than solving for $\mathbf{B}_3$ and $\mathbf{B}_4$ in $\mathbf{F'B}_3=\mathbf{Z'MZ}$ and $\mathbf{F'B}_4=\mathbf{B}_3'$. Decomposing $\mathbf{Z'MZ}$ can also be implemented to advantage in Method 1.

For most realistic situations, only Methods 1 and 3 are applicable. Whether Method 1 is superior to Method 3 depends on whether $\mathbf{A}$ is easier to compute than $\mathbf{A}^{-1}$. When evaluating $\mathbf{A}$, the methods of Hudson et al. (1982) can be applied. Method 2 can be used if an increase in the number of animals in the analysis causes no problem. Increasing the number of animals will make tridiagonalization, which is described next, harder.

10.4.3 Tridiagonalization and the EM Algorithm

Estimating σ_s^2 by the EM algorithm involves the iterative scheme

$$\sigma_s^{2[k+1]} = (\mathbf{b}_u^{[k]'}\mathbf{b}_u^{[k]} + \mathrm{tr}\mathbf{C}^{[k]})/q, \tag{10.11}$$

where the bracket defines the iteration number, q is the order of $\mathbf{b}_r$, $\mathbf{b}_u^{[k]}$ is a solution to the equations

$$\{\mathbf{L'Z'MZL}+\mathbf{I}/(\sigma_s^{2[k]})\}\mathbf{b}_u^{[k]} = \mathbf{L'Z'My}, \tag{10.12}$$

and $\mathbf{C}^{[k]}$ is the inverse of the coefficient matrix in (10.12).

Tridiagonalization can be used to speed up the EM algorithm (Smith and Graser 1986). A tridiagonal matrix, $\mathbf{V}$, can be found such that $\mathbf{V=P'L'Z'MZLP}$, where $\mathbf{P}$ is orthogonal (i.e., $\mathbf{P'P=PP'=I}$). System (10.12) can be transformed to

$$\{\mathbf{V}+\mathbf{I}/(\sigma_s^{2[k]})\}\mathbf{b}^{*[k]} = \mathbf{g}, \tag{10.13}$$

where $\mathbf{b}^{*[k]}=\mathbf{P}'\mathbf{b}_u^{[k]}$ and $\mathbf{g}=\mathbf{P}'\mathbf{L}'\mathbf{Z}'\mathbf{M}\mathbf{y}$.

Once **V** and **g** are determined, system (10.13) can be solved in linear time for any $\sigma_s^{2[k]}$. Moreover, the trace of the inverse of the coefficient matrix in (10.13) can be evaluated in linear time. Because

$$\mathbf{b}_u^{[k]'}\mathbf{b}_u^{[k]} = \mathbf{b}^{*[k]'}\mathbf{b}^{*[k]}$$

and

$$\mathrm{tr}\mathbf{C}^{[k]} = \mathrm{tr}\{\mathbf{V}+\mathbf{I}/(\sigma_s^{2[k]})\}^{-1},$$

(10.11) can be evaluated rapidly after tridiagonalization. Each round of the EM algorithm requires only linear time and, consequently, there is no practical limit to the number of rounds that might be conducted.

Tridiagonalization requires 2/3 q^3 operations, a little more than what is needed for matrix inversion. However, tridiagonalization is done only once, whereas alternative strategies can involve matrix inversion at each round of the EM algorithm.

Diagonalization can also be used to speed up the EM algorithm (Dempster et al. 1984). Diagonalization involves finding eigenvectors and eigenvalues of **L'Z'MZL**.

10.4.4 Remarks

In the previous discussion, methods were described for REML estimation of σ_s^2, given the constructed linear model. The practitioner must decide if the process is to be repeated for each "new" linear model. For example, following Harville and Mee (1984) a new model can be constructed using the maximum *a posteriori* approach given the new value of σ_s^2.

Numerical methods are designed for the case where the number of random effects is not so large as to disallow tridiagonalization. For some situations there may be too many random effects, and it may be undesirable to treat some of the random effects as fixed. For this case, the derivative-free algorithm of Graser et al. (1987) may be a more feasible way to perform REML estimation. This procedure involves calculating likelihoods directly and searching for the maximum. Likelihoods are evaluated using sparse-matrix absorption, and can also be computed by adapting the "QR algorithm" described in Hudson (1986).

10.5 A Preliminary Investigation

The data set described in Smith, Chapter 16, this Volume, was used to estimate heritability for a measure of survival in dairy cows. The data set contained 12 705 records and 74% of the observations were censored.

The model used in the analysis was:

$$h_{s(i)}(y_i) = b_o a_i + b_1 a_i^2 + s_{j(i)} + e_i, \tag{10.14}$$

where y_i is the i-th observation measured as the duration in days between first calving and "failure", s(i) is the herd-year group to which observation i belongs, $h_{s(i)}(.)$ is an arbitrary transformation defined for herd-year s(i), a_i is the age at first calving in years, and $s_{j(i)}$ is the j(i)-th sire effect. There were 395 sire effects and these were treated as random and unrelated. There were 1275 herd-year groups. Note that treating $h_{s(i)}(.)$ as unknown automatically accommodates herd-year effects.

The residual term, e_i, was assumed to have a log-exponential distribution, and this makes model (10.14) a Cox regression model. An equivalent model based on the hazard function is given in Smith, Chapter 16, this Volume.

Conditionally on b_o, b_1 and the sire effects, the likelihood employed was constructed by pooling the marginal information across herd-year groups as outlined by Smith, Chapter 16, this Volume. Marginal information is equivalent to rank information when there is no censoring. With censoring, marginal information corresponds to what is known about the rank information. The likelihood constructed from the pooled information is invariant to all $h_{s(i)}(.)$. As estimation and prediction can be performed without knowing or estimating the unknown transformations, model (10.14) is very attractive. The likelihood constructed for (10.14) is represented by the general likelihood (10.1) described for non-linear models.

The sire variance was estimated using the numerical techniques described earlier. That is, the likelihood was used to define constructed linear models, and for each case, σ_s^2 was estimated by REML using the EM algorithm. Linear models were constructed by the maximum *a posteriori* approach using a range of heritabilities. Also, a linear model was constructed using Pettitt's (1983) approximation (i.e., $\mathbf{b}^*=\mathbf{0}$).

Tridiagonalizing a matrix of order 395 required 11 min of computing time on a DEC2060. The EM algorithm required up to 500 rounds of iteration, but 500 rounds required only 29 s of computing.

The residual variance in (10.14) is standardized to $\pi^2/6$. Thus, a paternal half-sib analysis gives heritability on the underlying scale as $4\sigma_s^2/(\sigma_s^2+\pi^2/6)$.

The REML estimates of heritability and sire variance are displayed in Table 10.1. Each row in Table 10.1 represents a linear model constructed by using the maximum *a posteriori* approach. Each row is defined by the prior heritability. For the broad range of prior heritabilities considered, estimated heritabilities were in a narrow range (5.86 to 6.88%).

Pettitt's (1983) approximation produced a heritability of 6.73%, and this is also within the range.

Table 10.1 Estimates of genetic parameters obtained by REML using a constructed linear model

Prior heritability[a] (%)	Sire variance[b]	Estimated heritabilty (%)
5	0.02879	6.88
10	0.02879	6.88
20	0.02840	6.79
30	0.02785	6.66
40	0.02727	6.52
50	0.02671	6.39
60	0.02616	6.26
70	0.02564	6.14
80	0.02515	6.02
90	0.02468	5.91
95	0.02445	5.86

[a] Used to construct a linear model by the maximum *a posteriori* approach.
[b] Residual variance standardized to $\pi^2/6$.

The approach of Harville and Mee (1984) involves iterating by using the estimated heritability as a prior. Starting with a heritabiity of 10%, this approach gave an estimate of 6.89% after two rounds of REML. The results in Table 10.1 suggest that this estimate would be obtained in two rounds regardless of the starting value.

The results demonstrate that techniques designed for variance component estimation in linear models can be applied to non-linear models; estimates of genetic parameters can be insensitive to the approach taken to linearize a non-linear model. This last point is obviously data-dependent but it is worth considering for other data sets.

There were enough sires in the data set to suggest that heritability estimates reported are biologically significant. Unfortunately, the quality of this data set is in question because only 26% of the observations represented failures, and this is likely to be a gross underestimate. Thus, estimates reported here should be viewed with caution. Efficient methodology awaits a "clean" data set.

10.6 Conclusion

The goal of the present paper was to describe variance component estimation for a general non-linear model. There are methods of estimating heritability in non-linear models that have not been considered here. These are for discrete data, and they may involve the application of a correction for discontinuity such as Robertson et al. (1950), or an alternative strategy, e.g., Thompson et al. (1985).

There are plenty of unanswered questions in variance component estimation. This is true for both linear and non-linear models. Sometimes a solution to a problem in linear models, also applies to the non-linear domain. This was the case with tridiagonalization. In this paper emphasis was given entirely to univariate non-linear models. We have just begun to consider the multivariate case.

References

Berger JO (1980) Statistical decision theory. Springer-Verlag, Berlin Heidelberg New York

Dempster AP, Laird NM, Rubin DB (1977) Maximum likelihood from incomplete data via the EM algorithm. J R Stat Soc Ser B 38:1-38

Dempster AP, Selwyn MR, Patel CM, Roth AJ (1984) Statistical and computational aspects of mixed model analysis. Appl Stat 33:203-214

Foulley JL, Gianola D (1984) Estimation of genetic merit from bivariate "all or none" responses. Genet Sel Evol 16:285-306

Foulley JL, Gianola D, Thompson R (1983) Prediction of genetic merit from data on binary and quantiative variates with an application to calving difficulty, birth weight and pelvic opening. Genet Sel Evol 15:401-423

Gianola D, Foulley JL (1983) Sire evaluation for ordered categorical data with a threshold model. Genet Sel Evol 15:201-223

Gilmour AR, Anderson RD, Rae AL (1985) The analysis of binomial data by a generalized linear mixed model. Biometrika 72:593-599

Graser HU, Smith SP, Tier B (1987) A derivative-free approach for estimating variance components in animal models by restricted maximum likelihood. J Anim Sci 64:1362-1370

Harville DA (1977) Maximum likelihood approaches to variance component estimation and to related problems. J Am Stat Assoc 72:320-340

Harville DA, Mee RW (1984) A mixed model procedure for analyzing ordered categorical data. Biometrics 40:398-408

Henderson CR (1953) Estimation of variance and covariance components. Biometrics 9:226-252

Henderson CR (1973) Sire evaluation and genetic trends. In: Proc Anim Breed Genet Symp in Honor of Dr. J.L. Lush. ASAS and ADSA, Champaign, Illinois pp 10-41

Henderson CR (1976) A simple method for computing the inverse of a numerator relationship matrix used in prediction of breeding values. Biometrics 32:69-83

Hoschele I (1986) Estimation of breeding values and variance components with quasi-continuous data. PhD Diss, Hohenheim Univ, Stuttgart

Hudson GFS (1986) Computing genetic evaluations through application of generalized least squares to an animal model. Genet Sel Evol 18:31-40

Hudson GFS, Quaas RL, Van Vleck LD (1982) Computer algorithm for the recursive method of calculating large numerator relationship matrices. J Dairy Sci 65:2018-2022

Lindgren BW (1976) Statistical theory. 3rd edn. Macmillan, New York

Melsa JL, Cohn DL (1978) Decision and estimation theory. McGraw-Hill, New York

Meyer K (1986) Restricted maximum likelihood to estimate genetic parameters - in practice. In: Dickerson GE, Johnson RK (eds) Proc 3rd World Congr Genet Appl Lives Prod. Agric Commun, Univ Nebraska, Lincoln, Nebraska XII:454-459

Nelder JA, Wedderburn RWM (1972) Generalized linear models. J R Stat Soc Ser A 135:370-384

Patterson HD, Thompson R (1971). Recovery of interblock information when block sizes are unequal. Biometrika 58:545-554

Pettitt AN (1983) Approximate methods using ranks for regression with censored data. Biometrika 70:121-132

Quaas RL, Van Vleck LD (1980) Categorical trait sire evaluation by best linear unbiased prediction of future progeny category frequencies. Biometrics 36:117-122

Rao CR (1971) Minimum variance quadratic unbiased estimation of variance components. J Multivar Anal 1:257-275

Robertson A, Dempster ER, Lerner IM (1950) Heritability of threshold characters. Genetics 35:212-236

Smith SP (1987) Genetic parameters for type in Australian Holstein-Friesian dairy cattle. In: Proc 6th Conf Austr Assoc Anim Breed Genet, Perth pp 55-58

Smith SP, Allaire FR (1985) Durability of traits: non-linear association between initial and subsequent measures. J Dairy Sci 68:939-945

Smith SP, Allaire FR (1986) Analysis of failure times measured in dairy cows: theoretical considerations in animal breeding. J Dairy Sci 69:217-227

Smith SP, Graser HU (1986) Estimating variance components in a class of mixed models by restricted maximum likelihood. J Dairy Sci 69:1156-1165

Smith SP, Hammond K (1988) Rank regression with log-gamma residuals. Biometrika 75:741-751

Tallis GM (1962) The maximum likelihood estimation of correlation from contingency tables. Biometrics 18:342-353

Thompson R, McGuirk BJ, Gilmour AR (1985) Estimating the heritability of all-or-none and categorical traits by offspring-parent regression. Z Tierz Züchtungsbiol 102:342-354

Discussion Summary
PART III: ESTIMATION OF GENETIC PARAMETERS

R. Thompson[1]

8 Computational Aspects of Likelihood-Based Inference for Variance Components

It was commented jocularly that the EM algorithm would be the most widely used algorithm to estimate variance components if the number of iterates were used as a criterion. The EM algorithm was thought easy to compute, and the number of iterates is probably a minor part of the total calculations. It was worrying that the changes in each iteration can be so small and, therefore, it is difficult to know when to stop. Using a number of starting points and over relaxation methods was suggested. For large problems maximum likelihood estimation was thought to be simpler than restricted maximum likelihood, but the estimates of the genetic variances would be biased. The author referred to his 1977 paper for discussion of appropriate procedures. It was suggested that a plot of t_i^2 against Δ would be a useful graphical device, as REML estimation was equivalent to fitting a straight line.

9 Parameter Estimation in Variance Component Models for Binary Response Data

It was suggested that there would be no practical difference between using the probit and logit transformations. It was thought more difficult to interpret a flat prior for fixed effects in binary traits than in normally distributed traits. The author was referred to the paper of Thompson (Chapter 4) for an attempt to compare the methods described by Dr. Laird.

1 Institute of Animal Physiology and Genetics Research, Edinburgh Research Station, Edinburgh, Scotland.

10 Estimation of Genetic Parameters in Non-Linear Models

Several points clarifying the oral presentation were made. It was suggested that the estimated heritability would be appropriate to predict progress if selection were carried out on the survival traits.

Part IV: Prediction and Estimation of Genetic Merit

11 A Framework for Prediction of Breeding Value

D. Gianola[1], S. Im[2], and F.W. Macedo[3]

Inferences in a univariate normal linear model with two components of variance are discussed from a Bayesian perspective. When variances are known, three scenarios are considered: unknown "nuisance" parameters ($\boldsymbol{\beta}$) and breeding values ($\mathbf{u}$), unknown $\boldsymbol{\beta}$ and known $\mathbf{u}$, and unknown $\mathbf{u}$ and known $\boldsymbol{\beta}$. With unknown variances, several situations with corresponding predictors of breeding value are considered; the predictors differ in the degree of marginalization with respect to nuisance parameters and variance components, and in all these cases, inverted chi-square distributions are used as priors for the variance components. Prediction is discussed using the concept of a loss function, and it is argued that uniformly best predictors cannot be obtained in a classical framework, unless attention is restricted to the class of unbiased statistics. However, there is an unequivocal Bayesian solution. Model choice and prediction of future observations are also discussed.

11.1 Introduction

Production recording programs typically collect data on fertility, ease of birth, litter size, perinatal and early postnatal mortality, maternal ability, milk production, growth through slaughter, conformation, product quality and herdlife. Although the information is used primarily for management purposes or for quantifying output, the data are often employed for calculating breeding values, estimating genetic parameters from field records, or for assessing the potential importance of "nuisance" effects such as age of dam, parity or sex of the animal.

A theory for prediction of breeding value relying on normality is clearly insufficient because the traits listed above include variables for which at least some degree of non-linearity needs to be accommodated in the analysis, e.g., "all or none" characters or censored observations such as life span. It is also important to consider selection and assortative

[1] Department of Animal Sciences, University of Illinois, Urbana, Illinois, USA
[2] Laboratoire de Biometrie, Centre de Recherches de Toulouse, Institut National de la Recherche Agronomique, Castanet-Tolosan, France
[3] Department of Mathematics, University of Tras-os-Montes e Alto Douro, Vila Real, Portugal

mating because these are ubiquitous in animal breeding practice. Further, there may be genes of major effect segregating in the population, and these can cause multimodality so that the end result is a "mixture" of distributions. In addition to addressing these complications, a general theory should attempt to answer questions such as:

1. "Given that a certain distribution can be tentatively entertained, what model should be considered in the analysis?"
2. "How can prior information be formally (as opposed to algorithmically) incorporated in inferences or decisions?"
3. "How do we take into account the degree of uncertainty about unknowns, e.g., heritability, in the analysis?"
4. "Which predictors should be used in the context of a particular problem of decision?"
5. "What is the quality of the predictions obtained from a given model?"

The state of the art in prediction of breeding value is mixed model methodology (Henderson 1973, 1974). Application of mixed linear models to animal breeding has been justified on the grounds of best linear unbiased estimation and prediction. Their properties require knowledge of the variance-covariance structure and pertain to the performance of statistics in an infinite sequence of independent resamplings over the distributions of random variables involved in the model (Peixoto and Harville 1986). Minimization of prediction error variance does not seem to provide, at least *a prima facie*, a compelling reason for using best linear unbiased prediction (BLUP) as a tool for ranking and selecting breeding animals. A stronger justification for using BLUP is that under normality, selection with BLUP predictions maximizes expected merit in a selected sample of k individuals when m candidates are available. This has been shown by Bulmer (1980, 1982) and by Goffinet (1983) who restricted attention to the class of predictors that are linear, translation invariant functions of the data. These authors assumed that variances and covariances were known so at least one of the requirements for a general theory was not met. Also, this "optimal" ranking property of BLUP requires normality, and this would be violated at least with "all or none" traits (Gianola 1982), or with censored observations (Smith and Allaire 1986).

The purpose of this paper is to describe a framework which achieves the requirements outlined above in a **unified** and, we believe, clear manner. We consider a univariate mixed linear model with two components of variance, and assume multivariate normality. This allows detailed presentation of the derivations without getting involved in excessively technical details. It should not be construed that we advocate linear models and multivariate normality for **every** conceivable animal breeding problem. Although robustness of the assumption of normality is often invoked in animal breeding research (Harvey 1982), satisfactory results are not always obtained. Some of the concepts developed here with linear models and normality are applicable to other situations. A Bayesian approach is followed, and in some instances we resort to empirical Bayes techniques.

11.2 The Mixed Linear Model

The univariate model often employed in animal breeding is:

$$\mathbf{y} = \mathbf{X}\boldsymbol{\beta}+\mathbf{Z}\mathbf{u}+\mathbf{e}, \tag{11.1}$$

where $\mathbf{y}$ is an nx1 vector of observations, $\mathbf{X}$ and $\mathbf{Z}$ are known, fixed matrices, $\boldsymbol{\beta}$ and $\mathbf{u}$ are unknowns, and $\mathbf{e}$ is a vector of residuals reflecting the inability of the model to explain completely the variation of $\mathbf{y}$. The residuals are assumed independent of $\boldsymbol{\beta}$ and $\mathbf{u}$. Both vectors $\boldsymbol{\beta}$ and $\mathbf{u}$ are formally location parameters, but we distinguish between the two depending on the relative state of uncertainty prior to the realization of the data. It will be assumed that $\boldsymbol{\beta}$ is a vector on which prior knowledge is completely vague so, *a priori*, the investigator is indifferent with respect to the values it takes. On the other hand, $\mathbf{u}$ is a vector on which it is possible to make prior probability statements with some degree of sharpness. These two situations correspond to the "fixed" ($\boldsymbol{\beta}$) and "random" ($\mathbf{u}$) effects of the sampling school (Henderson 1973, 1974). Whereas it is possible that $\boldsymbol{\beta}$ has a "fixed" meaning (e.g., the difference in basal metabolic rate between two breeds of cattle), there is randomness stemming from subjective uncertainty about its value. The situation is similar with $\mathbf{u}$ but here, the investigator has past experience that can be represented by a distribution, with possibly unspecified parameters. If $\mathbf{u}$ in (1) is a vector of breeding values, under polygenic inheritance it is reasonable to assume a normal distribution for $\mathbf{u}$ (Bulmer 1980). More details on the Bayesian linear model are given by Lindley and Smith (1972), Dempfle (1977), Broemeling (1985), Smith and Allaire (1985) and Gianola and Fernando (1986).

The non-informative prior distribution for $\boldsymbol{\beta}$ is (Jeffreys 1961; Zellner 1971; Box and Tiao 1973):

$$p(\boldsymbol{\beta}) \propto \text{constant}. \tag{11.2}$$

Expression (11.2) can be interpreted assuming a sequence of normal distributions with increasing variance; these distributions become flatter as σ^2_β increases and, in the limit, all points have equal density over the real line. Equivalently, (11.2) can be viewed as a statement of the assertion that all $\boldsymbol{\beta}$'s are equally likely, *a priori*.

Using the argument suggested earlier, it will be assumed that

$$\mathbf{u} \sim N_q(\mathbf{0},\mathbf{A}\sigma^2_u), \tag{11.3}$$

so that $\mathbf{u}$ follows a q-variate normal distribution; in (11.3), $\mathbf{A}$ is a known matrix of numerator relationships (Henderson 1976) and σ^2_u is the variance of additive genetic effects or of transmitting abilities, depending on the context. In general, σ_u^2 is unknown so (11.3) states the form of the distribution but the values of all parameters are not necessarily specified.

It will be assumed that

$$\mathbf{y}|\boldsymbol{\beta},\mathbf{u},\sigma_e^2 \sim N_n(\mathbf{X}\boldsymbol{\beta}+\mathbf{Z}\mathbf{u},\mathbf{R}\sigma_e^2), \tag{11.4}$$

where $\mathbf{R}$ is a known matrix and σ_e^2 is the residual variance. In general, σ_e^2 will be unknown, so (11.4) is a statement of distributional form with all parameter values unspecified. Hereafter, and without loss of generality, it will be assumed that $\mathbf{R}=\mathbf{I}$.

The model would be incomplete without stating the degree of uncertainty about σ_u^2 and σ_e^2. In this paper, it will be assumed that *a priori* the variance components follow the independent inverted chi-square distributions (Lindley and Smith 1972; Broemeling 1985):

$$p(\sigma_u^2) \propto (\sigma_u^2)^{-1/2(\upsilon_u+2)}\exp(-1/2\upsilon_u s_u^2/\sigma_u^2) \tag{11.5}$$

and

$$p(\sigma_e^2) \propto (\sigma_e^2)^{-1/2(\upsilon_e+2)}\exp(-1/2\upsilon_e s_e^2/\sigma_e^2). \tag{11.6}$$

In (11.5) and (11.6), $s_u^2(s_e^2)$ can be interpreted as a prior value of $\sigma_u^2(\sigma_e^2)$, and $\upsilon_u(\upsilon_e)$ is a parameter analogous to degrees of freedom that conveys the "degree of belief" on the prior value (Chen 1979). These are "hyperparameters" of the prior distribution of the variances in the same way σ_u^2 is a parameter of the distribution of $\mathbf{u}$. The choice of an inverted chi-square distribution for a variance stems from its conjugate nature (Lindley and Smith 1972; Broemeling 1985) and because it appears as a posterior distribution of the appropriate parameter in certain settings (Zellner 1971; Box and Tiao 1973; Broemeling 1985). Their conjugate property simplifies considerably subsequent mathematical analysis.

In the absence of prior knowledge about the variances, one could set υ_u and υ_e to 0 in which case the prior distributions become

$$p(\sigma_u^2) \propto \sigma_u^{-2}\,;\; p(\sigma_e^2) \propto \sigma_e^{-2}. \tag{11.7}$$

The prior density for σ_e^2 corresponds to Jeffreys (1961) ignorance prior for a variance parameter; the prior for σ_u^2 will give essentially the same result as a "flat" or uniform prior when the number of elements (q) in $\mathbf{u}$ is large. Flat priors for variance parameters can cause technical difficulties in Bayesian analysis.

Specifying the value of "hyperparameters" of informative prior distributions such as (11.5) and (11.6) can be difficult in practice. If this is the case, one can resort to an empirical Bayes approach (Efron and Morris 1973; Casella 1985), and estimate "hyperparameters" from the data. A feasible alternative in the case of the prior distributions (11.5) and (11.6) would be to fit the inverted chi-square density function to a set of past estimates. A simple procedure could involve a method of moments fit. From past series of data one would calculate the mean and variance of estimates of σ_u^2, appropriately weighted. These statistics would be equated to the expected value and variance of an inverted chi-square random variable, which depend on s_u^2 and υ_u. This gives two equations on two unknowns that can be readily solved. While specifying prior knowledge is difficult, it is not impossible. The end result is a more complete model using pertinent information, which would be wasted otherwise. Malecot (1947) brilliantly described the role of prior knowledge in science, and emphasized the need of having a mechanism for incorporating it in the analysis of "current" data.

11.3 Joint Posterior Distribution

Bayes theorem gives (Zellner 1971; Box and Tiao 1973):

$$p(\boldsymbol{\beta},\mathbf{u},\sigma_u^2,\sigma_e^2|\mathbf{y},s_u^2,\upsilon_u,s_e^2,\upsilon_e) \propto p(\mathbf{y}|\boldsymbol{\beta},\mathbf{u},\sigma_e^2).p(\boldsymbol{\beta}).p(\mathbf{u}|\sigma_u^2)$$

$$.p(\sigma_e^2|s_e^2,v_e).p(\sigma_u^2|s_u^2,v_u) \qquad (11.8)$$

as posterior density of all unknowns; the first term on the right-hand side is the likelihood function, which depends on σ_e^2 but not on σ_u^2. Using (11.2)-(11.6) in (11.8) yields as joint posterior density:

$$p(\boldsymbol{\beta},\mathbf{u},\sigma_u^2,\sigma_e^2|\mathbf{y},s_u^2,\upsilon_u,s_e^2,\upsilon_e) \propto$$

$$(\sigma_e^2)^{-(n+\upsilon_e+2)/2}\exp\{-1/2[(\mathbf{y}-\mathbf{X}\boldsymbol{\beta}-\mathbf{Z}\mathbf{u})'(\mathbf{y}-\mathbf{X}\boldsymbol{\beta}-\mathbf{Z}\mathbf{u})+\upsilon_e s_e^2]/\sigma_e^2\}$$

$$.(\sigma_u^2)^{-(q+\upsilon_u+2)/2}\exp\{-1/2[\mathbf{u}'\mathbf{A}^{-1}\mathbf{u}+\upsilon_u s_u^2]/\sigma_u^2\} \qquad (11.9)$$

$$-\infty<\beta_i<\infty,\ i=1,\ldots,p$$

$$-\infty<u_i<\infty,\ i=1,\ldots,q$$

$$\sigma_u^2>0$$

$$\sigma_e^2>0.$$

We consider next several posterior distributions that give bases for inference under different states of knowledge.

11.4 Known Variance Components

Suppose σ_e^2 and σ_u^2 are known, i.e., the investigator is willing to assign all density to some specific values of these parameters. This implies that the variances can be treated as constants rather than as random variables. The only unknowns would then be $\boldsymbol{\beta}$ and $\mathbf{u}$, and their joint posterior density is obtained directly from (11.9) by regarding the variance components as constants:

$$p(\boldsymbol{\beta},\mathbf{u}|\sigma_u^2,\sigma_e^2,\mathbf{y}) \propto \exp\{-1/2[(\mathbf{y}-\mathbf{X}\boldsymbol{\beta}-\mathbf{Z}\mathbf{u})'(\mathbf{y}-\mathbf{X}\boldsymbol{\beta}-\mathbf{Z}\mathbf{u})+\mathbf{u}'\mathbf{A}^{-1}\mathbf{u}\alpha]/\sigma_e^2]\}, \quad (11.10)$$

where $\alpha=\sigma_e^2/\sigma_u^2$. As shown by several authors (Lindley and Smith 1972; Dempfle 1977; Robinson 1982; Gianola and Fernando 1986; Gianola et al., Chap. 2, this Vol.) this density is multivariate normal and can be written as

$$p(\boldsymbol{\beta},\mathbf{u}|\sigma_u^2,\sigma_e^2,\mathbf{y}) \propto \exp[-1/2(\theta-\hat{\theta})'(\mathbf{W}'\mathbf{W}+\Sigma)(\theta-\hat{\theta})/\sigma_e^2] \quad (11.11)$$

so

$$\theta|\sigma_u^2,\sigma_e^2,\mathbf{y} \sim N_{p+q}[\hat{\theta},(\mathbf{W}'\mathbf{W}+\Sigma)^{-1}\sigma_e^2], \quad (11.12)$$

where $\theta'=[\boldsymbol{\beta}',\mathbf{u}'],\mathbf{W}=[\mathbf{X},\mathbf{Z}]$,

$$\Sigma = \begin{bmatrix} \mathbf{0} & \mathbf{0} \\ \mathbf{0} & \mathbf{A}^{-1}\alpha \end{bmatrix}$$

and $\hat{\theta}$ is such that

$$\begin{bmatrix} \mathbf{X}'\mathbf{X} & \mathbf{X}'\mathbf{Z} \\ \mathbf{Z}'\mathbf{X} & \mathbf{Z}'\mathbf{Z}+\mathbf{A}^{-1}\alpha \end{bmatrix} \begin{bmatrix} \hat{\boldsymbol{\beta}} \\ \hat{\mathbf{u}} \end{bmatrix} = \begin{bmatrix} \mathbf{X}'\mathbf{y} \\ \mathbf{Z}'\mathbf{y} \end{bmatrix}. \quad (11.13)$$

These are the mixed model equations of Henderson (1973), known to yield $\hat{\boldsymbol{\beta}}$=BLUE($\boldsymbol{\beta}$) and $\hat{\mathbf{u}}$=BLUP($\mathbf{u}$), where BLUE denotes "best linear unbiased estimator" and BLUP was defined earlier. We note that BLUE($\boldsymbol{\beta}$) and BLUP($\mathbf{u}$) are the components of the mean vector of the posterior distribution (11.12); the Bayesian development, however, does not invoke the long-run performance over repeated sampling of the **y** vector. Because (11.12) is multivariate normal, the $\boldsymbol{\beta}$ and **u** components of the vector of joint means give directly the marginal means so:

$$\boldsymbol{\beta}|\sigma_u^2,\sigma_e^2,\mathbf{y} \sim N_p[\hat{\boldsymbol{\beta}},(\mathbf{X}'\mathbf{V}^{-1}\mathbf{X})^{-1}\sigma_e^2] \tag{11.14}$$

and

$$\mathbf{u}|\sigma_u^2,\sigma_e^2,\mathbf{y} \sim N_q[\hat{\mathbf{u}},(\mathbf{Z}'\mathbf{M}\mathbf{Z}+\mathbf{A}^{-1}\alpha)^{-1}\sigma_e^2]. \tag{11.15}$$

Above, $\mathbf{V}=\mathbf{I}+\mathbf{Z}\mathbf{A}\mathbf{Z}'\alpha^{-1}$, and $\mathbf{M}=\mathbf{I}-\mathbf{X}(\mathbf{X}'\mathbf{X})^{-1}\mathbf{X}'$. The distributions (11.14) and (11.15) can, of course, be obtained by integration of (11.11) with respect to the appropriate variable. This illustrates that in the context of the multivariate normal distribution, integration and "absorption" are equivalent operations. Inferences about $\boldsymbol{\beta}$ or **u**, posterior to the data, can be completed using (11.14) or (11.15), and standard multivariate normal theory. For example, the linear combination $\mathbf{K}'\boldsymbol{\beta}+\mathbf{P}'\mathbf{u}=\mathbf{L}'\theta$ has a normal posterior distribution with mean $\mathbf{L}'\hat{\theta}$ and variance $\mathbf{L}'\mathrm{Var}(\theta|\sigma_u^2,\sigma_e^2,\mathbf{y})\mathbf{L}$.

11.4.1 Posterior Distribution of $\boldsymbol{\beta}$ With Known u

For completeness, we also consider the situation where the density of **u** is a point mass so it can be treated as a constant in the analysis. This would occur, for example, in a trial involving different levels of concentrate feeding in daughters of proven sires with "repeatability" coefficients equal to 1. Because **u** is known, one can work with $\mathbf{v}=\mathbf{y}-\mathbf{Z}\mathbf{u}$ instead of **y**. The posterior distribution of $\boldsymbol{\beta}$ can be written as:

$$p(\boldsymbol{\beta}|\mathbf{u},\sigma_e^2,\mathbf{y}) \propto p(\mathbf{v}|\boldsymbol{\beta},\sigma_e^2).p(\boldsymbol{\beta}). \tag{11.16}$$

With $p(\boldsymbol{\beta})$ taken as "flat", the posterior distribution is strictly proportional to the likelihood of **v**. We have

$$\mathbf{v}|\boldsymbol{\beta},\sigma_e^2 \sim N_n(\mathbf{X}\boldsymbol{\beta},\mathbf{I}\sigma_e^2), \tag{11.17}$$

so (11.16) becomes:

$$p(\boldsymbol{\beta}|\mathbf{u},\sigma_e^2,\mathbf{y}) \propto \exp\{-1/2(\mathbf{v}-\mathbf{X}\boldsymbol{\beta})'(\mathbf{v}-\mathbf{X}\boldsymbol{\beta})/\sigma_e^2\}. \tag{11.18}$$

Using a standard decomposition (Zellner 1971; Gianola et al. 1986)

$$(\mathbf{v}-\mathbf{X}\boldsymbol{\beta})'(\mathbf{v}-\mathbf{X}\boldsymbol{\beta}) = (\mathbf{v}-\mathbf{X}\tilde{\boldsymbol{\beta}})'(\mathbf{v}-\mathbf{X}\tilde{\boldsymbol{\beta}})+(\boldsymbol{\beta}-\tilde{\boldsymbol{\beta}})\mathbf{X}'\mathbf{X}(\boldsymbol{\beta}-\tilde{\boldsymbol{\beta}}),$$

where $\tilde{\boldsymbol{\beta}}=(\mathbf{X}'\mathbf{X})^{-1}\mathbf{X}'\mathbf{v}$. Retaining only the quadratic form in $\boldsymbol{\beta}$, (11.16) can be expressed as:

$$p(\boldsymbol{\beta}|\mathbf{u},\sigma_e^2,\mathbf{y}) \propto \exp\{-1/2(\boldsymbol{\beta}-\tilde{\boldsymbol{\beta}})\mathbf{X}'\mathbf{X}(\boldsymbol{\beta}-\tilde{\boldsymbol{\beta}})/\sigma_e^2\}. \quad (11.19)$$

Thus

$$\boldsymbol{\beta}|\mathbf{u},\sigma_e^2,\mathbf{y} \sim N_p[\tilde{\boldsymbol{\beta}},(\mathbf{X}'\mathbf{X})^{-1}\sigma_e^2]. \quad (11.20)$$

Because this posterior distribution is multivariate normal, any linear combination $\mathbf{k}'\boldsymbol{\beta}$ is also normal, and any joint or marginal inferences about elements of $\boldsymbol{\beta}$ can be completed using standard least-squares analysis with $\mathbf{v}$ as data. The Bayesian interpretation would be, of course, radically different from the frequentist. Again, (11.20) does not refer to performance over a series of repetitions of the trial.

11.4.2 Posterior Distribution of u When β is Known

We address now the situation where the location vector $\boldsymbol{\beta}$ is known "without" error. This means that the prior density function of $\boldsymbol{\beta}$ is a point mass. This would occur, for example, in a sire evaluation problem where all "nuisance" parameters (e.g., herd and parity effects) are known, *a priori*. It should be noted that this is the setting of classical selection index theory (Smith 1936; Hazel 1943; Henderson 1963). Here, the vector $\mathbf{w}=\mathbf{y}-\mathbf{X}\boldsymbol{\beta}$ can be treated as data and the posterior density of $\mathbf{u}$ is

$$p(\mathbf{u}|\boldsymbol{\beta},\sigma_u^2,\sigma_e^2,\mathbf{y}) \propto \exp\{-1/2[(\mathbf{w}-\mathbf{Z}\mathbf{u})'(\mathbf{w}-\mathbf{Z}\mathbf{u})+\mathbf{u}'\mathbf{A}^{-1}\mathbf{u}\alpha]/\sigma_e^2\}. \quad (11.21)$$

Using an argument similar to the one leading to (11.11), the posterior density (11.21) can be expressed as:

$$p(\mathbf{u}|\boldsymbol{\beta},\sigma_u^2,\sigma_e^2,\mathbf{y}) \propto \exp[-1/2(\mathbf{u}-\bar{\mathbf{u}})(\mathbf{Z}'\mathbf{Z}+\mathbf{A}^{-1}\alpha)(\mathbf{u}-\bar{\mathbf{u}})/\sigma_e^2], \quad (11.22)$$

where

$$\bar{\mathbf{u}} = (\mathbf{Z}'\mathbf{Z}+\mathbf{A}^{-1}\alpha)^{-1}\mathbf{Z}'(\mathbf{y}-\mathbf{X}\boldsymbol{\beta}).$$

Hence

$$\mathbf{u}|\boldsymbol{\beta},\sigma_u^2,\sigma_e^2,\mathbf{y} \sim N_q[\bar{\mathbf{u}},(\mathbf{Z}'\mathbf{Z}+\mathbf{A}^{-1}\alpha)^{-1}\sigma_e^2] \tag{11.23}$$

so decisions or inferences about any linear combination **Pu** (**P** is typically a matrix of relative economic weights in the multiple trait situation) can be made by recourse to normal distribution theory.

11.5 Unknown Variance Components

When the dispersion parameters are unknown the problem (especially interval estimation) becomes difficult because the resulting distributions are much more complex. However, several options for point estimation are available. The point of view taken here is that given a series of distributions, the most appropriate one for inferences about a vector θ_1 is the distribution that has achieved a higher degree of marginalization with respect to the nuisance parameters θ_2. In other words, inferences about θ_1 from the distribution:

$$p(\theta_1|y) = \int_{R_{\theta_2}} p(\theta_1,\theta_2|y)\, d\theta_2 \tag{11.24}$$

would be "better" than those from:

$$p(\theta_1,\theta_2|y), \tag{11.25}$$

because point or interval estimators derived from the latter do not take into account uncertainty or "error" about θ_2.

11.5.1 Joint Inferences About Location Parameters and Variance Components

Using the Joint Posterior Distribution. An approach suggested by Lindley and Smith (1972) involves finding point estimators of $\boldsymbol{\beta},\mathbf{u},\sigma_u^2$ and σ_e^2 by maximizing (11.9) with respect to these variables. This was discussed by Gianola and Fernando (1986) and by Gianola et al (1986); the point estimates so obtained are known as "maximum a posteriori" (Beck and Arnold 1977), or the values which are most likely in the posterior density. Computing the estimates involves iteration with equations:

$$\begin{bmatrix} \mathbf{X'X} & \mathbf{X'Z} \\ \mathbf{Z'X} & \mathbf{Z'Z}+\mathbf{A}^{-1}\alpha^{[k]} \end{bmatrix} \begin{bmatrix} \boldsymbol{\beta}^{[k]} \\ \mathbf{u}^{[k]} \end{bmatrix} = \begin{bmatrix} \mathbf{X'y} \\ \mathbf{Z'y} \end{bmatrix} \tag{11.26}$$

$$\sigma_u^{2[k+1]} = \frac{[\mathbf{u'A}^{-1}\mathbf{u}]^{[k]}+\upsilon_u s_u^2}{q+\upsilon_u+2} \tag{11.27}$$

and

$$\sigma_e^{2[k+1]} = \frac{[\mathbf{e'e}]^{[k]}+\upsilon_e s_e^2}{n+\upsilon_e+2}\,. \tag{11.28}$$

where [k] is iterate number and $\mathbf{e}=\mathbf{y}-\mathbf{X}\boldsymbol{\beta}-\mathbf{Zu}$. Computations start setting a trial value of α, say $\alpha^{[0]}$, and solving (11.26). The values of $\boldsymbol{\beta}$ and $\mathbf{u}$ so obtained would then be used in (11.27) and (11.28). It should be noted that expressions (11.27) and (11.28) can be viewed as weighted averages of a variance component "obtained from the data" and the corresponding "hyperparameter". For example, one can write

$$\sigma_u^{2[k+1]} = [(q+2)\tilde{\sigma}_u^{2[k]}+\upsilon_u s_u^2]/(q+\upsilon_u+2),$$

where $\tilde{\sigma}^{2[k]}=[\mathbf{u'A}^{-1}\mathbf{u}]^{[k]}/(q+2)$ is the "data" variance component at the current iterate. The preceding expression illustrates that the estimate of the variance component will tend to the prior value s_u^2 either when $q<<\upsilon_u$ or when $[\mathbf{u'A}^{-1}\mathbf{u}]^{[k]}$ is close to 0. The latter can happen when the $\mathbf{u}$-solutions are strongly pulled towards 0, which occurs when α is very large (heritability is low), or when there is little information about $\mathbf{u}$ on the data, e.g., small progeny group sizes, so that the mean of the prior distribution of $\mathbf{u}(\mathbf{0})$ "dominates". This is the main reason why many times the estimate of σ_u^2 becomes null when the "hyperparameter" υ_u is set to zero. Thompson (1980) gave conditions for this to occur in a one-way random model. Harville (1977) attributed the problem to "dependencies" between $\mathbf{u}$ and σ_u^2 in the joint posterior distribution. As it will be seen later, the problem arises again when one searches for the mode of $p(\boldsymbol{\beta},\mathbf{u}|\mathbf{y})$ or of $p(\mathbf{u}|\mathbf{y})$: setting υ_u to 0 can yield modal values identical to 0. In any event, the procedure is seldom satisfactory (even with υ_u different from zero), and the only attractive feature is its relative ease of computation because the expressions calculated do not involve elements of the coefficient matrix in (11.26). This means that all computations can be done in a purely iterative manner; the "indirect" techniques described by Schaeffer and Kennedy (1986), Misztal (1986) and Misztal and Gianola (1987) could be used to advantage here. Hoschele et al. (1987), in a simulation study, found a poor mean squared error performance of the Lindley-Smith procedure when used for variance component estimation in threshold models.

Using the Posterior Distribution of u and of the Variance Components. Gianola et al. (1986) described an alternative to the Lindley and Smith (1972) procedure that takes into account uncertainty on $\boldsymbol{\beta}$. This is accomplished by integrating $\boldsymbol{\beta}$ out of the posterior density (11.9), and then maximizing with respect to $\mathbf{u}, \sigma_u^2$ and σ_e^2. Computations are carried out via iteration with:

$$\left(\mathbf{Z'MZ}+\mathbf{A}^{-1}\alpha^{[k]}\right)\left(\mathbf{u}^{[k]}\right) = \left(\mathbf{Z'My}\right) \tag{11.29}$$

$$\sigma_u^{2[k+1]} = \frac{[\mathbf{u'A}^{-1}\mathbf{u}]^{[k]}+\upsilon_u s_u^2}{q+\upsilon_u+2} \tag{11.30}$$

and

$$\sigma_e^{2[k+1]} = \frac{[\mathbf{e'e}]^{[k]}+\upsilon_e s_e^2}{n-p+\upsilon_e+2}. \tag{11.31}$$

The expression for $\mathbf{u}$ in (11.29) is (11.26) after absorption of $\boldsymbol{\beta}$, and Eqs. (11.30) and (11.31) are similar to (11.27) and (11.28), respectively; the latter has n rather than n-p in the denominator. The difference stems from the integration procedure or, equivalently, from taking into account uncertainty about $\boldsymbol{\beta}$. The procedure poses problems similar to the one described in the preceding section whenever a "flat" prior is used for σ_u^2. We conjecture that the mean squared error performance of this method will be generally unsatisfactory, both for variance components and breeding values, because of strong biases.

Strictly speaking, the procedure is not self-contained if one wishes to make inferences about $\boldsymbol{\beta}$. However, it is possible to approximate its marginal posterior density. In general, one can write:

$$p(\boldsymbol{\beta}|\mathbf{y},\upsilon_u,s_u^2,\upsilon_e,s_e^2) = \int p(\boldsymbol{\beta}|\mathbf{y},\mathbf{u},\sigma_u^2,\sigma_e^2,\upsilon_u,s_u^2,\upsilon_e,s_e^2)$$

$$\cdot p(\mathbf{u},\sigma_u^2,\sigma_e^2|\mathbf{y},\upsilon_u,s_u^2,\upsilon_e,s_e^2)\,d\mathbf{u}\,d\sigma_u^2\,d\sigma_e^2,$$

where the integration is over the space of $\mathbf{u}$, σ_u^2 and σ_e^2. If the joint density $p(\mathbf{u},\sigma_u^2,\sigma_e^2|\mathbf{y},\upsilon_u,s_u^2,\upsilon_e,s_e^2)$ is "peaked", one can use the approximation (Box and Tiao 1973; Gianola and Fernando 1986; Gianola et al. 1986):

$$p(\boldsymbol{\beta}|\mathbf{y},\upsilon_u,s_u^2,\upsilon_e,s_e^2) = p(\boldsymbol{\beta}|\mathbf{y},\mathbf{u}=\mathbf{u}^*,\sigma_u^2=\sigma_u^{2*},\sigma_e^2=\sigma_e^{2*},\upsilon_u,s_u^2,\upsilon_e,s_e^2),$$

where $\mathbf{u}^*, \sigma_u^{2*}$, and σ_e^{2*} are the converged solutions after iteration with (11.29)-(11.31). Now, given $\mathbf{u}$, we are in the situation of (11.20), so approximately

$$\boldsymbol{\beta}|\mathbf{y},\upsilon_u,s_u^2,\upsilon_e,s_e^2 \sim N_p[(\mathbf{X}'\mathbf{X})^{-1}\mathbf{X}'(\mathbf{y}-\mathbf{Z}\mathbf{u}^*),(\mathbf{X}'\mathbf{X})^{-1}\sigma_e^{2*}],$$

which yields approximate inferences about $\boldsymbol{\beta}$. This conditional density depends on the "hyperparameters" through the modal values $\mathbf{u}^*$ and σ_e^{2*}.

Using the Posterior Distribution of $\boldsymbol{\beta}$ and of the Variance Components. The density $p(\boldsymbol{\beta},\sigma_u^2,\sigma_e^2|\mathbf{y},\upsilon_u,s_u^2,\upsilon_e,s_e^2)$ allows one to make joint inferences about $\boldsymbol{\beta}$ and the variances after taking into account uncertainty on the values of the $\mathbf{u}$ vector. The "maximum a posteriori" estimates of $\boldsymbol{\beta}$ and of the variances from this density can be obtained taking advantage of the result:

$$\frac{\delta}{\delta\theta_1}\,\ell n\ f(\theta_1|\mathbf{y},\upsilon_u,s_u^2,\upsilon_e,s_e^2)$$

$$= E[\frac{\delta}{\delta\theta_1}\ \ell n\ f(\theta_1,\theta_2|\mathbf{y},\upsilon_u,s_u^2,\upsilon_e,s_e^2)], \tag{11.32}$$

where $\theta_1'=[\boldsymbol{\beta}',\sigma_u^2,\sigma_e^2],\theta_2=\mathbf{u}$ and the expectation is taken with respect to the distribution $\theta_2|\theta_1,\mathbf{y},\upsilon_u,s_u^2,\upsilon_e,s_e^2$ (Gianola et al. 1986; Foulley et al. 1987; Gianola et al. Chap. 2, this Vol.). This distribution is given in (11.23). Differentiating, taking expectations, setting to 0 and rearranging gives the iterative scheme:

$$\begin{bmatrix} \mathbf{X}'\mathbf{X} & \mathbf{X}'\mathbf{Z} \\ \mathbf{Z}'\mathbf{X} & \mathbf{Z}'\mathbf{Z}+\mathbf{A}^{-1}\alpha \end{bmatrix} \begin{bmatrix} \boldsymbol{\beta}^{[k]} \\ \mathbf{u}^{[k]} \end{bmatrix} = \begin{bmatrix} \mathbf{X}'\mathbf{y} \\ \mathbf{Z}'\mathbf{y} \end{bmatrix} \tag{11.33}$$

$$\sigma_u^{2[k+1]} = \frac{\mathbf{u}^{[k]'}\mathbf{A}^{-1}\mathbf{u}^{[k]}+\mathrm{tr}\mathbf{A}^{-1}\mathbf{C}^{[k]}\sigma_e^{2[k]}+\upsilon_u s_u^2}{q+\upsilon_u+2} \tag{11.34}$$

and

$$\sigma_e^{2[k+1]} = \frac{\mathbf{e}^{[k]'}\mathbf{e}^{[k]}+(q-\mathrm{tr}\mathbf{A}^{-1}\mathbf{C}^{[k]}\alpha^{[k]})\sigma_e^{2[k]}+\upsilon_e s_e^2}{n+\upsilon_e+2}, \tag{11.35}$$

where $\mathbf{C}=(\mathbf{Z'Z}+\mathbf{A}^{-1}\alpha)^{-1}$. When flat priors are employed for variance components, (11.33)-(11.34) define an algorithm for maximum likelihood estimation of $\boldsymbol{\beta}$ and of the variances (Harville 1974, 1977). Note that (11.34) and (11.35) can be viewed as weighted averages of "maximum likelihood" estimates of variances (the divisors being q+2 and n+2 rather than q and n) and of the "hyperparameters" s_u^2 and s_e^2. In other words, there is shrinkage towards the prior distribution. This phenomenon, typical in Bayesian analysis with informative prior distributions, may lead to an improved mean squared error performance over that obtained with maximum likelihood estimates.

While this procedure is not self-contained for prediction of $\mathbf{u}$, it is possible to make inferences about this vector using an argument similar to the one employed in the preceding section. Approximately:

$$p(\mathbf{u}|\mathbf{y},\upsilon_u,s_u^2,\upsilon_e,s_e^2) \simeq p(\mathbf{u}|\mathbf{y},\boldsymbol{\beta}=\boldsymbol{\beta}^{**},\sigma_u^{2**},\upsilon_u,s_u^2,\sigma_e^{2**},\upsilon_e,s_e^2),$$

where $\boldsymbol{\beta}^{**},\sigma_u^{2**},\sigma_e^{2**}$ are the solutions after iteration with (11.33)-(11.35). We are now in the situation of (11.23) so approximately

$$\mathbf{u}|\mathbf{y},\upsilon_u,s_u^2,\upsilon_e,s_e^2 \sim N_q[(\mathbf{Z'Z}+\mathbf{A}^{-1}\alpha^{**})^{-1}\mathbf{Z}'(\mathbf{y}-\mathbf{X}\boldsymbol{\beta}^{**}),(\mathbf{Z'Z}+\mathbf{A}^{-1}\alpha^{**})\sigma_e^{2**}].$$

It should be noted that the mean of the above distribution (the point predictor of $\mathbf{u}$) is given by the $\mathbf{u}$-component of (11.33) evaluated at the stabilized values of $\boldsymbol{\beta}$ and of the variances. The predictor obtained differs from the maximum likelihood estimator of the (classical) conditional mean of $\mathbf{u}$ given $\mathbf{y}$, $\boldsymbol{\beta}$ and the variances, because of the dependence on the "hyperparameters".

11.5.2 Marginal Inferences About Variance Components and Functions Thereof

Using the Joint Density of the Two Variance Components. The marginal density of the variance components when Jeffreys-type priors are employed was given by Gianola et al., Chapter 2, this Volume. With inverted chi-square densities taken as priors for σ_u^2 and σ_e^2, it can be shown that

$$p(\sigma_u^2,\sigma_e^2|\mathbf{y},\upsilon_u,s_u^2,\upsilon_e,s_e^2) \propto (\sigma_e^2)^{-(n-p-q+2+\upsilon_e)/2}$$

$$.\exp\{-1/2[\mathbf{y'y}-\hat{\boldsymbol{\theta}}'\mathbf{W'y}+\upsilon_e s_e^2]/\sigma_e^2\}.(\sigma_u^2)^{-(q+2+\upsilon_u)}$$

$$.\exp\{-1/2\upsilon_u s_u^2\}.|\mathbf{W'W}+\boldsymbol{\Sigma}|^{-1} \quad (11.36)$$

for σ_u^2 and σ_e^2 positive. The "maximum a posteriori" estimates of the variances in the above density can be found iterating with:

$$(\mathbf{Z'MZ}+\mathbf{A}^{-1}\alpha^{[k]})\mathbf{u}^{[k]}=\mathbf{Z'My} \tag{11.37}$$

$$\sigma_u^{2[k+1]}=\frac{[\mathbf{u'A}^{-1}\mathbf{u}+\mathrm{tr}\mathbf{A}^{-1}\mathbf{C}_{uu}\sigma_e^2]^{[k]}+\upsilon_u s_u^2}{q+\upsilon_u+2} \tag{11.38}$$

and

$$\sigma_e^{2[k+1]}=\frac{[\mathbf{e'e}+(p+q)-\mathrm{tr}\mathbf{A}^{-1}\mathbf{C}_{uu}]^{[k]}+\upsilon_e s_e^2}{n+\upsilon_e+2}\ , \tag{11.39}$$

where $\mathbf{C}_{uu}=(\mathbf{Z'MZ}+\mathbf{A}^{-1}\alpha)^{-1}$. When flat priors are employed for the variance components, (11.38) and (11.39) reduce to expressions appearing in the expectation-maximization algorithm for estimation by restricted maximum likelihood (Dempster et al. 1977). The estimates obtained with (11.38) and (11.39) are weighted averages (in the sense described previously) between restricted maximum likelihood estimates and the hyperparameters s_u^2 and s_e^2. Hoschele et al. (1987), working with threshold models, found that the mean squared error of estimators based on expressions similar to (11.38) and (11.39) was smaller than that of solutions calculated with flat priors for the variance components.

The procedure is not self-contained for prediction of **u**, but the marginal distribution of **u** can be approximated (Gianola et al. 1986) as:

$$p(\mathbf{u}|\mathbf{y},\upsilon_u,s_u^2,\upsilon_e,s_e^2)\simeq p(\mathbf{u}|\mathbf{y},\sigma_u^2=\tilde{\sigma}_u^2,\sigma_e^2=\tilde{\sigma}_e^2,\upsilon_u,s_u^2,\upsilon_e,s_e^2),$$

where $\tilde{\sigma}_u^2$ and $\tilde{\sigma}_e^2$ are estimates obtained with (11.38) and (11.39). This is precisely the situation of (11.15), so approximately

$$\mathbf{u}|\mathbf{y},\upsilon_u,s_u^2,\upsilon_e,s_e^2\sim N_q[\tilde{\mathbf{u}},(\mathbf{Z'MZ}+\mathbf{A}^{-1}\tilde{\alpha})^{-1}\tilde{\sigma}_e^2].$$

The mean of the above distribution, or point predictor of **u**, is obtained calculating (11.37) evaluated at the stabilized values $\tilde{\sigma}_e^2$ and $\tilde{\sigma}_u^2$ of the variance components. When flat priors are employed, this corresponds to the **u**-component of the solution to the mixed model equations with the variances replaced by their restricted maximum likelihood estimates (Harville 1974). In the general case, the hyperparameters will enter in a non-trivial manner, especially when there is not much information contained in the data about the variance components.

Approximate inferences about $\boldsymbol{\beta}$ can be obtained using a similar argument as for **u**. Here, one would use (11.14) with **V** and σ_e^2 evaluated at the modal values obtained with (11.38) and (11.39). Advantage should be taken of the mixed model equations when calculating the estimate of $\boldsymbol{\beta}$ and the (approximate) marginal posterior variance.

Using the Ratio of Variance Components. The joint posterior density in (11.9) can be written as

$$p(\boldsymbol{\beta},\mathbf{u},\sigma_u^2,\sigma_e^2|\mathbf{y},s_u^2,\upsilon_u,s_e^2,\upsilon_e) \propto (\sigma_e^2)^{-(n+\upsilon_e+2)/2}(\sigma_u^2)^{-(q+\upsilon_u+2)/2}$$

$$.\exp\{-1/2\upsilon_u s_u^2/\sigma_u^2\}.\exp\{-1/2\upsilon_e s_e^2/\sigma_e^2\}$$

$$.\exp\{-1/2(\mathbf{y}'\mathbf{y}-\hat{\theta}'\mathbf{W}'\mathbf{y})/\sigma_e^2\}.\exp\{-1/2(\theta-\hat{\theta})'(\mathbf{W}'\mathbf{W}+\Sigma)(\theta-\hat{\theta})/\sigma_e^2\}. \quad (11.40)$$

Making the change of variables $v=\sigma_e^2$ and $\alpha=\sigma_e^2/\sigma_u^2$, the determinant of the Jacobian of the transformation is σ_e^2/α^2. Using this and (11.40) gives:

$$p(\boldsymbol{\beta},\mathbf{u},\sigma_e^2,\alpha|\mathbf{y},\upsilon_u,s_u^2,\upsilon_e,s_e^2) \propto (\sigma_e^2)^{-(n+q+\upsilon_e+\upsilon_u+2)/2}\alpha^{(q+\upsilon_u-2)/2}$$

$$.\exp\{-1/2[(\theta-\hat{\theta})'(\mathbf{W}'\mathbf{W}+\Sigma)(\theta-\hat{\theta})+\upsilon_u s_u^2\alpha+\upsilon_e s_e^2+\mathbf{y}'\mathbf{y}-\hat{\theta}'\mathbf{W}'\mathbf{y}]/\sigma_e^2\}.(11.41)$$

The vector θ appears in a normal kernel so it can be integrated readily to obtain:

$$p(\sigma_e^2,\alpha|\mathbf{y},\upsilon_u,s_u^2,\upsilon_e,s_e^2) \propto (\sigma_e^2)^{-(n+\upsilon_e+\upsilon_u+2-p)/2}\alpha^{(q+\upsilon_u-2)/2}$$

$$.\exp\{-1/2(\mathbf{y}'\mathbf{y}-\hat{\theta}'\mathbf{W}'\mathbf{y}+\upsilon_u s_u^2\alpha+\upsilon_e s_e^2)/\sigma_e^2\}|\mathbf{W}'\mathbf{W}+\Sigma|^{-1/2}. \quad (11.42)$$

One can also integrate σ_e^2 out using expression A2.1.6 in Box and Tiao (1973) so

$$p(\alpha|\mathbf{y},\upsilon_u,s_u^2,\upsilon_e,s_e^2)\propto\alpha^{(q+\upsilon_u-2)/2}[\mathbf{y}'\mathbf{y}+\upsilon_e s_e^2+\upsilon_u s_u^2\alpha-\hat{\theta}'\mathbf{W}'\mathbf{y}]^{-(n+\upsilon_e+\upsilon_u-p)/2}$$

$$.|\mathbf{W}'\mathbf{W}+\Sigma|^{-1/2}$$

$$\propto \alpha^{(q+\upsilon_u-2)/2}\left[\prod_{i=1}^{q}(\gamma_i+\alpha)\right]^{-1/2}\left[S_\alpha+\upsilon_u s_u^2\alpha+\upsilon_e s_e^2\right]^{-(n+\upsilon_e+\upsilon_u-p)/2}, \quad (11.43)$$

where $S_\alpha = y'y - \hat{\theta}'W'y$ is the "mixed model" residual sum of squares and the γ_i's are the eigenvalues of a matrix related to **Z'MZ** as described by Gianola et al., Chapter 2, this Volume.

It does not seem possible to find the constant of integration of (11.43) by analytical means, but the density function can be integrated numerically. The range of α depends on the model: in a "sire" model, α varies between 3 and ∞; in an animal model, it can take values between 0 and ∞. The posterior distribution of heritability or of the intra-class coefficient of correlation can be made by transformation of α. For example, in a sire model $\alpha = 4h^{-2} - 1$, where h^2 is heritability, a parameter that varies between 0 and 1. Parameterization in terms of h^2 or of intra-class correlation may facilitate numerical integration because the ranges of integration (in a sire model) are 0-1 and 0-1/4, respectively.

Equation (11.41) also gives the basis for deriving the posterior distribution of $\boldsymbol{\beta}$ and **u**, given α. Regarding the ratio α in (11.41) as a constant, and integrating σ_e^2 out gives

$$p(\boldsymbol{\beta},\mathbf{u}|\mathbf{y},\alpha,\upsilon_u,s_u^2,\upsilon_e,s_e^2) \propto \left[1 + \frac{(\theta-\hat{\theta})'(\mathbf{W}'\mathbf{W}+\Sigma)(\theta-\hat{\theta})}{d(n-p+\upsilon_e+\upsilon_u)}\right]^{-(n-p+\upsilon_e+\upsilon_u+p+q)/2} \quad (11.44)$$

where $d = (\mathbf{y}'\mathbf{y} - \hat{\theta}'\mathbf{W}'\mathbf{y} + \upsilon_u s_u^2 \alpha + \upsilon_e s_e^2)/(n-p+\upsilon_e+\upsilon_u)$. The above is a multivariate t-distribution with mean $\hat{\theta}$, variance-covariance matrix $(\mathbf{W}'\mathbf{W}+\Sigma)^{-1}dk$, where $k = (n-p+\upsilon_e+\upsilon_u)/(n-p+\upsilon_e+\upsilon_u-2)$, and $n-p+\upsilon_e+\upsilon_u$ degrees of freedom. Hence, it follows that $\boldsymbol{\beta}|\mathbf{y},\alpha,\upsilon_u,s_u^2,\upsilon_e,s_e^2$ and $\mathbf{u}|\mathbf{y},\alpha,\upsilon_u,s_u^2,\upsilon_e,s_e^2$ are also distributed as multivariate t (Zellner 1971; Box and Tiao 1973).

It is possible to make inferences about breeding values, marginalized completely with respect to the variance components, by noting that

$$p(\mathbf{u}|\mathbf{y},\upsilon_u,s_u^2,\upsilon_e,s_e^2) = \int_0^1 p(\mathbf{u}|\mathbf{y},h^2,\upsilon_u,s_u^2,\upsilon_e,s_e^2)$$

$$\cdot p(h^2|\mathbf{y},\upsilon_u,s_u^2,\upsilon_e,s_e^2)dh^2. \quad (11.45)$$

The two densities required are (11.43), after transformation to h^2, and (11.44). Hence,

$$E(\mathbf{u}|\mathbf{y},\upsilon_u,s_u^2,\upsilon_e,s_e^2) = \int_0^1 E(\mathbf{u}|\mathbf{y},h^2,\upsilon_u,s_u^2,\upsilon_e,s_e^2)$$

$$\cdot p(h^2|\mathbf{y},\upsilon_u,s_u^2,\upsilon_e,s_e^2)dh^2 \quad (11.46)$$

can be evaluated via numerical integration. The expectation in the integrand of (11.46) is given by the **u**-component of the solution to the mixed model equations evaluated at the "current" (in the course of integration) value of h^2. In a large animal breeding problem the computations are probably formidable, but the current developments in supercomputing suggest that these calculations will be feasible in the near future. However, the principles discussed clearly indicate that there is an exact solution to the problem of prediction of breeding value when there is uncertainty about variance components, at least for the model considered here. This result supersedes the approximation suggested by Gianola et al. (1986): estimate variances by restricted maximum likelihood, and then use the estimates in the mixed model equations as if these were the true parameters.

The problem of interval estimation can be also attacked using the principles outlined above. Suppose one wishes to compute

$$\text{Prob}\{a<\mathbf{k}'\boldsymbol{\beta}+\mathbf{m}'\mathbf{u}<b|\mathbf{y},\upsilon_u,s_u^2,\upsilon_e,s_e^2\},$$

where a and b are arbitrary numbers and **k** and **m** are known, fixed vectors. Let $\mathbf{k}'\hat{\boldsymbol{\beta}}+\mathbf{m}'\hat{\mathbf{u}}$ be the linear combination evaluated via the mixed model equations at a given value of h^2 so

$$\text{Prob}\{a<\mathbf{k}'\boldsymbol{\beta}+\mathbf{m}'\mathbf{u}<b|\mathbf{y},h^2,\upsilon_u,s_u^2,\upsilon_e,s_e^2\}=\text{Prob}\{a^*<t<b^*|\mathbf{y},h^2,\upsilon_u,s_u^2,\upsilon_e,s_e^2\},$$

where a^*,b^* and t are studentized values of a, b and $\mathbf{k}'\boldsymbol{\beta}+\mathbf{m}'\mathbf{u}$, respectively, based on (11.44). The above probability statement is **conditional** on h^2, and it should be noted that a^*,b^* and t are also functions of h^2. An unconditional probability can be obtained integrating the two sides of the above expression with respect to the marginal posterior distribution of h^2. Then

$$\text{Prob}\{a<\mathbf{k}'\boldsymbol{\beta}+\mathbf{m}'\mathbf{u}<b|\mathbf{y},\upsilon_u,s_u^2,\upsilon_e,s_e^2\} = \int_0^1 \text{Prob}\{a^*<t<b^*|\mathbf{y},h^2,\upsilon_u,s_u^2,\upsilon_e,s_e^2\}$$

$$.p(h^2|\mathbf{y},\upsilon_u,s_u^2,\upsilon_e,s_e^2).dh^2,$$

and this can be evaluated by numerical means. This provides the basis for Bayesian "confidence" intervals with uncertainty on h^2 taken into account.

11.5.3 Marginal Inferences About Location Parameters

Joint Inferences About $\boldsymbol{\beta}$ and u. Consider the joint posterior distribution of all parameters in Eq. (11.9). Integrating σ_u^2 and σ_e^2 yields

$$p(\boldsymbol{\beta},\mathbf{u}|\mathbf{y},\upsilon_u,s_u^2,\upsilon_e,s_e^2) \propto \left[1+\frac{(\mathbf{y}-\mathbf{X}\boldsymbol{\beta}-\mathbf{Z}\mathbf{u})'(\mathbf{y}-\mathbf{X}\boldsymbol{\beta}-\mathbf{Z}\mathbf{u})}{\upsilon_e s_e^2}\right]^{-(n+\upsilon_e)/2}$$

$$\cdot\left[1+\frac{\mathbf{u}'\mathbf{A}^{-1}\mathbf{u}}{\upsilon_u s_u^2}\right]^{-(q+\upsilon_u)/2}. \tag{11.47}$$

This density is in the form of a poly-t distribution (Broemeling 1985) and does not lend itself to analytical treatment. One possibility is to derive "maximum a posteriori" estimators of $\boldsymbol{\beta}$ and $\mathbf{u}$ by joint maximization of (11.47) with respect to these variables. Differentiating the log of (11.47), setting to 0 and rearranging gives the scheme:

$$\begin{bmatrix} \mathbf{X}'\mathbf{X} & \mathbf{X}'\mathbf{Z} \\ \mathbf{Z}'\mathbf{X} & \mathbf{Z}'\mathbf{Z}+\mathbf{A}^{-1}w_e/w_u \end{bmatrix} \begin{bmatrix} \tilde{\boldsymbol{\beta}} \\ \tilde{\mathbf{u}} \end{bmatrix} = \begin{bmatrix} \mathbf{X}'\mathbf{y} \\ \mathbf{Z}'\mathbf{y} \end{bmatrix}, \tag{11.48}$$

where

$$w_e = (\tilde{\mathbf{e}}'\tilde{\mathbf{e}}+\upsilon_e s_e^2)/(n+\upsilon_e)$$

$$w_u = (\tilde{\mathbf{u}}'\mathbf{A}^{-1}\tilde{\mathbf{u}}+\upsilon_u s_u^2)/(q+\upsilon_u)$$

can be viewed as "variances". The system (11.48) is not explicit in $\tilde{\boldsymbol{\beta}}$ and $\tilde{\mathbf{u}}$, so it must be solved iteratively. There is a similarity with the Lindley and Smith (1972) setting described in (11.26)-(11.28), but there are no equations for the variances here because these parameters have been integrated out. Therefore, the same problems posed by the Lindley and Smith procedure should be expected here whenever υ_u is set to 0. Because σ_e^2 and σ_u^2 were formally integrated out, it is not possible to argue in this case that the problems stem from "severe" dependencies between $\mathbf{u}$ and σ_u^2, as Harville (1977) conjectured. The difficulty seems to be connected with the behavior of the function log(x) when x tends to 0; this is discussed in more detail in the following section. The performance of estimates obtained with (11.48) has not been studied.

Marginal Inferences About u. With the decomposition of a quadratic form employed previously, and letting $\boldsymbol{\beta}^{\#}=(\mathbf{X}'\mathbf{X})^{-1}\mathbf{X}'\mathbf{w}$, where $\mathbf{w}=\mathbf{y}-\mathbf{Z}\mathbf{u}$, one can write:

$$(\mathbf{y}-\mathbf{X}\boldsymbol{\beta}-\mathbf{Z}\mathbf{u})'(\mathbf{y}-\mathbf{X}\boldsymbol{\beta}-\mathbf{Z}\mathbf{u}) = (\mathbf{w}-\mathbf{X}\boldsymbol{\beta}^{\#})'(\mathbf{w}-\mathbf{X}\boldsymbol{\beta}^{\#})+(\boldsymbol{\beta}-\boldsymbol{\beta}^{\#})'\mathbf{X}'\mathbf{X}(\boldsymbol{\beta}-\boldsymbol{\beta}^{\#}).$$

Using this, the joint posterior density function (11.9) can be expressed as

$$p(\boldsymbol{\beta},\mathbf{u},\sigma_u^2,\sigma_e^2|\mathbf{y},\upsilon_u,s_u^2,\upsilon_e,s_e^2) \propto (\sigma_e^2)^{-(n+\upsilon_e+2)/2}.(\sigma_u^2)^{-(q+\upsilon_u+2)/2}$$

$$.\exp\{-1/2[(\mathbf{w}-\mathbf{X}\boldsymbol{\beta}^{\#})'(\mathbf{w}-\mathbf{X}\boldsymbol{\beta}^{\#})+\upsilon_e s_e^2]/\sigma_e^2\}$$

$$.\exp\{-1/2(\mathbf{u}'\mathbf{A}^{-1}\mathbf{u}+\upsilon_u s_u^2)/\sigma_u^2\}$$

$$.\exp\{-1/2(\boldsymbol{\beta}-\boldsymbol{\beta}^{\#})'\mathbf{X}'\mathbf{X}(\boldsymbol{\beta}-\boldsymbol{\beta}^{\#})/\sigma_e^2\}. \quad (11.49)$$

The vector $\boldsymbol{\beta}$ appears now in a normal kernel so it can be integrated as

$$\int_{R_{\boldsymbol{\beta}}} \exp\{-1/2(\boldsymbol{\beta}-\boldsymbol{\beta}^{\#})'\mathbf{X}'\mathbf{X}(\boldsymbol{\beta}-\boldsymbol{\beta}^{\#})/\sigma_e^2\}d\boldsymbol{\beta} = (2\pi)^{p/2}|(\mathbf{X}'\mathbf{X})^{-1}\sigma_e^2|^{1/2}.$$

Using this in (11.49) allows integrating σ_e^2 and σ_u^2 out to obtain:

$$p(\mathbf{u}|\mathbf{y},\upsilon_u,s_u^2,\upsilon_e,s_e^2) \propto [(\mathbf{y}-\mathbf{Z}\mathbf{u})'\mathbf{M}(\mathbf{y}-\mathbf{Z}\mathbf{u})+\upsilon_e s_e^2]^{-(n-p+\upsilon_e)/2}$$

$$.[\mathbf{u}'\mathbf{A}^{-1}\mathbf{u}+\upsilon_u s_u^2]^{-(q+\upsilon_u)/2}. \quad (11.50)$$

It is difficult to go further without making approximations. Calculation of the optimal selection rule, $E(\mathbf{u}|\mathbf{y},\upsilon_u,s_u^2,\upsilon_e,s_e^2)$, needs to be done by numerical methods, although the integration may not be feasible in most animal breeding situations where q is very large. An approximation to the posterior mean would be the "maximum a posteriori" estimate, or posterior mode of (11.50). The logarithm of (11.50) is

$$L(\mathbf{u}) = \text{constant}-1/2(n-p+\upsilon_e)\ell n[(\mathbf{y}-\mathbf{Z}\mathbf{u})'\mathbf{M}(\mathbf{y}-\mathbf{Z}\mathbf{u})+\upsilon_e s_e^2]$$

$$-1/2(q+\upsilon_u)\ell n[\mathbf{u}'\mathbf{A}^{-1}\mathbf{u}+\upsilon_u s_u^2].$$

Differentiating this expression with respect to $\mathbf{u}$ and setting to $\mathbf{0}$ gives the non-linear system:

$$\left(\mathbf{Z}'\mathbf{M}\mathbf{Z}+\mathbf{A}^{-1}v_e/v_u\right)\left(\tilde{\mathbf{u}}\right)=\left(\mathbf{Z}'\mathbf{M}\mathbf{y}\right), \quad (11.51)$$

where v_u is as w_u in the preceding section, and v_e is as w_e except that its denominator is now $n-p+v_e$. Whereas the solution to this system may, in principle, give a reasonably close approximation to the posterior mean, the equations are expected to behave in a similar fashion to those arising in the procedure of Lindley and Smith (1972), i.e., often giving the trivial solution $\mathbf{u}=0$ if $\upsilon_u=0$. The problem cannot be attributed to a "dependence" between $\mathbf{u}$ and σ_u^2, as mentioned earlier. Some insight can be obtained by examination of L($\mathbf{u}$) above. The argument of the first logarithmic expression cannot be 0 even if υ_e and $\mathbf{u}$ are both zero, reducing to $\mathbf{y'My}$ in this case. However, the argument of the second expression can be 0 only if $\upsilon_u s_u^2$ is null. In this case $\ell n[\mathbf{u'A^{-1}u}]$ tends to $-\infty$ as $\mathbf{u}$ tends to $\mathbf{0}$, so L($\mathbf{u}$) becomes very large and is maximum at $\mathbf{u}=\mathbf{0}$. In general, $\mathbf{u'A^{-1}u}$ will be close to $\mathbf{0}$ when there is little information about the u's (e.g., small progeny group sizes) or when heritability is low.

It is sensible to discuss whether the procedures described in this and in the preceding section represent an improvement over their logical contenders, e.g., best linear unbiased prediction assuming the variances are known, or "best linear unbiased prediction" with variances replaced by restricted maximum likelihood estimates obtained from the data from which predictions are to be made. If variances are known with complete certainty we would be in the situation of (11.10) with $\sigma_e^2=s_e^2$ and $\sigma_u^2=s_u^2$, so these "hyperparameters" would be employed in the prediction problem. If the variances are estimated from the data via restricted maximum likelihood, i.e., ignoring prior knowledge, the procedure would be only approximately optimal (Gianola et al. 1986) but paying a (perhaps severe) penalty for not using information external to the data. Malecot (1947) criticized this approach as a form of scientific paradigm. Alternatively, the procedures described in the last two sections combine current data with past data based knowledge. Clearly, the "weights" $w_e(v_e)$ and $w_u(v_u)$ can be viewed as weighted averages of the hyperparameters s_e^2 and s_u^2, and of "natural" statistics stemming from the data. For example

$$v_u = \frac{q\tilde{s}_u^2+\upsilon_u s_u^2}{q+\upsilon_u}$$

is a weighted averaged of the "natural" estimate of variance $\tilde{s}_u^2 = \tilde{\mathbf{u}}'\mathbf{A}^{-1}\tilde{\mathbf{u}}/q$ and s_u^2. If there is little information in the data about the variance in question, the "hyperparameter" will dominate v_u. Incorporation of this prior knowledge should result in improved prediction of breeding value and, thus, in faster genetic gain.

Marginal Inferences About $\boldsymbol{\beta}$. Deriving the marginal distribution $p(\boldsymbol{\beta}|\mathbf{y},\upsilon_u,s_u^2,\upsilon_e,s_e^2)$ by successive integration of "nuisance" parameters is technically difficult. Alternatively, consider:

$$p(\boldsymbol{\beta}|\mathbf{y},\upsilon_u,s_u^2,\upsilon_e,s_e^2) = \int_{R_\alpha} p(\boldsymbol{\beta}|\mathbf{y},\alpha,\upsilon_u,s_u^2,\upsilon_e,s_e^2)$$

$$\cdot p(\alpha|\mathbf{y},\upsilon_u,s_u^2,\upsilon_e,s_e^2)d\alpha. \qquad (11.52)$$

From (11.44), the first density under the integral sign is a p-variate t-distribution with its mean being the $\boldsymbol{\beta}$-component of $\hat{\theta}$. It is well known (Henderson 1973, 1974) that this component can be calculated using (11.13), which strictly depends on α. Hence, by virtue of the double expectation theorem:

$$E(\boldsymbol{\beta}|\mathbf{y},\upsilon_u,s_u^2,\upsilon_e,s_e^2)=E_\alpha[E(\boldsymbol{\beta}|\mathbf{y},\alpha,\upsilon_u,s_u^2,\upsilon_e,s_e^2)]$$

with the outside expectation taken with respect to the posterior distribution of α given in (11.43). This can be done numerically by "averaging" values of $\boldsymbol{\beta}$ calculated at specific values of α, with the weights obtained from (11.43), which describes the plausibility of different values of α, posterior to the data.

11.6 Choosing a Predictor

The search for "optimal" predictors in animal breeding (Henderson, 1973, 1984; Thompson 1979) has emphasized statistical properties holding over repeated sampling, e.g., unbiasedness, with little reference to utility considerations. Along these lines, Harville (1984), in connection with estimation of variances and covariances observed:

> "... it would seem that alternative variance-component estimators should be compared in terms of the goodness of the corresponding predictors rather than as estimators per se".

Smith and Allaire (1985) and Smith and Hammond (1986) presented a discussion of mate and dairy sire selection, with reference to utility considerations. These authors stressed that the "goodness" of a predictor should consider the rewards (or penalties) stemming from taking a particular action.

In general, the formulation of the problem should start with a definition of "merit", "utility" or "loss". Let θ be an unobservable vector, e.g., breeding values, and $\hat{\theta}$ be a function of the data used to predict θ. From a decision theoretic viewpoint (Judge et al. 1985), one must define a loss function, $L(\theta,\hat{\theta})$, reflecting the consequences (perhaps monetary) of choosing $\hat{\theta}$ as a predictor when θ is the "true" value. The sampling school solution to the problem is to find $\hat{\theta}$ such that the average loss or "risk":

$$E_{y|\theta}[L(\theta,\hat{\theta})] = \int_{R_y} L(\theta,\hat{\theta})\, f(\mathbf{y}|\theta)\, d\mathbf{y} \tag{11.53a}$$

is minimum over repeated sampling. Clearly this is a function of θ, so it is not possible to find a $\hat{\theta}$ minimizing (11.53a) for all $\theta\varepsilon\Omega$. One way around the problem is to average (11.53a) with respect to a prior distribution, $p(\theta)$, and choose $\hat{\theta}$ such that

$$E_{\theta}\{E_{y|\theta}[L(\theta-\hat{\theta})]\} = \int_{R_\theta}\int_{R_y} L(\theta,\hat{\theta})\, f(\mathbf{y}|\theta)\, d\mathbf{y}p(\theta)d\theta \tag{11.53b}$$

is minimum. When the integrals exist, (11.53b) is equal to

$$\int_{R_y}[\int_{R_\theta} L(\theta,\hat{\theta}).f(\mathbf{y}|\theta)p(\theta)/p(\mathbf{y})d\theta]p(\mathbf{y})d\mathbf{y}. \tag{11.53c}$$

Because $f(\mathbf{y}|\theta).p(\theta)/p(\mathbf{y})$ is the posterior density of θ, we see that the term within brackets in (11.53c) is the expected posterior loss. Hence, any $\hat{\theta}$ minimizing expected posterior loss, would also minimize (11.53c) and, thus, average "risk" over repeated sampling. However, the change in order of integration from (11.53b) to (11.53c) can be done only when the integral over R_θ converges. If this is not so, an estimator (predictor) minimizing average "risk" does not exist (Judge et al. 1985). However, one can always obtain an estimator minimizing expected posterior loss provided the posterior density exists. If average "risk" is finite (this in general implies propriety of the prior distribution of θ), the estimator minimizing expected posterior loss will be admissible (Berger 1980) so that there is no "better" estimator. However, this is not so when $p(\theta)$ is improper, where it has been shown that estimates with lower average "risk" (e.g., James-Stein rule) than the Bayesian estimator minimizing expected posterior loss exist.

Consider now the animal breeding situation, model (11.1) with multivariate normality and the quadratic loss function $L(\theta,\hat{\theta}) = (\hat{\theta}-\theta)'(\hat{\theta}-\theta)$. Let $\theta'=[\boldsymbol{\beta}',\mathbf{u}']$ and assume, for simplicity, that the variances are known. If the prior distribution of θ is normal (so both $\boldsymbol{\beta}$

and **u** are normal, *a priori*), BLUP(θ)=E(θ|y) minimizes expected posterior loss and average "risk". Here, BLUP is the best predictor both in the Bayesian and frequentist settings. However, if **β** is "fixed", the corresponding prior distribution of θ is improper and BLUP is only best in the sense of minimizing expected posterior loss; this will not hold, in general, over repeated sampling. One possibility is to transform **y** to a statistic (**w**) not depending on **β**, i.e., a "translation invariant" function. In this class, BLUP(**u**)=E(**u**|**w**) is best in both settings because **β** has been "eliminated" and only the prior distribution of **u** (which is proper), needs to be specified in the computation of average "risk". Unfortunately, **β** is not always a "nuisance" parameter; for example, one may be interested in predicting functions of **β** and **u** ("group" plus sire effects) and here it is unreasonable to "eliminate" **β** from the analysis. In this case all one can say is that BLUP minimizes expected posterior loss but that there may be predictors with better performance over repeated sampling. One class of statistics would be that of biased predictors based on "shrinkage" of **β**. There is no good frequentist theory for this (at least in the mixed model), and most developments in this area are expected to depend on heuristic arguments. The conditional or posterior mean is a sensible criterion to use when loss is quadratic or when interest is on the **β** or **u**'s themselves. As stated earlier, ranking individuals with conditional means maximizes expected merit in a sample having a fixed number of selected individuals; this involves a repeated sampling argument (Bulmer 1980; Goffinet 1983; Fernando and Gianola 1986). Based on the above considerations, BLUP will accomplish so only in the class of linear, translation invariant functions of the data because the conditional mean (in a classical sense) can be calculated only when **β** is known or when it has been "eliminated". The problem is even more complicated when variances are unknown. However, if their prior distributions are proper, there is an unequivocal Bayesian solution along the lines described earlier.

What happens when selection has occurred? We deal with this briefly because other contributors to this volume discuss the problem in detail. If selection is based on any function of the data, and if these data are included in the analysis, the posterior distribution of θ is unaffected by such selection (Gianola and Fernando 1986). Hence, predictors would be constructed ignoring selection, and these would be best in the Bayesian sense. However, these predictors are probably not optimum over repeated sampling.

11.7 Choosing a Model

Selecting a model or functional form for data analysis is a fundamental and difficult issue in animal breeding practice. A model is an abstraction of reality and probably there is no such thing as the "true" model. However, to the extent that events can be described reasonably, models are useful to study complex phenomena in terms of just a few parameters. In animal breeding there is considerable empiricism in model building, because many simplifying assumptions need to be made in order to render the model computable. However, relatively little attention has been given to the problem of how to choose between competing models. The title of this paper would be unjustified without at least a reference to the issue.

Consider choosing between two "grouping" strategies, M_1 and M_2, in a sire evaluation problem with variance components known. A purpose of "grouping" is to account for genetic differences among "subpopulations" of sires due to "physical" origin, e.g., Australia vs. New Zealand, or due to genetic trend. One can use model (11.1) but with assumption (11.3) replaced by

$$\mathbf{u} \sim N(\eta, \mathbf{A}\sigma_u^2), \tag{11.54}$$

where the vector η allows for the possibility that elements of $\mathbf{u}$ may have different means. Suppose there are two alternative hypotheses about the structure of η: M_1, with $\eta=\mathbf{Q}_1\mathbf{g}_1$ and M_2, with $\eta=\mathbf{Q}_2\mathbf{g}_2$; the $\mathbf{Q}$'s are known matrices relating "group" effects $\mathbf{g}$ to η. The "groups" are to be defined arbitrarily and the question is which of the two definitions is more plausible in the light of the data. Suppose further that under M_1, the covariance matrix of $\mathbf{u}$ is $\mathbf{I}\sigma_u^2$, and that under M_2 is as in (11.54). The two models consist then of a grouping strategy without relationships vs. another grouping form that allows for relationships.

Using (11.1), the two models, with $\mathbf{u}=\mathbf{Q}_i\mathbf{g}_i+\varepsilon_i$, can be written as:

$$\mathbf{y} = \mathbf{X}\boldsymbol{\beta}+\mathbf{Z}\mathbf{Q}_i\mathbf{g}_i+\mathbf{Z}\varepsilon_i + \mathbf{e}; \qquad i=1,2$$

with $\varepsilon_1 \sim N(\mathbf{0},\mathbf{I}\sigma_u^2)$ and $\varepsilon_2 \sim N(\mathbf{0},\mathbf{A}\sigma_u^2)$.

Following ideas in Zellner (1971), let x be a discrete random variable taking the value 1 if M_1 is true, or 2 if M_2 is true. Also, let the prior probabilities for each of the models be $p(M_1)= p(x=1)$ and $p(M_2)=p(x=2)$, such that $p(M_1)+p(M_2)=1$. From Bayes theorem:

$$p(x=i,\boldsymbol{\beta},\mathbf{g}_i,\varepsilon_i|\mathbf{y}) = p(\mathbf{y}|x=i,\boldsymbol{\beta},\mathbf{g}_i,\varepsilon_i).p(\boldsymbol{\beta},\mathbf{g}_i,\varepsilon_i|x=i).p(x=i)/p(\mathbf{y}) \quad i=1,2 \tag{11.55}$$

gives the posterior density of $x=i,\boldsymbol{\beta}$, $\mathbf{g}_i$, and ε_i. Above, $p(\boldsymbol{\beta},\mathbf{g}_i,\varepsilon_i|x=i)$ is the joint prior density under M_i, and $p(\mathbf{y}|x=i,\boldsymbol{\beta}, \mathbf{g}_i,\varepsilon_i)$ is the density function of the observations given $M_i,\boldsymbol{\beta},\mathbf{g}_i$ and ε_i. Further,

$$p(\mathbf{y}) = \sum_i \int_{R_\beta} \int_{R_{g_i}} \int_{R_{\varepsilon_i}} p(\mathbf{y}|x=i,\boldsymbol{\beta},\mathbf{g}_i,\varepsilon_i)p(\boldsymbol{\beta},\mathbf{g}_i,\varepsilon_i|x=i)d\boldsymbol{\beta}d\mathbf{g}_i d\varepsilon_i$$

is the marginal density of the data. Integrating $\boldsymbol{\beta}$, $\mathbf{g}_i$ and ε_i out of (11.55), one obtains:

$$p(x{=}i|y) = \frac{p(y|x{=}i).p(x{=}i)}{p(y)}; \quad i{=}1,2 \tag{11.56}$$

as posterior probabilities for each of the models, where $p(y|x{=}i)$ is the marginal likelihood under M_i. The posterior odds (Zellner 1971) in favor of M_1 are then:

$$K_{12} = \frac{p(M_1|y)}{p(M_2|y)} = \frac{p(y|M_1)}{p(y|M_2)} \cdot \frac{p(M_1)}{p(M_2)}, \tag{11.57}$$

so we see that if $p(M_1)=p(M_2)$, the posterior odds in favor of M_1 are proportional to the ratio of the corresponding marginal likelihoods. In this case, the model chosen would be the one which produces the largest marginal likelihood. When the variance components are known, the conditional likelihoods are:

$$p(\mathbf{y}|M_1,\boldsymbol{\beta},\mathbf{g}_1) \propto \exp\{-1/2(\mathbf{y}-\mathbf{X}\boldsymbol{\beta}-\mathbf{Z}\mathbf{Q}_1\mathbf{g}_1)'\mathbf{V}_1^{-1}(\mathbf{y}-\mathbf{X}\boldsymbol{\beta}-\mathbf{Z}\mathbf{Q}_1\mathbf{g}_1)\}$$

and

$$p(\mathbf{y}|M_2,\boldsymbol{\beta},\mathbf{g}_2) \propto \exp\{-1/2(\mathbf{y}-\mathbf{X}\boldsymbol{\beta}-\mathbf{Z}\mathbf{Q}_2\mathbf{g}_2)'\mathbf{V}_2^{-1}(\mathbf{y}-\mathbf{X}\boldsymbol{\beta}-\mathbf{Z}\mathbf{Q}_2\mathbf{g}_2)\}$$

with $\mathbf{V}_1=\mathbf{ZZ}'\sigma_u^2+\mathbf{I}\sigma_e^2$ and $\mathbf{V}_2=\mathbf{ZAZ}'\sigma_u^2+\mathbf{I}\sigma_e^2$. The corresponding marginal likelihoods can be obtained (approximately) evaluating the respective conditional likelihoods at the maximum likelihood estimates of $(\boldsymbol{\beta},\mathbf{g}_i)$. Note that $-2\log\hat{K}_{12}$ is the difference in "residual" sums of squares between the two models. Referring this criterion to a χ^2 distribution with degrees of freedom equal to $|\text{rank}(\mathbf{X},\mathbf{ZQ}_1)-\text{rank}(\mathbf{X},\mathbf{ZQ}_2)|$ can be viewed as a sampling test for the "implicit hypothesis" that $K_{12}=1$. The posterior odds (11.57) can be used as prior odds in analyses of subsequent data sets. It is possible to obtain a more general result but this is beyond the scope of the present paper.

11.8 Prediction of Future Records

The ultimate objective when assessing breeding values is to predict future performance, e.g., new records of a given animal, mean performance of the progeny of a sire, etc. The Bayesian approach uses the concept of predictive density function to make inferences about yet unobserved records, given past data (Zellner 1971). Let $\mathbf{y}$ and $\tilde{\mathbf{y}}$ be past and future observations, respectively. The joint density of $\tilde{\mathbf{y}},\theta'=[\boldsymbol{\beta}',\mathbf{u}']$ and of the variance components σ_u^2, σ_e^2 is

$$p(\tilde{y},\theta,\sigma_u^2,\sigma_e^2|y,\upsilon_u,s_u^2,\upsilon_e,s_e^2) \propto p(\tilde{y}|y,\theta,\sigma_u^2,\sigma_e^2)$$

$$\cdot p(\theta,\sigma_e^2,\sigma_u^2|y,\upsilon_u,s_u^2,\upsilon_e,s_e^2).$$

The predictive density function is

$$p(\tilde{y}|y,\upsilon_u,s_u^2,\upsilon_e,s_e^2) \propto \int_{R_\theta}\int_0^\infty\int_0^\infty p(\tilde{y}|y,\theta,\sigma_u^2,\sigma_e^2)$$

$$\cdot p(\theta,\sigma_e^2,\sigma_u^2|y,\upsilon_u,s_u^2,\upsilon_e,s_e^2)\, d\theta d\sigma_e^2 d\sigma_u^2. \tag{11.58}$$

Thus, the predictive density function is a weighted average of "conditional" predictive densities, the weighting function being the posterior density of the unknown parameters.

For simplicity, so as to illustrate concepts without further algebra, suppose the model for the future observations is:

$$\tilde{\mathbf{y}} = \tilde{\mathbf{X}}\beta+\tilde{\mathbf{Z}}\mathbf{u}+\tilde{\mathbf{e}}, \tag{11.59}$$

where the incidence matrices relate future observations to $\boldsymbol{\beta}$ and $\mathbf{u}$, and $\tilde{\mathbf{e}}\sim N(\mathbf{0},\mathbf{I}\sigma_e^2)$. It will also be assumed that α, the variance ratio, is known, and that the future and past observations are **conditionally** independent. Then using (11.44) in (11.59)

$$E(\tilde{\mathbf{y}}|\mathbf{y},\alpha,\upsilon_u,s_u^2,\upsilon_e,s_e^2) = \tilde{\mathbf{W}}\hat{\theta} \tag{11.60}$$

and

$$Var(\tilde{\mathbf{y}}|\mathbf{y},\alpha,\upsilon_u,s_u^2,\upsilon_e,s_e^2) = [\,\tilde{\mathbf{W}}(\mathbf{W}'\mathbf{W}+\Sigma)^{-1}\,\tilde{\mathbf{W}}'+\mathbf{I}]dk, \tag{11.61}$$

where dk is as defined after (11.44). The derivation of (11.60)-(11.61) has been omitted, but it can be shown that the predictive density function is a multivariate t-distribution with degrees of freedom as in (11.44). With this result, joint or marginal inferences about future observations, or linear combinations thereof, can be completed. Having established a basis to predict future outcomes, the next logical step is to contrast the predictions with the realized values of the future observations, and to decide how much departure from the predictions one is willing to accept. If the discrepancy is excessive, the entire construct should be revised, including models (11.1) and (11.59), all prior distributions, corresponding "hyperparameters", and all assumptions made. As suggested by Box and Tiao (1973) one can change subsets of assumptions and calculate the predictions so obtained. In this manner, it is possible to examine systematically discrepancies between predictions and specific aspects of the model.

Acknowledgements. The support of the Illinois Agriculture Experiment Station and of Grant US-805-84 from BARD-The United States-Israel Binational Agricultural Research and Development Fund for conducting the research reported here is acknowledged. Dr. F.W. Macedo was supported by Contract AID/NE-C-1701 between Purdue University and the University of Tras-os-Montes e Alto Douro. Drs. R.L. Fernando and C.R. Henderson are thanked for helpful discussions.

References

Beck JV, Arnold KJ (1977) Parameter estimation in engineering and science. J Wiley, New York

Berger JO (1980) Statistical decision theory: foundations, concepts and methods. Springer-Verlag, Berlin Heidelberg New York

Box GEP, Tiao GC (1973) Bayesian inference in statistical analysis. Adison-Wesley, Reading

Broemeling LD (1985) Bayesian analysis of linear models. M Dekker, New York

Bulmer MG (1980) The mathematical theory of quantitative genetics. Clarendon Press, Oxford

Bulmer MG (1982) Sire evaluation with best linear unbiased predictors. Biometrics 38:1086-1088

Casella G (1985) An introduction to empirical Bayes data analysis. Am Statistician 39:83-87

Chen CF (1979) Bayesian inference for a normal dispersion matrix and its application to stochastic multiple regression analysis. J R Stat Soc Ser B 41:235-248

Dempfle L (1977) Relation entre BLUP (best linear unbiased prediction) et estimateurs bayesiens. Ann Genet Sel Anim 9:27-32

Dempster AP, Laird NM, Rubin DB (1977) Maximum likelihood from incomplete data via the EM algorithm. J R Stat Soc Ser B 39:1-38

Efron B, Morris C (1973) Stein's estimation rule and its competitors: an empirical Bayes approach. J Am Stat Assoc 68:17-130

Fernando RL, Gianola D (1986) Optimal properties of the conditional mean as a selection criterion. Theor Appl Genet 72:822-825

Foulley JL, Im S, Gianola D, Hoschele I (1987) Empirical Bayes estimation of parameters for n polygenic binary traits. Genet Sel Evol 19:197-224

Gianola D (1982) Theory and analysis of threshold characters. J Anim Sci 54:1079-1096

Gianola D, Fernando RL (1986) Bayesian methods in animal breeding theory. J Anim Sci 63:217-244

Gianola D, Foulley JL, Fernando RL (1986) Prediction of breeding values when variances are not known. In: Dickerson GE, Johnson RK (eds) Proc 3rd World Congr Genet Appl Livest Prod. Agric Commun, Univ Nebraska, Lincoln, Nebraska XII:356-370

Goffinet B (1983) Selection on selected records. Genet Sel Evol 15:91-97

Harvey WR (1982) Least-squares analysis of discrete data. J Anim Sci 54:1067-1071

Harville DA (1974) Bayesian inference for variance components using only error contrasts. Biometrika 61:383-385

Harville DA (1977) Maximum likelihood approaches to variance component estimation and related problems. J Am Stat Assoc 72:320-338

Harville DA (1984) Discussion on interpolation and estimation. In: David HA, David HT (eds) Statistics: an appraisal. Iowa State Univ Press, Ames, pp 281-286

Hazel LN (1943) The genetic basis for constructing selection indexes. Genetics 28:476-490

Henderson CR (1963) Selection index and expected genetic advance. In: Hanson WD, Robinson HR (eds) Statistical genetics and plant breeding. National Academy of Sciences, Washington DC, Publication 982:141-153

Henderson CR (1973) Sire evaluation and genetic trends. In: Proc Anim Breed Genet Symp in Honor of Dr. J.L. Lush. ASAS and ADSA, Champaign, Illinois, pp 10-41

Henderson CR (1974) General flexibility of linear model techniques for sire evaluation. J Dairy Sci 57:963-972

Henderson CR (1976) A simple method for computing the inverse of a numerator relationship matrix used in prediction of breeding values. Biometrics 32:69-83

Henderson CR (1984) Applications of linear models in animal breeding. Univ Guelph, Guelph

Hoschele I, Gianola D, Foulley JL (1987) Estimation of variance components with quasi-continuous data using Bayesian methods. Z Tierz Züchtungsbiol 104:334-349

Jeffreys H (1961) Theory of probability. Oxford, Clarendon Press

Judge GC, Griffiths WE, Hill RC, Lutkepol H, Lee TC (1985) The theory and practice of econometrics. Wiley, New York

Lindley DV, Smith AFM (1972) Bayes estimates for the linear model. J R Stat Soc B 34:1-18

Malecot G (1947) Les criteres statistiques et la subjectivite de la connaisance scientifique. Ann Univ Lyon X:42-74

Misztal I (1986) Survey of some computing methods in BLUP sire evaluation. In: Mixed models-applications and analysis, pp 8-9, Institut für Mathematik, Univ Augsburg, Augsburg

Misztal I, Gianola D (1987) Indirect solution of mixed model equations. J Dairy Sci 70:716-723

Peixoto JL, Harville DA (1986) Comparisons of alternative predictors under the balanced one-way random model. J Am Stat Assoc 81:431-436

Robinson G (1982) That BLUP is a good thing. CSIRO, Division of Mathematics and Statistics, Clayton, Victoria, mimeo 13, pp

Schaeffer LR, Kennedy BW (1986) Computing solutions to mixed model equations. In: Dickerson GE, Johnson RK (eds) Proc 3rd World Congr Genet Appl Livest Prod. Agric Commun, Univ Nebraska, Lincoln, Nebraska, XII:382-393

Smith HF (1936) A discriminant function for plant selection. Ann Eugen 7:240-250

Smith SP, Allaire FR (1985) Efficient selection rules to increase non-linear merit: application in mate selection. Genet Sel Evol 17:387-406

Smith SP, Allaire FR (1986) Analysis of failure times measured in dairy cows: theoretical considerations in animal breeding. J Dairy Sci 69:217-227

Smith SP, Hammond K (1986) The ramifications of portfolio and utility theories for dairy breeding programs. In: Dickerson GE, Johnson RK (eds) Proc 3rd World Congr Genet Appl Livest Prod. Agric Commun, Univ Nebraska, Lincoln, Nebraska IX:101-105

Thompson R (1979) Sire evaluation. Biometrics 35:339-353

Thompson R (1980) Maximum likelihood estimation of variance components. Math Operationsforsch Stat 11:545-561

Zellner A (1971) An introduction to Bayesian inference in econometrics. J Wiley, New York

12 BLUP (Best Linear Unbiased Prediction) and Beyond

D.A. Harville[1]

The problem considered is that of predicting the value of an unobservable random variable w from the value of an observable random vector **y**. This problem is considered under each of four states of knowledge about the joint distribution of w and **y**, ranging from complete knowledge to "no" knowledge. Point predictors, estimators of the mean squared error of prediction, and interval predictors are presented for each case. Both frequentist and Bayesian approaches are discussed and relationships between the two are pointed out. Specifics are given for the prediction of a linear combination of the fixed and random effects in a mixed linear model. The results are illustrated by applying them to some animal breeding data.

12.1 Introduction

Animal breeders and other practitioners are often faced with the problem of making predictions, for example, about the milk production records of the future offspring of a dairy bull. Many such problems can be formulated as special cases of the general problem of predicting the values of unobservable random variables from the value (called the data vector) of an observable random vector. Statistical approaches to this general problem and to its special cases have been considered in many publications.

The purpose of this paper is to present and illustrate a unified approach to statistical prediction. The approach taken is one that reflects the following points of view:

1. Statistical prediction is best understood if it is considered in a rather general framework, with features specific to particular applications being introduced only as necessary.
2. It is informative to regard statistical inference for parameters as a special case of statistical prediction.
3. It is instructive to consider statistical prediction under various states of knowledge about the joint distribution of an unobservable random variable (whose value is to be

[1] Department of Statistics, Iowa State University, Ames, Iowa, USA

predicted) and the observable random vector (on whose value the prediction is to be based).

4. There has been too much emphasis on point prediction and not enough on prediction intervals and on the estimation of the mean squared error of prediction. Prediction intervals and estimates of the mean squared error of prediction are often obtained by acting (naively) as though the estimated values of various unknown parameters were true values.
5. A more extensive use of Bayesian ideas by animal breeders and other practitioners is desirable and is more feasible from a computational standpoint than commonly thought. Many currently used prediction procedures have Bayesian interpretations (as discussed, in an animal breeding context, e.g., by Dempfle 1977 and Gianola and Fernando 1986). The Bayesian approach can be used to devise prediction procedures that are more sensible from both a Bayesian and frequentist perspective than those in current use.

In Section 12.2, the prediction problem is formulated in general terms and in the context of a mixed linear model (that contains a single set of uncorrelated and homoscedastic random effects), and an example is introduced. Then, in Sections 12.3-12.6, respectively, the prediction problem is discussed under each of four successive states of knowledge (about the underlying joint distribution). Both Bayesian and frequentist approaches are considered, and the coverage includes prediction intervals and the estimation of the mean squared error of prediction, as well as point prediction.

12.2 Formulation of the Prediction Problem

12.2.1 Mixed Model

In many animal breeding (and other) applications, it is assumed that the data vector is the value of an $n \times 1$ observable random vector $\mathbf{y}$ that follows a mixed linear model.

One relatively simple, yet important, class of mixed linear models comprises models of the general form

$$\mathbf{y} = \mathbf{X}\boldsymbol{\beta}+\mathbf{Z}\mathbf{s}+\mathbf{e}, \tag{12.1}$$

where $\boldsymbol{\beta}$ is a $p \times 1$ vector of parameters, $\mathbf{X}$ and $\mathbf{Z}$ are known matrices, and $\mathbf{s}$ and $\mathbf{e}$ are unobservable random vectors of dimensions $m \times 1$ and $n \times 1$ with $E(\mathbf{s})=\mathbf{0}$, $E(\mathbf{e})=\mathbf{0}$, $\text{var}(\mathbf{s})=\sigma_s^2\mathbf{I}$, $\text{cov}(\mathbf{s},\mathbf{e})=\mathbf{0}$, and $\text{var}(\mathbf{e})=\sigma_e^2\mathbf{I}$. Here, σ_s^2 and σ_e^2 are scalar-valued parameters, called variance components. The elements $s_1,\ldots, s_m$ of $\mathbf{s}$ consist of a single set of "random effects", while the elements of $\mathbf{e}$ are random "errors".

In some applications (including many animal breeding applications), it is easy to conceive of the values of $s_1,\ldots,s_m$ as being one set in a sequence of m-dimensional sets generated by repeated sampling from the same distribution, in which case the distribution of **s** has a frequentist interpretation. In other applications, it may be difficult to envision the values of $s_1,\ldots,s_m$ as one set in such a sequence, in which case the distribution of **s** might be regarded simply as a prior "opinion". In the latter case, the distribution of **s** might be referred to as a prior distribution, and the variance component σ_s^2 might be referred to as a prior variance.

Let $\gamma=\sigma_s^2/\sigma_e^2$ represent the variance ratio. Assume that $\sigma_e^2>0$. Since σ_s^2 is, by definition, a variance, there is an implicit assumption that σ_s^2 and hence γ are non-negative. Moreover, in some applications (including many animal breeding applications), there is (as a consequence of the interpretation assigned to the elements of **s** and **e**) a known upper bound on γ. Assume then that $0\le\gamma\le u$, where either $u=\infty$ or u is a known, finite constant, or equivalently that $0\le\sigma_s^2\le u\sigma_e^2$.

Define $p^*=\text{rank}(\mathbf{X})$, $r=\text{rank}(\mathbf{X},\mathbf{Z})-\text{rank}(\mathbf{X})$, and $f=n-\text{rank}(\mathbf{X},\mathbf{Z})$. It is assumed that $r>0$ and $f>0$.

Each random variable whose value is to be predicted is typically (in animal breeding applications) of the form $w=\lambda'\boldsymbol{\beta}+\delta'\mathbf{s}$, where λ ($=\mathbf{X}\mathbf{k}$ for some vector **k**) and δ are vectors of specified constants. Note that inference for the parametric function $\lambda'\boldsymbol{\beta}$ is a special case of the prediction of the value of the random variable w [a special case where var(w)=0], though it is somewhat unconventional to use the term prediction in referring to inference for a parametric function.

Observe that

$$E(\mathbf{y}) = \mathbf{X}\boldsymbol{\beta}, \quad E(\lambda'\boldsymbol{\beta}+\delta'\mathbf{s}) = \lambda'\boldsymbol{\beta}, \tag{12.2}$$

$$\text{var}(\mathbf{y}) = \sigma_e^2\mathbf{I}+\sigma_s^2\mathbf{Z}\mathbf{Z}' = \sigma_e^2(\mathbf{I}+\gamma\mathbf{Z}\mathbf{Z}'), \tag{12.3}$$

$$\text{cov}(\mathbf{y}, \lambda'\boldsymbol{\beta}+\delta'\mathbf{s}) = \sigma_s^2\mathbf{Z}\delta = \sigma_e^2\gamma\mathbf{Z}\delta, \tag{12.4}$$

$$\text{var}(\lambda'\boldsymbol{\beta}+\delta'\mathbf{s}) = \sigma_s^2\delta'\delta = \sigma_e^2\gamma\delta'\delta. \tag{12.5}$$

Let $\bar{\boldsymbol{\beta}}$ and $\bar{\mathbf{s}}$ represent the two parts of an arbitrary solution to the linear system

$$\begin{bmatrix} \mathbf{X}'\mathbf{X} & \mathbf{X}'\mathbf{Z} \\ \mathbf{Z}'\mathbf{X} & \mathbf{Z}'\mathbf{Z} \end{bmatrix} \begin{bmatrix} \bar{\boldsymbol{\beta}} \\ \bar{\mathbf{s}} \end{bmatrix} = \begin{bmatrix} \mathbf{X}'\mathbf{y} \\ \mathbf{Z}'\mathbf{y} \end{bmatrix}, \tag{12.6}$$

which consists of the normal equations (NE) obtained by treating both **s** and $\boldsymbol{\beta}$ as vectors of unknown parameters and applying ordinary least-squares. Note that defining $\bar{\mathbf{s}}$ in this way is equivalent to defining it to be an arbitrary solution to the reduced NE obtained by absorbing the first set of NE (12.6) into the second set, i.e., equivalent to the equations

$$\mathbf{C}\bar{\mathbf{s}} = \mathbf{q}, \tag{12.7}$$

where $\mathbf{C}=\mathbf{Z}'(\mathbf{I}-\mathbf{P}_X)\mathbf{Z}$ and $\mathbf{q}=\mathbf{Z}'(\mathbf{I}-\mathbf{P}_X)\mathbf{y}$ with $\mathbf{P}_X=\mathbf{X}(\mathbf{X}'\mathbf{X})^{-}\mathbf{X}'$. ($\mathbf{A}^{-}$ represents an arbitrary generalized inverse of a matrix **A**.)

Following Harville and Fenech (1985, Sect. 3), observe that rank(**C**)=r, let $\Delta_1,\ldots,\Delta_r$ represent the non-zero (and hence positive) characteristic values of **C**, and define **D**=diag $(\Delta_1,\ldots,\Delta_r)$. Take **R** and **U** to be matrices of dimensions $m\times r$ and $m\times(m-r)$ whose columns are orthonormal characteristic vectors of **C** corresponding to the values $\Delta_1,\ldots,\Delta_r,0,\ldots,0$, respectively. That is, **R** is an $m\times r$ matrix such that **R'R=I** and **CR=RD**, and **U** is an $m\times(m-r)$ matrix such that **U'U=I** and **CU=0** (and **R'U=0**).

If **s**, as well as $\boldsymbol{\beta}$, were treated as a vector of unknown parameters, there would exist an $r\times 1$ vector of linearly independent estimable functions of **s**. In fact, one such vector is

$$\mathbf{t} = (t_1,\ldots,t_r)' = \mathbf{D}^{1/2}\mathbf{R}'\mathbf{s} = \mathbf{D}^{-1/2}(\mathbf{RD})'\mathbf{s} = \mathbf{D}^{-1/2}\mathbf{R}'\mathbf{C}\mathbf{s}$$

(Harville and Fenech 1985, Sect. 3). Let

$$\bar{\mathbf{t}} = (\bar{t}_1,\ldots,\bar{t}_r)' = \mathbf{D}^{1/2}\mathbf{R}'\bar{\mathbf{s}} = \mathbf{D}^{-1/2}\mathbf{R}'\mathbf{C}\bar{\mathbf{s}} = \mathbf{D}^{-1/2}\mathbf{R}'\mathbf{q},$$

so that, if **s** (as well as $\boldsymbol{\beta}$) were a vector of unknown parameters, $\bar{t}_1,\ldots,\bar{t}_r$ would be the best linear unbiased estimators of $t_1,\ldots,t_r$, respectively.

Corresponding to NE (12.6) is the residual sum of squares

$$S_e = \mathbf{y}'\mathbf{y}-\bar{\boldsymbol{\beta}}'\mathbf{X}'\mathbf{y}-\bar{\mathbf{s}}'\mathbf{Z}'\mathbf{y} = \mathbf{y}'(\mathbf{I}-\mathbf{P}_X)\mathbf{y}-\bar{\mathbf{s}}'\mathbf{q} = \mathbf{y}'(\mathbf{I}-\mathbf{P}_X)\mathbf{y}-\Sigma_i\bar{t}_i^2.$$

12.2.2 Example

Consider the data of Harville and Fenech (1985), which are reproduced in Table 12.1. These data consist of the weights (at birth) of 62 single-birth male lambs, and come from five distinct population lines. Each lamb was the progeny of 1 of 23 rams, and each lamb had a different dam. Age of dam was recorded as belonging to one of three categories, numbered 1 (1-2 years), 2 (2-3 years), and 3 (over 3 years).

Table 12.1 Birth weights (in pounds) of lambs

Sire	Dam age	Wt.	Sire	Dam age	Wt.	Sire	Dam age	Wt.
Line 1			Line 3			Line 5		
1	1	6.2	1	2	9.0	1	1	11.7
2	1	13.0		3	9.5			12.6
3	1	9.5			12.6	2	1	9.0
		10.1	2	1	11.0		3	11.0
		11.4		2	10.1	3	3	9.0
	2	11.8			11.7			12.0
	3	12.9		3	8.5	4	3	9.9
		13.1			8.8	5	2	13.5
4	1	10.4			9.9	6	2	10.9
	2	8.5			10.9		3	5.9
					11.0	7	2	10.0
Line 2					13.9			12.7
1	3	13.5	3	1	11.6		3	13.2
2	2	10.1		3	13.0			13.3
	3	11.0	4	2	12.0	8	1	10.7
		14.0						11.0
		15.5	Line 4					12.5
3	1	12.0	1	1	9.2		3	9.0
4	1	11.5			10.6			10.2
	3	10.8			10.6			
				3	7.7			
					10.0			
					11.2			
			2	1	10.2			
					10.9			
			3	1	11.7			
				3	9.9			

Let y_{ijkd} represent the weight (at birth) of the dth of those lambs that are the offspring of the kth sire in the jth population line and of a dam belonging to the ith age category. Following Harville and Fenech (1985), assume that y_{ijkd} follows the mixed linear model

$$y_{ijkd} = \mu+\delta_i+\pi_j+s_{jk}+e_{ijkd},$$

where the age effects $(\delta_1,\delta_2,\delta_3)$ and the line effects $(\pi_1,\ldots,\pi_5)$ are fixed effects, where the sire (within line) effects $(s_{11},s_{12},\ldots,s_{58})$ are uncorrelated random effects having mean 0 and common variance σ_s^2, and where the random errors $e_{1111}, e_{1121},\ldots,e_{3582}$ have mean 0 and common variance σ_e^2 and are uncorrelated with each other and with the sire effects. Clearly, this model is expressible as a special case of mixed linear model (12.1).

Let $h^2=4\sigma_s^2/(\sigma_s^2+\sigma_e^2)=4\gamma/(1+\gamma)$. The parametric function h^2 is interpretable (in the context of the example) as a heritability. Since a heritability is inherently less than or equal to one, there is an implicit assumption that $0\le h^2\le 1$ or equivalently that $0\le\gamma\le 1/3$.

One objective in analyzing the lamb-weight data might be the prediction (for each j and k) of the value of the random variable

$$w_{jk} = \mu+\delta_*+\pi_j+s_{jk},$$

where $\delta_*=(22/62)\delta_1+(11/62)\delta_2+(29/62)\delta_3$ is a weighted average of the age effects δ_1, δ_2, and δ_3 (with weights proportional to frequency of representation). The value of w_{jk} can be interpreted as the average birth weight of an infinite number of male offspring of the kth sire in the jth line.

12.2.3 General Prediction Problem

Consider now the general problem of predicting the value of an unobservable random variable w based on the value of an nx1 observable random vector $\mathbf{y}$. Assume the existence of the first- and second-order moments of the joint distribution of w and $\mathbf{y}$, and adopt the following notation: μ_w=E(w), μ_y=E($\mathbf{y}$), v_w =var(w), $\mathbf{v}_{yw}$= cov($\mathbf{y}$,w), and $\mathbf{V}_y$=var($\mathbf{y}$). For simplicity, assume that $\mathbf{V}_y$ is non-singular.

In some applications, the joint distribution of w and $\mathbf{y}$ may be interpretable in terms of long-run frequencies. In other applications, it may be best regarded as a prior "opinion".

Note that, if v_w=0 (in which case $\mathbf{v}_{yw}$=$\mathbf{0}$), then w=μ_w (with probability one), and the problem of predicting the value of w reduces (in essence) to that of inference for μ_w.

The general prediction problem is to be considered under each of the following four states of knowledge:

1. The joint distribution of w and $\mathbf{y}$ is known.
2. Only the first- and second-order moments, or equivalently μ_w, μ_y, v_w, $\mathbf{v}_{yw}$, and $\mathbf{V}_y$, are known.
3. Only v_w, $\mathbf{v}_{yw}$, and $\mathbf{V}_y$ are known, while μ_y=$\mathbf{X}\boldsymbol{\beta}$ and μ_w=$\lambda'\boldsymbol{\beta}$, where $\boldsymbol{\beta}$ is a px1 vector of unknown parameters, $\mathbf{X}$ is a known matrix of rank p*, and λ (=$\mathbf{X}'\mathbf{k}$ for some $\mathbf{k}$) is a vector of specified constants.
4. The only knowledge is that v_w and the elements of $\mathbf{v}_{yw}$ and $\mathbf{V}_y$ are specified functions of an unknown parameter vector $\theta=(\theta_1,...,\theta_q)'$ whose value is restricted to a specified space Ω, while μ_w and μ_y are of the same form as in the third state (with the joint parameter space of the two vectors $\boldsymbol{\beta}$ and θ being the product of the individual spaces).

In considering prediction under the fourth state of knowledge, resort is sometimes made to one of the following two assumptions:

S. The joint distribution of $e_w = w - \mu_w$ and $\mathbf{e}_y = \mathbf{y} - \boldsymbol{\mu}_y$ is a symmetric distribution (symmetric in the sense that the joint distribution of $-e_w$ and $-\mathbf{e}_y$ is the same as that of e_w and $\mathbf{e}_y$) that is known up to the value of θ.

N. The joint distribution of e_w and $\mathbf{e}_y$ is MVN (multivariate normal).

Note that Assumption N implies (and hence is stronger than) Assumption S.

Clearly, the problem of predicting the value of a random variable of the form $w = \boldsymbol{\lambda}'\boldsymbol{\beta} + \boldsymbol{\delta}'\mathbf{s}$ under mixed linear model (12.1) can be formulated as a special case of the general prediction problem. [In so doing, it is assumed (for definiteness) that $\theta = (\gamma, \sigma_e^2)'$]. In this special case, the first through fourth states of knowledge correspond respectively to the following four assumptions:

1. The vector $\mathbf{X}\boldsymbol{\beta}$ is known, and the joint distribution of $\mathbf{s}$ and $\mathbf{e}$ is known.
2. The quantities $\mathbf{X}\boldsymbol{\beta}$, γ and σ_e^2 are known.
3. Only γ and σ_e^2 are known.
4. The parameters γ and σ_e^2 are unknown, as are $\mathbf{X}\boldsymbol{\beta}$ and the general form of the joint distribution of $\mathbf{s}$ and $\mathbf{e}$.

Terminology. Let $d(\mathbf{y})$ represent an arbitrary (point) predictor of the value of w. The difference $d(\mathbf{y})-w$ is called the **prediction error**. Following, for example, Goldberger (1962) and Henderson (1963), the predictor $d(\mathbf{y})$ is said to be **unbiased** if $E[d(\mathbf{y})-w] \equiv 0$, that is, if the expected value of the prediction error is identically equal to zero. (Unless otherwise indicated, expectations are defined with respect to the joint distribution of w and $\mathbf{y}$ and are unconditional. Further, identically equal means equal for all possible joint distributions, what distributions are possible depends on the assumed state of knowledge and on any additional assumptions about the joint distribution.) Note that $d(\mathbf{y})$ is unbiased if and only if $E[d(\mathbf{y})] \equiv \mu_w$, that is, if and only if the expected value of $d(\mathbf{y})$ is identically equal to the expected value of the random variable w whose value is to be predicted. Note also that

$$E[d(\mathbf{y})-w] = E\{E[d(\mathbf{y})-w|\mathbf{y}]\}. \tag{12.8}$$

The quantity $E\{[d(\mathbf{y})-w]^2\}$ is called the **mean squared error** (MSE) of the predictor $d(\mathbf{y})$. Note that if $d(\mathbf{y})$ is an unbiased predictor, then

$$E[(d(\mathbf{y})-w)^2] = \mathrm{var}[d(\mathbf{y})-w],$$

that is, the MSE equals the variance of the prediction error. Note also that

$$E[(d(\mathbf{y})-w)^2] = E\{E[(d(\mathbf{y})-w)^2|\mathbf{y}]\}. \tag{12.9}$$

Let $S(\mathbf{y})$ represent an arbitrary set (which may vary with the value of $\mathbf{y}$) of values of w. If $\Pr[w \varepsilon S(\mathbf{y})] \equiv 1-\alpha$, then $S(\mathbf{y})$ is said to be an **(exact) 100(1-α)% prediction set**. If $\Pr[w \varepsilon S(\mathbf{y})]$ only "approximates", rather than equals, $1-\alpha$ for some or all of the possible joint distributions of w and **y**, then $S(\mathbf{y})$ is called an **approximate 100(1-α)% prediction set**. Note that

$$\Pr[w \varepsilon S(\mathbf{y})] = E\{\Pr[w \varepsilon S(\mathbf{y}) | \mathbf{y}]\}. \tag{12.10}$$

12.3 State 1: Joint Distribution Known

Suppose that the joint distribution of w and **y** is known. Then, results on point and interval prediction are obtained as immediate consequences of well-known results on Bayesian statistics.

12.3.1 Point Prediction

When the joint distribution of w and **y** is viewed as a prior opinion, the conditional MSE $E[(d(\mathbf{y})-w)^2 | \mathbf{y}]$ of a (point) predictor $d(\mathbf{y})$ is interpretable as a Bayes expected loss and the (unconditional) MSE $E[(d(\mathbf{y})-w)^2]$ is interpretable as a Bayes risk (Berger 1985, Sect. 1.3).

The conditional MSE $E[(d(\mathbf{y})-w)^2 | \mathbf{y}]$ is minimized uniquely by setting $d(\mathbf{y})=E(w|\mathbf{y})$, as is well-known (e.g., Berger 1985, Sect. 4.4.2) and as can be readily verified by observing that for "any" function $d(\cdot)$,

$$\begin{aligned} E[(d(\mathbf{y})-w)^2 | \mathbf{y}] &= E[(d(\mathbf{y})-E(w|\mathbf{y})+E(w|\mathbf{y})-w)^2 | \mathbf{y}] \\ &= [d(\mathbf{y})-E(w|\mathbf{y})]^2+E[(E(w|\mathbf{y})-w)^2 | \mathbf{y}] \qquad (12.11) \\ &\geq E[(E(w|\mathbf{y})-w)^2 | \mathbf{y}] \end{aligned}$$

with equality holding if and only if $d(\mathbf{y})=E(w|\mathbf{y})$. Further, it follows from results (12.11) and (12.9) that

$$E[(d(\mathbf{y})-w)^2] = E\{[d(\mathbf{y})-E(w|\mathbf{y})]^2\}+E[(E(w|\mathbf{y})-w)^2] \tag{12.12}$$

$$\geq E[(E(w|\mathbf{y})-w)^2] \tag{12.13}$$

[with equality holding if and only if $d(\mathbf{y})=E(w|\mathbf{y})$ with probability one], so that the (unconditional) MSE $E[(d(\mathbf{y})-w)^2]$ is also minimized "uniquely" by setting $d(\mathbf{y})=E(w|\mathbf{y})$.

Thus, the conditional mean E(w|y) is the optimum predictor in the sense that it has minimum MSE (both conditionally and unconditionally).

Obviously, E[E(w|y)-w|y]=0, implying [in light of result (12.8)] that E(w|y) is an unbiased predictor. Further, the conditional MSE of the predictor E(w|y) is

$$E[(E(w|\mathbf{y})-w)^2|\mathbf{y}] = \mathrm{var}(w|\mathbf{y}),$$

and, according to result (12.9), its (unconditional) MSE is

$$E[(E(w|\mathbf{y})-w)^2] = E[\mathrm{var}(w|\mathbf{y})]. \tag{12.14}$$

If the joint distribution of w and **y** is MVN, then

$$E(w|\mathbf{y}) = \mu_w+\mathbf{v}_{yw}'\mathbf{V}_y^{-1}(\mathbf{y}-\mu_y),$$

$$\mathrm{var}(w|\mathbf{y}) = E[\mathrm{var}(w|\mathbf{y})] = v_w-\mathbf{v}_{yw}'\mathbf{V}_y^{-1}\mathbf{v}_{yw} \tag{12.15}$$

(e.g., Searle 1971, Sect. 2.4).

The predictor E(w|y) is uncorrelated with its prediction error. In fact, for an "arbitrary" function d(·),

$$\mathrm{cov}[d(\mathbf{y}),E(w|\mathbf{y})-w] = E[d(\mathbf{y})(E(w|\mathbf{y})-w)]$$

$$= E\{E[d(\mathbf{y})(E(w|\mathbf{y})-w)|\mathbf{y}]\} = 0.$$

12.3.2 Interval Prediction

For each value of **y**, take $S_1(\mathbf{y})$ to be any set (of values of w) such that $\Pr[w\in S_1(\mathbf{y})|\mathbf{y}]=1-\alpha$. The set $S_1(\mathbf{y})$ is an exact 100(1-α)% prediction set, as is evident from result (12.10). A Bayesian would refer to $S_1(\mathbf{y})$ as a 100(1-α)% credible set (Berger 1985, Sect. 4.3.2).

Suppose now that, for each value of **y**, the conditional distribution of w given **y**, is absolutely continuous, in which case there exists a function f(w|y), known as the p.d.f. (probability density function) of the distribution, such that

$$\Pr[w\in S(\mathbf{y})|\mathbf{y}] = \int_{S(\mathbf{y})}f(w|\mathbf{y})dw$$

for "any" set S(**y**). Take $S_1^h(\mathbf{y})$ to be a set such that (1) $\Pr[w\in S_1^h(\mathbf{y})|\mathbf{y}]=1-\alpha$ and (2) $f(w^*|\mathbf{y})\le f(w|\mathbf{y})$ for every value of $w\in S_1^h(\mathbf{y})$ and for every $w^*\notin S_1^h(\mathbf{y})$. By definition, the conditional probability of coverage of $S_1^h(\mathbf{y})$ given $\mathbf{y}$ equals $1-\alpha$ [implying, in particular, that it is a $100(1-\alpha)\%$ prediction set], and its "length" is less than or equal to that of any other prediction set whose conditional probability of coverage equals $1-\alpha$ (Box and Tiao 1973, Sect. 2.8). A Bayesian would refer to $S_1^h(\mathbf{y})$ as a $100(1-\alpha)\%$ HPD (highest posterior density) credible set (Berger 1985, Sect. 4.3.2).

Let z_α represent the upper-α point of the standard normal distribution, and take $\alpha(1)$ and $\alpha(2)$ to be any two non-negative constants such that $\alpha(1)+\alpha(2)=\alpha$. If the conditional distribution of w given $\mathbf{y}$ is normal, then every interval of the form

$$E(w|\mathbf{y})-z_{\alpha(2)}[\mathrm{var}(w|\mathbf{y})]^{1/2} \le w \le E(w|\mathbf{y})+z_{\alpha(1)}[\mathrm{var}(w|\mathbf{y})]^{1/2} \tag{12.16}$$

is a $100(1-\alpha)\%$ prediction interval. In fact, the conditional probability of coverage of interval (12.16) (given $\mathbf{y}$) equals $1-\alpha$. Moreover, if the conditional distribution of w (given $\mathbf{y}$) is normal, then the prediction set $S_1^h(\mathbf{y})$ reduces to the interval

$$E(w|\mathbf{y})-z_{\alpha/2}[\mathrm{var}(w|\mathbf{y})]^{1/2} \le w \le E(w|\mathbf{y})+z_{\alpha/2}[\mathrm{var}(w|\mathbf{y})]^{1/2}$$

obtained by setting $\alpha(2)=\alpha(1)=\alpha/2$ in interval (12.16).

12.3.3 Special Case: Mixed Linear Model

Suppose that $\mathbf{y}$ follows the mixed linear model (12.1) and that $w=\lambda'\boldsymbol{\beta}+\delta'\mathbf{s}$. Suppose further that $\mathbf{X}\boldsymbol{\beta}$, γ and σ_e^2 are known and that the form of the joint distribution of $\mathbf{s}$ and $\mathbf{e}$ is known to be MVN, in which case the joint distribution of w and $\mathbf{y}$ is known; it is MVN and μ_w, μ_y, v_w, $\mathbf{v}_{yw}$, and $\mathbf{V}_y$ [which are given by (12.2)-(12.5)] are known.

The conditional mean and variance of w are

$$E(w|\mathbf{y}) = \lambda'\boldsymbol{\beta}+\gamma\delta'\mathbf{Z}'(\mathbf{I}+\gamma\mathbf{Z}\mathbf{Z}')^{-1}(\mathbf{y}-\mathbf{X}\boldsymbol{\beta}), \tag{12.17}$$

$$\mathrm{var}(w|\mathbf{y}) = \sigma_e^2\gamma\delta'[\mathbf{I}-\gamma\mathbf{Z}'(\mathbf{I}+\gamma\mathbf{Z}\mathbf{Z}')^{-1}\mathbf{Z}]\delta. \tag{12.18}$$

Representations more suitable for computational purposes than (12.17) and (12.18) are obtained by exploiting the identity

$$\mathbf{Z}'(\mathbf{I}+\gamma\mathbf{Z}\mathbf{Z}')^{-1} = (\mathbf{I}+\gamma\mathbf{Z}'\mathbf{Z})^{-1}\mathbf{Z}', \tag{12.19}$$

which can be derived by pre- and post-multiplying both sides of the identity $(\mathbf{I}+\gamma\mathbf{Z}'\mathbf{Z})\mathbf{Z}'=\mathbf{Z}'(\mathbf{I}+\gamma\mathbf{Z}\mathbf{Z}')$ by $(\mathbf{I}+\gamma\mathbf{Z}'\mathbf{Z})^{-1}$ and $(\mathbf{I}+\gamma\mathbf{Z}\mathbf{Z}')^{-1}$, respectively. Substitution of expression (12.19) in (12.17) and (12.18) gives

$$E(w|\mathbf{y}) = \boldsymbol{\lambda}'\boldsymbol{\beta}+\gamma\boldsymbol{\delta}'(\mathbf{I}+\gamma\mathbf{Z}'\mathbf{Z})^{-1}\mathbf{Z}'(\mathbf{y}-\mathbf{X}\boldsymbol{\beta}) = \boldsymbol{\lambda}'\boldsymbol{\beta}+\gamma\boldsymbol{\delta}'\mathbf{u}, \tag{12.20}$$

where $\mathbf{u}$ is the (unique) solution to $(\mathbf{I}+\gamma\mathbf{Z}'\mathbf{Z})\mathbf{u}=\mathbf{Z}'(\mathbf{y}-\mathbf{X}\boldsymbol{\beta})$, and

$$\begin{aligned} \mathrm{var}(w|\mathbf{y}) &= \sigma_e^2\gamma\boldsymbol{\delta}'[\mathbf{I}-(\mathbf{I}+\gamma\mathbf{Z}'\mathbf{Z})^{-1}(\gamma\mathbf{Z}'\mathbf{Z})]\boldsymbol{\delta} \\ &= \sigma_e^2\gamma\boldsymbol{\delta}'(\mathbf{I}+\gamma\mathbf{Z}'\mathbf{Z})^{-1}\boldsymbol{\delta}. \end{aligned} \tag{12.21}$$

The advantage of expressions (12.20) and (12.21) over (12.17) and (12.18) is that the dimensions of the matrix $\mathbf{I}+\gamma\mathbf{Z}'\mathbf{Z}$, whose inverse enters in (12.20) and (12.21), are only $m \times m$, whereas the dimensions of the matrix $\mathbf{I}+\gamma\mathbf{Z}\mathbf{Z}'$, whose inverse appears in (12.17) and (12.18), are $n \times n$.

12.4 State 2: Only First and Second Moments Known

Suppose that μ_w, $\boldsymbol{\mu}_y$, v_w, $\mathbf{v}_{yw}$, and $\mathbf{V}_y$ are known but that there is no further knowledge about the joint distribution of w and $\mathbf{y}$. Then, $E(w|\mathbf{y})$ and the set $S_1(\mathbf{y})$ are not determinable from the available information (and hence can no longer be used for prediction).

12.4.1 Best Linear (Point) Prediction

A possible (point) predictor is

$$\eta(\mathbf{y}) = \mu_w+\mathbf{v}_{yw}'\mathbf{V}_y^{-1}(\mathbf{y}-\boldsymbol{\mu}_y) = \tau+\mathbf{v}_{yw}'\mathbf{V}_y^{-1}\mathbf{y}, \tag{12.22}$$

where $\tau=\mu_w-\mathbf{v}_{yw}'\mathbf{V}_y^{-1}\boldsymbol{\mu}_y$. Note that if the joint distribution of w and $\mathbf{y}$ is MVN, then $\eta(\mathbf{y})=E(w|\mathbf{y})$.

Clearly, $E(\eta(\mathbf{y})-w)=0$, that is, $\eta(\mathbf{y})$ is an unbiased predictor. Moreover, $\eta(\mathbf{y})$ is the best (minimum MSE) linear predictor (e.g., Rao 1965, Sect. 4a.11). To see this, observe that

$$\mathrm{cov}[\mathbf{y},\eta(\mathbf{y})-w] = E[\mathbf{y}(\eta(\mathbf{y})-w)] = \mathbf{0}, \tag{12.23}$$

and hence that for any scalar c and any $n \times 1$ vector $\mathbf{a}$,

$$E[(c+\mathbf{a}'\mathbf{y}-w)^2] = E[(c+\mathbf{a}'\mathbf{y}-\eta(\mathbf{y})+\eta(\mathbf{y})-w)^2]$$

$$= E\{[c+\mathbf{a}'\mathbf{y}-\eta(\mathbf{y})]^2\}+E[(\eta(\mathbf{y})-w)^2] \geq E[(\eta(\mathbf{y})-w)^2],$$

with equality holding if and only if $c=\tau$ and $\mathbf{a}'=\mathbf{v}'_{yw}\mathbf{V}_y^{-1}$.

Define $v_e=v_w-\mathbf{v}'_{yw}\mathbf{V}_y^{-1}\mathbf{v}_{yw}$. Then, the MSE of the predictor $\eta(\mathbf{y})$ is

$$E[(\eta(\mathbf{y})-w)^2] = \text{var}[\eta(\mathbf{y})-w] = v_e, \tag{12.24}$$

as is easily verified. The predictor $\eta(\mathbf{y})$ and its MSE v_e coincide with Hartigan's (1969) definitions of the linear expectation and linear variance of w given $\mathbf{y}$.

It follows from result (12.12) that for any scalar c and any $n \times 1$ vector $\mathbf{a}$,

$$E\{[c+\mathbf{a}'\mathbf{y}-E(w|\mathbf{y})]^2\} = E[(c+\mathbf{a}'\mathbf{y}-w)^2]-E[(E(w|\mathbf{y})-w)^2]. \tag{12.25}$$

Thus, since $E[(E(w|\mathbf{y})-w)^2]$ does not involve c or $\mathbf{a}$ and since $E[(c+\mathbf{a}'\mathbf{y}-w)^2]$ is minimized (uniquely) by setting $c=\tau$ and $\mathbf{a}'=\mathbf{v}'_{yw}\mathbf{V}_y^{-1}$, $\eta(\mathbf{y})$ is the BLA (best linear approximation) to $E(w|\mathbf{y})$ in the sense that the unique values of c and $\mathbf{a}$ that minimize the expected squared difference $E\{[c+\mathbf{a}'\mathbf{y}-E(w|\mathbf{y})]^2\}$ are $c=\tau$ and $\mathbf{a}=\mathbf{V}_y^{-1}\mathbf{v}_{yw}$.

Observe that as a consequence of results (12.24), (12.13), and (12.14),

$$v_e = E[(\eta(\mathbf{y})-w)^2] \geq E[(E(w|\mathbf{y})-w)^2] = E[\text{var}(w|\mathbf{y})],$$

implying [in light of result (12.15)] that $E[\text{var}(w|\mathbf{y})]$ attains its maximum value when the joint distribution of w and $\mathbf{y}$ is MVN. Observe also that it follows from result (12.23) that $\text{cov}[\eta(\mathbf{y}), \eta(\mathbf{y})-w]=0$, that is, $\eta(\mathbf{y})$ is uncorrelated with its prediction error.

Note that for $\eta(\mathbf{y})$ to be determinable, complete knowledge of μ_w, $\boldsymbol{\mu}_y$, v_w, $\mathbf{v}_{yw}$, and $\mathbf{V}_y$ is unnecessary. Rather, knowledge of the scalar τ and the vector $\mathbf{v}'_{yw}\mathbf{V}_y^{-1}$ suffices. Similarly, to determine v_e, it is not necessary to have complete knowledge of μ_w, $\boldsymbol{\mu}_y$, v_w, $\mathbf{v}_{yw}$, and $\mathbf{V}_y$.

12.4.2 Interval Prediction (Frequentist Approach)

The mean and variance of the distribution of the error $\eta(\mathbf{y})-w$ of the predictor $\eta(\mathbf{y})$ are known to be 0 and v_e, respectively. However, nothing else is known about the distribution of $\eta(\mathbf{y})-w$.

One approach to obtaining an approximate 100(1-α)% prediction set (for the value of w) is based on assigning some distribution (having mean 0 and variance v_e) to $\eta(\mathbf{y})$-w and acting as though the assigned distribution is the true distribution. Let S_2^* represent any set of values of the prediction error $\eta(\mathbf{y})$-w such that (under the assigned distribution) $\Pr[\eta(\mathbf{y})-w \in S_2^*]=1-\alpha$. For each value of $\mathbf{y}$, define $S_2(\mathbf{y})$ to be the set of values of w such that $\eta(\mathbf{y})-w \in S_2^*$. The set $S_2(\mathbf{y})$ can be regarded as an approximate 100(1-α)% set for the value of w, if the true distribution of $\eta(\mathbf{y})$-w coincides with the assigned distribution, then the probability $\Pr[w\in S_2(\mathbf{y})]$ that the set $S_2(\mathbf{y})$ will include the value of w is exactly 1-α.

Suppose now that the distribution assigned to $\eta(\mathbf{y})$-w is $N(0,v_e)$ (normal with mean 0 and variance v_e). Then, choosing S_2^* to be the interval

$$-z_{\alpha(1)}v_e^{1/2} \le \eta(\mathbf{y})-w \le z_{\alpha(2)}v_e^{1/2},$$

$S_2(\mathbf{y})$ is the interval

$$\eta(\mathbf{y})-z_{\alpha(2)}v_e^{1/2} \le w \le \eta(\mathbf{y})+z_{\alpha(1)}v_e^{1/2}. \tag{12.26}$$

Note that when the joint distribution of w and $\mathbf{y}$ is MVN, interval (12.16) reduces to interval (12.26).

If the joint distribution of w and $\mathbf{y}$ is not MVN, then, at least in the case where the conditional distribution of w given $\mathbf{y}$ is normal, interval (12.26) tends to be conservative. To see this, observe that

1. $\eta(\mathbf{y})$ is the BLA to $E(w|\mathbf{y})$ $\{$and $E[\eta(\mathbf{y})]=\mu_w=E[E(w|\mathbf{y})]\}$, so that intervals (12.26) and (12.16) tend to be "centered" at about the same point;
2. $v_e \ge E[\mathrm{var}(w|\mathbf{y})]$, so that interval (12.26) tends to be longer than interval (12.16); and
3. when the conditional distibution of w and $\mathbf{y}$ is normal, the probability of coverage of interval (12.16) equals 1-α.

12.4.3 Bayesian Prediction

The Bayesian approach is to complete the specification of the joint distribution of w and $\mathbf{y}$ (by using prior information), in which case the results of Section 12.3 are applicable.

12.5 State 3: Only Variances and Covariances Known

Suppose that v_w, $\mathbf{v}_{yw}$, and $\mathbf{V}_y$ are known and that μ_y and μ_w are of the general form $\mu_y=\mathbf{X}\boldsymbol{\beta}$ and $\mu_w=\lambda'\boldsymbol{\beta}$ ($=\mathbf{k}'\mathbf{X}\boldsymbol{\beta}$), but that there is no further information about the joint distribution of w and $\mathbf{y}$. Then,

$$\tau = (\lambda'-\mathbf{v}_{yw}'\mathbf{V}_y^{-1}\mathbf{X})\boldsymbol{\beta} = (\mathbf{k}'-\mathbf{v}_{yw}'\mathbf{V}_y^{-1})\mathbf{X}\boldsymbol{\beta},$$

and in general τ, and hence $\eta(\mathbf{y})$ and the set $S_2(\mathbf{y})$, are not determinable from the available information (and, as a consequence, can no longer be used for prediction).

12.5.1 Best Linear Unbiased (or Location-Equivariant) Prediction

It is well-known that the BLUE [best (minimum variance) linear unbiased estimator] of τ is

$$\tilde{\tau}(\mathbf{y}) = (\lambda'-\mathbf{v}_{yw}'\mathbf{V}_y^{-1}\mathbf{X})\tilde{\boldsymbol{\beta}},$$

where $\tilde{\boldsymbol{\beta}}$ is any solution to the Aitken equations

$$\mathbf{X}'\mathbf{V}_y^{-1}\mathbf{X}\tilde{\boldsymbol{\beta}} = \mathbf{X}'\mathbf{V}_y^{-1}\mathbf{y} \qquad (12.27)$$

(e.g., Rao 1965, Sect. 4a), and that

$$\text{var}[\tilde{\tau}(\mathbf{y})] = (\lambda'-\mathbf{v}_{yw}'\mathbf{V}_y^{-1}\mathbf{X})(\mathbf{X}'\mathbf{V}_y^{-1}\mathbf{X})^-(\lambda-\mathbf{X}'\mathbf{V}_y^{-1}\mathbf{v}_{yw}).$$

Note that $\tilde{\tau}(\mathbf{y})$ is expressible as $\tilde{\tau}(\mathbf{y})=\boldsymbol{\ell}'\mathbf{y}$, where

$$\boldsymbol{\ell}' = (\lambda'-\mathbf{v}_{yw}'\mathbf{V}_y^{-1}\mathbf{X})(\mathbf{X}'\mathbf{V}_y^{-1}\mathbf{X})^-\,\mathbf{X}'\mathbf{V}_y^{-1} = (\mathbf{k}'-\mathbf{v}_{yw}'\mathbf{V}_y^{-1})\mathbf{X}(\mathbf{X}'\mathbf{V}_y^{-1}\mathbf{X})^-\,\mathbf{X}'\mathbf{V}_y^{-1}.$$

A possible (point) predictor

$$\tilde{\eta}(\mathbf{y}) = \tilde{\tau}(\mathbf{y})+\mathbf{v}_{yw}'\mathbf{V}_y^{-1}\mathbf{y} = \lambda'\tilde{\boldsymbol{\beta}}+\mathbf{v}_{yw}'\mathbf{V}_y^{-1}(\mathbf{y}-\mathbf{X}\tilde{\boldsymbol{\beta}}) = (\boldsymbol{\ell}'+\mathbf{v}_{yw}'\mathbf{V}_y^{-1})\mathbf{y}$$

(of the value of w) is obtained by replacing τ in formula (12.22) for $\eta(\mathbf{y})$ with its BLUE $\tilde{\tau}(\mathbf{y})$. The predictor $\tilde{\eta}(\mathbf{y})$ is unbiased. In fact, $\tilde{\eta}(\mathbf{y})$ is the BLUP [best (minimum MSE) linear unbiased predictor] (e.g., Goldberger 1962; Henderson 1963; Harville 1976).

To see that $\tilde{\eta}(\mathbf{y})$ is the BLUP, observe that a predictor (of the value of w) is linear and unbiased if and only if it is expressible in the form

$$c+\mathbf{r}'\mathbf{y}+\mathbf{v}_{yw}'\mathbf{V}_y^{-1}\mathbf{y} = c+(\mathbf{r}'+\mathbf{v}_{yw}'\mathbf{V}_y^{-1})\mathbf{y}, \tag{12.28}$$

where $c+\mathbf{r}'\mathbf{y}$ is a (linear) unbiased estimator of τ, that is, where $c=0$ and $\mathbf{r}'\mathbf{X}=\lambda'-\mathbf{v}_{yw}'\mathbf{V}_y^{-1}\mathbf{X}$. Moreover, if $c+\mathbf{r}'\mathbf{y}$ is an unbiased estimator of τ, then the MSE of predictor (12.28) is

$$E[(c+\mathbf{r}'\mathbf{y}+\mathbf{v}_{yw}'\mathbf{V}_y^{-1}\mathbf{y}-w)^2] = \text{var}(c+\mathbf{r}'\mathbf{y}+\mathbf{v}_{yw}'\mathbf{V}_y^{-1}\mathbf{y}-w)$$

$$= \text{var}(c+\mathbf{r}'\mathbf{y})+\text{var}(\mathbf{v}_{yw}'\mathbf{V}_y^{-1}\mathbf{y}-w)$$

$$= \text{var}(c+\mathbf{r}'\mathbf{y})+v_e,$$

implying [since $\tilde{\tau}(\mathbf{y})=\ell'(\mathbf{y})$ is the BLUE of τ] that $\tilde{\eta}(\mathbf{y})$ is the BLUP. A further implication is that the MSE of the BLUP is $\text{var}[\tilde{\eta}(\mathbf{y})-w]= v_e^*$, where

$$v_e^* = v_e+\text{var}[\tilde{\tau}(\mathbf{y})] = v_e+(\lambda'-\mathbf{v}_{yw}'\mathbf{V}_y^{-1}\mathbf{X})(\mathbf{X}'\mathbf{V}_y^{-1}\mathbf{X})^-(\lambda-\mathbf{X}'\mathbf{V}_y^{-1}\mathbf{v}_{yw}).$$

In conjunction with states of knowledge 3 and 4, a predictor $d(\mathbf{y})$ [or, more generally, any function $d(\mathbf{y})$ of $\mathbf{y}$] is said to be **location equivariant** if $d(\mathbf{y}+\mathbf{X}\mathbf{b})=d(\mathbf{y})+\lambda'\mathbf{b}$ for every vector $\mathbf{b}$ (and every value of $\mathbf{y}$). (This usage of the term location equivariant is consistent with that employed in estimation theory; refer to, e.g., Lehmann 1983, Chap.3).

A linear predictor $c+\mathbf{a}'\mathbf{y}$ is location equivariant if and only if $c+\mathbf{a}'(\mathbf{y}+\mathbf{X}\mathbf{b})=c+\mathbf{a}'\mathbf{y}+\lambda'\mathbf{b}$ (for every $\mathbf{b}$), that is, if and only if

$$\mathbf{a}'\mathbf{X} = \lambda' \tag{12.29}$$

or, equivalently, if and only if $\mathbf{a}'\mathbf{y}$ is a (linear) unbiased predictor (of the value of w). Further, if condition (12.29) is satisfied, then the MSE of the predictor $c+\mathbf{a}'\mathbf{y}$ is

$$E[(c+\mathbf{a}'\mathbf{y}-w)^2] = c^2+E[(\mathbf{a}'\mathbf{y}-w)^2]. \tag{12.30}$$

Since $\tilde{\eta}(\mathbf{y})=(\ell'+\mathbf{v}_{yw}'\mathbf{V}_y^{-1})\mathbf{y}$ is the BLUP, $E[(\mathbf{a}'\mathbf{y}-w)^2]$ is minimized, subject to the constraint (12.29), by taking $\mathbf{a}'=\ell'+\mathbf{v}_{yw}'\mathbf{V}_y^{-1}$ and consequently expression (12.30) is minimized [subject to the constraint (12.29)] by taking $c=0$ and $\mathbf{a}'=\ell'+\mathbf{v}_{yw}'\mathbf{V}_y^{-1}$. Thus, the BLUP $\tilde{\eta}(\mathbf{y})$ is location equivariant and has minimum MSE among all linear location-equivariant predictors.

The BLUP $\tilde{\eta}(\mathbf{y})=(\ell'+\mathbf{v}_{yw}'\mathbf{V}_y^{-1})\mathbf{y}$ is the BLUA (best linear unbiased approximation) to $E(w|\mathbf{y})$ in the sense that the unique values of c and $\mathbf{a}$ that minimize the expected squared difference $E\{[c+\mathbf{a}'\mathbf{y}-E(w|\mathbf{y})]^2\}$, subject to the constraint $E[c+\mathbf{a}'\mathbf{y}-E(w|\mathbf{y})]=0$, are $c=0$ and $\mathbf{a}'=\ell'+\mathbf{v}_{yw}'\mathbf{V}_y^{-1}$ {as is evident from equality (12.25), upon observing that $E[(E(w|\mathbf{y})-w)^2]$ does not involve c or $\mathbf{a}$ and that $E[E(w|\mathbf{y})]= E(w)$}.

Note that for the BLUP $\tilde{\eta}(\mathbf{y})$ to be determinable, complete knowledge of v_w, $\mathbf{v}_{yw}$, and $\mathbf{V}_y$ is unnecessary. Rather, it suffices to know the vector $\boldsymbol{\ell}'+\mathbf{v}_{yw}'\mathbf{V}_y^{-1}$. Similarly, to determine v_e^*, it is not necessary to have complete knowledge of v_w, $\mathbf{v}_{yw}$, and $\mathbf{V}_y$.

12.5.2 Interval Prediction (Frequentist Approach)

An approximate 100(1-α)% prediction interval for the value of w can be obtained via an approach analogous to that described (in Sect. 12.4.2) for prediction under state of knowledge 2. Some distribution (having mean 0 and variance v_e^*) is assigned to $\tilde{\eta}(\mathbf{y})-w$, and a set S_3^* of values of $\tilde{\eta}(\mathbf{y})-w$ is determined such that (under the assigned distribution) $\Pr[\tilde{\eta}(\mathbf{y})-w \in S_3^*]= 1-\alpha$. Then, the set $S_3(\mathbf{y})$, consisting of values of w such that $\tilde{\eta}(\mathbf{y})-w \in S_3^*$, can be regarded as an approximate 100(1-α)% prediction set for the value of w. If the distribution assigned to $\tilde{\eta}(\mathbf{y})-w$ is $N(0,v_e^*)$, then, choosing S_3^* to be the interval

$$-z_{\alpha(1)}v_e^{*\,1/2} \le \tilde{\eta}(\mathbf{y})-w \le z_{\alpha(2)}v_e^{*\,1/2},$$

$S_3(\mathbf{y})$ is the interval

$$\tilde{\eta}(\mathbf{y})-z_{\alpha(2)}v_e^{*\,1/2} \le w \le \tilde{\eta}(\mathbf{y})+z_{\alpha(1)}v_e^{*\,1/2}.$$

12.5.3 Special Case: Mixed Linear Model

Suppose that $\mathbf{y}$ follows mixed linear model (12.1) and that $w=\boldsymbol{\lambda}'\boldsymbol{\beta}+\boldsymbol{\delta}'\mathbf{s}$. Suppose further that γ and σ_e^2 are known, in which case v_w, $\mathbf{v}_{yw}$, and $\mathbf{V}_y$ are known.

Then,

$$\tilde{\eta}(\mathbf{y}) = \tau+\gamma\boldsymbol{\delta}'\mathbf{Z}'(\mathbf{I}+\gamma\mathbf{ZZ}')^{-1}\mathbf{y},$$

with $\tau=[\boldsymbol{\lambda}'-\gamma\boldsymbol{\delta}'\mathbf{Z}'(\mathbf{I}+\gamma\mathbf{ZZ}')^{-1}\mathbf{X}]\boldsymbol{\beta}$,

$$v_e = \sigma_e^2\gamma\boldsymbol{\delta}'[\mathbf{I}-\gamma\mathbf{Z}'(\mathbf{I}+\gamma\mathbf{ZZ}')^{-1}\mathbf{Z}]\boldsymbol{\delta}, \tag{12.31}$$

and the Aitken equations (12.27) are equivalent to the equations

$$\mathbf{X}'(\mathbf{I}+\gamma\mathbf{ZZ}')^{-1}\mathbf{X}\tilde{\boldsymbol{\beta}} = \mathbf{X}'(\mathbf{I}+\gamma\mathbf{ZZ}')^{-1}\mathbf{y}. \tag{12.32}$$

Further,

$$\tilde{\tau}(\mathbf{y}) = [\lambda'-\gamma\delta'\mathbf{Z}'(\mathbf{I}+\gamma\mathbf{ZZ}')^{-1}\mathbf{X}]\tilde{\boldsymbol{\beta}},$$

$$\tilde{\eta}(\mathbf{y}) = \tilde{\tau}(\mathbf{y})+\gamma\delta'\mathbf{Z}'(\mathbf{I}+\gamma\mathbf{ZZ}')^{-1}\mathbf{y}$$

$$= \lambda'\tilde{\boldsymbol{\beta}}+\gamma\delta'\mathbf{Z}'(\mathbf{I}+\gamma\mathbf{ZZ}')^{-1}(\mathbf{y}-\mathbf{X}\tilde{\boldsymbol{\beta}}), \tag{12.33}$$

$$\mathrm{var}[\tilde{\tau}(\mathbf{y})] = \sigma_e^2[\lambda'-\gamma\delta'\mathbf{Z}'(\mathbf{I}+\gamma\mathbf{ZZ}')^{-1}\mathbf{X}]$$

$$\cdot[\mathbf{X}'(\mathbf{I}+\gamma\mathbf{ZZ}')^{-1}\mathbf{X}]^{-}[\lambda-\gamma\mathbf{X}'(\mathbf{I}+\gamma\mathbf{ZZ}')^{-1}\mathbf{Z}\delta], \tag{12.34}$$

and (as in the general case)

$$\mathrm{var}\,[\tilde{\eta}(\mathbf{y})-w] = v_e^* = v_e+\mathrm{var}\,[\tilde{\tau}(\mathbf{y})].$$

Note that the BLUP $\tilde{\eta}(\mathbf{y})$ depends on γ and σ_e^2 only through the value of γ.

Representations better suited for computational purposes than (12.31)-(12.34) are obtained by exploiting identity (12.19) and the related identity

$$(\mathbf{I}+\gamma\mathbf{ZZ}')^{-1} = \mathbf{I}-\gamma\mathbf{Z}(\mathbf{I}+\gamma\mathbf{Z}'\mathbf{Z})^{-1}\mathbf{Z}'.$$

Specifically,

$$v_e = \sigma_e^2\gamma\delta'(\mathbf{I}+\gamma\mathbf{Z}'\mathbf{Z})^{-1}\delta;$$

equations (12.32) are re-expressible as

$$[\mathbf{X}'\mathbf{X}-\gamma\mathbf{X}'\mathbf{Z}(\mathbf{I}+\gamma\mathbf{Z}'\mathbf{Z})^{-1}\mathbf{Z}'\mathbf{X}]\tilde{\boldsymbol{\beta}} = \mathbf{X}'\mathbf{y}-\gamma\mathbf{X}'\mathbf{Z}(\mathbf{I}+\gamma\mathbf{Z}'\mathbf{Z})^{-1}\mathbf{Z}'\mathbf{y}; \tag{12.35}$$

the BLUP $\tilde{\eta}(\mathbf{y})$ is re-expressible as

$$\tilde{\eta}(\mathbf{y}) = \lambda'\tilde{\boldsymbol{\beta}}+\gamma\delta'\tilde{\mathbf{u}}, \tag{12.36}$$

where $\tilde{\mathbf{u}}=(\mathbf{I}+\gamma\mathbf{Z}'\mathbf{Z})^{-1}\mathbf{Z}'(\mathbf{y}-\mathbf{X}\tilde{\boldsymbol{\beta}})$ or equivalently where

$$(\mathbf{I}+\gamma\mathbf{Z}'\mathbf{Z})\tilde{\mathbf{u}} = \mathbf{Z}'(\mathbf{y}-\mathbf{X}\tilde{\boldsymbol{\beta}}); \tag{12.37}$$

and

$$\text{var}[\tilde{\tau}(\mathbf{y})] = \sigma_e^2[\boldsymbol{\lambda}'-\gamma\boldsymbol{\delta}'(\mathbf{I}+\gamma\mathbf{Z}'\mathbf{Z})^{-1}\mathbf{Z}'\mathbf{X}]$$

$$\cdot[\mathbf{X}'\mathbf{X}-\gamma\mathbf{X}'\mathbf{Z}(\mathbf{I}+\gamma\mathbf{Z}'\mathbf{Z})^{-1}\mathbf{Z}'\mathbf{X}]^{-}[\boldsymbol{\lambda}-\gamma\mathbf{X}'\mathbf{Z}(\mathbf{I}+\gamma\mathbf{Z}'\mathbf{Z})^{-1}\boldsymbol{\delta}].$$

Equations (12.35) and (12.37), which define $\tilde{\boldsymbol{\beta}}$ and $\tilde{\mathbf{u}}$, are equivalent to the equations

$$\begin{bmatrix} \mathbf{X}'\mathbf{X} & \gamma\mathbf{X}'\mathbf{Z} \\ \mathbf{Z}'\mathbf{X} & \mathbf{I}+\gamma\mathbf{Z}'\mathbf{Z} \end{bmatrix}\begin{bmatrix} \tilde{\boldsymbol{\beta}} \\ \tilde{\mathbf{u}} \end{bmatrix} = \begin{bmatrix} \mathbf{X}'\mathbf{y} \\ \mathbf{Z}'\mathbf{y} \end{bmatrix} \tag{12.38}$$

as is well-known (Henderson 1963) and as is easily verified. Moreover,

$$v_e^* = \sigma_e^2[\boldsymbol{\lambda}'\mathbf{G}_{11}\boldsymbol{\lambda}+\boldsymbol{\lambda}'\mathbf{G}_{12}\boldsymbol{\delta}+\gamma\boldsymbol{\delta}'\mathbf{G}_{21}\boldsymbol{\lambda}+\gamma\boldsymbol{\delta}'\mathbf{G}_{22}\boldsymbol{\delta}], \tag{12.39}$$

where

$$\mathbf{G} = \begin{bmatrix} \mathbf{G}_{11} & \mathbf{G}_{12} \\ \mathbf{G}_{21} & \mathbf{G}_{22} \end{bmatrix}$$

is any generalized inverse of the coefficient matrix of linear system (12.38). To confirm this, observe that one choice for $\mathbf{G}$ is

$$\mathbf{G}_{11} = [\mathbf{X}'\mathbf{X}-\gamma\mathbf{X}'\mathbf{Z}(\mathbf{I}+\gamma\mathbf{Z}'\mathbf{Z})^{-1}\mathbf{Z}'\mathbf{X}]^{-}, \tag{12.40}$$

$$\mathbf{G}_{12} = -\gamma\mathbf{G}_{11}\mathbf{X}'\mathbf{Z}(\mathbf{I}+\gamma\mathbf{Z}'\mathbf{Z})^{-1}, \quad \mathbf{G}_{21} = -(\mathbf{I}+\gamma\mathbf{Z}'\mathbf{Z})^{-1}\mathbf{Z}'\mathbf{X}\mathbf{G}_{11}, \tag{12.41}$$

$$\mathbf{G}_{22} = (\mathbf{I}+\gamma\mathbf{Z}'\mathbf{Z})^{-1}+\gamma(\mathbf{I}+\gamma\mathbf{Z}'\mathbf{Z})^{-1}\mathbf{Z}'\mathbf{X}\mathbf{G}_{11}\mathbf{X}'\mathbf{Z}(\mathbf{I}+\gamma\mathbf{Z}'\mathbf{Z})^{-1}, \tag{12.42}$$

and that, for this particular choice for $\mathbf{G}$, formula (12.39) is valid. Formula (3.1) of Urquhart (1969) can be used to show that expression (12.39) does not vary with the choice of $\mathbf{G}$.

It is now clear that in computing the BLUP $\tilde{\eta}(\mathbf{y})$ and its MSE v_e^*, the direct inversion of the $n \times n$ matrix $\mathbf{I}+\gamma\mathbf{Z}\mathbf{Z}'$ can be circumvented. In particular, by making use of formulae (12.36) and (12.39), the problem of computing $\tilde{\eta}(\mathbf{y})$ and v_e^* can be reduced to that of "solving" linear system (12.38).

If (as in the lamb-weight example) the matrix $\mathbf{I}+\gamma\mathbf{Z}'\mathbf{Z}$ is diagonal or otherwise easy to invert or if the matrix $\mathbf{X}'\mathbf{X}$ is easy to "invert", then, in solving linear system (12.38), the use of "absorption" may be helpful. The absorption of the equations for $\tilde{\mathbf{u}}$ into those for $\tilde{\boldsymbol{\beta}}$ gives

Eq. (12.35) for solving for $\tilde{\boldsymbol{\beta}}$ and Eq. (12.37) for "backsolving" for $\tilde{\mathbf{u}}$ Alternatively, the absorption of the equations for $\tilde{\boldsymbol{\beta}}$ into those for $\tilde{\mathbf{u}}$ gives equation

$$(\mathbf{I}+\gamma\mathbf{C})\tilde{\mathbf{u}} = \mathbf{q}$$

for solving for $\tilde{\mathbf{u}}$ and equation

$$\mathbf{X}'\mathbf{X}\tilde{\boldsymbol{\beta}} = \mathbf{X}'\mathbf{y}-\gamma\mathbf{X}'\mathbf{Z}\tilde{\mathbf{u}} \tag{12.43}$$

for backsolving for $\tilde{\boldsymbol{\beta}}$. Choice (12.40)-(12.42) for $\mathbf{G}$ is that associated with the absorption of the equations for $\tilde{\mathbf{u}}$ into those for $\tilde{\boldsymbol{\beta}}$, while that associated with the absorption of the equations for $\tilde{\boldsymbol{\beta}}$ into those for $\tilde{\mathbf{u}}$ is

$$\mathbf{G}_{11} = (\mathbf{X}'\mathbf{X})^{-}+\gamma(\mathbf{X}'\mathbf{X})^{-}\mathbf{X}'\mathbf{Z}(\mathbf{I}+\gamma\mathbf{C})^{-1}\mathbf{Z}'\mathbf{X}(\mathbf{X}'\mathbf{X})^{-}, \tag{12.44}$$

$$\mathbf{G}_{12} = -\gamma(\mathbf{X}'\mathbf{X})^{-}\mathbf{X}'\mathbf{Z}(\mathbf{I}+\gamma\mathbf{C})^{-1},\quad \mathbf{G}_{21} = -(\mathbf{I}+\gamma\mathbf{C})^{-1}\mathbf{Z}'\mathbf{X}(\mathbf{X}'\mathbf{X})^{-}, \tag{12.45}$$

$$\mathbf{G}_{22} = (\mathbf{I}+\gamma\mathbf{C})^{-1}. \tag{12.46}$$

The BLUPs of the elements of $\mathbf{s}$ are the corresponding elements of the vector $\tilde{\mathbf{s}}=\gamma\tilde{\mathbf{u}}$. By using the identity

$$\gamma(\mathbf{I}+\gamma\mathbf{C})^{-1}\mathbf{C} = \mathbf{I}-(\mathbf{I}+\gamma\mathbf{C})^{-1}, \tag{12.47}$$

$\tilde{\mathbf{s}}$ can be re-expressed, in terms of the solution $\bar{\mathbf{s}}$ to Eqs. (12.7), as follows:

$$\tilde{\mathbf{s}} = \gamma(\mathbf{I}+\gamma\mathbf{C})^{-1}\mathbf{q} = \gamma(\mathbf{I}+\gamma\mathbf{C})^{-1}\mathbf{C}\bar{\mathbf{s}} = \bar{\mathbf{s}}-(\mathbf{I}+\gamma\mathbf{C})^{-1}\bar{\mathbf{s}}. \tag{12.48}$$

Also, the BLUE OF $\boldsymbol{\lambda}'\boldsymbol{\beta}$ is expressible as

$$\boldsymbol{\lambda}'\tilde{\boldsymbol{\beta}} = \boldsymbol{\lambda}'(\mathbf{X}'\mathbf{X})^{-}\mathbf{X}'\mathbf{y}-\boldsymbol{\lambda}'(\mathbf{X}'\mathbf{X})^{-}\mathbf{X}'\mathbf{Z}\tilde{\mathbf{s}}, \tag{12.49}$$

as is evident from Eq. (12.43). The first term $\boldsymbol{\lambda}'(\mathbf{X}'\mathbf{X})^{-}\mathbf{X}'\mathbf{y}$ of expression (12.49) is the estimator of $\boldsymbol{\lambda}'\boldsymbol{\beta}$ that would have been the BLUE had the term $\mathbf{Zs}$ been omitted from model (12.1).

Let $\tilde{t}_i$ represent the BLUP of t_i (i=1,...,r), and define $\tilde{\mathbf{t}}= (\tilde{t}_1,...,\tilde{t}_r)'$. According to Harville and Fenech (1985, expression A.1)

$$(\mathbf{I}+\gamma\mathbf{C})^{-1} = \mathbf{R}(\mathbf{I}+\gamma\mathbf{D})^{-1}\mathbf{R}'+\mathbf{U}\mathbf{U}' \tag{12.50}$$

$$= \mathbf{R}(\mathbf{I}+\gamma\mathbf{D})^{-1}\mathbf{R}'+\mathbf{I}-\mathbf{R}\mathbf{R}' = \mathbf{I}-\gamma\mathbf{R}(\mathbf{I}+\gamma\mathbf{D})^{-1}\mathbf{D}\mathbf{R}'. \tag{12.51}$$

Result (12.50) implies, in particular, that

$$\mathbf{R}'(\mathbf{I}+\gamma\mathbf{C})^{-1} = (\mathbf{I}+\gamma\mathbf{D})^{-1}\mathbf{R}'. \tag{12.52}$$

Substitution from (12.48) and (12.52) gives

$$\tilde{\mathbf{t}} = \gamma\mathbf{D}^{1/2}\mathbf{R}'\tilde{\mathbf{u}} = \mathbf{D}^{1/2}\mathbf{R}'\tilde{\mathbf{s}}$$

$$= \mathbf{D}^{1/2}\mathbf{R}'\bar{\mathbf{s}}-\mathbf{D}^{1/2}\mathbf{R}'(\mathbf{I}+\gamma\mathbf{C})^{-1}\bar{\mathbf{s}} = \mathbf{D}^{1/2}\mathbf{R}'\bar{\mathbf{s}}-\mathbf{D}^{1/2}(\mathbf{I}+\gamma\mathbf{D})^{-1}\mathbf{R}'\bar{\mathbf{s}}$$

$$= \bar{\mathbf{t}}-(\mathbf{I}+\gamma\mathbf{D})^{-1}\bar{\mathbf{t}} = \gamma\mathbf{D}(\mathbf{I}+\gamma\mathbf{D})^{-1}\bar{\mathbf{t}}. \tag{12.53}$$

Thus, there is a simple relationship between the BLUP $\tilde{t}_i$ of t_i and the "least-squares estimator" $\bar{t}_i$ (introduced in Sect. 12.2.1), namely,

$$\tilde{t}_i = \bar{t}_i-(1+\gamma\Delta_i)^{-1}\bar{t}_i = [\gamma\Delta_i/(1+\gamma\Delta_i)]\bar{t}_i.$$

In light of results (12.50) and (12.53),

$$\tilde{\mathbf{s}} = \gamma(\mathbf{I}+\gamma\mathbf{C})^{-1}\mathbf{q} = \gamma\mathbf{R}(\mathbf{I}+\gamma\mathbf{D})^{-1}\mathbf{R}'\mathbf{q}+\gamma\mathbf{U}\mathbf{U}'\mathbf{q}$$

$$= \gamma\mathbf{R}(\mathbf{I}+\gamma\mathbf{D})^{-1}\mathbf{R}'\mathbf{q}+\gamma\mathbf{U}\mathbf{U}'\mathbf{C}\bar{\mathbf{s}} = \gamma\mathbf{R}(\mathbf{I}+\gamma\mathbf{D})^{-1}\mathbf{R}'\mathbf{q}$$

$$= \gamma\mathbf{R}(\mathbf{I}+\gamma\mathbf{D})^{-1}\mathbf{D}^{1/2}\bar{\mathbf{t}} = \gamma\mathbf{R}\mathbf{D}^{1/2}(\mathbf{I}+\gamma\mathbf{D})^{-1}\bar{\mathbf{t}} \tag{12.54}$$

$$= \mathbf{R}\mathbf{D}^{-1/2}\tilde{\mathbf{t}}.$$

Substitution from (12.49) and (12.54) gives the following expression for the BLUP of $\lambda'\boldsymbol{\beta}+\delta'\mathbf{s}$:

$$\tilde{\eta}(\mathbf{y}) = \lambda'(\mathbf{X}'\mathbf{X})^{-}\mathbf{X}'\mathbf{y}+[\delta'-\lambda'(\mathbf{X}'\mathbf{X})^{-}\mathbf{X}'\mathbf{Z}]\tilde{\mathbf{s}}$$

$$= \lambda'(\mathbf{X}'\mathbf{X})^{-}\mathbf{X}'\mathbf{y}+\gamma[\delta'-\lambda'(\mathbf{X}'\mathbf{X})^{-}\mathbf{X}'\mathbf{Z}]\mathbf{R}(\mathbf{I}+\gamma\mathbf{D})^{-1}\mathbf{D}^{1/2}\bar{\mathbf{t}}. \tag{12.55}$$

Similarly, substitution from (12.44)-(12.46) and (12.51) gives the following expression for the MSE of the BLUP:

$$v_e^* = \sigma_e^2\{\lambda'(\mathbf{X}'\mathbf{X})^{-}\lambda+\gamma[\delta'-\lambda'(\mathbf{X}'\mathbf{X})^{-}\mathbf{X}'\mathbf{Z}](\mathbf{I}+\gamma\mathbf{C})^{-1}[\delta-\mathbf{Z}'\mathbf{X}(\mathbf{X}'\mathbf{X})^{-}\lambda]\}$$

$$= \sigma_e^2\{\lambda'(\mathbf{X'X})^-\lambda+\gamma[\delta'-\lambda'(\mathbf{X'X})^-\mathbf{X'Z}][\delta-\mathbf{Z'X}(\mathbf{X'X})^-\lambda]$$

$$-\gamma^2[\delta'-\lambda'(\mathbf{X'X})^-\mathbf{X'Z}]\mathbf{R}(\mathbf{I}+\gamma\mathbf{D})^{-1}\mathbf{DR}'[\delta-\mathbf{Z'X}(\mathbf{X'X})^-\lambda]\}. \qquad (12.56)$$

In expressions (12.55) and (12.56), the only inverse matrix that involves γ is $(\mathbf{I}+\gamma\mathbf{D})^{-1}$, which is diagonal. As a consequence, these expressions are very useful if the values of $\tilde{\eta}(\mathbf{y})$ and v_e^* are to be computed for many values of γ.

Example. Consider, for the lamb-weight example, the prediction of the value of the random variable w_{jk}. The BLUP $\tilde{w}_{jk}$ of w_{jk} is plotted in Fig. 12.1 (as a function of γ) for j=1 and k=3 (the third sire in the first population line) and for j=5 and k=6 (the sixth sire in the fifth line). The BLUP of the difference $w_{13}-w_{56}=\pi_1-\pi_5+s_{13}-s_{56}$ in "breeding value" between these two sires can, of course, be obtained by subtracting $\tilde{w}_{56}$ from $\tilde{w}_{13}$. Also plotted in Fig. 12.1 is the root MSE $\{\text{var}[\tilde{w}_{13}-\tilde{w}_{56}-(w_{13}-w_{56})]\}^{1/2}$ of the BLUP of $w_{13}-w_{56}$. (In plotting the root MSE, the value of σ_e^2 was taken to be 2.962, which is an estimated value to be discussed in Sect. 12.6.1). Note that the sign of the difference $\tilde{w}_{13}-\tilde{w}_{56}$, as well as its magnitude, depends on γ.

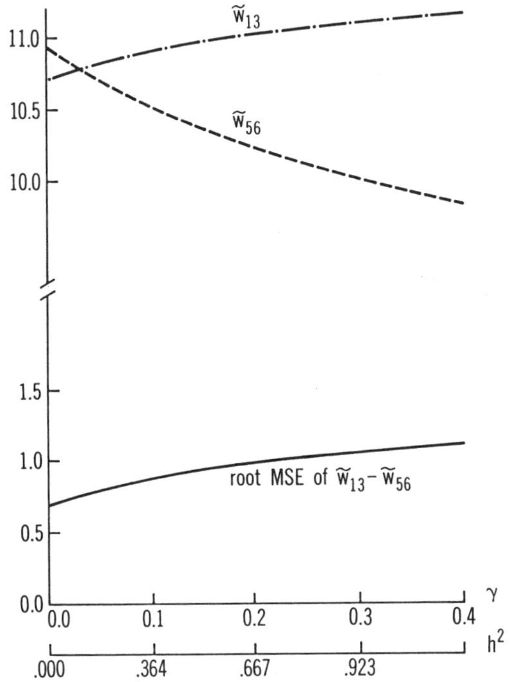

Fig. 12.1 The BLUPS of w_{13} and w_{56} and the root MSE of the BLUP of $w_{13}-w_{56}$

12.5.4 Linear-Bayes Prediction

The BLUP $\tilde{\eta}(\mathbf{y})$ is interpretable in terms of Hartigan's (1969) "linear-Bayes" approach. In this approach, $\boldsymbol{\beta}$ is regarded as a random vector having known mean vector α and known variance-covariance matrix Σ; think of the distribution of $\boldsymbol{\beta}$ as a prior opinion and the joint distribution of w and $\mathbf{y}$ (under the third state of knowledge) is regarded as the conditional distribution given $\boldsymbol{\beta}$.

Assume now that $p^*=p$ and that $\Sigma=\varepsilon\mathbf{I}$, where ε is a (known) positive scalar, these assumptions can always be satisfied via reparameterization, and let $E^*(\cdot)$, $var^*(\cdot)$, and $cov^*(\cdot,\cdot)$ denote an expectation, variance, and covariance, respectively, defined with respect to the unconditional joint distribution of w and $\mathbf{y}$. Then, $E^*(w)=\lambda'\alpha$, $E^*(\mathbf{y})=\mathbf{X}\alpha$, $var^*(w)=v_w+\varepsilon\lambda'\lambda$, $cov^*(\mathbf{y},w)=\mathbf{v}_{yw}+\varepsilon\mathbf{X}\lambda$, and $var^*(\mathbf{y})=\mathbf{V}_y+\varepsilon\mathbf{X}\mathbf{X}'$. Note that the first- and second-order moments of the unconditional joint distribution of w and $\mathbf{y}$ are known.

The linear-Bayes approach to state-3 prediction consists, in effect, of applying the results of Sections 12.4.1 and 12.4.2 to the unconditional joint distribution of w and $\mathbf{y}$. The (point) predictor obtained in this way is

$$\eta^{(\varepsilon)}(\mathbf{y}) = \lambda'\alpha+(\mathbf{v}_{yw}+\varepsilon\mathbf{X}\lambda)'(\mathbf{V}_y+\varepsilon\mathbf{X}\mathbf{X}')^{-1}(\mathbf{y}-\mathbf{X}\alpha),$$

and its MSE (under the unconditional distribution of w and $\mathbf{y}$) is

$$v_e^{(\varepsilon)} = v_w+\varepsilon\lambda'\lambda-(\mathbf{v}_{yw}+\varepsilon\mathbf{X}\lambda)'(\mathbf{V}_y+\varepsilon\mathbf{X}\mathbf{X}')^{-1}(\mathbf{v}_{yw}+\varepsilon\mathbf{X}\lambda).$$

By making use of the identities

$$\lim_{\varepsilon\to\infty}[\varepsilon\mathbf{X}'(\mathbf{V}_y+\varepsilon\mathbf{X}\mathbf{X}')^{-1}] = (\mathbf{X}'\mathbf{V}_y^{-1}\mathbf{X})^{-1}\mathbf{X}'\mathbf{V}_y^{-1},$$

$$\lim_{\varepsilon\to\infty}[\varepsilon\mathbf{I}-\varepsilon^2\mathbf{X}'(\mathbf{V}_y+\varepsilon\mathbf{X}\mathbf{X}')^{-1}\mathbf{X}] = (\mathbf{X}'\mathbf{V}_y^{-1}\mathbf{X})^{-1},$$

$$\lim_{\varepsilon\to\infty}[(\mathbf{V}_y+\varepsilon\mathbf{X}\mathbf{X}')^{-1}] = \mathbf{V}_y^{-1}-\mathbf{V}_y^{-1}\mathbf{X}(\mathbf{X}'\mathbf{V}_y^{-1}\mathbf{X})^{-1}\mathbf{X}'\mathbf{V}_y^{-1}$$

(e.g., Sallas and Harville 1981, thm. 2; Dempster et al. 1981, Sect. 2), it can be shown that

$$\lim_{\varepsilon\to\infty}[\eta^{(\varepsilon)}(\mathbf{y})] = \tilde{\eta}(\mathbf{y}), \quad \lim_{\varepsilon\to\infty}[v_e^{(\varepsilon)}] = v_e^*. \tag{12.57}$$

Thus, as $\varepsilon\to\infty$ (i.e., as the distribution of $\boldsymbol{\beta}$ becomes increasingly diffuse), the linear-Bayes predictor $\eta^{(\varepsilon)}(\mathbf{y})$ and its MSE $v_e^{(\varepsilon)}$ (under the unconditional distribution of w and $\mathbf{y}$) converge to the BLUP $\tilde{\eta}(\mathbf{y})$ and its MSE v_e^*.

12.5.5 Bayesian Prediction

The Bayesian approach consists of

1. Regarding $\boldsymbol{\beta}$ as a random vector whose distribution is known (on the basis of prior information) and regarding the joint distribution of w and $\mathbf{y}$ (under the third state of knowledge) as their conditional joint distribution given $\boldsymbol{\beta}$;
2. Completing the specification of the conditional joint distribution of w and $\mathbf{y}$ given $\boldsymbol{\beta}$ (on the basis of prior information), so that this distribution is known up to the value of $\boldsymbol{\beta}$; and
3. Applying the results of Section 12.3 to the unconditional joint distribution of w and $\mathbf{y}$ (which, in light of points 1 and 2 is known).

If the distribution of $\boldsymbol{\beta}$ and the conditional joint distribution of w and $\mathbf{y}$ given $\boldsymbol{\beta}$ are both taken to be MVN, then the Bayesian approach is essentially the same as the linear-Bayes approach.

12.6 State 4: No Information

Suppose that the only knowledge about the joint distribution of w and $\mathbf{y}$ is that μ_y and μ_w are of the general form $\mu_y=\mathbf{X}\boldsymbol{\beta}$ and $\mu_w=\boldsymbol{\lambda}'\boldsymbol{\beta}$ and that v_w and the elements of $\mathbf{v}_{yw}$ and $\mathbf{V}_y$ are specified functions of the unknown parameter vector θ ($\theta \in \Omega$). Then, $\tilde{\eta}(\mathbf{y})$ and the set $S_3(\mathbf{y})$ may be functionally dependent on θ, in which case they can no longer be used for prediction.

In what follows, $v_w(\theta)$, $\mathbf{v}_{yw}(\theta)$, $\mathbf{V}_y(\theta)$, $\tilde{\eta}(\mathbf{y};\theta)$, and $v_e^*(\theta)$ may sometimes be written for v_w, $\mathbf{v}_{yw}$, $\mathbf{V}_y$, $\tilde{\eta}(\mathbf{y})$, and v_e^*. Doing so emphasizes the dependence of these quantities on θ and provides a means for identifying the values of these quantities corresponding to particular values of θ.

One frequently adopted approach to prediction consists (in the context of predicting the value of w from the value of $\mathbf{y}$) of the following two steps: (1) estimate θ; and (2) act as though the estimated value of θ is the true value and use the prediction procedures described in Sections 12.5.1 and 12.5.2. In the more sophisticated versions of this approach, some modifications may be introduced in step (2) to account for the estimation of θ. An alternative approach to prediction is the Bayesian approach.

12.6.1 Estimation of θ

To implement the two-step approach to prediction, a (point) estimator of θ must be chosen. Here, attention is restricted to estimators of θ that are translation invariant and that are even functions of $\mathbf{y}$. [A possibly vector-valued function $\mathbf{f}(\mathbf{y})$ of $\mathbf{y}$ (in particular, an estimator of θ)

is said to be **translation invariant** if **f(y+Xb)=f(y)** for every $p \times 1$ vector **b** (and every value of **y**). A possibly vector-valued function **h(t)** of a vector **t** is said to be **even** if **h(-t)=h(t)** for every **t** in the domain of **h** and is said to be **odd** if **h(-t)=-h(t)** for every such **t**.] In the following, $\hat{\theta}$ represents an arbitrary even, translation-invariant estimator of θ.

Let **z=L'y**, where **L** is an $n \times (n-p^*)$ matrix such that **L'X=0** and rank(**L**)=n-p*. (For example, the rows of **L'** could be chosen to be any n-p* linearly independent rows of the matrix $\mathbf{I}-\mathbf{P}_\mathbf{X}$.) The vector **z** satisfies the definition of a maximal invariant (given, e.g., by Ferguson 1967, Sect. 5.6) with respect to transformations of the general form T(**y**)=**y**+**Xb**. Hence, a function f(**y**) of **y** is translation invariant if and only if it is expressible as a function of **z**, that is, if and only if f(**y**)=h(**L'y**) for some function h(·) (e.g., Ferguson 1967). It follows that $\hat{\theta}$ is expressible as a function of **z**, say $\hat{\theta}(\mathbf{z})$. Moreover, $\hat{\theta}(\mathbf{z})$ is an even function of **z**.

One way to obtain an estimate of θ is to act as though the distribution of **y** is MVN and to adopt the ML (maximum likelihood) approach. An appealing variation on this approach was described by Thompson (1962) and Patterson and Thompson (1971). In their approach, which has come to be known as REML (for restricted maximum likelihood), θ is estimated by applying the ML approach to the likelihood function that results from regarding **z**, rather than **y**, as the data vector. The form of the likelihood equations suggests that the REML approach may produce a sensible estimate of θ even if the underlying distribution deviates from the MVN (e.g., Harville 1977, Sect. 8.1). Both the REML and (ordinary) ML estimators of θ are even, translation-invariant estimators, as is well-known (e.g., Harville 1977, Sect. 4.3) and as is easily verified.

Special Case: Mixed Linear Model. In the special case of a mixed linear model (12.1), θ has two components, γ and σ_e^2. REML estimates of γ and σ_e^2 are obtained by maximizing the function

$$\ell(\gamma,\sigma_e^2;\mathbf{z}) = (2\pi\sigma_e^2)^{-(n-p^*)/2}[\prod_i(1+\gamma\Delta_i)]^{-1/2}$$

$$\cdot\exp\{-(2\sigma_e^2)^{-1}[S_e+\sum_i \bar{t}_i^2/(1+\gamma\Delta_i)]\},$$

which is equal (or proportional) to the likelihood function that results from regarding **z** as the data vector (Harville and Callanan, Chap. 8, this Vol.). Harville and Callanan discussed the computation of the REML estimates of γ and σ_e^2 and found that for the lamb-weight data, these estimates are 0.175 and 2.962, respectively.

12.6.2 Point Prediction

Substitution of $\hat{\theta}(\mathbf{z})$ for θ in $\tilde{\eta}(\mathbf{y};\theta)$ gives the point predictor

$$\hat{\eta}(\mathbf{y}) = \hat{\eta}[\mathbf{y};\hat{\theta}(\mathbf{z})].$$

Since (for any particular value of θ) $\tilde{\eta}(\mathbf{y};\theta)$ is a location-equivariant function of $\mathbf{y}$,

$$\hat{\eta}(\mathbf{y}+\mathbf{Xb}) = \tilde{\eta}[\mathbf{y}+\mathbf{Xb};\hat{\theta}(\mathbf{L}'\mathbf{y}+\mathbf{L}'\mathbf{Xb})] = \tilde{\eta}[\mathbf{y}+\mathbf{Xb};\, \hat{\theta}(\mathbf{L}'\mathbf{y})]$$

$$= \tilde{\eta}[\mathbf{y};\hat{\theta}(\mathbf{L}'\mathbf{y})]+\lambda'\mathbf{b} = \hat{\eta}(\mathbf{y})+\lambda'\mathbf{b}.$$

Further, since (for any particular value of θ) $\tilde{\eta}(\mathbf{y};\theta)$ is an odd function of $\mathbf{y}$ (as is easily verified) and $\hat{\theta}(\mathbf{z})$ is an even function of $\mathbf{z}$,

$$\hat{\eta}(-\mathbf{y}) = \tilde{\eta}[-\mathbf{y};\hat{\theta}(-\mathbf{z})] = \tilde{\eta}[-\mathbf{y};\hat{\theta}(\mathbf{z})] = -\tilde{\eta}[\mathbf{y};\hat{\theta}(\mathbf{z})] = -\hat{\eta}(\mathbf{y}).$$

Thus, $\hat{\eta}(\mathbf{y})$ is an odd, location-equivariant predictor.

The following proposition is a special case of a result proved by Jeske and Harville (1988).

Proposition 1. Let

$$t(w,\mathbf{y}) = [d(\mathbf{y})-w]/[m(\mathbf{z})]^{1/2},$$

where $d(\mathbf{y})$ is any odd, location-equivariant predictor and $m(\mathbf{z})$ is any non-negative, even, translation-invariant statistic. [If $m(\mathbf{z})= 0$, set $t(w,\mathbf{y})=0$.] Then, under Assumption S,

1. The distribution of $t(w,\mathbf{z})$ depends on $\boldsymbol{\beta}$ and θ only through the value of θ, and
2. The distribution of $t(w,\mathbf{y})$ is symmetric about $\mathbf{0}$.

Consideration of the special case of Proposition 1 where $m(\mathbf{z})\equiv 1$ leads to the following conclusion.

Corollary. Under Assumption S, the distribution of the prediction error $d(\mathbf{y})-w$ of any odd, location-equivariant predictor $d(\mathbf{y})$ is symmetric about 0. In particular, the distribution of $\hat{\eta}(\mathbf{y})-w$ is symmetric about 0.

It follows from the corollary that, under Assumption S, the predictor $\hat{\eta}(\mathbf{y})$ (or any other odd, location-equivariant predictor) is unbiased (provided that the expected value of the predictor exists).

Example. Consider the prediction of the value of w_{jk} from the lamb-weight data. Let $\hat{w}_{jk}$ represent the point prediction given by $\hat{\eta}(y)=\tilde{\eta}(y;\hat{\theta})$ when $\hat{\theta}$ is taken to be the REML estimate of θ. Recalling that the REML estimate of γ is 0.175 and referring to Fig. 12.1, the predictions of the values of w_{13} and w_{56} are $\hat{w}_{13}=10.98$ and $\hat{w}_{56}=10.29$. Further, the prediction of the value of the difference $w_{13}-w_{56}$ is $\hat{w}_{13}-\hat{w}_{56}=0.69$.

12.6.3 MSE of Prediction

Let $\varepsilon(w,y)=\hat{\eta}(y)-w$ represent the prediction error of $\hat{\eta}(y)$. This error can be decomposed into two parts in accordance with the identity

$$\varepsilon(w,y) = \varepsilon_1(w,y;\theta)+\varepsilon_2(y;\theta). \tag{12.58}$$

Here, $\varepsilon_1(w,y;\theta)=\tilde{\eta}(y;\theta)-w$ is the prediction error of the BLUP, and $\varepsilon_2(y;\theta)=\hat{\eta}(y)-\tilde{\eta}(y;\theta)=\tilde{\eta}(y;\hat{\theta})-\tilde{\eta}(y;\theta)$ is the difference between the predictor $\hat{\eta}(y)$ and the BLUP.

The following result was proved by Kackar and Harville (1984) and Harville (1985) and in a special case by Khatri and Shah (1981).

Proposition 2. Under Assumption N,

1. $\varepsilon_1(w,y;\theta)$ and $\varepsilon_2(y;\theta)$ are statistically independent, and
2. The MSE $E[(\hat{\eta}(y)-w)^2]$ of $\hat{\eta}(y)$ equals

$$v_e^+(\theta) = v_e^*(\theta)+\text{var}[\varepsilon_2(y;\theta)] \tag{12.59}$$

[provided the expected value of $\hat{\eta}(y)$ exists].

Result (12.59) indicates that the effect of not knowing θ, and thereby having to use the predictor $\hat{\eta}(y)$ instead of the BLUP $\tilde{\eta}(y;\theta)$, is to inflate the MSE of prediction by an amount equal to $\text{var}[\varepsilon_2(y;\theta)]$. This amount is "large" if (1) $\tilde{\eta}(y;\theta)$ is "sensitive" to changes in θ and (2) the variability of the estimator $\hat{\theta}$ of θ is "substantial".

12.6.4 Approximating the MSE

In general, closed-form expressions for the second term of sum (12.59), and hence for the MSE of $\hat{\eta}(y)$, are not possible. Even in the special case of the prediction of a group "mean" under a balanced one-way random-effects model, the simplest possible expression for the second term involves incomplete beta functions (Kackar and Harville 1984). More seriously, even when resort is made to numerical methods, the evaluation of the second term presents (except for relatively simple special cases) a difficult, if not intractable, problem.

Kackar and Harville (1984) proposed an approximation for the second term of sum (12.59). Letting $\delta(\mathbf{z};\theta)=\partial\eta(\mathbf{y};\theta)/\partial\theta$, the proposed approximation is

$$\mathrm{var}[\varepsilon_2(\mathbf{y};\theta)] \doteq \mathrm{tr}(\mathbf{AB}), \tag{12.60}$$

where $\mathbf{A}=\mathrm{var}[\delta(\mathbf{z};\theta)]$ and where $\mathbf{B}$ is a matrix that equals or approximates the MSE matrix $E[(\hat{\theta}-\theta)(\hat{\theta}-\theta)']$ of $\hat{\theta}$. When $\hat{\theta}$ is taken to be the REML estimator of θ, one choice for $\mathbf{B}$ is the inverse of the (REML) information matrix, which is given by, for example, Harville (1977, Sect. 5).

Substitution of approximation (12.60) for the second term of sum (12.59) gives, as an approximation for the MSE of $\hat{\eta}(\mathbf{y})$,

$$v_e^{@}(\theta) = v_e^{*}(\theta)+\mathrm{tr}[\mathbf{A}(\theta)\mathbf{B}(\theta)].$$

[To emphasize functional dependence on θ, $\mathbf{A}(\theta)$ and $\mathbf{B}(\theta)$ are written for $\mathbf{A}$ and $\mathbf{B}$.]

It can be shown that the ijth element of the matrix $\mathbf{A}$ is $a_{ij}= \mathbf{c}_i'\mathbf{V}_y^{-1}(\mathbf{I}-\mathbf{P})\mathbf{c}_j$, where $\mathbf{P}=\mathbf{X}(\mathbf{X}'\mathbf{V}_y^{-1}\mathbf{X})^-\mathbf{X}'\mathbf{V}_y^{-1}$ and $\mathbf{c}_i'=(\partial\mathbf{v}_{yw}'/\partial\theta_i)-\mathbf{v}_{yw}'\mathbf{V}_y^{-1}(\mathbf{I}-\mathbf{P})(\partial\mathbf{V}_y/\partial\theta_i)-\lambda'(\mathbf{X}'\mathbf{V}_y^{-1}\mathbf{X})^-\mathbf{X}'\mathbf{V}_y^{-1}$ $(\partial\mathbf{V}_y/\partial\theta_i)$ (Harville 1985).

Special Case: Mixed Linear Model. Consider now approximation (12.60) as applied to the prediction of $\lambda'\boldsymbol{\beta}+\delta'\mathbf{s}$ under mixed linear model (12.1).

Differentiation of expression (12.55) gives

$$\partial\tilde{\eta}(\mathbf{y};\theta)/\partial\gamma = [\delta'-\lambda'(\mathbf{X}'\mathbf{X})^-\mathbf{X}'\mathbf{Z}]\mathbf{R}(\mathbf{I}+\gamma\mathbf{D})^{-2}\mathbf{D}^{1/2}\mathbf{t} \tag{12.61}$$

Result (12.61), when combined with various results given by Harville and Fenech (1985), can be used to show that

$$\begin{aligned} a_{11} &= \mathrm{var}[\partial\tilde{\eta}(\mathbf{y};\theta)/\partial\gamma] \\ &= \sigma_e^2[\delta'-\lambda'(\mathbf{X}'\mathbf{X})^-\mathbf{X}'\mathbf{Z}]\mathbf{RD}(\mathbf{I}+\gamma\mathbf{D})^{-3}\mathbf{R}'[\delta-\mathbf{Z}'\mathbf{X}(\mathbf{X}'\mathbf{X})^-\lambda] \qquad (12.62) \\ &= \sigma_e^2[\delta'-\lambda'(\mathbf{X}'\mathbf{X})^-\mathbf{X}'\mathbf{Z}]\mathbf{C}(\mathbf{I}+\gamma\mathbf{C})^{-3}[\delta-\mathbf{Z}'\mathbf{X}(\mathbf{X}'\mathbf{X})^-\lambda]. \end{aligned}$$

Clearly, $\partial\tilde{\eta}(\mathbf{y};\theta)/\partial\sigma_e^2=0$, implying that $a_{12}=a_{21}=a_{22}=0$ and hence (letting b_{ij} represent the ijth element of $\mathbf{B}$) that

$$\mathrm{tr}(\mathbf{AB}) = a_{11}b_{11}.$$

Moreover, when **B** is taken to be the inverse of the (REML) information matrix,

$$b_{11} = 2\{\Sigma_i\Delta_i^2/(1+\gamma\Delta_i)^2 - [\Sigma_i\Delta_i/(1+\gamma\Delta_i)]^2/(n-p^*)\}^{-1} \tag{12.63}$$

$$= 2[\mathrm{tr}\{[(\mathbf{I}+\gamma\mathbf{C})^{-1}\mathbf{C}]^2\} - (n-p^*)^{-1}\{\mathrm{tr}[(\mathbf{I}+\gamma\mathbf{C})^{-1}\mathbf{C}]\}^2]^{-1},$$

as can be verified by applying the results of Harville and Callanan (Chap. 8, this Vol., Sects. 8.11 and 8.12).

12.6.5 Estimating the MSE

In estimating the MSE of the predictor $\hat{\eta}(\mathbf{y})$, it is common practice to ignore the contribution of the second term of sum (12.59) and to use the estimator $v_e^*(\hat{\theta})$. A more conservative, and perhaps better, estimator of the MSE is the estimator $v_e^@(\hat{\theta})$ obtained by substituting $\hat{\theta}$ for θ in the MSE approximation $v_e^@(\theta)$.

Jeske (1985, Sect. 3.3) gave some results relevant to the estimation of the MSE of $\hat{\eta}(\mathbf{y})$. His results are valid for the special case where the following two assumptions (as well as Assumption N) are satisfied:

C. $\hat{\theta}$ is a complete sufficient statistic (for the distribution of **z**);

U. $E[v_w(\hat{\theta})]=v_w$, $E[\mathbf{v}_{yw}(\hat{\theta})]=\mathbf{v}_{yw}$, and $E[\mathbf{V}_y(\hat{\theta})]=\mathbf{V}_y$.

It can be shown that, in the context of mixed linear model (12.1), Assumption C is equivalent to the assumption that $\Delta_1=...=\Delta_r$ and Assumption U is equivalent to the assumption that the two elements $\hat{\gamma}$, and $\hat{\sigma}_e^2$ of $\hat{\theta}$ are such that $\hat{\gamma}\hat{\sigma}_e^2$ and $\hat{\sigma}_e^2$ estimate σ_s^2 and σ_e^2 unbiasedly. Strictly speaking, Assumption C is satisfied only in relatively simple special cases, and (because $\hat{\theta}$ is confined to the parameter space Ω) Assumption U is generally not satisfied either. Nevertheless, results derived under these assumptions can have far-reaching implications.

The following result is among those given by Jeske (1985).

Proposition 3. Under Assumptions N, C, and U,

$$E[v_e^*(\hat{\theta})] = v_e^*(\theta) - \mathrm{var}[\varepsilon_2(\mathbf{y};\theta)]. \tag{12.64}$$

The asymptotic results of Prasad and Rao (1986) suggest that even if Assumption C is violated (but Assumptions N and U are satisfied, at least approximately), then

$$E[v_e^*(\hat{\theta})] \doteq v_e^*(\theta)\text{-var}[\varepsilon_2(\mathbf{y};\theta)], \tag{12.65}$$

that is, result (12.64) holds as an approximation. Their results further suggest that if Assumptions N and U hold exactly or approximately, then

$$E\{\text{tr}[\mathbf{A}(\hat{\theta})\mathbf{B}(\hat{\theta})]\} \doteq \text{var}[\varepsilon_2(\mathbf{y};\theta)], \tag{12.66}$$

that is, an approximately unbiased estimator of the second term of expression (12.59) for the MSE of $\hat{\eta}(\mathbf{y})$ is obtained simply by substituting $\hat{\theta}$ for θ in approximation (12.60).

Observe that, in light of result (12.59), result (12.65) suggests that

$$E[v_e^*(\hat{\theta})] \doteq v_e^+(\theta)\text{-2var}[\varepsilon_2(\mathbf{y};\theta)], \tag{12.67}$$

that is, the "conventional" estimator $v_e^*(\hat{\theta})$ tends to underestimate the MSE of $\hat{\eta}(\mathbf{y})$ by approximately twice the amount that might have been anticipated. Moreover, result (12.67), when combined with result (12.66), suggests that

$$E[v_e^@(\hat{\theta})] \doteq v_e^+(\theta)\text{-var}[\varepsilon_2(\mathbf{y};\theta)],$$

that is, $v_e^@(\hat{\theta})$ also tends to underestimate the MSE, though the extent of the underestimation is only about one-half that of $v_e^*(\hat{\theta})$. A further implication of results (12.66) and (12.67) is that the estimator $\hat{v}_e^@$ defined by

$$\hat{v}_e^@ = v_e^*(\hat{\theta})+2\text{tr}[\mathbf{A}(\hat{\theta})\mathbf{B}(\hat{\theta})]$$

might be a better estimator [of the MSE of $\hat{\eta}(\mathbf{y})$] than either $v_e^*(\hat{\theta})$ or $v_e^@(\hat{\theta})$.

Example. Consider, for the lamb-weight example, the various MSE estimators as applied to the predictor $\hat{w}_{13}$-$\hat{w}_{56}$ of the value of w_{13}-w_{56}. When **B** is taken to be the inverse of the (REML) information matrix (and the estimates of γ and σ_e^2 are taken to be the REML estimates 0.175 and 2.962), the estimates of the root MSE are $[v_e^*(\hat{\theta})]^{1/2}$=0.955, $[v_e^@(\hat{\theta})]^{1/2}$=1.017, and $(\hat{v}_e^@)^{1/2}$=1.076.

12.6.6 Interval Prediction (Frequentist Approach)

One approach to obtaining an approximate 100(1-α)% prediction interval (for the value of w) is based on acting as though the joint distribution of e_w and $\mathbf{e}_y$ is MVN or, more generally, on assigning to e_w and $\mathbf{e}_y$ some symmetric distribution (that is specified up to the value of θ) and acting as though the assigned distribution is the true distribution.

Let

$$t(w,\mathbf{y}) = [\hat{\eta}(\mathbf{y})-w]/[m(\mathbf{z})]^{1/2},$$

where $m(\mathbf{z})$ is a non-negative, even, translation-invariant estimator of the MSE of $\hat{\eta}(\mathbf{y})$. [One choice for $m(\mathbf{z})$ is $\hat{v}_e^{@}$]. According to Proposition 1, the distribution of $t(w,\mathbf{y})$, as derived from the joint distribution assigned to e_w and $\mathbf{e}_y$, is symmetric about 0 and depends on $\boldsymbol{\beta}$ and θ only through the value of θ. Define $k_{\alpha/2}(\theta)$ to be the upper-$\alpha/2$ point of this distribution. [For simplicity, attention is restricted to the case where the distribution function of $t(w,\mathbf{y})$ is continuous.]

Let

$$\ell_{\alpha/2}(\mathbf{y};\theta) = \hat{\eta}(\mathbf{y})-k_{\alpha/2}(\theta)[m(\mathbf{z})]^{1/2},$$

$$u_{\alpha/2}(\mathbf{y};\theta) = \hat{\eta}(\mathbf{y})+k_{\alpha/2}(\theta)[m(\mathbf{z})]^{1/2}.$$

Note that under the joint distribution assigned to e_w and $\mathbf{e}_y$,

$$\Pr[-k_{\alpha/2}(\theta) \le t(w,\mathbf{y}) \le k_{\alpha/2}(\theta)] = 1-\alpha$$

or, equivalently,

$$\Pr[\ell_{\alpha/2}(\mathbf{y};\theta) \le w \le u_{\alpha/2}(\mathbf{y};\theta)] = 1-\alpha$$

In those special cases where $k_{\alpha/2}(\theta)$, and hence $\ell_{\alpha/2}(\mathbf{y};\theta)$ and $u_{\alpha/2}(\mathbf{y};\theta)$, are not functionally dependent on θ, the interval

$$\ell_{\alpha/2}(\mathbf{y};\theta) \le w \le u_{\alpha/2}(\mathbf{y};\theta) \tag{12.68}$$

can be regarded as an approximate $100(1-\alpha)\%$ prediction interval, if the true joint distribution of e_w and $\mathbf{e}_y$ coincides with the assigned distribution, the probability of coverage of this interval is exactly $1-\alpha$. More generally, the interval

$$\ell_{\alpha/2}(\mathbf{y};\hat{\theta}) \le w \le u_{\alpha/2}(\mathbf{y};\hat{\theta}), \tag{12.69}$$

obtained from interval (12.68) by substituting $\hat{\theta}$ for θ, can be regarded as an approximate $100(1-\alpha)\%$ prediction interval. Interval (12.69) belongs to a general class of approximate $100(1-\alpha)\%$ prediction intervals considered by Jeske and Harville (1988).

Except for relatively simple special cases, the analytical or numerical determination of the value of $k_{\alpha/2}(\theta)$ (corresponding to a specified value of θ) presents an intractable problem. Thus, as a practical matter, it may be necessary to modify the definitions of $\ell_{\alpha/2}(\mathbf{y};\theta)$ and $u_{\alpha/2}(\mathbf{y};\theta)$ by replacing $k_{\alpha/2}(\theta)$ with an approximation. In particular, $k_{\alpha/2}(\theta)$ could be approximated by the upper-$\alpha/2$ point of a t-distribution (as discussed, in detail, by Jeske 1985, Sect. 5.3).

An alternative way to approximate $k_{\alpha/2}(\theta)$ is by simulation. In this approach, a random sample $(w^{(1)},\mathbf{y}^{(1)}),...,(w^{(s)},\mathbf{y}^{(s)})$ is generated from the joint distribution assigned to e_w and $\mathbf{e}_y$ (which is equivalent to setting $\boldsymbol{\beta}=\mathbf{0}$ and generating a random sample from the joint distribution of w and $\mathbf{y}$), the empirical distribution of the absolute values $|t(w^{(1)},\mathbf{y}^{(1)})|,...,|t(w^{(s)},\mathbf{y}^{(s)})|$ of the sample values $t(w^{(1)},\mathbf{y}^{(1)}),...,t(w^{(s)},\mathbf{y}^{(s)})$ is formed, and $k_{\alpha/2}(\theta)$ is approximated by the upper-α point of this empirical distribution. In various special cases, $t(w,\mathbf{y})$ depends on w and $\mathbf{y}$ only through a relatively small number of functions of w and $\mathbf{y}$, in which case it suffices to generate a random sample from the joint distribution of these functions rather than from that of w and $\mathbf{y}$. When the approximation to $k_{\alpha/2}(\theta)$ obtained by simulation is substituted for $k_{\alpha/2}(\theta)$ in the definitions of $\ell_{\alpha/2}(\mathbf{y};\theta)$ and $u_{\alpha/2}(\mathbf{y};\theta)$, prediction interval (12.69) has a parametric bootstrap (Efron 1982, Sect. 5.2) interpretation.

Special Case: Mixed Linear Model. Consider now the approximation of $k_{\alpha/2}(\theta)$ by simulation in the special case where $\mathbf{y}$ follows mixed linear model (12.1) and where $w=\boldsymbol{\lambda}'\boldsymbol{\beta}+\boldsymbol{\delta}'\mathbf{s}$. For definiteness, suppose that e_w and $\mathbf{e}_y$ are assigned the MVN distribution, that the two elements $\hat{\gamma}$ and $\hat{\sigma}_e^2$ of $\hat{\theta}$ are taken to be the REML estimators (of γ and σ_e^2), that $m(\mathbf{z})=\hat{v}_e^{@}$, and that $\mathbf{B}$ is taken to be the inverse of the (REML) information matrix.

It follows from results (12.58) and (12.55) that

$$t(w,\mathbf{y}) = [\varepsilon_1(w,\mathbf{y};\theta)+\varepsilon_2(\mathbf{y};\theta)]/[\hat{v}_e^{@}]^{1/2},$$

$$\varepsilon_2(\mathbf{y};\theta) = [\boldsymbol{\delta}'-\boldsymbol{\lambda}'(\mathbf{X}'\mathbf{X})^-\mathbf{X}'\mathbf{Z}]\mathbf{R}[\hat{\gamma}(\mathbf{I}+\hat{\gamma}\mathbf{D})^{-1}-\gamma(\mathbf{I}+\gamma\mathbf{D})^{-1}]\mathbf{D}^{1/2}\bar{\mathbf{t}}. \tag{12.70}$$

The REML estimators $\hat{\gamma}$ and $\hat{\sigma}_e^2$ depend on $\mathbf{y}$ only through the values of $\bar{\mathbf{t}}$ and S_e (Harville and Callanan, Chap. 8, this Vol., Sect. 8.4). Thus, in light of result (12.70), $\varepsilon_2(\mathbf{y};\theta)$ and $\hat{v}_e^{@}$ depend on $\mathbf{y}$ only through the values of $\bar{\mathbf{t}}$ and S_e. Accordingly, for any particular value of θ, $t(w,\mathbf{y})$ can be regarded as a function, say $t^*(\varepsilon_1,\bar{\mathbf{t}},S_e;\theta)$ of $\varepsilon_1(w,\mathbf{y};\theta)$, $\bar{\mathbf{t}}$, and S_e.

Let $x_0,x_1,...,x_r,\chi^2$ represent statistically independent random variables, and take the distribution of x_i to be standard normal $(i=0,...,r)$ and that of χ^2 to be chi-square with f degrees of freedom. Define $\mathbf{x}=(x_1,x_2,...,x_r)'$. It can be shown that the joint distribution of $\varepsilon_1(w,\mathbf{y};\theta)$, $\bar{\mathbf{t}}$, and S_e is the same as that of $v_e^{*1/2}x_0$, $\sigma_e(\mathbf{I}+\gamma\mathbf{D})^{1/2}\mathbf{x}$, and $\sigma_e^2\chi^2$. Thus, to approximate $k_{\alpha/2}(\theta)$, it suffices to generate a random sample $(x_0{}^{(1)},\mathbf{x}^{(1)},\chi^{2(1)}),...,$

$(x_0^{(s)}, \mathbf{x}^{(s)}, \chi^{2(s)})$ from the joint distribution of $x_0, \mathbf{x}$, and χ^2 and to form the empirical distribution of $|t^*[v_e^{*1/2}x_0^{(1)}, \sigma_e(\mathbf{I}+\gamma\mathbf{D})^{1/2}\mathbf{x}^{(1)}, \sigma_e^2\chi^{2(1)};\theta]|, \ldots, |t^*[v_e^{*1/2}x_0^{(s)}, \sigma_e(\mathbf{I}+\gamma\mathbf{D})^{1/2}\mathbf{x}^{(s)}, \sigma_e^2\chi^{2(s)};\theta]|$. The approximation is taken to be the upper-α point of this empirical distribution.

It can be shown that the distribution of $t(w,\mathbf{y})$, and hence its upper-$\alpha/2$ point $k_{\alpha/2}(\theta)$, depend on $\theta=(\gamma, \sigma_e^2)'$ only through the value of γ. Thus, in carrying out the simulation, the value of σ_e^2 can be arbitrarily set equal to 1 (and the value of $\boldsymbol{\beta}$ to $\mathbf{0}$). Moreover, to keep the computations to a minimum, advantage should be taken of representations (12.70), (12.56), (12.62), and (12.63) [for $\varepsilon_2(\mathbf{y};\theta)$, v_e^*, a_{11}, and b_{11}] and various related (recursive) representations given by Harville and Callanan, Chapter 8, this Volume, for $\hat{\gamma}$ and $\hat{\sigma}_e^2$.

12.6.7 Bayesian Prediction

The Bayesian approach consists of

1. Regarding $\boldsymbol{\beta}$ and θ as random vectors whose joint distribution is known (on the basis of prior information) and regarding the joint distribution of w and $\mathbf{y}$ (under the fourth state of knowledge) as their conditional joint distribution given $\boldsymbol{\beta}$ and θ);
2. Completing the specification of the conditional joint distribution of w and $\mathbf{y}$ given $\boldsymbol{\beta}$ and θ (on the basis of prior information), so that this distribution is known up to the values of $\boldsymbol{\beta}$ and θ; and
3. Applying the results of Section 12.3 to the unconditional joint distribution of w and $\mathbf{y}$ (which, in light of points 1 and 2 is known).

Consider now the specifics of the Bayesian approach when (1) $\boldsymbol{\beta}$ and θ are statistically independent; (2) $p^*=p$; (3) the distribution of $\boldsymbol{\beta}$ is MVN with known mean vector α and with variance-covariance matrix $\varepsilon\mathbf{I}$ (where ε is a known positive scalar); (4) the distribution of θ has a p.d.f. (probability density function) $\pi(\theta)$; and (5) the conditional joint distribution of w and $\mathbf{y}$ given $\boldsymbol{\beta}$ and θ is MVN.

The conditional joint distribution of w and $\mathbf{y}$ given θ is MVN with mean vector and variance-covariance matrix

$$\begin{bmatrix} \lambda'\alpha \\ \mathbf{X}\alpha \end{bmatrix} \quad \text{and} \quad \begin{bmatrix} v_w+\varepsilon\lambda'\lambda & \mathbf{v}_{yw}'+\varepsilon\lambda'\mathbf{X}' \\ \mathbf{v}_{yw}+\varepsilon\mathbf{X}\lambda & \mathbf{V}_y+\varepsilon\mathbf{X}\mathbf{X}' \end{bmatrix},$$

respectively. This conditional distribution has a p.d.f. $g(w,\mathbf{y}|\theta)$, which is expressible in the form

$$g(w,\mathbf{y}|\theta) = g_1(w|\mathbf{y},\theta)g_2(\mathbf{y}|\theta), \tag{12.71}$$

where $g_1(w|\mathbf{y},\theta)$ is the p.d.f. of the conditional distribution of w given $\mathbf{y}$ and θ, which (in the notation of Sect. 12.5.4) is normal with mean $\eta^{(\varepsilon)}(\mathbf{y})$ and variance $v_e^{(\varepsilon)}$, and where $g_2(\mathbf{y}|\theta)$ is the p.d.f. of the conditional distribution of $\mathbf{y}$ given θ (which is MVN with mean vector $\mathbf{X}\alpha$ and variance-covariance matrix $\mathbf{V}_y+\varepsilon\mathbf{XX}'$).

The prediction of the value of w is based on the conditional distribution of w given $\mathbf{y}$. This distribution has a p.d.f. $f(w|\mathbf{y})$, which [in light of result (12.71)] is expressible as

$$\begin{aligned} f(w|\mathbf{y}) &= \int_\Omega g(w,\mathbf{y}|\theta)\pi(\theta)d\theta / \int_\Omega g_2(\mathbf{y}|\theta)\pi(\theta)d\theta \\ &\propto \int_\Omega g(w,\mathbf{y}|\theta)\pi(\theta)d\theta = \int_\Omega g_1(w|\mathbf{y},\theta)g_2(\mathbf{y}|\theta)\pi(\theta)d\theta \\ &\propto \int_\Omega g_1(w|\mathbf{y},\theta)(2\pi\varepsilon)^{p/2}|\mathbf{X}'\mathbf{X}|^{1/2}g_2(\mathbf{y}|\theta)\pi(\theta)d\theta. \end{aligned} \tag{12.72}$$

It follows from result (12.57) that

$$\lim_{\varepsilon\to\infty} [g_1\ (w|\mathbf{y},\theta)] = g_1^*(w|\mathbf{y},\theta), \tag{12.73}$$

where $g_1^*(w|\mathbf{y},\theta)$ is the p.d.f. of a normal distribution with mean $\tilde{\eta}(\mathbf{y};\theta)$ and variance $v_e^*(\theta)$. Moreover,

$$\lim_{\varepsilon\to\infty}[(2\pi\varepsilon)^{p/2}|\mathbf{X}'\mathbf{X}|^{1/2}g_2(\mathbf{y}|\theta)] \propto g_2^*(\mathbf{L}'\mathbf{y}|\theta), \tag{12.74}$$

where $g_2^*(\mathbf{z}|\theta)$ is the p.d.f. of the conditional distribution of $\mathbf{z}$ given θ, which is MVN with mean vector $\mathbf{0}$ and variance-covariance matrix $\mathbf{L}'\mathbf{V}_y(\theta)\mathbf{L}$ (e.g., Sallas 1979, pp. 93-94; Dempster et al. 1981, Sect. 5.4).

Together, results (12.72)-(12.74) suggest that if the (prior) distribution of $\boldsymbol{\beta}$ is "diffuse", that is, if ε is "large", then the conditional distribution of w given $\mathbf{y}$ can be approximated by the distribution with p.d.f. $f^*(w|\mathbf{y})$, where

$$f^*(w|\mathbf{y}) \propto \int_\Omega g_1^*(w|\mathbf{y},\theta)g_2^*(\mathbf{z}|\theta)\pi(\theta)d\theta$$

or equivalently where

$$f^*(w|\mathbf{y}) = \int_\Omega g_1^*(w|\mathbf{y},\theta)h(\theta|\mathbf{z})d\theta.$$

Here,

$$h(\theta|\mathbf{z})=g_2^*(\mathbf{z}|\theta)\pi(\theta)/\int_\Omega g_2^*(\mathbf{z}|\omega)\pi(\omega)d\omega$$

represents the p.d.f. of the conditional distribution of θ given $\mathbf{z}$.

Let $E_B(\cdot)$ and $\text{var}_B(\cdot)$ denote an expectation and variance (or variance-covariance matrix), respectively, defined with respect to the distribution whose p.d.f. is $h(\theta|\mathbf{z})$. Then, the mean and variance of the distribution with p.d.f. $f^*(w|\mathbf{y})$ are respectively

$$\eta_B(\mathbf{y}) = E_B[\tilde{\eta}(\mathbf{y};\theta)], \tag{12.75}$$

$$v_B = E_B[v_e^*(\theta)]+\text{var}_B[\tilde{\eta}(\mathbf{y};\theta)]. \tag{12.76}$$

To obtain approximations to $\eta_B(\mathbf{y})$ and v_B, let $\bar{\theta}(\mathbf{z})=E_B(\theta)$, define $\mathbf{s}(\theta)=\partial v_e^*(\theta)/\partial\theta$, and take $\mathbf{S}(\theta)$ to be the qxq matrix with ijth element $\partial^2 v_e^*(\theta)/\partial\theta_i\partial\theta_j$. Expanding $\tilde{\eta}(\mathbf{y};\theta)$ in a Taylor series (in θ) around $\bar{\theta}(\mathbf{z})$ gives the following first-order approximation:

$$\tilde{\eta}(\mathbf{y};\theta) \doteq \tilde{\eta}[\mathbf{y};\bar{\theta}(\mathbf{z})]+\{\delta[\mathbf{z};\bar{\theta}(\mathbf{z})]\}'[\theta-\bar{\theta}(\mathbf{z})]. \tag{12.77}$$

Similarly, expanding $v_e^*(\theta)$ in a Taylor series gives the following first- and second-order approximations:

$$v_e^*(\theta) \doteq v_e^*[\bar{\theta}(\mathbf{z})]+\{\mathbf{s}[\bar{\theta}(\mathbf{z})]\}'[\theta-\bar{\theta}(\mathbf{z})] \tag{12.78}$$

$$\doteq v_e^*[\bar{\theta}(\mathbf{z})]+\{\mathbf{s}[\bar{\theta}(\mathbf{z})]\}'[\theta-\bar{\theta}(\mathbf{z})]$$

$$+(1/2)[\theta-\bar{\theta}(\mathbf{z})]'\mathbf{S}[\bar{\theta}(\mathbf{z})][\theta-\bar{\theta}(\mathbf{z})]. \tag{12.79}$$

Substitution of approximation (12.77) in (12.75) results in the approximation

$$\eta_B(\mathbf{y}) \doteq \tilde{\eta}[\mathbf{y};\bar{\theta}(\mathbf{z})].$$

Substitution of approximations (12.77) and (12.78) or (12.79) in (12.76) results in the approximations

$$v_B \doteq v_e^*[\bar{\theta}(\mathbf{z})]+\{\delta[\mathbf{z};\bar{\theta}(\mathbf{z})]\}'\text{var}_B(\theta)\delta[\mathbf{z};\bar{\theta}(\mathbf{z})]$$

$$\doteq v_e^*[\bar{\theta}(\mathbf{z})]+\{\delta[\mathbf{z};\bar{\theta}(\mathbf{z})]\}'\text{var}_B(\theta)\delta[\mathbf{z};\bar{\theta}(\mathbf{z})]$$

$$+(1/2)\mathrm{tr}\{S[\bar{\theta}(\mathbf{z})]\mathrm{var}_B(\theta)\}.$$

Frequentist Properties of the Bayesian (Point) Predictor. Both $\eta_B(\mathbf{y})$ and its approximation $\tilde{\eta}[\mathbf{y};\bar{\theta}(\mathbf{z})]$ are odd, location-equivariant predictors, as is easily verified. Thus, it follows from the corollary to Proposition 1 that, under Assumption S, the distributions of their prediction errors $\eta_B(\mathbf{y})$-w and $\tilde{\eta}[\mathbf{y};\bar{\theta}(\mathbf{z})]$-w are symmetric about 0, in which case they are unbiased predictors (provided their expected values exist).

Special Case: Mixed Linear Model. Consider now the distribution with p.d.f. $f^*(w|\mathbf{y})$, and the mean $\eta_B(\mathbf{y})$ and variance v_B of this distribution, in the special case where $\mathbf{y}$ follows mixed linear model (12.1) and where $w=\lambda'\boldsymbol{\beta}+\delta'\mathbf{s}$. Set $\pi(\theta)\equiv 1$, corresponding to a "non-informative", improper distribution. Then, $h(\theta|\mathbf{z})=c^{-1}\ell(\gamma,\sigma_e^2;\mathbf{z})$, where $c=\int_0^u\int_0^\infty \ell(\gamma,\sigma_e^2;\mathbf{z})\,d\sigma_e^2 d\gamma$ [and where $\ell(\gamma,\sigma_e^2;\mathbf{z})$ is as defined in Sect. 12.6.1].

For $k < (n-p^*)/2$,

$$\int_0^\infty (\sigma_e^2)^{(k-1)}\ell(\gamma,\sigma_e^2;\mathbf{z})d\sigma_e^2 = 2^{-k}\pi^{-(n-p^*)/2}\Gamma[(n-p^*-2k)/2].[\prod_i(1+\gamma\Delta_i)]^{-1/2}$$

$$\cdot[S_e+\sum_i \tilde{t}_i^2/(1+\gamma\Delta_i)]^{-(n-p^*-2k)/2} \qquad (12.80)$$

(e.g., Box and Tiao 1973, Appendix A2.1). It follows, in particular, that the p.d.f. of the conditional distribution of γ given $\mathbf{z}$ is

$$h_1(\gamma|\mathbf{z}) = \int_0^\infty h(\theta|\mathbf{z})d\sigma_e^2$$

$$= c_1^{-1}[\prod_i(1+\gamma\Delta_i)]^{-1/2}[S_e+\sum_i \tilde{t}_i^2/(1+\gamma\Delta_i)]^{-(n-p^*-2)/2},$$

where

$$c_1 = \int_0^u[\prod_i(1+\gamma\Delta_i)]^{-1/2}[S_e+\sum_i \tilde{t}_i^2/(1+\gamma\Delta_i)]^{-(n-p^*-2)/2}d\gamma.$$

Note that [since $\tilde{\eta}(\mathbf{y};\theta)$ does not depend on σ_e^2]

$$\eta_B(\mathbf{y}) = \int_0^u \tilde{\eta}(\mathbf{y};\theta)h_1(\gamma|\mathbf{z})d\gamma, \qquad (12.81)$$

$$\mathrm{var}_B[\tilde{\eta}(\mathbf{y};\theta)] = \int_0^u[\tilde{\eta}(\mathbf{y};\theta)-\eta_B(\mathbf{y})]^2h_1(\gamma|\mathbf{z})d\gamma. \qquad (12.82)$$

Moreover, observing that $v_e^*(\theta)=\sigma_e^2\psi(\gamma)$ for a suitably defined function $\psi(\gamma)$ of γ, it follows from result (12.80) that

$$E_B[v_e^*(\theta)] = c_1^{-1}(n-p^*-4)^{-1}\int_0^u \psi(\gamma)[\prod_i(1+\gamma\Delta_i)]^{-1/2}$$

$$\cdot[S_e+\sum_i \bar{t}_i^2/(1+\gamma\Delta_i)]^{-(n-p^*-4)/2}d\gamma. \tag{12.83}$$

Thus, by taking advantage of expressions (12.81)-(12.83), the problem of computing $\eta_B(\mathbf{y})$ and v_B can be reduced to one of numerically evaluating one-dimensional integrals. A similar approach can be used to reduce the problem of evaluating the p.d.f. $f^*(w|\mathbf{y})$ to one of numerically evaluating one-dimensional integrals.

Example. Consider, for the lamb-weight example, the prediction of the value of $w_{13}-w_{56}$. The values of $\eta_B(\mathbf{y})$ and $v_B^{1/2}$ are found to be 0.55 and 1.016, respectively. The value of $\eta_B(\mathbf{y})$ is in reasonably good agreement with the value (0.69) reported in Section 12.6.2 for the corresponding frequentist quantity $\hat{\eta}(\mathbf{y})$. Suppose, however, that the data were modified in such a way that the value of S_e were increased to 138 (from its original value of 102.235) but $\bar{\mathbf{t}}$ and $\mathbf{X'y}$ were left unchanged. As depicted in Fig. 12.2, the conditional distribution of γ given $\mathbf{z}$ is much more severely skewed for the modified data than the original. For the modified data, the value of $\eta_B(\mathbf{y})$ is found to be 0.47. This value differs considerably from that of $\hat{\eta}(\mathbf{y})$, which is found to be -0.18 (corresponding to an REML estimate of γ of 0.010).

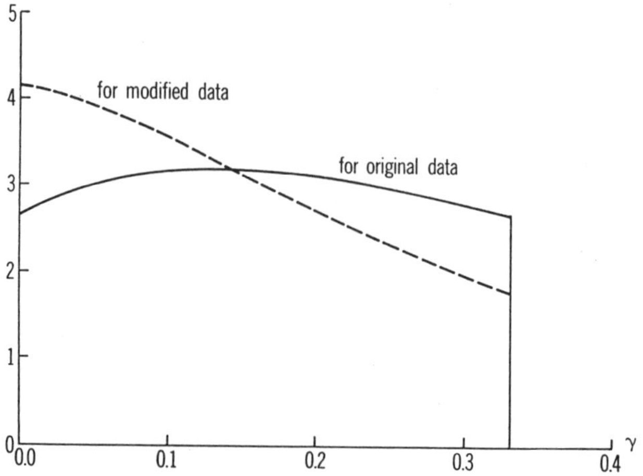

Fig. 12.2 P.d.f. of the conditional distribution of γ given $\mathbf{z}$

Acknowledgements. Thanks are due Alicia L. Carriquiry for assisting with the computations and with the preparation of the figures. The work was supported in part by the Office of Naval Research Contract N00014-85-K-0418.

References

Berger, JO (1985) Statistical decision theory and Bayesian analysis. 2nd edn. Springer-Verlag, Berlin Heidelberg New York Tokyo

Box, GEP, Tiao GC (1973) Bayesian inference in statistical analysis. Addison-Wesley, Reading

Dempfle L (1977) Relation entre BLUP (best linear unbiased prediction) et estimateurs bayesiens. Ann Genet Sel Anim 9:27-32

Dempster AP, Rubin DB, Tsutakawa, RK (1981) Estimation in covariance components models. J Am Stat Assoc 76:341-353

Efron B (1982) The jackknife, the bootstrap, and other resampling plans. Society for Industrial and Applied Mathematics, Philadelphia

Ferguson TS (1967) Mathematical statistics: a decision theoretic approach. Academic Press, New York

Gianola D, Fernando RL (1986) Bayesian methods in animal breeding theory. J Anim Sci 63:217-244

Goldberger AS (1962) Best linear unbiased prediction in the generalized regression model. J Am Stat Assoc 57:369-375

Hartigan JA (1969) Linear Bayesian methods. J R Stat Soc Ser B 31:446-454

Harville DA (1976) Extension of the Gauss-Markov theorem to include the estimation of random effects. Ann Stat 4:384-395

Harville DA (1977) Maximum likelihood approaches to variance component estimation and to related problems. J Am Stat Assoc 72:320-340

Harville DA (1985) Decomposition of prediction error. J Am Stat Assoc 80:132-138

Harville DA, Fenech AP (1985) Confidence intervals for a variance ratio, or for heritability, in an unbalanced mixed linear model. Biometrics 41:137-152

Henderson CR (1963) Selection index and expected genetic advance. In: Hanson WD, Robinson HF (eds) Statistical genetics and plant breeding. NAS-NRC Publ 982, Washington DC, pp 141-163

Jeske DR (1985) Prediction intervals for the realization of a random variable under a general mixed linear model. PhD Diss, Iowa State Univ, Ames

Jeske DR, Harville DA (1988) Prediction-interval procedures and (fixed-effects) confidence-interval procedures for mixed linear models. Commun Stat Theor Meth 17:1053-1087

Kackar RN, Harville DA (1984) Approximations for standard errors of estimators of fixed and random effects in mixed linear models. J Am Stat Assoc 79:853-862

Khatri CG, Shah KR (1981) On the unbiased estimation of fixed effects in a mixed model for growth curves. Commun Stat Theor Meth 10:401-406

Lehmann EL (1983) Theory of point estimation. Wiley, New York

Patterson HD, Thompson R (1971) Recovery of interblock information when block sizes are unequal. Biometrika 58:545-554

Prasad NGN, Rao JNK (1986) On the estimation of mean square error of small area predictors. In: 1986 Proc Section Survey Res Meth, Am Stat Assoc, pp 108-116

Rao CR (1965) Linear statistical inference and its applications. Wiley, New York

Sallas WM (1979) Recursive mixed model estimation. PhD Diss, Iowa State Univ, Ames

Sallas WM, Harville DA (1981) Best linear recursive estimation for mixed linear models. J Am Stat Assoc 76:860-869

Searle SR (1971) Linear models. Wiley, New York

Thompson WA Jr (1962) The problem of negative estimates of variance components. Ann Math Stat 33:273-289

Urquhart NS (1969) The nature of the lack of uniqueness of generalized inverse matrices. SIAM Review 11:268-271

13 Connectedness in Genetic Evaluation

J.L. Foulley[1], J. Bouix[2], B. Goffinet[3], and J.M. Elsen[2]

The problem of connectedness in genetic evaluation is part of the more general one of "adjusting" for nuisance parameters (e.g., herd x year x season) and of estimating genetic population means (group effects). After a review of approaches to the study of connectedness, its impact is discussed in relation to the type of model used (purely fixed or mixed; with or without genetic group) and to the properties sought from an estimator of genetic merit (unbiasedness vs. mean square error). Consideration of connectedness is especially important in the choice of alternative models. The degree of connectedness is likely to be a factor limiting the effectiveness of adjusting genetic evaluations for sources of bias. A coefficient to assess the degree of connectedness in practice is suggested. This coefficient involves pairs of levels of factors, and it varies between 0 (non-estimability) and 1 (balanced distribution according to the factors involved in disconnectedness). Small numerical examples are given to illustrate theoretical aspects of this problem. Practical applications based on the use of artificial insemination sires to connect herds over space and time are also presented.

13.1 Introduction

Artificial insemination (AI) is sufficiently extended in dairy cattle to permit progeny testing of males and to disseminate genetic progress in the population as a whole. In other species of livestock as well as in other systems, e.g., production of beef in suckling herds, natural service and AI coexist. In these "mixed" systems, it is necessary to include natural service sires in the base for selection to obtain reasonable genetic progress. This requires an effective system of comparison and evaluation of sires based on an extended on-the-farm performance recording program. One of the main limitations to taking full advantage of the data obtained is the insufficient use of the same sires in several herds. The intra-herd deviation of the progeny of a given sire from contemporary animals is free of biases due to

1 Institut National de la Recherche Agronomique (INRA), Station de Génétique Quantitative et Appliquée 78350 Jouy-en-Josas, France
2 INRA, Station d'Amélioration Génétique des Animaux, Toulouse, France.
3 INRA Laboratoire de Biométrie BP 27, Auzeville 31326 Castanet-Tolosan Cedex, France

environmental effects peculiar to the herd and the period in question. However, this deviation also includes a component due to the average genetic level of the sires of the contemporaries (Spike and Freeman 1977), and this can vary considerably among herds either due to sampling or to systematic genetic differences among herds. Artificial insemination provides a useful tool for creating a network of connections among herds which permits accounting for these differences in genetic level (Foulley and Clerget-Darpoux 1978; Foulley and Menissier 1978).

There is a great diversity of situations with respect to the feasibility of creating such a network, and these are related to the species, state of the AI technique, production system, degree of use of AI, and extent of collective participation in organized selection schemes. For example, certain schemes in beef cattle require intensive planning of matings and of recording to achieve a reasonable degree of success (Foulley and Sapa 1982).

This paper, without pretending to be exhaustive, aims to address theoretical and practical issues arising in the development of connections in animal improvement programs. The concept of connection presents difficulties of interpretation so a precise definition is needed. After a brief review of statistical models used to predict the genetic value of sires, the effects of connection on properties of the estimators such as bias and mean square error will be considered.

13.2 The Models

Attention will be restricted to situations where the model relating the vector of sire effects $\mathbf{w}$ to observations $\mathbf{y}$ is linear. An important problem in the study of connection is the nature of the effects and the form to model them. Two situations will be considered.

13.2.1 Classical Model

If the distribution of sires across herds is random, i.e., if there is no association between characteristics of the sires (breed, age) and those of the herds, the genetic values w_{ij} of male i in herd j are independent (at least in the large numbers) random variables with the same distribution. Using asymptotic arguments, it can be often assumed that the random variables are normally distributed and this leads to the mixed linear model:

$$\mathbf{y} = \mathbf{Xb}+\mathbf{Zw}+\mathbf{e}, \qquad (13.1)$$

where **b** is a vector of nuisance parameters, taken in this paper as herd effects, to simplify developments; **w** is the vector of genetic effects to be predicted; **X** and **Z** are known incidence matrices, and **e** is a vector of random residuals. Assumptions are

$$E\begin{bmatrix}\mathbf{y}\\ \mathbf{w}\\ \mathbf{e}\end{bmatrix} = \begin{bmatrix}\mathbf{Xb}\\ \mathbf{0}\\ \mathbf{0}\end{bmatrix} \tag{13.2a}$$

and

$$\text{var}\begin{bmatrix}\mathbf{y}\\ \mathbf{w}\\ \mathbf{e}\end{bmatrix} = \begin{bmatrix}\mathbf{V} & \mathbf{ZG} & \mathbf{0}\\ \mathbf{GZ'} & \mathbf{G} & \mathbf{0}\\ \mathbf{0} & \mathbf{0} & \mathbf{R}\end{bmatrix} \tag{13.2b}$$

with

$$\mathbf{V} = \mathbf{ZGZ'}+\mathbf{R}. \tag{13.2c}$$

In view of the assumptions, **G** can be written as

$$\mathbf{G} = \mathbf{I}\sigma_w^2. \tag{13.3a}$$

For reasons of simplicity, we will take

$$\mathbf{R} = \mathbf{I}\sigma_e^2. \tag{13.3b}$$

However, if the variance of the residuals is not homogeneous, or if these are correlated, it suffices to make the transformations

$$\mathbf{y}^* = \mathbf{S}^{-1}\mathbf{y} \quad \mathbf{X}^* = \mathbf{S}^{-1}\mathbf{X} \quad \mathbf{Z}^* = \mathbf{S}^{-1}\mathbf{Z}, \tag{13.4}$$

where **S** is a square matrix of full rank such that **R=SS'**, to return to the situation of (13.3b) (Gianola 1986).

It often happens that the assumption of a random distribution of sires across herds does not hold, as frequently the better sires are used in the better herds. If this is so, the best linear unbiased predictor of **w** obtained with model (13.1) might be adequate to the extent that herd effects **b** are treated as fixed (Henderson 1973; Pollak and Quaas 1982). A simple reason to explain this is that the best linear predictor (BLUP) is invariant to translations on the effects considered fixed (Goffinet 1983). In other words, advantage is taken of the fact that BLUP is a function of n-rank(**X**) contrasts free of herd effects. This is accomplished via the projector

$$\mathbf{M} = \mathbf{I}-\mathbf{X}(\mathbf{X}'\mathbf{X})^{-}\mathbf{X}'. \tag{13.5}$$

Alternatively, this might be explained by invoking a Bayesian argument. The joint prior distribution f(**b**,**w**) can be written as f(**w**)·f(**b**|**w**). Under multivariate normality, the prior covariance matrix of **b**|**w** is

$$\Gamma_{bb}^{1/2}(\mathbf{R}_{bb}-\mathbf{R}_{bw}\mathbf{R}_{ww}^{-1}\mathbf{R}_{wb})\Gamma_{bb}^{1/2},$$

where Γ_{bb}=Diag[Var($\mathbf{b}_j$)], and $\mathbf{R}_{bw}$ is a correlation matrix between **b** and **w**, *a priori*. Taking Γ_{bb} to be very large, f(**b**|**w**) tends towards a uniform distribution, which does not depend on the prior associations between **b** and **w**. The BLUP of **w** using (13.1) with **b** fixed corresponds to a type of Bayesian estimator using a flat prior for **b** (Dempfle 1977; Gianola and Fernando 1986).

13.2.2 Certain Characteristics of the Males Known

Suppose that in addition to **y**, there is information available on, e.g., line of the males, age or genetic relationship to other males. Let **z** be a vector denoting this additional information. The more natural evaluation criterion to consider, to best utilize the new information, is (Bulmer 1980; Goffinet and Elsen 1984; Fernando and Gianola 1986):

$$\mathrm{E}(\mathbf{w}|\mathbf{y},\mathbf{z}). \tag{13.6}$$

When **z** is a quantitative variable, and if one assumes joint multivariate normality of **w**,**y**,**z** with known covariance matrices, (13.6) can be readily calculated using the technique of BLUP. When **z** is a qualitative variate such as the line or the year of birth of the sire, the model to consider is

$$\mathbf{y} = \mathbf{Xb}+\mathbf{Zw}+\mathbf{e} \tag{13.7a}$$

with

$$\mathbf{w} = \mathbf{Pg}+\mathbf{u}, \tag{13.7b}$$

where $\mathbf{g}$ is an $mx1$ vector of expectations of the values of the sires conditionally to the qualitative factor $\mathbf{z}$, which is often designed as containing "group" effects. The matrix $\mathbf{P}$, of order qxm, allocates the q males to the m groups, and $\mathbf{u}$ is a $qx1$ vector of sire effects intra-group with $\mathbf{u} \sim (\mathbf{0},\mathbf{G})$.

Thus, $\mathbf{Pg}$ is interpretable as $E(\mathbf{w}|\mathbf{z})$, and $\mathbf{g}$ is a hyper-parameter which in a Bayesian framework can be specified. Then, conditionally to $\mathbf{g}$ and $\mathbf{G}$, the predictor $\hat{\mathbf{w}}$ of $\mathbf{w}$ given $\mathbf{y}$ and $\mathbf{z}$, is

$$\hat{\mathbf{w}} = E(\mathbf{w}|\mathbf{y},\mathbf{z}) = \mathbf{Pg}+E(\mathbf{u}|\mathbf{y},\mathbf{g}), \tag{13.8a}$$

so the last two terms of (13.8a) depend on $\mathbf{g}$. Under the assumption of multivariate normality, and taking $\mathbf{G}=\mathbf{I}\sigma^2$, $k=\sigma_e^2/\sigma_u^2$, $\tilde{\mathbf{u}}=E(\mathbf{u}|\mathbf{y},\mathbf{g})$ is the solution to the system

$$(\mathbf{Z'MZ} + k\mathbf{I})\tilde{\mathbf{u}} = \mathbf{Z'M}(\mathbf{y}-\mathbf{ZPg}).$$

In practice, we are forced to calculate $\hat{\mathbf{w}}$ of (13.8a) conditionally to a value $\hat{\mathbf{g}}$ estimated from the data so the criterion of selection becomes

$$\tilde{\mathbf{w}} = \mathbf{P}\hat{\mathbf{g}}+E(\mathbf{u}|\mathbf{y},\mathbf{g}=\hat{\mathbf{g}}). \tag{13.8b}$$

In an empirical Bayes approach, $\hat{\mathbf{g}}$ is estimated from the marginal distribution $f(\mathbf{y}|\mathbf{g})$ using, for example, the method of maximum likelihood (ML). Under normality, this ML estimator is identical to the generalized least-squares estimator so the criterion $\tilde{\mathbf{w}}$ corresponds to BLUP of Henderson.

In contrast to (13.8b) one could consider the estimator $\mathbf{w}^*$ derived from model (13.1) ignoring the information contributed by $\mathbf{z}$, so

$$\mathbf{w}^* = E(\mathbf{w}|\mathbf{y}, \text{Model 13.1}), \tag{13.8c}$$

noting that $\mathbf{w}$ in model (13.1) does not contain a "group" effect. Hence, the choice between (13.8b) and (13.8c) is essentially a choice between models with or without groups. This is related to:

1. The goodness of the approximation of $\hat{\mathbf{w}}$ by $\tilde{\mathbf{w}}$, which depends highly on the precision with which $\mathbf{g}$ is estimated from the data; and
2. The difference between $\hat{\mathbf{w}}$ and $\mathbf{w}^*$, and this is primarily a function of the real importance of the "group" effects.

As it will be seen later, the problems of connection enter in point 1 as they may limit the effective correction of systematic genetic differences. However, it is always important to consider points 1 and 2 simultaneously, to the extent that the information used in estimating group effects is not really employed to advantage unless differences between groups are not negligible.

In (13.8b), **g** is estimated almost always using generalized least-squares, i.e., regarding this vector as fixed. However, it is possible to consider calculating $E(\mathbf{g}|\mathbf{y})$ using an informative prior density in the analysis, which implies regarding **g** as a random factor in the model. There is a well-known situation where one proceeds as if the group effects were random. This is the case where the qualitative information used to define the groups is the knowledge of ancestors or, more generally, of genetic relationships among sires. Describing the *a priori* relations among males via a matrix of variances and covariances (**A** replacing **I** in 13.3a), it is possible to considerably reduce the number of parameters to be estimated in a model where, for example, each common ancestor would furnish a level of the vector **g**. Including the relationship matrix, however, can pose other modelling problems when the fixed group effects are included in the model. In theory, when all data are available and the relationship matrix is complete, including the latter, this eliminates the need to group sires based on year of birth or year of entering service. Unfortunately, this is rarely the case and one is led to consider both relationships and groups in the model. An interesting approach in this case has been proposed by Thompson (1979) and studied by Quaas and Pollak (1982). Recently, Westell (1984) developed a grouping procedure of individuals in the base population for an animal model which uses Thompson, Quaas and Pollak's accumulated group model. The procedure takes into account the theory of quantitative genetics by accumulating effects of ancestral groups in the genetic value of an individual. This is done with transformations $\mathbf{w}^*$, $\mathbf{g}^*$, $\mathbf{u}^*$, letting

$$\begin{aligned}\mathbf{w} &= \mathbf{T}\mathbf{w}^* \\ &= \mathbf{T}(\mathbf{P}\mathbf{g}^*+\mathbf{u}^*) \\ &= \mathbf{T}\mathbf{P}\mathbf{g}^*+\mathbf{u},\end{aligned} \tag{13.9a}$$

where **T** is a lower triangular matrix such that

$$\mathbf{A} = \mathbf{T}\mathbf{D}\mathbf{T}' \tag{13.9b}$$

with **D** diagonal.

13.3 The Unbiasedness Constraint

13.3.1 Models without Group Effects

It is important to observe that in a model without group effects such as (13.1), BLUP always gives a unique solution regardless of the state of connectedness between sires and herds. After absorption of the fixed effects, the mixed model equations are

$$(\mathbf{Z'MZ}+k\mathbf{A}^{-1})\hat{\mathbf{u}} = \mathbf{Z'My}, \tag{13.10}$$

where from now on, following common usage, the genetic values in the absence of group effects are designated as $\mathbf{u}$ (and not $\mathbf{w}$), $k=\sigma_e^2/\sigma_u^2$, and $\mathbf{M}$ is the projection matrix defined in (13.5). Because the coefficient matrix in (13.10) is the sum of a positive semi-definite and of a positive-definite matrix, it must be positive-definite and, therefore, invertible (Rao 1952).

The system of equations in $\hat{\mathbf{u}}$ can then always be solved even in the trivial case of a sire x herd layout with a single sire per herd; here, the BLUP solution is identical to the mean of the prior distribution of the sire effects $\mathbf{u}$. Nevertheless, whereas in the case of connectedness there is a single relationship between the BLUP solutions, $\mathbf{1'A}^{-1}\hat{\mathbf{u}}=0$, the existence of disconnection results in the appearance of q-1-rank($\mathbf{Z'MZ}$) relations of the form

$$\mathbf{d}_i'\mathbf{A}^{-1}\hat{\mathbf{u}} = 0, \tag{13.11}$$

where $\mathbf{d}_i'$ is a row vector of q elements defining a relationship other than the trivial one "sum of the rows=0", between the rows or columns of $\mathbf{Z'MZ}$. This property follows from the fact that the least-squares equations obtained by putting k=0 in (13.10), are consistent.

13.3.2 Models with Group Effects

Estimability of w. Consider the estimability of the vector $\mathbf{w}=\mathbf{Pg}+\mathbf{u}$. In general, a linear combination of fixed effects ($\mathbf{k'b}$) and of random effects ($\mathbf{m'u}$) is predictable if and only if $\mathbf{k'b}$ is estimable. With the estimator of $\mathbf{b}$ being that of generalized least-squares, $\mathbf{k'b}$ is an estimable function if

$$\mathbf{k}' = \mathbf{k}'(\mathbf{X'V}^{-1}\mathbf{X})^{-}(\mathbf{X'V}^{-1}\mathbf{X}), \tag{13.12}$$

where $\mathbf{V}$=Var($\mathbf{y}$) is the matrix defined in (13.2b). There is considerable simplification, as an estimable function $\mathbf{k'b}$ is estimable in a generalized least-squares model, if it is also estimable in the corresponding ordinary least-squares model. In fact, if $\mathbf{k'b}$ is estimable under the model

$$\mathbf{y} = \mathbf{Xb}+\mathbf{e} \tag{13.13}$$

with

$$\text{Var}(\mathbf{y}) = \mathbf{I}\sigma_e^2, \tag{13.14}$$

there exists a column vector **t** such that E(**t'y**)=**k'b**, with **k'**=**t'X** (Searle 1971). The same linear combination is centered on **k'b** in the generalized least-squares model because, under that model, E(**t'y**)=**t'Xb**=**k'b**. The problem of connection between factors in relation to prediction of **k'b**+**m'u**, reduces then to the study of connection between factors in the simpler model (13.13). This will be considered in the following section.

Estimability of Fixed Effects. An experimental design is said to be disconnected when certain parameters or functions thereof which are normally estimable under the complete design are no longer estimable when there are empty cells. A simple two-factor layout with "sires" and "herds" (Table 13.1) illustrates the problem. In this example, there are only ten estimable (linearly independent) parameters instead of the 11 that could be estimated were the layout connected. Letting μ,s_i and h_j be a mean, the effect of sire i and of herd j, respectively, the estimable parameters are:

$$\mu_1, s_1-s_4, s_4-s_6, h_1-h_2, h_2-h_5$$

$$\mu_2, s_2-s_3, s_3-s_5, h_3-h_4, h_4-h_6.$$

These two sets of estimable functions correspond to two disconnected sets which appear very clearly in the information matrix after permutation of rows and columns (Table 13.2). Disconnectedness creates a problem of lack of estimability which, in matrix terms, leads to a reduction in the rank of the information matrix **X'X** (Chakrabarti 1963; Eccleston 1972; Eccleston and Russell 1975; Rhagavarao and Federer 1975). Shah and Yadolah (1977) published a remarkable review of this type of statistical approach to the study of connectedness and introduced a number of definitions which we utilize here because of their clarity.

Let $\phi=(F_{i_1},F_{i_2},...,F_{i_s})$ and $\theta=(F_{i_{s+1}},F_{i_{s+2}},...,F_{i_{s+k}})$ be two disjoint sets of s and k factors, respectively, among the m factors included in a cross-classified, additive layout, with $i_1,i_2,..., i_{s+k}$ designating an ordered subset of 1,2,...,m. Also, let $r(\phi|\theta)$ be the rank of the space of estimable functions relative to the effects of factors ϕ after eliminating the effects of factors θ, and ignoring factors not included in $\phi U\theta$ (U:union). With this notation taken from the above authors, it is said that:

1. Factors ϕ in the complete layout are connected if

$$r(\phi) = \sum_{j=1}^{s} f_j - s + 1, \tag{13.15}$$

where f_j is the number of levels of factor j in θ;

2. Factors ϕ are connected after having eliminated factors θ, and ignoring the remaining factors if

$$r(\phi|\theta) = \sum_{j=1}^{s} f_j - s. \tag{13.16}$$

Table 13.1 Example of a disconnected "sire x herd" design

	Herd[a]					
Sire	1	2	3	4	5	6
1	X (7)	X (10)			X (15)	
2			X (2)			X (7)
3			X (2)	X (5)		X (8)
4	X (3)				X (10)	
5				X (7)		
6		X (4)			X (4)	

[a] In parentheses are the number of progeny per subclass.

Relationship (13.15) applies particularly to the case of a complete layout with m factors which is, thus, connected if the rank of the estimable functions is equal to $\sum_{j=1}^{m}$ f-m+1. In the example given in Table 13.1, if one is interested in ϕ="sire" and θ="herd", it should hold (in case of connectedness) that $r(\phi)=6-1+1=6$, $r(\phi|\theta)= 6-1=5$ and $r(\phi,\theta)=6+6-2+1=11$. These definitions can also be expressed in terms of matrix ranks because

$$r(\phi) = \text{rank}(\mathbf{X}'_{\phi}\mathbf{X}_{\phi}) \qquad (13.17a)$$

$$r(\phi|\theta) = \text{rank}(\mathbf{C}_{\phi|\theta}), \qquad (13.17b)$$

where

$$\mathbf{C}_{\phi|\theta} = \mathbf{X}'_{\phi}[\mathbf{I}-\mathbf{X}_{\theta}(\mathbf{X}'_{\theta}\mathbf{X}_{\theta})^{-}\mathbf{X}'_{\theta}]\mathbf{X}_{\phi}. \qquad (13.17c)$$

Table 13.2 Information matrix for the design given in Table 13.1 under an additive sire(s) and herd(h) model[a]

A.

32	0	0	0	0	0	7	10	0	0	15	0
0	9	0	0	0	0	0	0	2	0	0	7
0	0	15	0	0	0	0	0	2	5	0	8
0	0	0	13	0	0	3	0	0	0	10	0
0	0	0	0	7	0	0	0	0	7	0	0
0	0	0	0	0	8	0	4	0	0	4	0
7	0	0	3	0	0	10	0	0	0	0	0
10	0	0	0	0	4	0	14	0	0	0	0
0	2	2	0	0	0	0	0	4	0	0	0
0	0	5	0	7	0	0	0	0	12	0	0
15	0	0	10	0	4	0	0	0	0	29	0
0	7	8	0	0	0	0	0	0	0	0	15

B.

32	0	0	7	10	15	0	0	0	0	0	0
0	13	0	3	0	10	0	0	0	0	0	0
0	0	8	0	4	4	0	0	0	0	0	0
7	3	0	10	0	0	0	0	0	0	0	0
10	0	4	0	14	0	0	0	0	0	0	0
15	10	4	0	0	29	0	0	0	0	0	0
0	0	0	0	0	0	9	0	0	2	0	7
0	0	0	0	0	0	0	15	0	2	5	8
0	0	0	0	0	0	0	0	7	0	7	0
0	0	0	0	0	0	2	2	0	4	0	0
0	0	0	0	0	0	0	5	7	0	12	0
0	0	0	0	0	0	7	8	0	0	0	15

[a] A: Matrix pertaining to the vector $[S_1S_2S_3S_4S_5S_6;\ h_1h_2h_3h_4h_5h_6]$.

B: Matrix obtained after the same permutation of rows and columns of A and corresponding to the vector: $(S_1S_4S_6h_1h_2h_5;\ S_2S_3S_5h_3h_4h_6]$.

Note the two following relationships of rows and columns:

(1) + (2) + (3) = (4) + (5) + (6)
(7) + (8) + (9) = (10) + (11) + (12)

The notation $C_{\phi|\theta}$ of Chakrabarti (1963) corresponds to the projector (13.5) to the information matrix $X'_{\phi}M_{\theta}X_{\phi}$ relative to factors ϕ after absorption of factors θ. The structure of this matrix is important in problems of connection. In practice, one is often interested in assessing estimability of contrasts between levels of a single factor T (e.g., between sire groups) after eliminating the effects of additional factors θ (e.g., herd-year-season). It is known that the levels of the factor are connected if $r(\phi_T|\theta)$ is equal to f_T-1, where f_T is the number of levels of factor T. A necessary and sufficient condition for this property to hold is that the matrix $C_{T|\theta}+J$ be non-singular, where J is a square matrix of order f_T with all its elements equal to 1. In a layout with two disconnected, fixed factors, the matrix $C_{T|\theta}$ can be put, by permutation of rows and columns, in a block diagonal form, with each of the blocks corresponding to a set of connected levels. Furthermore, Petersen (1978) proposed a test for connectedness among sires based on this property of the matrix **C** after absorption of "herd-year-season" effects, when this is a large matrix.

Definitions (13.15) and (13.16) in terms of rank of the space of estimable functions have the advantage of providing a general and rigorous approach to the problems of connection, irrespective of the number of factors involved in the layout. In fact, with a two-factor design with "treatments" and "blocks", e.g., there is an equivalence between the properties of connection between treatments, connection between blocks, connection between treatments and blocks, and connection of data. Unfortunately, this does not extend beyond two factors. For example, with three additive factors, there can be a two-factor connection jointly with disconnection between levels of the remaining factor after absorbing the first two. This causes the entire design to be disconnected. The example in Table 13.3, due to Eccleston and Russell (1975) and analyzed by taking two factors at a time, illustrates the problem (Table 13.4). Such a layout is an example of what Shah and Yadolah (1977) termed a non-trivially disconnected[a] design, and this can only be detected by checking the rank of the matrix **C** in (13.17c) (see also Table 13.5). In this case, after absorption of the effects of "rows" and "columns" the matrix **C** relative to the treatment effects should have rank 8-1=7, provided that the preceding effects are connected; however, its rank is 8-2=6, because of a relationship among rows (columns), i.e., (2)+(5)=(3)+(8). Thus, three subsets of treatment effects which are pairwise comparable are obtained: (B,E), (C,H), (A,D,F,H); this allows one to form five linearly independent, pairwise differences, plus a more complex, estimable contrast such as A+B+C-D-F-G.

In this simple case where the model has only two cross-classified factors, following Bose (1947), cited by Eccleston and Hedayat (1974), it is possible to assess connection in terms of a "physical" approach via the establishment of links between objects. Two treatments or a treatment and a block are said to be connected if they can be linked by an alternative chain

---

[a] According to these authors, an additive m-factor design is said to be trivially disconnected if there is at least one set of (m-1) factors disconnected. In contrast, in a non-trivially disconnected design, every subset of (m-1) factors is connected but the design is disconnected for each factor after eliminating (absorbing) the others.

of treatments and blocks associated by contiguous pairs (a treatment and a block are associated if the treatment appears in the block). In the example of Table 13.1, sire 1 is connected to sire 4 by the chain "sire 1, herd 1, sire 4". Likewise, sire 2 is connected with sire 5 by the chain "sire 2, herd 3, sire 3, herd 4, sire 5". In the second case, the connection is "indirect" because sires 2 and 5 appear in two different herds. Also, herds 1 and 2, 2 and 5, and consequently 1 and 5 are directly connected. This physical approach to connection was developed by Weeks and Williams (1964), particularly through a test for connection suggested by these authors for an additive model with N cross-classified factors. The test is based on defining and constituting sets of points, "N-tuples", which are adjacent ("nearly identical"). The data in a given cell can be viewed as a point with N coordinates represented by the ordered sequence of levels of factors $(Z_1,Z_2,...,Z_i,...,Z_N)$, where Z_i can take any value among $1,2,...,m_i$, and m_i is the number of levels of factor i. Two N-tuples are said to be nearly identical if these are equal componentwise except for one component. Using this property systematically in a design, the N-tuples which are nearly identical form a connected subset of design points which may be analyzed separately. The whole data set is then partitioned into subsets, each of which is a collection of connected design points. In a two-way (rows and columns) layout, the procedure reduces to joining cells with observations via horizontal and vertical segments. The networks that do not have common segments constitute sets of data points that are connected intra-set and disconnected inter-set. In the example of Table 13.1, there are two such sets: (11,12,15,41,45,62,65) and (23,26,33,34,36,54), respectively. Beyond two factors, this test only provides a sufficient but not necessary condition for connectedness.

Table 13.3 Example of a non-trivially disconnected design with 3 additive factors: treatment (A,B,...,G,H), rows, and columns (1,2,3,4) (Eccleston and Russell 1975)

Row	Column[a]			
	1	2	3	4
1	$A_{(5)}$	$B_{(8)}$	$E_{(5)}$	$F_{(3)}$
2	$C_{(3)}$	$D_{(4)}$	$G_{(2)}$	$H_{(5)}$
3	$H_{(9)}$	$F_{(10)}$	$A_{(1)}$	$C_{(10)}$
4	$G_{(6)}$	$E_{(6)}$	$B_{(4)}$	$D_{(10)}$

[a] In parenthesis the number of observations in a subclass.

Table 13.4 Diagrams showing that factors[a] in the design of Table 13.3 are pairwise connected, although the entire design is disconnected

R \ T	A	B	C	D	E	F	G	H
1	X (5)	X (8)			X (5)	X (3)		
2			X (3)	X (4)			X (2)	X (5)
3	X (1)		X (10)			X (10)		X (9)
4		X (4)		X (10)	X (6)		X (6)	

C \ T	A	B	C	D	E	F	G	H
1	X (5)		X (3)				X (6)	X (9)
2		X (8)		X (4)	X (6)	X (10)		
3	X (1)	X (4)			X (5)		X (2)	
4			X (10)	X (10)		X (3)		X (5)

R \ C	1	2	3	4
1	X (5)	X (8)	X (5)	X (3)
2	X (3)	X (4)	X (2)	X (5)
3	X (9)	X (10)	X (1)	X (10)
4	X (6)	X (6)	X (4)	X (10)

[a] T,R,C: treatment, row, column, respectively.
Number of observations in a subclass are in parentheses.

Table 13.5 Information matrices for the design in Table 13.3: (I) complete matrix, (II) after absorbing row and column effects

I

	Treatment (sire)								Row				Column			
	1	2	3	4	5	6	7	8	9	10	11	12	13	14	15	16
	6	0	0	0	0	0	0	0	5	0	1	0	5	0	1	0
	0	12	0	0	0	0	0	0	8	0	0	4	0	8	4	0
	0	0	13	0	0	0	0	0	0	3	10	0	3	0	0	10
Treatment	0	0	0	14	0	0	0	0	0	4	0	10	0	4	0	10
	0	0	0	0	11	0	0	0	5	0	0	6	0	6	5	0
	0	0	0	0	0	13	0	0	3	0	10	0	0	10	0	3
	0	0	0	0	0	0	8	0	0	2	0	6	6	0	2	0
	0	0	0	0	0	0	0	14	0	5	9	0	9	0	0	5
	5	8	0	0	5	3	0	0	21	0	0	0	5	8	5	3
Row	0	0	3	4	0	0	2	5	0	14	0	0	3	4	2	5
	1	0	10	0	0	10	0	9	0	0	30	0	9	10	1	10
	0	4	0	10	6	0	6	0	0	0	0	26	6	6	4	10
	5	0	3	0	0	0	6	9	5	3	9	6	23	0	0	0
Column	0	8	0	4	6	10	0	0	8	4	10	6	0	28	0	0
	1	4	0	0	5	0	2	0	5	2	1	4	0	0	12	0
	0	0	10	10	0	3	0	5	3	5	10	10	0	0	0	28

II

1	2	3	4	5	6	7	8
4.016	-1.217	-0.379	0.659	-0.630	-0.088	-0.892	-1.468
-1.217	6.919	1.213	-1.242	-4.247	-2.352	-0.533	1.459
-0.379	1.213	7.134	-2.003	1.191	-2.501	0.075	-4.730
0.659	-1.242	-2.003	7.342	-1.630	-0.662	-1.595	-0.869
-0.630	-4.247	1.191	-1.630	6.764	-1.508	-1.266	1.326
-0.088	-2.352	-2.501	-0.662	-1.508	7.426	1.042	-1.358
-0.892	-0.533	0.075	-1.595	-1.266	1.042	5.044	-1.875
-1.468	1.459	-4.730	-0.869	1.326	-1.358	-1.875	7.515

Note the following relationship of rows and columns:

. (1) + (2) + (3) + (4) + (5) + (6) + (7) + (8) = 0
. (2) + (5) = (3) + (8)

Conclusions. With the effects treated as random being always predictable, it would appear that connectedness between sires is not an issue. In view of the property of estimability presented previously, under the model

$$\mathbf{y} = \mathbf{Hg}+\mathbf{Xb}+\mathbf{Zu}+\mathbf{e} \tag{13.18a}$$

with

$$E(\mathbf{y}) = \mathbf{Hg}+\mathbf{Xb} \tag{13.18b}$$

and

$$\text{var}\begin{bmatrix} \mathbf{u} \\ \mathbf{e} \end{bmatrix} = \begin{bmatrix} \mathbf{A}\sigma_u^2 & 0 \\ 0 & \mathbf{I}\sigma_e^2 \end{bmatrix}, \tag{13.18c}$$

the function **k'Pg** is estimable if and only if it is also estimable in the simple model

$$\mathbf{y} = \mathbf{Hg}+\mathbf{Xb}+\mathbf{e} \tag{13.19}$$

with Var(**e**)=$\mathbf{I}\sigma_e^2$. Under (13.19), the equations for $\hat{\mathbf{g}}$ have the form

$$\mathbf{H'MH}\hat{\mathbf{g}} = \mathbf{H'My}, \tag{13.20}$$

where **M** is as in (13.5). Because the factor "group" precedes in hierarchy the factor "sires", **H** can be written as **H**=**ZP** (Quaas and Pollak 1981), where **P** is a qxm matrix describing which of the m groups the q sires belong to ($\mathbf{P}_{ij}$=1 if sire j belongs to group i, or 0 otherwise). The system to solve when calculating "group" effects and transmitting ability of sires, intra-group, is

$$\begin{bmatrix} \mathbf{P'CP} & \mathbf{P'C} \\ \mathbf{CP} & \mathbf{C}+k\mathbf{A}^{-1} \end{bmatrix} \begin{bmatrix} \hat{\mathbf{g}} \\ \hat{\mathbf{u}} \end{bmatrix} = \begin{bmatrix} \mathbf{P'R} \\ \mathbf{R} \end{bmatrix}, \tag{13.21a}$$

where

$$\mathbf{C} = \mathbf{Z'MZ} \text{ and } \mathbf{R} = \mathbf{Z'My}. \tag{13.21b}$$

This system has rank [q+rank(**P'CP**)] and the groups are connected only if rank (**P'CP**)= m-1. This holds (Foulley et al. 1983) if every vector $\mathbf{d}_i$ in (13.11) defining a non-trivial relation on **C**, is linearly independent of the columns of **P**. In other words, the matrix $[\mathbf{d}_i,\mathbf{P}]'[\mathbf{d}_i,\mathbf{P}]$ must be non-singular (Searle 1971). In the example of Table 13.3, beyond the trivial relationship we have $\mathbf{d}'$=[0 1 -1 0 1 0 0 -1]. If one creates the factor "population" with three groups constituted by sires ADFG,BE, and CH, respectively, we have

$$\mathbf{P}' = \begin{bmatrix} 1 & 0 & 0 & 1 & 0 & 1 & 1 & 0 \\ 0 & 1 & 0 & 0 & 1 & 0 & 0 & 0 \\ 0 & 0 & 1 & 0 & 0 & 0 & 0 & 1 \end{bmatrix}.$$

Here, there is an evident dependence between $\mathbf{d}'$ and rows 2 and 3 of $\mathbf{P}'$; this makes the contrasts between the groups so defined to be unestimable. If, on the contrary, the groups were constituted by sires ABE,CH, and DFG, there would no longer be dependence other than the trivial relationship $\mathbf{P1}=\mathbf{1}$, and pairwise differences between groups would be estimable. Fernando et al. (1983) proposed an algorithm to search for connected groups in a layout with fixed herd-year-seasons and random sires intra-group.

As mentioned earlier in connection with (13.8b) one could consider omitting group effects from the model. In this case, the predictors of sire effects would be the quantities $\hat{\mathbf{u}}_2$ obtained applying the BLUP procedure to model (II) $\mathbf{y}=\mathbf{Xb}+\mathbf{Zu}+\mathbf{e}$, when the true model is (I), $\mathbf{y}=\mathbf{Hg}+\mathbf{Xb}+\mathbf{Zu}+\mathbf{e}$. The linear combination $\delta'\hat{\mathbf{u}}_2$ has as expectation under the true model:

$$E(\delta'\hat{\mathbf{u}}_2) = \delta'\mathbf{FHg},$$

where $\mathbf{F}$ is a matrix such that $\mathbf{Fy}$ is BLUP($\mathbf{u}$) under model II. If the true values of sires is as in (13.7), $\mathbf{w}=\mathbf{Pg}+\mathbf{u}$, the bias can be written

$$B=E(\delta'\hat{\mathbf{u}}_2)-\delta'\mathbf{w},$$

and it has expectation:

$$E(B) = \delta'(\mathbf{FH}-\mathbf{P})\mathbf{g}$$

$$= \delta'(\mathbf{FZ}-\mathbf{I})\mathbf{Pg}, \qquad (13.22)$$

obtained by putting $\mathbf{H}=\mathbf{ZP}$. The matrix $\mathbf{FZ}$ is explicitly

$$\mathbf{FZ} = \Omega\mathbf{C}\ ;\ \Omega = (\mathbf{C}+k\mathbf{A}^{-1})^{-1}. \qquad (13.23)$$

Thus, in the example of Table 13.3, the difference between sires E and H predicted under Model II is affected by a bias (given that Model I holds) of $0.9795(g_2-g_1)+0.0118(g_2-g_3)$, with the grouping ABE,CH,DFG. The bias can be important depending on the extent of the true differences between groups.

Using models with group effects implicitly poses the question of how the groups should be formed. This can be difficult, especially when artificial insemination and natural service coexist, such as in beef cattle and sheep. It is necessary to find a compromise between the extreme cases of assuming that there are no genetic differences between herds, on the one hand (see Foulley and Clerget-Darpoux 1978), and that all such differences can be accounted for via subclasses constituted by natural service sires used in a particular herd in a given year, on the other; this was considered in a study by Foulley et al. (1983). A general model for this AI-natural service situation could be:

$$y_{ijhkl} = h_{ij}+g_h+s_{hk}+e_{ijkl}, \tag{13.24}$$

where h_{ij} is the fixed effect of herd i and period j, g_h is the fixed effect of the group h of sires, s_{hk} is the effect of sire k in group h, and e_{ijkl} is a residual. Often, the nature of sires used in AI service will suggest allocating them to a particular group. Also, the effect of a particular sire in an AI group could be treated in the usual manner as random, or fixed as suggested by Henderson (1973). In this latter case, the difference between two natural service sires in different groups will be estimable only if group and herd-period effects are connected (the condition for random AI sires), and if the AI sires are connected among themselves after absorption of the h_{ij} effects. This is only possible if AI is used to a considerable extent in a well-planned manner. With this model, on the other hand, differences between groups of natural service sires no longer depend on values of AI sires contrary to what occurs in unbalanced layouts with sires treated as random. One can write:

$$E(g_h-g_{h'})=\begin{cases} g_h-g_{h'} & \text{if sires in AI groups } (s_{ok}) \text{ are fixed} \\ \\ g_h-g_{h'}+\sum\limits_k \alpha_{k,hh'}E(s_{ok}) & s_{ok} \text{ random} \end{cases}$$

where the coefficients α are such that $\sum\limits_k \alpha_{k,hh'}=0$ with $\alpha_{k,hh'}=0$ in a balanced distribution of AI sires. Above, $E(s_{ok})$ is the prior expectation of AI sire k. A non-negligible bias can result and its extent depends on the unbalancedness and on how large systematic differences between AI sires are. A compromise could be made forming some "realistic" groups with AI sires and treating these as random, intra-group.

13.4 Minimum Mean Square Error

13.4.1 Models without Group Effects

In the context of linear models, BLUP is an unbiased predictor, in the sense that $E(\hat{u}_i)=E(u_i)$, which minimizes the mean square error $E(\hat{u}_i-u_i)^2$. The prediction error variance is

$$PEV_i = E(\hat{u}_i-u_i)^2 = (\mathbf{A})_{ii}\sigma_u^2-(\Omega)_{ii}\sigma_e^2, \tag{13.25}$$

where $(\mathbf{A})_{ii}$ and $(\Omega)_{ii}$ indicate the diagonal terms of $\mathbf{A}$ in $\mathbf{G}=\mathbf{A}\sigma_u^2$, and of the matrix Ω defined in (13.23), respectively.

It is common to express the precision of that prediction in terms of a coefficient of determination (CD_i) such that

$$CD_i = [Var(u_i)-PEV_i]/Var(u_i),$$

which is the coefficient of correlation between u_i and $\hat{u}_i$. In the case of unrelated animals ($\mathbf{A}=\mathbf{I}$), this reduces to

$$CD_i = 1-k(\Omega)_{ii} \tag{13.26}$$

with k being the ratio σ_e^2/σ_u^2.

There is an information matrix and a coefficient of determination for the sires being tested associated with every connected layout. In a balanced design, all CD's are identical and, thus, provide a measure peculiar to the layout and connection situation. Foulley and Clerget-Darpoux (1978) gave an expression for the CD when in each herd there are p natural service sires, each having n progeny, and with the connection provided by an AI sire which has m progeny in each herd. Putting N=m+pn,

$$CD = [n/(n+k)][1-kn/(mn+kN)]. \tag{13.27}$$

With unbalancedness, there is not a single criterion for evaluating a plan of connections. A possible way around the problem would be to use a weighted average of the CD's; the weights could be, for example, progeny group sizes.

13.4.2 Models with Group Effects

This situation was considered by Foulley et al. (1983) in relation to the planning of a progeny test program for young beef AI sires, using reference bulls. These authors studied the average of the prediction error variances between and within groups, weighted by their respective frequencies. The groups consisted of reference sires (treated as fixed), and "AI unit-year of test" groups of young sires. Also, Tong et al. (1980) compared by simulation different plans of connection using exchanges of semen between two regions; the end point assessed was also the mean prediction error variance.

In models with "groups" where the value of a sire is estimated as $\hat{\mathbf{w}}=\mathbf{P}\hat{\mathbf{g}}+\hat{\mathbf{u}}$, the CD can be measured by the criterion

$$[\mathrm{Var}(w_i)-\mathrm{Var}(w_i-\hat{w}_i)]/\mathrm{Var}(w_i)$$

(Ufford et al. 1979; Foulley et al. 1983), although this is no longer the coefficient of correlation between w_i and $\hat{w}_i$. The CD is often used in practice as a selection variable. In fact, many breeders base selection decisions both on estimated breeding values and on their precision. In natural service with limited genetic exchanges between herds, it is tempting to generalize the above indicator to assess connection among sires. The formula would be applied to the difference $w_i-w_{i'}$ rather than to w_i (Foulley et al. 1983).

As noted earlier, connection is an "all-or-none" property: disconnectedness appears as an extreme case of unbalance resulting in lack of estimability. It would be useful to measure connection on a continuous scale where the two extremes represent complete balance and disconnectedness, respectively.

In (13.16) and (13.17b) it was seen that the factors ϕ would be disconnected after taking into account factors θ, if the rank of $\mathbf{C}_{\phi|\theta}$ in (13.17c) is less than the sum of levels of factors ϕ minus the number of factors θ. Using the notion of Co-rank[b] (Delattre 1983), this condition holds if

$$\text{Co-rank}(\mathbf{C}_{\phi|\theta}) > s.$$

In the case where ϕ represents a single factor (treatments, groups, or fixed sires) the above is equivalent to the statement that there is lack of connection among sires, for example, after taking into account the other nuisance factors if Co-rank ($\mathbf{C}$)>1, where $\mathbf{C}=\mathbf{Z}'\mathbf{M}\mathbf{Z}$. Now, Co-rank ($\mathbf{C}$) gives the order of multiplicity of the null eigenvalues of $\mathbf{C}$, and the remaining eigenvalues are equal if the distribution of observations for a given level of ϕ over θ levels does not change from one level to the other. Unbalancedness translates into a dispersion of the q-1 eigenvalues; in the extreme case of lack of connection, some of these q-1 eigenvalues are null. This illustrates why there is no problem with disconnectedness if the factor is treated as random. Because $\mathbf{C}$ changes to $\mathbf{C}^*=\mathbf{C}+k\mathbf{A}$, the eigenvalues $\lambda_i(\mathbf{C})$ change to $\lambda_i(\mathbf{C}^*)=\lambda_i(\mathbf{C})+k\varepsilon_i$, where $0<\min\lambda_i(\mathbf{A})\leq\varepsilon_i\leq\max\lambda_i(\mathbf{A})$ (Wilkinson 1965), and become all positive. Then unbalancedness is reflected in dispersion of the eigenvalues of $\mathbf{C}$. In this

---

[b] Note: If B is the square of order n with rank r, its co-rank is n-r.

respect Chakrabarti (1963) suggested for a connected design, to calculate the coefficient of variation of the q-1 non-null eigenvalues of **C**. Their harmonic mean is equally interesting because it is inversely proportional to the mean variance of pairwise differences between treatments. Chakrabarti (1963) also gave an expression for the extreme values (λ_{LP},λ_{MAX}) of the positive eigenvalues of **C** in a "treatment x block" design. This is

$$\lambda_{LP} \leq \{[tr^2(\mathbf{C})-N^2(\mathbf{C})]/[r(\mathbf{C}).(r(\mathbf{C})-1)]\}^{1/2} \tag{13.28}$$

$$tr(\mathbf{C})/r(\mathbf{C}) \leq \lambda_{MAX} \leq 2M$$

or if r(**C**)=q-1

$$qM/(q-1) \leq \lambda_{MAX} \leq 2M.$$

Above, tr and N designate the trace and the Euclidean norm, respectively, and M=Max$(\mathbf{C})_{ii}$. This permits one to give a general indication of the degree of disconnectedness between treatments and, in particular, the limits for the largest eigenvalue λ_{MAX} allow the assessment of the minimum level of precision for pairwise comparisons between treatments.

Because calculating the eigenvalues is prohibitively expensive in large **C** matrices, Foulley et al. (1984) proposed an alternative approach to measure balance-connectedness. The principle consists in expressing the precision of a contrast between treatments, i.e., $\mathbf{L'C^-L}\sigma_e^2$, in the full model (F) relative to its value in a model (R) reduced by the factors being responsible for the unbalancedness-disconnectedness.

Let the full (F) and reduced (R) models be respectively:

$$\mathbf{y} = \mathbf{X}_\phi\phi+\mathbf{X}_\theta\theta+\mathbf{e} \tag{13.29a}$$

and

$$\mathbf{y} = \mathbf{X}_\phi\phi+\mathbf{e} \tag{13.29b}$$

with $\phi'=[\phi_1',\phi_2']$, and ϕ_1 denotes the effects of the factor "treatment" on which there is main interest. The coefficient of balance-connectedness between levels i and i' of the factor ϕ_1 in the full model (ϕ,θ) and due to the incidence of θ is

$$\gamma_{ii'}=[\mathbf{L}_{ii'}'\mathbf{C}_R^-(\phi_1)\mathbf{L}_{ii'}]/[\mathbf{L}_{ii'}'\mathbf{C}_F^-(\phi_1)\mathbf{L}_{ii'}], \tag{13.30}$$

where $\mathbf{C}_R$ and $\mathbf{C}_F$ are information matrices as in (13.17c) for treatments ϕ_1 after absorption of the remaining factors in models R and F, respectively, and $\mathbf{L}_{ii'}'\phi_1$ is the contrast between treatments i and i'. The coefficient varies between 0 and 1; a value of 0 indicates lack of connection between the levels of the fixed treatment factor under the full model. The value 1 indicates that the factors θ do not affect the accuracy of comparisons between treatments.

We now illustrate the concept by considering the following models:

$$\mathbf{y} = \mathbf{Hg}+\mathbf{Xb}+\mathbf{Zu}+\mathbf{e} \tag{13.31}$$

$$\mathbf{y} = \mathbf{Zu}+\mathbf{e} \tag{13.32}$$

$$\mathbf{y} = \mathbf{Xb}+\mathbf{Zu}+\mathbf{e} \tag{13.33}$$

$$\mathbf{y} = \mathbf{Hg}+\mathbf{Zu}+\mathbf{e}, \tag{13.34}$$

where $\mathbf{g}$, $\mathbf{u}$, and $\mathbf{b}$ indicate "group", "sire", "herd", or other nuisance parameters. The "global" degree of connection among sires can be deduced from models (13.31) and (13.32). The degree of connection among sires due to the incidence of nuisance parameters $\mathbf{b}$ can be deduced from models (13.31) and (13.34), if one reasons with a model with "group" effects, or from (13.32) and (13.33) in the absence of "group" effects. In this last case, one could calculate the γs assuming that sires are fixed (formula 13.30), or random with $(\mathbf{C}+k\mathbf{A}^{-1})$ replacing $\mathbf{C}$ in (13.30); treating sires as fixed would exacerbate the effects of unbalancedness. The mixed model analysis of (13.33) is much more flexible and allows the study of connectedness when sires are related. An example is given in Tables 13.6 and 13.7 for the two layouts considered throughout this presentation. In the case of the layout in Table 13.1, the combination of sires 2 and 5 gives a weaker coefficient of connection (γ=0.46 and 0.75 for ratios k of 0 and 9, respectively) than for other estimable contrasts. This can be explained by the fact that the connection between sires 2 and 5 is indirect, as they do not have progeny in the same herd. Likewise, for the layout in Table 13.3, the coefficient γ is smaller for the combinations AD and FG because the two sires never appear together in the same combination of the other two factors. The utilization of related sires from groups initially disconnected improves the average level of connection among sires (Tables 13.6 and 13.7). However, as shown by the comparisons of 1-4 and B-E in the two layouts considered, this practice could cause a reduction in the γ-values of well-connected sires due to an inappropriate distribution of their progeny.

When the "treatment" factor is considered random, the coefficients γ defined in (13.30) are within the range of the solutions of the equation

$$|\Omega_R-\gamma^*\Omega_F| = 0,$$

where Ω_R and Ω_F are matrices as in (13.23) (Ω^{-1}=$\mathbf{C}+k\mathbf{A}^{-1}$) for models R and F, respectively. The average value γ^* of these solutions, which can be calculated as $\bar{\gamma}^*=\mathrm{tr}[(\mathbf{C}_F+k\mathbf{A}^{-1})\Omega_R]/q$, could provide an index of connection between treatments. As suggested by Chaterjee and

Hadi (1986) in the analysis of influential observations, one could also consider the ratio $\eta=|\Omega_R| / |\Omega_F|$; this can be also calculated as $\eta=|C_F+kA^{-1}| / |C_R+kA^{-1}|$. Thus, in the case of the layout of Table 13.1 and for the situations a,b,c,d considered in Table 13.6 $\bar{\gamma}^*$ takes values of 0.548, 0.840, 0.851 and 0.855, respectively, whereas the values of η are 0, 0.263, 0.305, and 0.324, respectively.

Table 13.6 γ coefficients among sires for the design given in Table 13.1[a]

a \ b	1	2	3	4	5	6
1	1	0.578	0.508	0.930	0.606	0.950
2	0	1	0.950	0.675	0.751	0.738
3	0	0.842	1	0.627	0.851	0.704
4	0.903	0	0	1	0.692	0.927
5	0	0.458	0.611	0	1	0.749
6	0.951	0	0	0.872	0	1

d \ c	1	2	3	4	5	6
1	1	0.611	0.585	0.894	0.630	0.954
2	0.625	1	0.953	0.762	0.751	0.758
3	0.619	0.953	1	0.733	0.859	0.772
4	0.892	0.775	0.767	1	0.755	0.931
5	0.641	0.751	0.861	0.764	1	0.763
6	0.928	0.804	0.822	0.931	0.795	1

[a] a: $k=\sigma_e^2/\sigma_u^2=0$; b: $k=9$, $\mathbf{A}=\mathbf{I}$; c: $k=9$, $a_{34}=0.5$; d: $k=9$, $a_{34}=0.5$, $a_{36}=0.25$. Average values of γ coefficients are (diagonal excluded) 0.309, 0.749, 0.781, and 0.795 in a, b, c, d, respectively.

The information represented by the matrix of pairwise connection indexes γ can be reduced by applying cluster analysis to the values $1-\gamma$, which can be viewed as measures of disconnectedness. A cluster analysis using Johnson's algorithm was applied to data from the French program for evaluation of natural service sires using reference AI bulls (Foulley and Sapa 1982). A small example involving 53 Charolais sires is presented in Fig. 13.1. At a level of connection of $\gamma=0.42$, eight groups can be distinguished. The first four groups (starting from the top of the figure) correspond to sires utilized in areas different from the main region in which the program is applied. The last group discriminates producers from that region using a fall calving season rather than the usual spring calving.

In the same way that there is interest in the precision of comparisons between sires, it is possible to consider the precision of contrasts between herds associated with such sires, as shown by

$$[\mathrm{Var}(\tilde{\mathbf{u}}-\mathbf{u})|33]^{-1}[\mathrm{Var}(\hat{\mathbf{u}}-\mathbf{u})|32] = [\mathbf{I}+\mathbf{Z}'\mathbf{X}\mathrm{Var}(\hat{\mathbf{b}})\mathbf{X}'\mathbf{Z}(\mathbf{Z}'\mathbf{Z}+k\mathbf{I})^{-1}]^{-1}, \qquad (13.35)$$

where $\hat{\mathbf{u}}$ and $\tilde{\mathbf{u}}$ are the BLUP's of $\mathbf{u}$ under the reduced (13.32) and full (13.33) models, respectively, and assuming $\mathrm{Var}(\mathbf{u})=\mathbf{I}\sigma_u^2$.

Table 13.7 γ coefficients among treatments (e.g., sires) for the design given in Table 13.3 (effect of row and column factors on unbalancedness)[a]

b / a	A	B	C	D	E	F	G	H
A	1	0.830	0.779	0.764	0.820	0.808	0.891	0.826
B	0	1	0.582	0.735	0.978	0.794	0.766	0.568
C	0	0	1	0.746	0.595	0.786	0.731	0.924
D	0.529	0	0	1	0.772	0.744	0.839	0.700
E	0	0.950	0	0	1	0.778	0.815	0.585
F	0.573	0	0	0.539	0	1	0.723	0.726
G	0.720	0	0	0.650	0	0.454	1	0.803
H	0	0	0.830	0	0	0	0	1

c / d	A	B	C	D	E	F	G	H
A	1	0.843	0.831	0.764	0.864	0.807	0.892	0.844
B	0.872	1	0.682	0.751	0.962	0.815	0.775	0.606
C	0.833	0.722	1	0.805	0.717	0.847	0.788	0.910
D	0.764	0.770	0.809	1	0.830	0.745	0.839	0.715
E	0.867	0.972	0.730	0.833	1	0.848	0.850	0.683
F	0.807	0.839	0.852	0.744	0.850	1	0.723	0.739
G	0.892	0.804	0.789	0.840	0.855	0.724	1	0.825
H	0.871	0.653	0.921	0.737	0.723	0.769	0.845	1

[a] a: $k = 0$; b: $k = 9$, $\mathbf{A} = \mathbf{I}$; c: $k = 9$, $a_{CE} = 0.5$; d: $k = 9$, $a_{CE} = 0.5$, $a_{BH} = 0.25$.

Average values of γ coefficients are (diagonal excluded) 0.187, 0.765, 0.796, and 0.810 in a, b, c, and d, respectively.

In particular, if only natural service sires ($\mathbf{u}_i$) are considered, having descendants (n_i) in a single herd (h_i), one can write for contrasts $\Delta_{ii'}=u_i-u_{i'}$

$$\mathrm{Var}(\tilde{\Delta}_{ii'}-\Delta_{ii'})-\mathrm{Var}(\tilde{\Delta}_{ii'}-\Delta_{ii'})$$

$$= \mathrm{Var}\{[n_i h_i/(n_i+k)]-[n_{i'} h_{i'}/(n_{i'}+k)]\} .$$

Fig. 13.1 Example of a cluster analysis (Johnson's algorithm) performed on the matrix of disconnectedness indices (1-γ) among 53 Charolais sires (Eight groups can be discriminated at the 0.42 level of aggregation)

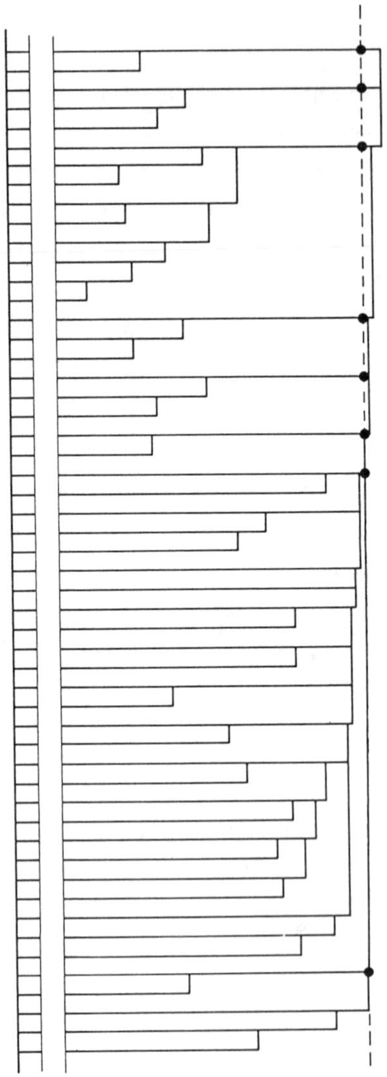

Thus, in the balanced case the increase in variance of prediction errors of contrasts between sires due to disconnectedness between herds is proportional to the sampling variance of contrasts between the herds where the sires are utilized.

These observations suggest avenues for further research. For example, when using animal models it will probably be simpler to examine connections between herds rather than between animals. A numerical example of the calculation of coefficients of connection between levels of a fixed "nuisance" factor is given in Table 13.8.

Table 13.8 γ coefficients among column effects due to unbalancedness over treatments (sires) in a "column, row, treatment" model[a]

a \ b	1	2	3	4
1	1	0.822	0.769	0.775
2	0.760	1	0.919	0.837
3	0.689	0.919	1	0.659
4	0.773	0.790	0.612	1

[a] a: $k = \sigma_e^2/\sigma_u^2 = 9$; $\mathbf{A} = \mathbf{I}$; b: $k = 9$; $a_{CE} = 0.5$; $a_{BH} = 0.25$.

Another possibility, for fixed effects ϕ such as herd or group effects, would be to use the concept of the Kullback-Leibler distance (Kullback 1968) which quantifies the discrepancy between the density p_F and p_R of ϕ under full(F) and reduced(R) models (see 13.29a,b)

$$I(p_F,p_R) = E_{p_F}[\log(p_F/p_R)]. \tag{13.36a}$$

With a Gaussian mixed linear model, and assuming a full-rank parameterization (r) in ϕ, this distance can be expressed for the GLS estimator as

$$I(p_F,p_R) = 1/2[\log|\mathbf{C}_F|-\log|\mathbf{C}_R|+\mathrm{tr}(\mathbf{C}_F^{-1}\mathbf{C}_R)-r+\mathrm{tr}(\mathbf{C}_R\Delta\Delta')], \tag{13.36b}$$

where $\mathbf{C}_F$ and $\mathbf{C}_R$ are the information matrices of ϕ under the full and reduced models, respectively.

If the last term involving bias $\Delta=E_F(\hat{\phi}_R)-E_F(\hat{\phi}_F)$ is ignored, this formula might serve as a structural index to quantify the overall degree of connectedness among ϕ levels.

13.5 Other Objectives and Constraints

13.5.1 Relaxing the Unbiasedness Requirement for Group Effects

It should be stated at the onset that removing this requirement is mandatory when certain differences between groups are not estimable. A formula for the bias when estimating the contrast $\delta'\mathbf{w}$ between sires via the BLUP $\hat{\mathbf{u}}_2$ under model (13.33) when the true model (13.31) includes group effects, was given by (13.22). This bias can be important. In fact, it is maximum when progeny group sizes tend to 0 and the ratio σ_e^2/σ_u^2 tends to infinity. It is minimum in the opposite situation (large progeny groups, small ratio) because in that case the models (13.31) and (13.33) lead to similar least-squares equations and the hierarchical "group" effects are not estimable and, therefore, can be ignored.

The requirement of unbiasedness is classical in animal breeding. While it is rarely questioned in the case of nuisance effects (herd-year-season, parity, sex, etc.), it is legitimate to question its significance vis a vis "group" effects, as they are often defined in an arbitrary manner. Using biased criteria of the type $\hat{\mathbf{u}}_2$ when predicting $\mathbf{w}=\mathbf{Pg}+\mathbf{u}$, was addressed by Tong et al. (1980) and Kennedy (1981). The point of departure of this approach is to consider mean square error (MSE) of comparisons between sires according to whether the full (13.31) or reduced (13.33) models are considered; these models will be denoted using subscripts (1) and (2) to simplify notation. These MSE's are written as

$$MSE_1 = Var[\delta'(\hat{\mathbf{w}}-\mathbf{w})] \tag{13.37a}$$

$$MSE_2 = Var[\delta'(\hat{\mathbf{u}}_2-\mathbf{w})]+E^2[\delta'(\hat{\mathbf{u}}_2-\mathbf{w})]$$

$$= Var[\delta'(\hat{\mathbf{u}}_2-\mathbf{w})]+E^2(B), \tag{13.37b}$$

where E(B) is the expectation of the bias as given in (13.22). Because the variances of $\hat{\mathbf{u}}_2-\mathbf{w}$ and of $\hat{\mathbf{u}}_2-\mathbf{u}$ are equal, MSE_2 can be decomposed as the sum of the prediction error variance of $\delta'\mathbf{u}$ taken over the model used for calculation, and of the squared bias with respect to the true model. Thus, for a given contrast (k), the MSE's for the true (1) and for the "operational" model (2) differ as:

$$MSE_{1,k}-MSE_{2,k}=Var[\delta_k'(\hat{\mathbf{w}}-\mathbf{w})]-Var[\delta_k'(\hat{\mathbf{u}}_2-\mathbf{u})]-E^2(B_k). \tag{13.38}$$

The variance of prediction errors is always higher than or equal to that obtained with the "operational" model, the difference being mainly due to the sampling variance of the estimated group effects. The effectiveness of the operational model will depend on the balance between the "cost" of estimating group effects and the real differences between them. The expression (13.38) can also be put in relative terms:

$$(MSE_{2,k}-MSE_{1,k})/MSE_{1,k}=\{E^2(B_k)/MSE_{1,k}\}-[1-\gamma_k(\mathbf{g})], \qquad (13.39)$$

where $\gamma_k(\mathbf{g})$ is the index of connection between sires due to taking into account "groups", as stemming from (13.30). Formula (13.39) shows the roles played by the index of disconnection $(1-\gamma)$ and the bias in the relative difference between MSE's. This expression also suggests that the bias removal ability of a model cannot be discussed irrespective of the degree of connection.

In a general discussion on the need and the extent of using alternative grouping strategies in on-the-farm selection programs, McClintock and Taylor (1982) have suggested not to group unless there is a clear genetic differentiation between prospective groups and a good structure of connections.

Kennedy (1981) argued similarly in the context of the relevance of grouping dairy sires into two subpopulations when there is exchange of semen between two regions. In this case, it appears that such grouping is justified when the exchange program allows the estimation of the genetic difference between regions with sufficient precision. This author showed that this "critical" genetic difference is larger as the semen exchange is weaker.

When testing is discontinuous, e.g., the French case for beef cattle AI sires, the genetic difference between series is sufficiently large to justify (in terms of MSE of comparison between sires of different series) the systematic use of control sires and a grouping of the series. Foulley et al. (1983), in a theoretical study, stated that the condition to be met is:

$$(\Delta\mu/\sigma_G)^2 \geq (k/2)[(1/pn)+(1-CD)/qm], \qquad (13.40)$$

where

$\Delta\mu$: genetic difference between series as measured in the progeny of sires;

σ_G : the genetic standard deviation for the trait considered;

k : the ratio between sire and residual variances, σ_e^2/σ_s^2, also equal to $(4/h^2)-1$, where h^2 is heritability;

CD : the intra-series coefficient of determination, equal to $n/(n+k)$;

p,q: the number of control and tested sires;

n,m: number of progeny per control and tested sire, respectively.

Foulley et al. (1983) gave values for progeny group sizes which optimize the distribution of animals among descendants of tested and control sires for a given precision of the comparison between sires of different series. Using these results gives the following values for $\Delta\mu$:

1. 0.27, 0.23, and 0.21 σ_G for 10, 20, and 30 bulls tested per series and a precision of 0.5;
2. 0.20, 0.17, and 0.15 σ_G for the above number of bulls and a precision of 0.7.

In the present French context for selection of beef cattle, the above results justify the use of control bulls primarily in comparisons between sires in different "selection units".

The main difficulty with the mean square error criterion is the need to know the differences between levels of treatments, e.g., groups. However, if such information exists, it would be reasonable to treat these effects as random, centered on a common mean.

13.5.2 Maximum Genetic Progress

We have seen that a possible solution to plan connections in an unbalanced layout or with grouping is to assess mean square error. However, this will not assure that for a given model such error will be minimum.

Maximization of genetic progress would be an alternative way, satisfactory from a theoretical viewpoint because it relates directly to the objective of the breeding program. We will consider the expected value of selected animals for different selection protocols. Goffinet and Elsen (1984) discussed optimal selection criteria in this framework. These authors considered two types of selection: (C_1) number of selected animals R fixed given a total number T of candidates, and (C_2) fixing the expected value of the selected animals.

Under C_2, which has been used extensively following Cochran (1951), the best selection rule is to retain every candidate whose conditional mean $E(w_i|\mathbf{y})$ is larger than a threshold λ which is a function of the selection intensity E(R/T). Applying this method, it can be found that for a model without "groups" the expectation of the value of the selected animals depends only on the individual quadratic error and not on the covariances between errors. Under normality, the genetic gain can be written (Foulley et al. 1984) as:

$$\bar{w} = \sum_i (2\pi)^{-1/2} \exp[-(\lambda/\sigma_i)^2/2]\sigma_i \tag{13.41a}$$

with

$$E(R/T) = \sum_i \int_{\lambda/\sigma_i}^{\infty} (2\pi)^{-1/2} \exp(-t^2/2)dt \tag{13.41b}$$

and $\sigma_i=[\mathrm{Var}(\hat{w}_i)]^{1/2}$. A global optimization of the connection plan may be possible: for any plan, there is a σ_i and, therefore, for a given selection intensity E(R/T) there must be a threshold λ and a gain $\bar{w}$.

Under C_1, the optimum strategy is to retain the R candidates having the largest conditional means. In this situation, the expression for expected genetic progress is more complex and depends on covariances between errors of prediction. In general, it would appear that planning the connection layout such that expected genetic progress is maximum is more difficult.

13.6 Discussion and Conclusions

It appears that the problems of connection in genetic evaluation are related to the more general issue of elimination of nuisance parameters (herd-year-season) and to the estimation of eventual "population" or genetic "group" effects.

Strictly speaking, connectedness can be viewed statistically as a problem of estimability of fixed effects, so it can be formulated in terms of matrix rank. Connectedness between genetic values to be predicted is formally insured in models without population effects. Thus, if the model is additive and it has the two factors "herd-year-season" and "sire", the BLUP solutions are always computable and satisfy for each subset of q_i connected sires the condition $\sum_j \hat{u}_{ij}=0$. The contribution of the data to the BLUP of a given sire stems only from the connection group to which the sire belongs. This underlines the crucial importance of the assumption $E(\mathbf{u})$=constant, which implicitly states that the sires belong to the same population.

This assumption is relaxed when group effects are included in the model; these must be connected for comparisons between sires to be valid. If the differences between groups are really unimportant or if the state of connectedness is such that these differences cannot be determined, it may be tempting to exclude group effects from the model or to consider biased predictors of breeding values. The same reasoning can be applied to any fixed "nuisance" parameter; a poor connection between its levels always increases the prediction error variance of breeding values by an important amount. This is a situation which will merit further examination as the animal model is gradually introduced in genetic evaluation. We suggested a continuous index to measure the state of "connection-balancedness". It is based on the value of the prediction error variance relative to a reference value in the model without the factors being assessed for connectedness. The index, which characterizes the model and the structure of the data, should be considered jointly with the squared bias deduced from using a reduced model, as shown in Eq. (13.39). Only by making this type of analysis will it be possible to reach judicious decisions when planning genetic evaluation programs.

Some minimum planning is often required to take advantage of the data collected in recording programs. The process comprises several essential phases: (1) analysis of the demographic and breeding structure of the population; (2) analytical or simulation study of alternative protocols making simplifying assumptions; (3) formulation of practical recommendations; (4) minimum *a posteriori* control of the quality of the protocol. This approach has been employed in France (Foulley and Menissier 1978; Foulley and Sapa 1982) in the Charolais and Limousin breeds to simultaneously plan matings to natural service sires and a connection plan involving a small number of AI sires (three, of which one is replaced every year). From 1980 to 1985, 139 Charolais and 226 Limousin bulls were evaluated for pre-weaning growth traits. This was possible because of inseminations with semen from nine and seven AI sires, respectively, which were chosen from the best for maternal traits via a station progeny test (Sapa and Menissier 1986).

Artificial insemination plays the central role in this scheme because it allows the distribution of progeny among many herds, regions, and years, and creates connections among all factors. The planned use of AI for connection between herds and years is also being undertaken in sheep and dairy goats, especially in large flocks which use natural service extensively. Certain connection plans for sheep have been described by Foulley et al. (1984).

In general, the planning poses theoretical problems (model, criterion for assessing the quality of a given plan) as well as practical ones (number, nature, distribution of reference sires), which are far from being resolved. This paper may make a contribution towards solving these problems.

Acknowledgements. This paper draws widely from material presented at a conference entitled "Sire evaluation and connectedness", given by the author at an INRA seminar on artificial insemination and genetic improvement. The conference was held in Toulouse-Auzeville, November 23-24, 1983, and the proceedings were published by J.M. Elsen and J.L. Foulley (eds) in "Les colloques de l'INRA", 1984, No.29. Thanks are expressed to Dr. D. Gianola who translated the present version of the paper from French into English.

References

Bose RC (1947) Presidential address. In: Proc 34th Indian Sci Congr.

Bulmer MG (1980) The mathematical theory of quantitative genetics. Clarendon Press, Oxford

Chakrabarti C (1963) On the C matrix in design of experiments. J Indian Stat Assoc 1:8-23

Chatterjee S, Hadi AS (1986) Influential observations, high leverage points and outliers in linear regression. J Stat Sci 1:379-405

Cochran WG (1951) Improvement by means of selection. In: Neyman J (ed) Proc 2nd Berkeley Symp Math Stat Probab. Univ California Press, Berkeley, pp 449-470

Delattre P (1983) Matrices, determinants, formes quadratiques. Ecole d'automne de biologie theorique, Solignac, 19 Sept-7 Oct 1983, 130 pp Mimeo

Dempfle L (1977) Relation entre BLUP (Best linear unbiased prediction) et estimateurs bayesiens. Ann Genet Sel Anim 9:27-32

Eccleston JA (1972) On the theory of connected designs. PhD Thesis, Cornell Univ

Eccleston JA, Hedayat A (1974) On the theory of connected designs; characterization and optimality. Ann Stat 2:1238-1255

Eccleston JA, Russell K (1975) Connectedness and orthogonality in multifactor designs. Biometrika 62:341-345

Fernando RL, Gianola D (1986) Optimal properties of the conditional mean as a selection criterion. Theor Appl Genet 72:822-825

Fernando RL, Gianola D, Grossman M (1983) Identifying all connected subsets in a two-way classification without interaction. J Dairy Sci 66:1399-1402

Foulley JL, Clerget-Darpoux F (1978) Progeny group size for evaluating natural service bulls using AI reference sires. Ann Genet Sel Anim 10:541-556

Foulley JL, Menissier F (1978) Arguments in favour of an evaluation system for natural service bulls belonging to beef breeds, by taking into account the diffusion of the best AI bulls. XVth World Charolais convention, Sept 16-29, 1978, Charolais 52

Foulley JL, Sapa J (1982) The French evaluation program for natural service beef bulls using AI sire progeny as herd ties. British Cattle Breeders Club, Winter Conference, Cambridge, January 11-12, 1982, Digest 37:64-67

Foulley JL, Schaeffer LR, Song H, Wilton JW (1983) Progeny group size in an organized progeny test program of AI beef bulls using reference sires. Can J Anim Sci 63:17-26

Foulley JL, Bouix J, Goffinet B, Elsen JM (1984) Comparaison de peres et connexions. In: Elsen JM, Foulley JL (eds) Insemination artificielle et amelioration genetique: bilan et perspectives critiques. Colloques de l'INRA 29:131-176, INRA Service des publication, Versailles

Gianola D (1986) On selection criteria and estimation of parameters when the variance is heterogeneous. Theor Appl Genet 72:671-677

Gianola D, Fernando RL (1986) Bayesian methods in animal breeding theory. J Anim Sci 63:217-244

Goffinet B (1983) Selection on selected records. Genet Sel Evol 15:91-98

Goffinet B, Elsen JM (1984) Critere optimal de selection quelques resultats generaux. Genet Sel Evol 16:307-317

Henderson CR (1973) Sire evaluation and genetic trends. In: Proc Anim Breed Genet Symp in Honor of Dr. J.L. Lush. ASAS and ADSA, Champaign, Illinois, pp 10-41

Kennedy BW (1981) Bias and mean square error from ignoring genetic groups in mixed model sire evaluation. J Dairy Sci 62:689-697

Kullback S (1968) Information theory and statistics. Dover Publications, New York

McClintock AE, Taylor JF (1982) Developments in the use of BLUP for estimating genetic merit. In: Barker JSF, McClintock AE (eds) Future developments in the genetic improvement of animals. Academic Press, Aust, pp 157-178

Petersen PH (1978) A test for connectedness fitted for the two-way BLUP sire evaluation. Acta Agric Scand 28:360-362

Pollak EJ, Quaas RL (1982) A comparison of sire evaluations when sires are nonrandomly used across herds. J Dairy Sci suppl 1 (Abstr) 65:102-103

Quaas RL, Pollak EJ (1981) Modified equations for sire models with groups. J Dairy Sci 64:1868-1870

Quaas RL, Pollak EJ (1982) Thompson's accumulated group model for sire evaluation. J Anim Sci 55 suppl 1 (Abstr):160

Raghavarao D, Federer WT (1975) On connectedness in two-way elimination of heterogeneity designs. Ann Stat 3:730-735

Rao CR (1952) Advanced statistical methods in biometric research. Wiley, New York

Sapa J, Menissier F (1986) Beef sire evaluation for natural service in France. II. On-farm progeny-testing. 37th annu meet EAAP, Budapest, Hungary, Sept 1-4, 1986. Mimeo, 11 pp

Searle, SR (1971) Linear models. Wiley, New York

Shah KR, Yadolah D (1977) On the connectedness of designs. Sankhya 39:284-287

Spike PL, Freeman AE (1977) Effect of genetic differences among herds on accuracy of selection and expected genetic change. J Dairy Sci 60:967-974

Thompson R (1979) Sire evaluation. Biometrics 35:339-353

Tong AKW, Kennedy BW, Moxley JE (1980) Potential errors in sire evaluation from regional genetic differences. J Dairy Sci 63:627-633

Ufford GE, Henderson CR, Keown JF, Van Vleck LD (1979) Accuracy of first lactation versus all lactations sire evaluations by best linear unbiased prediction. J Dairy Sci 63:603-612

Weeks DL, Williams DR (1964) A note on the determination of connectedness in an N-way cross-classification. Technometrics 6:319-324

Westell RA (1984) Simultaneous evaluation of sires and cows for a large population. PhD Thesis, Cornell Univ, Ithaca, NY

Wilkinson JH (1965) The algebraic eigenvalue problem. Clarendon Press, Oxford

Discussion Summary
PART IV: PREDICTION AND ESTIMATION OF GENETIC MERIT

H.-U. Graser[1]

11 A Framework for Prediction of Breeding Value

The view was expressed that BLUP is not the procedure which gives necessarily minimum Mean Squared Error (MSE). However, other methods with smaller MSE are mostly unknown for a particular data set.

The pessimistic view expressed in the talk about the computational aspects of a full Bayesian analysis is not justified. The problem can be reduced to a single dimension, with requirements comparable to those of REML. Supercomputers, therefore, may not be needed.

A question was raised on the relationship between MINQUE and Bayesian methods. Dr. Gianola stated that there is none. MINQUE is unbiased and minimizes a norm which depends on a "prior" value. However, one can obtain negative estimates with non-null probability, and this is "illegal" in a Bayesian approach. The so-called prior is, in effect, a fixed value and no knowledge about its distribution is incorporated in the corresponding analysis.

With respect to bias in estimates, a Bayesian is often interested in minimizing loss functions, where biased estimates may do better. We seem happy using REML, a biased method for variance component estimation, but express concern about whether breeders would accept biased estimates of breeding value, even if these have smaller MSE. This would seem to be a contradictory attitude.

Concern was expressed about the refusal of Bayesians to allow for negative variance components when establishing the prior, particularly for traits which have zero heritability. A combination of positive prior plus data will always be positive, and wrong conclusions with respect to the design of breeding programs might be drawn.

[1] Animal Genetics and Breeding Unit, University of New England, Armidale, Australia

12 BLUP (Best Linear Unbiased Prediction) and Beyond

The proposed approach for predicting unobserved random variables without knowledge of their distributions involves computing a one-dimensional integral. The mixed model equations can be represented in terms of eigenvalues and eigenvectors. The computational requirements are then similar to those of an REML estimation. However, if this is not feasible, "crude" averaging over the distribution of variance components leads to a more sensible predictor than BLUP using a single estimated value. Bayesian estimates are more robust against poor variance component estimates.

The Bayesian approach has an analogy to REML. In the latter one integrates over β and, in the proposed method, integration is over the variance components.

13 Connectedness in Genetic Evaluation

There is no "magic" value for the coefficients of the proposed method at which one would reject connectedness. The values depend on the data and on the number of factors in the model. Treating nuisance parameters as fixed or random does not make much difference for the coefficients because only the mean rather than the spread of eigenvalues is affected. However, randomness implies connection through prior knowledge.

The possible reduction of the coefficients when including the inverse of a numerator relationship matrix is an artifact of a data set if very close relatives (e.g., identical twins) are included.

The method is computationally feasible, and has been applied to sire reference schemes in France. It can be used to develop recommendations for across-herd use of animals by the industry via simulation of designs and relationships. A pragmatic approach has to be chosen to develop these criteria.

The relationship between degree of connectedness and effectiveness of selection (e.g., increased genetic trend) is not completely clear, as suggested by simulation studies.

Part V: Prediction and Estimation in Non-Linear Models

14 Generalized Linear Models and Applications to Animal Breeding

R. Thompson[1]

Generalized linear models (GLM), a flexible extension of linear models, are described. Specific examples of their application to animal breeding problems are given. These include the estimation of variance components, for normal and non-normal data, the estimation of heritability by offspring-parent regression for binary traits, and the estimation of gene frequencies. Unresolved problems are highlighted. Animal breeding studies that have used GLM methods are reviewed.

14.1 Introduction

Linear models are often used by animal breeders in theoretical studies of different strategies and when making summaries and inferences from experimental and field data. At the simplest level we may wish to explain a response variable y, of length n, in terms of r explanatory variables $x_1,...,x_r$. Then a linear model for the expectation of y, μ, can be written as

$$\eta = \sum_{i=1}^{r} x_i\beta_i$$

or

$$E(y) = \mu = \eta = \sum_{i=1}^{r} x_i\beta_i \tag{14.1}$$

with the discrepancy between y and η, the error, being normally distributed with variance $\mathbf{I}\sigma^2$, where $\mathbf{I}$ is the identity matrix and σ^2 is the variance of individual errors. In animal breeding more complicated variance models are often needed to take account of genetic and environmental covariances between animals. For instance, a variance matrix $\mathbf{V}=\mathbf{I}\sigma^2+\mathbf{ZZ}'\sigma_1^2$ might be appropriate, with σ_1^2 being a variance parameter and $\mathbf{Z}$ an $n \times s$ matrix. If the columns

[1] Institute of Animal Physiology and Genetics Research, Edinburgh Research Station, Edinburgh, Scotland

of $\mathbf{Z}$ are $\mathbf{z}_1,...,\mathbf{z}_s$ then the variance structure can be mimicked (if $\sigma_1^2>0$) by thinking of the model:

$$y = \sum_{i=1}^{r} x_i\beta_i + \sum_{j=1}^{s} z_j u_j + e \tag{14.2}$$

with u_j and e normally distributed with means zero and $\mathrm{var}(u_j)=\sigma_1^2$, $\mathrm{cov}(u_j,u_k)=0$ and $\mathrm{var}(e)=\mathbf{I}\sigma^2$. As β_i are fixed effects and u_j and e random effects, (14.2) is often called a mixed model.

Least-squares estimates of the parameters of interest in model (14.1), β_i, are, in a well-known matrix notation:

$$\mathbf{X'X}\tilde{\beta} = \mathbf{X'y}. \tag{14.3}$$

For model (14.2),

$$\mathbf{X'V^{-1}X}\hat{\beta} = \mathbf{X'V^{-1}y}. \tag{14.4}$$

The inversion of $\mathbf{V}$ in (14.4) can be avoided by noting that $\hat{\beta}$ in (14.4) satisfies:

$$\begin{bmatrix} \mathbf{X'X} & \mathbf{X'Z} \\ \mathbf{Z'X} & \mathbf{Z'Z}+\mathbf{I}\gamma^{-1} \end{bmatrix} \begin{bmatrix} \hat{\beta} \\ \hat{\mathbf{u}} \end{bmatrix} = \begin{bmatrix} \mathbf{X'y} \\ \mathbf{Z'y} \end{bmatrix}, \tag{14.5}$$

where $\gamma=\sigma_1^2/\sigma^2$.

Although (14.1) and (14.2) are a fairly flexible set of models, they are not necessarily appropriate for all data. For example, the response variable might be discrete so that a normal distribution would be inappropriate. Alternatively, the explanatory effects might operate in a multiplicative rather than in an additive manner. The responses might be naturally bounded, but the fitted values, being a linear function of the explanatory variables x, might not be. Nelder and Wedderburn (1972) introduced the idea of generalized linear models to allow more flexibility. Their generalization came in two directions. They allowed the expectation of y, μ, to be a monotone differentiable function of the linear predictor $\eta=\Sigma x_i\beta_i$ so that $\mu=f(\eta)=f(\Sigma x_i\beta_i)$, or by inversion $g(\mu)=\eta$. As g links the mean to the linear predictor, it is called the link function. Link functions include the log function, so that $\log(\mu)=\eta$, giving rise to log-linear models in which the effects multiply together. For binomial data, $o<\mu<1$, and three link functions are often used that satisfy this natural constraint. The first is the probit function $\eta=\Phi^{-1}(\mu)$, where Φ is the normal cumulative distribution function so

$$\mu = \Phi(\eta) = [1/(2\pi)^{1/2}] \int_{-\infty}^{\eta} \exp(-x^2/2)dx.$$

The second function is the logit function $\eta=\log[\mu/(1-\mu)]$ so that the log-odds is linearly related to the effects, and the third function is the complementary log-log function $\eta=\log[-\log(1-\mu)]$. Other useful link functions include the power family $\eta=\mu^{\alpha}$, if $\alpha\neq0$ and $\eta=\log\mu$, if $\alpha=0$, and of course the identity link $\eta=\mu$.

The other extension that Nelder and Wedderburn (1972) introduce relaxes the assumption that y follows a normal distribution and allows y to follow a distribution that belongs to the exponential family. The probability density function for y then can be written as

$$f_y(y,\theta,\phi) = \exp\{[y\theta-b(\theta)]/a(\phi)+c(y,\theta)\}$$

for some functions $a(\phi)$, $b(\theta)$ and $c(y,\theta)$, and known ϕ. This exponential family is quite wide ranging. In addition to the normal distribution, it includes the Poisson distribution which is often appropriate for count or frequency data, and the gamma distribution, which can model data with a constant coefficient of variation. The binomial distribution, sometimes useful for percentages, and the inverse Gaussian distribution are also included.

Nelder and Wedderburn (1972) showed that the estimation of $\boldsymbol{\beta}$ by maximum likelihood could be fitted into an iterative weighted least-squares procedure of the form

$$\mathbf{X'DV^{-1}DX}\,\hat{\boldsymbol{\beta}} = \mathbf{X'DV^{-1}y^*}, \tag{14.6}$$

where $\mathbf{D}$ is a diagonal matrix with i-th diagonal element $(d\eta_i/d\mu_i)$, so that $\mathbf{D}$ depends on the link function. $\mathbf{V}$ is a diagonal matrix with i-th element proportional to the variance of y_i, and so $\mathbf{V}$ depends on the distribution used. The vector $\mathbf{y^*}$ is now a working variate $\mathbf{X}\boldsymbol{\beta}+\mathbf{D}^{-1}(y-\mu)$, and so is a linear combination of the linear predictor, η, and of the discrepancy between the observed and fitted value.

With normal data, an analysis of variance table, based on residual sums of squares for a sequence of models, can be used to investigate the fit of various models and the effects of the explanatory variates, factors and their interactions. For generalized linear models, Nelder and Wedderburn (1972) suggested an analogous procedure, using twice the log-likelihood instead of residual sums of squares, to construct an analysis of deviance.

Since their introduction, generalized linear models have been found to be useful in many areas. There are books on generalized linear models by Dobson (1983) and McCullagh and Nelder (1983) and an illustrative review using animal production data is given by Crosbie and Hinch (1985). The algorithms are relatively simple to program and are included in the widely distributed programs GENSTAT and GLIM.

In the rest of the paper I discuss the application of generalized linear model (GLM) ideas to animal breeding data. I will consider parent-offspring regression for binary data, estimation of gene frequencies, and estimation of variance components with normal and non-normal data.

14.2 Estimation of Heritability of Binary Traits by Offspring-Parent Regression

If a trait takes one of only two possible values it can be called a binary trait. Some economically important traits such as dystocia in cattle, leg weakness in pigs and fleece-rot in sheep are scored in this way.

I consider parent-offspring heritability estimation for binary traits, motivated by fleece-rot data of McGuirk and Atkins (1984). This example shows that GLM can be a convenient framework to model genetic ideas.

Because means and variances of binary traits are related, the genetic analysis of such traits can cause statistical problems. Heritability estimates are often related to the frequency of the condition (Robertson and Lerner 1949). For a population represented in several years in which incidences differ, this causes complications in specifying the degree of genetic variation and in comparing estimates from populations with differing incidences.

Sometimes these difficulties can be avoided if we consider "liability" to a condition rather than its incidence. We define liability as the sum of environmental and genetic effects influencing susceptibility, and assume it to be normally distributed (Falconer 1965). Animals with liability above a notional threshold are assumed to exhibit the condition so that expression is an all-or-none or binary phenomenon. Falconer (1965) proposed that the usual laws of quantitative genetic theory might follow if examined at the level of liability. Suppose the incidence in the parent generation is q. If liability is assumed to be normally distributed with mean 0 and variance 1, the mean liability of affected parents, those that exceed the threshold T, is equal to z/q, where z is the ordinate of the normal distribution at x where $x=\Phi^{-1}(p)$ and p=1-q, and the mean liability of unaffected parents is -z/p. The difference between the two groups is then z/pq.

If the mean liability in the offspring generation is L, then the mating of affected and unaffected sires to unselected dams gives offspring with mean liability $L+1/2h^2z/q$ and $L+1/2h^2(-z/p)$, respectively, and proportions not affected, p_1 and p_0. If liability is normally distributed then $p_i=\Phi(x_i)$, $x_0=u+1/2h^2z/p$, and $x_1=u-1/2h^2z/q$ with u=T-L. The difference in mean liability between offspring of the two parental types is x_o-x_1 or $1/2h^2(z/p+z/q)$ so that $h^2=2(x_o-x_1)pq/z$. As $z'(x_o-x)=(p_o-p_1)$, where z' is the ordinate corresponding to the mean incidence in the offspring generation, then $h^2=2[(p_o-p_1)/z']\cdot(pq/z)$ as suggested by Falconer (1965) and James and McGuirk (1982). Thompson et al. (1985) consider the case when there is a number of groups of sires as well as groups of progeny which differ in incidence. Then

$$x_{jkl} = u_{jk}-(1/2)h^2z_j/q_j$$

$$x_{jko} = u_{jk}+(1/2)h^2z_j/p_j,$$

where z_j and q_j relate to parental incidences in the j-th sire group, and u_{jk} measures the distance between the threshold and the mean liability of the offspring group.

This data fits into the GLM framework with the dependent variable being the number affected, A, out of N, the number scored. Independent variables include a factor COHORT and a variate MNL containing the mean liability of the parental groups (affected and unaffected). The regression coefficients associated with COHORT and MNL measure cohort effects (u_{jk}) and $(1/2)h^2$, respectively. The number affected A out of N is assumed to follow a binomial distribution. The link function is the probit transformation linking the x scale to the proportion (A/N) scale.

The method of Thompson et al. (1985) is approximate in that it takes no account of the reduction of variance in the offspring generation, and is limited to cases when the mean liability of parents can be assumed known. In some circumstances, one might want to use GLM's to describe the parental data, or the parents and offspring might have animals in common. Then it might be conceptually easier to think in terms of the variance matrix of all observations or in terms of a mixed model and variance components. Then, the methods discussed later might be more appropriate, although they will be more difficult to apply and interpret.

14.3 Estimation of Gene Frequencies

The estimation of gene frequencies by maximum likelihood has a long history and a comprehensive review is in Elandt-Johnson (1971). In some cases I have found it tedious and time-consuming to implement these algorithms; for example, for the inheritance of color in North Ronaldsay sheep (Ryder et al. 1974). The genetic models are highly structured but this is ignored in the algorithms. I wondered if GLM could be used.

Let us consider data on the O, A, B system for blood groups. Suppose the respective gene frequencies are r, p and q and that there is random mating with respect to this locus. This gives rise to the following frequency table:

		A	O	B
		p	r	q
A	p	p^2	pr	pq
O	r	rp	r^2	rq
B	q	qp	qr	q^2

and the frequencies in the four observed groups O, A, B and AB formed from the genotypes OO, (AA, OA and AO), (BB, OB and BO) and (AB and BA) are r^2, p^2+2pr, q^2+2qr and 2pq, respectively.

GLM could be used directly with contingency tables. For example, for a two-way contingency with independence, data in the form of counts can be thought of arising from a Poisson distribution. With independence, the mean of each observation, μ_i, can be viewed as a row effect multiplied by a column effect. This means that a logarithmic link function $\log(\mu_i)=\eta_i$ allows a linear structure to be defined linking the mean of an observation to explanatory variables representing rows and columns of the table. The nine genotypic frequencies fit into such a framework but the frequencies in the four observed groups do not.

One can think of the frequencies as arising from a "mixed-up" contingency table in which AA, AO and OA cells are combined together and, similarly, the BB, BO and OB cells, and the AB and BA cells. In the GLM, each observation is linked to a single linear predictor. If we could link each observation to several linear predictors we can embed these mixed-up contingency tables into the GLM framework. Suppose we have a linear predictor $\eta=\Sigma x_i\beta$ of length m to specify our linear model and that $\Upsilon=f(\mu)$. We suppose that the expected value of $y(\mu)$ can be written as $\mu=\mathbf{C}\Upsilon$, where **C** is a known *n*x*m* matrix. **C** links the m transformed linear predictors to the n observations and we call it a composite link (Thompson and Baker 1981). For this model, a simple extension of the procedure of Nelder and Wedderburn can be used to estimate β. In each iteration working dependent and independent variables $\mathbf{y}^*=\mathbf{CDX}\boldsymbol{\beta}+(\mathbf{y}-\mu)$ and $\mathbf{X}^*=\mathbf{CDX}$ are used with weight **V**, where **D** is a diagonal matrix with elements $d\mu_k/d\eta_k$. Obviously when **C**=**I** and $\mu_k=\Upsilon_k$ this algorithm is equivalent to that used by Nelder and Wedderburn (1972) for GLM.

For the O, A, B example, the model is $\mu=\mathbf{C}\exp(\mathbf{X}\boldsymbol{\beta})$, where **C** and **X**' are $4x9$ matrices and $\boldsymbol{\beta}$ is a $4x1$ vector, with

$$\mathbf{C} = \begin{bmatrix} 1 & 1 & 0 & 1 & 0 & 0 & 0 & 0 & 0 \\ 0 & 0 & 1 & 0 & 0 & 0 & 1 & 0 & 0 \\ 0 & 0 & 0 & 0 & 1 & 0 & 0 & 0 & 0 \\ 0 & 0 & 0 & 0 & 0 & 1 & 0 & 1 & 1 \end{bmatrix}$$

$$\mathbf{X}' = \begin{bmatrix} 1 & 1 & 1 & 1 & 1 & 1 & 1 & 1 & 1 \\ 2 & 1 & 1 & 1 & 0 & 0 & 1 & 0 & 0 \\ 0 & 1 & 0 & 1 & 2 & 1 & 0 & 1 & 0 \\ 0 & 0 & 1 & 0 & 0 & 1 & 1 & 1 & 2 \end{bmatrix}$$

and $\boldsymbol{\beta}=\ell n(\alpha)$, where α'=(N, p, r, q), and N is the total number of observations. For example, the first element of $\mathbf{X}\boldsymbol{\beta}$ is logN+2logp=log(Np^2), and the first element of **C** exp($\mathbf{X}\boldsymbol{\beta}$)=N($p^2$+pr+pr)=N($p^2$+2pr).

Thompson and Baker (1981) have given other examples of composite links. Rodger (1982) has shown how to take account of the binary nature of the composite matrix **C** in implementing the procedure in the GLIM program. Green (1982) has described the composite models as an ingenious but misguided small subset of models that can be fitted by iterative re-weighted least-squares. I have, however, found this additional flexibility useful in several situations. One example is grouped normal data when it is easier to describe cumulative probabilities in terms of linear predictors, but the grouped probabilities are (quasi) independent (Thompson and Baker 1981).

14.4 Variance Components for Normal Data

For orthogonal designs, an analysis of variance is often a convenient summary of the data. Variance components can then be found by equating mean squares to their expectation. If there are more sums of squares than components, Nelder and Wedderburn (1972) point out that GLM ideas can be used as the i-th mean square with df_i degrees of freedom distributed as a χ^2, or gamma variable and the expectation of this mean square is a linear function of variance parameters. For example, consider a trait with phenotypic variance $\sigma^2+\sigma_1^2$ and covariances between full sibs=$\sigma_1^2/2$, and between half-sibs $\sigma_1^2/4$, i.e., a trait with additive variance σ_1^2 and no dominance variance. Then a hierarchical mating design with each of s sires mated to d dams and n offspring raised gives rise to within-dam, between-dam and between-sire mean squares. These have expectations σ^2-1/2σ_1^2, σ^2-1/2σ_1^2+nσ_1^2/4=σ^2+(n/4-1/2) σ_1^2, and σ^2-1/2σ_1^2+nσ_1^2/4+ndσ_1^2/4=σ^2+[n(d-1)/4-1/2]σ_1^2, respectively. Hence, explanatory variates for the three mean squares are (1,1,1) and [-1/2, n/4-1/2, n(d-1)/4-1/2], with the effects represented by σ^2 and σ_1^2. In the iterative algorithm, each mean square is weighted by its variance (2*x*Expectation^2/degrees of freedom).

Restricted Maximum Likelihood (REML) (Patterson and Thompson 1971) is a popular method of estimating variance components, reducing to analysis of variance methods for orthogonal designs. For model (14.2) it can be conveniently thought of in terms of the mixed model equations (14.5). Eliminating β from (14.5) gives

$$(\mathbf{Z'SZ}+\mathbf{I}\gamma^{-1})\,\hat{\mathbf{u}} = \mathbf{Z'S\,y} \tag{14.7}$$

or

$$\hat{\mathbf{u}} = \mathbf{Q}^{-1}\mathbf{Z'Sy}, \tag{14.8}$$

giving an equation for the predicted value $\hat{\mathbf{u}}$.

Estimation of σ^2 and σ_1^2 is iterative, and at each stage can be thought of as finding two sums of squares, and equating them to their expectation. One sum of squares is based on residuals $(\mathbf{y}-\mathbf{X}\hat{\boldsymbol{\beta}}-\mathbf{Z}\hat{\mathbf{u}})$ and the other is based on the predicted values $\hat{\mathbf{u}}$. The estimating equations for σ^2 and σ_1^2 can be written as

$$(\mathbf{y}-\mathbf{X}\boldsymbol{\beta}-\mathbf{Z}\mathbf{u})'(\mathbf{y}-\mathbf{X}\boldsymbol{\beta}-\mathbf{Z}\mathbf{u}) = (n-r)\sigma^2$$

and

$$\mathbf{u'u}+\sigma^2 \,\mathrm{tr}\, \mathbf{Q}^{-1} = \sigma_1^2. \tag{14.9}$$

To avoid the inversion of $\mathbf{Q}$ at each stage, we could replace $\mathbf{Q}$ by $\mathbf{P}(\Lambda+\mathbf{I}\gamma^{-1})\mathbf{P}'$, with $\mathbf{P}$ orthogonal and Λ a diagonal matrix with i-th diagonal element λ_i. Then $\mathbf{Q}^{-1}$ can be found from $\mathbf{P}(\Lambda+\mathbf{I}\gamma^{-1})\mathbf{P}'$. The estimation procedure can be thought of (Patterson and Thompson 1971; Dempster et al. 1984) as using a residual sum of squares with expectation $(n-r-s+1)\sigma^2$, and using $\mathbf{P'Z'S}$y to form (s-1) independent sums of squares, with expectation of the i-th sum of squares being $\lambda_i\sigma_1^2+\sigma^2$. In one sense, an analysis of variance has been constructed with a residual mean square and the sire sum of squares split into (s-1) independent sums of squares. Again a linear model is being fitted to the sums of squares with the explanatory variables being functions of Λ and the linear effects being σ^2 and σ_1^2.

14.5 Variance Components with Generalized Linear Models

One restriction on the GLMs is that the theory has been developed for a single random component. There are, of course, many circumstances when one might wish models with more than one random component to fit into the GLM framework. For example, the heritability model in the second section is one such model. There have been several methods suggested for dealing with binomial and Poisson data. In some the main interest has been in fixed effects, for example giving adjustment factors to standard errors (McCullagh and Nelder 1983; Nelder 1985), and giving more efficient estimators (Williams 1982; Breslow 1984; Diaconis and Efron 1985). In others, the emphasis was on the random effects; for example, obtaining more efficient predictors of random effects (or combining information from data with prior information) (Laird 1978; Foulley and Gianola 1984) and estimating variance components (Gilmour et al. 1985).

The methods suggested generally resemble those of GLM models for fixed effects and variance components for normal data. The details differ partly because of differing statistical methods used including (empirical) Bayesian methods, maximum (quasi) likelihood procedures and the method of moments, and partly because of the various different numerical approximations which have been suggested. Except in simple cases with a small number of parameters (Tallis 1962; Im 1984) exact answers can be computationally expensive.

I will illustrate some of the methods for a probit model with binary data, such as the second section. I suppose a model $E(y_i)= \mu_i=\Phi(x_i\beta+u_i)$ is appropriate for given u_i, e.g., offspring of the i-th sire (i=1,...,s), and that genetic ideas suggest that the sire effects u are normally distributed with mean 0 and variance $\mathbf{I}\sigma_1^2$. The variance σ_1^2 represents the covariance between half-sibs or one-quarter of the additive genetic variance.

Then the likelihood of y_i can be written as:

$$\int L(y_i,\beta,u_i).\exp[-u_i^2/2\sigma_1^2]/(2\pi\sigma_1^2)^{1/2}du_i,$$

where $L(y_i,\beta_i,u_i)$ is the likelihood of y_i, given β and u_i.

We rewrite this as:

$$\frac{(2V_i^2)^{1/2}}{(2\sigma_1^2)^{1/2}}\int L(y_i,\beta,u_i).\exp\{[-u_i^2/\sigma_1^2+s_i^2]/2\}[(1/2\pi V_i^2)^{1/2}].\exp[-s_i^2/2]du_i$$

where $V_is_i=u_i-u_{ci}$, or

$$\frac{(2V_i^2)^{1/2}}{2\sigma_1^2}\int L(y_i,\beta,u_{ci}+V_is_i)\cdot\exp\{[-(u_{ci}+V_is_i)^2/\sigma_1^2+s_i^2]/2\}$$

$$\cdot(1/2\pi^{1/2})\cdot\exp[-s_i^2/2]ds_i.$$

We can approximate this integral using G Gaussian quadrature points and weights, c_g and w_g (g=1,...,G). Terms πw_g and $c_g/2^{1/2}$ are given in Abramowitz and Stegun (1972). This gives for the integral

$$\frac{(2V_i^2)^{1/2}}{(2\sigma_1^2)^{1/2}} \sum_{g=1}^{G} L(y_i,\beta,u_{ci}+V_ic_g)\cdot\exp\{[-(u_{ci}+V_ic_g)^2/\sigma_1^2+c_g^2]/2\}w_g. \qquad (14.10)$$

This approximation has been used, implicity or explicitly, in several forms. Anderson and Aitkin (1985) suggested using the approximation, with u_{ci}=0, $V_i^2=\sigma_i^2$ in (14.10) so that it reduces to

$$\sum_{g=1}^{G} w_g.L(y_i,\beta,\sigma_1c_g).$$

In a sense, we have now an extra explanatory covariate c_g in the linear model with regression coefficient σ_1. Anderson and Aitkin (1985) show that the maximum likelihood estimation of β and σ_1 can be thought of as a weighted GLM problem, with the observations and systematic effects repeated G times, and a covariate with value c_g for the g-th copy of the observation. Each copy has weight w^*_{gi}, obtained from

$$w^*_{gi}[\Sigma w_gL(y_i,\beta,\sigma_1c_g)] = w_gL(y_i,\beta,\sigma_1c_g)\ ,$$

such that w^*_{gi} is a weight for the g-th grid point taking into account prior information (w_g) and data information (L). This could be fitted directly using GLM algorithms with an extra stage of iteration as w^*_{gi} depends on β and σ_1. However, it might be useful to take advantage of the repeated nature of **X**, and form weights for the fixed effects and covariates and for the working dependent variate for the n observations (each being a group of G pseudo-observations). For model (14.2) with normally distributed data the estimation equations are an approximation to

$$\begin{bmatrix} \mathbf{X'X} & (1/\sigma_1)\mathbf{X'Zu} \\ (1/\sigma_1)\mathbf{u'Z'X} & \mathbf{A} \end{bmatrix} \begin{bmatrix} \boldsymbol{\beta} \\ \sigma_1 \end{bmatrix} = \begin{bmatrix} \mathbf{X'y} \\ (1/\sigma_1)\mathbf{u'Z'y} \end{bmatrix} \qquad (14.11)$$

with

$$\mathbf{A} = (1/\sigma_1^2)\mathbf{u'Z'Z'u}+\mathrm{tr}[(\mathbf{Z'Z}+\mathbf{I}\gamma^{-1})^{-1}\mathbf{Z'Z}],$$

and

$$\mathbf{u} = [\mathbf{Z'Z}+\mathbf{I}\gamma^{-1}]^{-1}\mathbf{Z}'(\mathbf{y}-\mathbf{X}\boldsymbol{\beta}).$$

Essentially, this is using $(1/\sigma_1)\mathbf{Zu}$ as a covariate, and taking into account that it is not known precisely. The first equation is a rearrangement of (14.5), and the second of (14.9), with **S**=**I**. This suggests that the speed of convergence depends on the degree of confounding between **X** and **Z** and, that schemes based on second differentials might be quicker for unbalanced designs.

Leonard (1972) uses a Bayesian argument incorporating prior knowledge for $\boldsymbol{\beta}$ and σ_1^2, and if there is vague knowledge for these parameters, then the expression maximized is equivalent to (14.10) with G=1 and V_i=1 and u_{ci} thought of as parameters satisfying GLM-augmented equations of the form:

$$\begin{bmatrix} \mathbf{X'DV^{-1}DX} & \mathbf{X'DV^{-1}DZ} \\ \mathbf{Z'DV^{-1}DX} & \mathbf{Z'DV^{-1}DZ}+\mathbf{I}/\sigma_1^2 \end{bmatrix} \begin{bmatrix} \boldsymbol{\beta} \\ \mathbf{u}_c \end{bmatrix} = \begin{bmatrix} \mathbf{X'y} \\ \mathbf{Z'y} \end{bmatrix} \qquad (14.12)$$

Eliminating $\boldsymbol{\beta}$ can give:

$$\mathbf{Qu}_c = \mathbf{Z'Sy} \qquad (14.13)$$

with σ_1^2 satisfying

$$q\sigma_1^2 = \mathbf{u}_c'\mathbf{u}_c. \qquad (14.14)$$

Although (14.14) is simpler than the REML estimators, it can often lead to zero estimates for σ_1^2 (Lindley and Smith 1972). Thompson (1980) quantified when this occurs for normal data. In particular, for well-designed animal breeding experiments we can expect these estimates of σ_1^2 to be zero. The problem statistically is that (14.14) assumes that the $\mathbf{u}_c$ are known. I would like to take account of prediction errors in $\mathbf{u}_c$, or alternatively, one should look at a function after integrating out the random effects u (O'Hagan 1976).

Consideration of model (14.2) shows that an exact likelihood follows from (14.10) with G=1 and estimating $\mathbf{u}_c$ if we use s log σ^2-log$|\mathbf{Z'Z}+\mathbf{I}\gamma^{-1}|$ to replace log V_i^2. Alternatively, if we wish to consider just the likelihood of error contrasts, we could use s log σ^2-log$|\mathbf{Q}|$ with **Q** given by (14.7) and (14.8). For GLM this naturally suggests using -log$|\mathbf{Q}|$ based on (14.13). If $(\mathbf{Q}-\mathbf{I}/\sigma_1^2)$ is thought to be independent of σ_1^2, then maximizing (14.10) will lead to estimating equations for variance components mimicking normal data methods. This type of estimating equation has been suggested using other arguments. Thompson (1979) and

Foulley et al. (1983) argue by direct analogy with normal data equations. Laird (1978) notes that the u_i's are missing data and that the EM algorithm could be used. This author assumes a flat prior for $\boldsymbol{\beta}$, and approximates the variance matrix by the inverse of the information matrix based on the posterior distributions, and evaluated at the modes. Similar approximations are used by Harville and Mee (1984) and Stiratelli et al. (1984). For convenience we denote this as Method A.

Gilmour et al. (1985), denoted as GAR, suggest that method A could be biased if the family size is small. For example, for a balanced one-way analysis of binary data with s families of size f if the mean frequency is p, then the approximations in the second section suggest an analysis of variance of the form

	M.S.	E.M.S.
Between families	B	$\sigma^2+f\sigma_1^2$
Within families	W	σ^2

with $\sigma^2=pq-z^2t$, $\sigma_1^2=z^2t$, where z is the ordinate corresponding to p and t is the intra-class correlation on the underlying scale. One iterate of the REML-type estimation procedure starting with $\beta=\Phi(p)$, u=0 gives rise to an estimate $\sigma_1^2=(B-pq)/f$ rather than $\sigma_1^2=(B-W)/f$ or $\sigma_1^2=(B-pq)/(f-1)$ suggested by the analysis of variance. Limited simulation suggests that with further iterates the estimate of t becomes smaller.

The efficiency of the GAR scheme can be investigated for this balanced design. If the probability of r observations of 1 in a family of size f is defined to be

$$p_r^* = p_r+(1/2)z^2t[d^2p_r/dp] \tag{14.15}$$

with $p_r=C_r^f p^r(1-p)^{f-r}$, then this model is consistent with the covariance between observations in the same family being z^2t, as assumed in the GAR scheme. Then, using (14.15) to work out the variances and covariances of B and W, it can be found that the GAR estimates of z^2t have a high asymptotic efficiency compared with maximum likelihood for model (14.15). In particular GAR is greater than 0.95 efficient provided t<0.25, and f<30 and greater than 0.90 efficient provided t<0.25 and f<40.

GAR suggest another method which perhaps explains why this bias is introduced. Essentially they integrate out the random effects **u** and determine the mean and approximate variance structure. They suggest that $E(y)=\mu=\Phi\{[1/(1+\sigma_1^2)^{1/2}]\mathbf{X}\boldsymbol{\beta}\}=\Phi(\mathbf{X}\boldsymbol{\beta}^*)$ and that the variance of **y** is approximately **V+DZZ'Dt**, where **V** is chosen so that the diagonal elements of **V+DZZ'Dt** are $\mu_i(1-\mu_i)$. The approximation enters because cov (y_i,y_j) is found by taking a Taylor series expansion in t about t=0. The approximation is more appropriate for μ near to 0.5, and small t. GAR suggest estimating $\boldsymbol{\beta}$ by generalized least-squares methods and note that they can be written in the form (14.12), where **D** and **V** are calculated as a function of $\boldsymbol{\beta}$ but not of $\mathbf{u}_c$. GAR use $\mathbf{u}_c'\mathbf{u}_c$ to estimate σ_1^2 employing a REML normal distribution argument. This scheme is computationally simpler than method A, and for the balanced one-way analysis leads to $t=(B-pq)/z^2(f-1)$ as suggested by a line of the analysis of variance.

GAR essentially use the $\mathbf{DZZ'D}t$ variance information to estimate t. Other authors have suggested similar schemes for estimating $\boldsymbol{\beta}$, with alternatives for estimating σ_1^2 using weighted sums of squares of residuals (Williams 1982; Breslow 1984), or deviances (Nelder 1985).

There has been very little experimental comparison of these and other competing methods (e.g., Haseman and Kupper 1979), apart from studies of Gaussian methods by Crowder (1985) and simulation by Hoschele (1986). The last author found that method A gave satisfactory estimates of heritability for one-way classifications when the family size was greater than 40 and the incidences were in the range 0.05-0.95. When the incidence was below 0.10 the true values were slightly underestimated for low or moderate heritabilities, and high heritabilities were substantially underestimated when the incidence was 0.01. For mixed models with large family sizes and small (average <10) herd-year-season classes Hoschele (1986) found that both A and GAR tended to overestimate heritability, and suggested that this might be due to the approximation involved in the marginal density.

There are two other possibilities. Both procedures use a residual variance $\mu(1-\mu)$ derived from linear estimates. This estimate can have a bias of the order of z^2t/f in the balanced one-way example. Also, note that the GAR parameterization gives parameters $\boldsymbol{\beta}^*$ relevant to averaging a distribution of genetic values. This suggests that rather than thinking of $\boldsymbol{\beta}$ as defined for a given value of σ_1^2 and, hence, depending on σ_1^2, that $\boldsymbol{\beta}/(1+\sigma_1^2)^{1/2}=\boldsymbol{\beta}^*$ might be a more stable parameterization. This poses the question about the appropriate scale if we use prior knowledge for $\boldsymbol{\beta}$. There is a similar, and perhaps more extreme case, when one tries to model prior information to find an appropriate power transformation for y (Box and Cox 1964). Cox and Reid (1987) suggest taking an orthogonal transformation of the parameters to make it easier to correct for the effects of nuisance parameters. This again suggests using $\boldsymbol{\beta}^*$ as parameters. Patterson and Thompson (1971) were careful to set up error contrasts orthogonal to linear contrasts used to estimate treatment effects. It is less clear how to implement a similar strategy for these non-linear models, and for the analysis of selected data (Henderson 1975; Thompson 1979).

14.6 Discussion

I have attempted to show the flexibility of GLMs and their application to animal breeding. In many cases, the animal breeding models and the distribution of the data provide compelling arguments for the use of GLMs. A GLM can also lead to simpler interpretation.

Atkins and Thompson (1986) examined the correlated response in litter size to selection on bone length in Blackface sheep, where litter size was 1 or 2. They found that using GLM, with a binomial distribution and probit link function, allowed a simpler interpretation. For example, the regressions of transformed litter size on selection differential for bone length at different parities were more homogeneous than the corresponding regressions on the original litter size scale.

Woolliams et al. (1986) have recently successfully modelled susceptibility to death in lambs by various causes including "Swayback". They used the hazard function, $\lambda_v(t).\exp[f_v(t)]$, such that the probability that a lamb alive at the beginning of day t succumbs to trait v on day t, $f_v(t)$, was a linear function of lamb attributes including genetic type and copper supplementation. Similarly, in an analysis of calving ease in Simmental cattle, Zhao (1986) found that dam age and sex effects were essentially additive when a threshold model was used.

We might also ask if the use of the GLM can lead to faster genetic progress. The answer is less clear-cut. Meijering and Gianola (1985a) compared linear (BLUP) with non-linear (GLM) methods of sire evaluation for categorical data. When a one-way model was tenable or responses were tetrachotomous, the differences between the methods were negligible. When the responses were binary, the layout highly unbalanced and a mixed model was appropriate to describe the underlying variate, the responses to selection using GLM were greater than those from BLUP at heritabilities of 0.20 or 0.50 for incidences less than 0.25. The largest difference in selection efficiency was 12%. Meijering and Gianola (1985b) have also shown that just correcting for residual variation and using linear prediction is unsatisfactory if a threshold model is appropriate.

Meijering (1985) analyzed Netherlands birth recording data considering dystocia (four categories) and stillbirth (two categories). He found that GLM and BLUP solutions were highly correlated (r=0.99) both for dystocia and stillbirth, with minor differences in sire ranking. This and the high computational cost, led Meijering to conclude that there was no advantage in applying the GLM procedure under the conditions of the national birth recording program in the Netherlands. Similarly, Jensen (1986) found very high correlations (>0.99) between linear and non-linear predictions of breeding values for type traits. In his data, however, fixed effects only explained about 2% of the variation.

Harville and Mee (1984) gave an example of calving difficulty with Holstein cattle. The first and last bulls on a threshold model were also ranked first and last on a linear model approach. Further, in 60 of the 89 bulls, the rankings using the two procedures differed by no more than ten positions. However, there were seven bulls with substantial information (at least 20 calvings) that changed ranking by more than 20 places.

Acknowledgements. I am very grateful to B.J. McGuirk for useful comments.

References

Abramowitz M, Stegun I (1972) Handbook of mathematical functions. Dover, New York

Anderson DA, Aitkin M (1985) Variance component models with binary response: interviewer variability. JR Stat Soc B 47:203-210

Atkins KD, Thompson R (1986) Predicted and realised responses to selection for an index of bone length and body size in Scottish Blackface sheep. 2. Correlated responses in lifetime productivity. Anim Prod 43:437-445

Breslow NE (1984) Extra-Poisson variation in log-linear models. Appl Stat 33:38-44

Box GEP, Cox DR (1964) An analysis of transformations (with discussion). JR Stat Soc B 26:211-252

Cox DR, Reid N (1987) Parameter orthogonality and approximate conditional inference. JR Stat Soc B 49:1-39

Crosbie SF, Hinch GN (1985) An intuitive explanation of generalised linear models. New Zealand Ag Res 28:19-29

Crowder M (1985) Gaussian estimation for correlated binomial data. JR Stat Soc 47:229-237

Dempster AP, Selwyn MR, Patel CM, Roth AJ (1984) Statistical and computational aspects of mixed model analysis. Appl Stat 33:203-214

Diaconis P, Efron B (1985) Testing for independence in a two-way table: new interpretations of the Chi-square statistic. Ann Stat 13:845-874

Dobson AJ (1983) An introduction to statistical modelling. Chapman and Hall, London

Elandt-Johnson RC (1971) Probability models and statistical methods in genetics. Wiley, New York

Falconer DS (1965) The inheritance of liability to certain diseases estimated from the incidence among relatives. Ann Hum Genet 29:51-76

Foulley JL, Gianola D (1984) Estimation of genetic merit from bivariate "all or none" responses. Genet Sel Evol 16:285-306

Foulley JL, Gianola D, Thompson R (1983) Prediction of genetic merit from data on categorical and quantitative variates with an application to calving difficulty, birth weight and pelvic opening. Genet Sel Evol 15:407-424

Gilmour AR, Anderson RD, Rae AL (1985) The analysis of binomial data in a generalised linear mixed model. Biometrika 72:593-599

Green PJ (1982) The APL alternative. GLIM: 82. In: Gilschrist R(ed) Proc Int Conf Gen Linear Models. Springer-Verlag, Berlin Heidelberg New York

Harville DA, Mee RW (1984) A mixed-model procedure for analyzing ordered categorical data. Biometrics 40:393-408

Haseman JD, Kupper LL (1979) Analysis of dichotomous response data from certain toxicological experiments. Biometrics 35:281-292

Henderson CR (1975) Best linear unbiased estimation and prediction under a selection model. Biometrics 31:423-449

Hoschele I (1986) Estimation of breeding values and variance components with quasi continuous data. PhD Thesis, Hohenheim Univ

Im S (1984) Plan mixte equilibre a deux facteurs hierarchiques pour les donnees binaires. Biometrics 40:283-292

James JW, McGuirk BJ (1982) Regression of offspring on parent for all-or-none traits. Z Tierz Züchtungsbiol 99:308-314

Jensen J (1986) Sire evaluation for type traits with linear and non-linear procedures. Livest Prod Science 15:165-171

Laird NM (1978) Empirical Bayes methods for two-way contingency tables. Biometrika 65:581-590

Leonard T (1972) Bayesian methods for binomial data. Biometrika 59:581-589

Lindley DV, Smith AFM (1972) Bayes estimates for the linear model. JR Stat Soc B 34:1-41

McCullagh P, Nelder JA (1983) Generalized linear models. Chapman and Hall, London

McGuirk BJ, Atkins KD (1984) Fleece rot in merino sheep. 1. The inheritance of fleece rot in a flock of medium-wool Peppin Merinos. Austr J Agric Res 35:425-434

Meijering A (1985) Sire evaluation for calving traits by best linear unbiased prediction and non-linear methodology. Z Tierz Züchtungsbiol 102:95-105

Meijering A, Gianola D (1985a) Linear versus non-linear methods of sire evaluation for categorical traits: a simulation study. Genet Sel Evol 17:115-132

Meijering A, Gianola D (1985b) Observations on sire evaluation with categorical data using heteroscedastic mixed linear models. J Dairy Sci 68:1226-1232

Nelder JA (1985) Quasi-likelihood and GLIM. In: Gilchrist R, Francis B, Whittaker J (eds) Generalized linear models, Proceedings. Springer-Verlag, Berlin Heidelberg New York Tokyo

Nelder JA, Wedderburn RWM (1972) Generalized linear models. JR Stat Soc A 135:370-384

O'Hagan A (1976) On posterior joint and marginal modes. Biometrika 63:329-333

Patterson HD, Thompson R (1971) Recovery of inter-block information when block sizes are unequal. Biometrika 58:545-554

Robertson A, Lerner IM (1949) The heritability of all-or-none traits: viability of poultry. Genetics 34:395-411

Rodger JH (1982) Composite link functions with linear log link and Poisson error. GLIM Newsletter, December, 1982

Ryder ML, Land RB, Ditchburn R (1974) Colour inheritance in Soay, Orkney and Shetland sheep. J Zool 173:477-485

Stiratelli R, Laird N, Ware JH (1984) Random-effects models for serial observations with binary response. Biometrics 40:961-971

Tallis GM (1962) The maximum likelihood estimation of correlation from contingency tables. Biometrics 18:342-353

Thompson R (1979) Sire evaluation. Biometrics 35:339-353

Thompson R (1980) Maximum likelihood estimation of variance components. Math Operationforsch Stat 11:545-561

Thompson R, Baker RJ (1981) Composite link functions in generalised linear models. Appl Stat 30:125-131

Thompson R, McGuirk BJ, Gilmour AR (1985) Estimating the heritability of all-or-none and categorical traits by offspring-parent regression. Z Tierz Züchtungsbiol 102:342-354

Williams DA (1982) Extra binomial variation in logistic linear models. Appl Stat 31:144-148

Woolliams C, Suttle NF, Woolliams JA, Jones DG, Wiener G (1986) Studies on lambs from lines genetically selected for low and high copper status. 1. Differences in mortality. Anim Prod 43:293-301

Zhao Y (1986) Applications of the threshold model to calving ease in Simmental cattle. MS Thesis, Cornell Univ

15 Analysis of Linear and Non-Linear Growth Models with Random Parameters

N.M. Laird[1]

This paper discusses the analysis of growth data using mixed models. Both Bayes and frequentist approaches are outlined for the linear model. The use of the EM algorithm is shown to offer a flexible and straightforward computational approach to the analysis. Extensions to non-linear models are described.

15.1 Introduction

This paper describes the use of emerging statistical methods for fitting both linear and non-linear growth curves to a sample of individuals. Methods for estimating both individual and population parameters will be discussed, as well as the relationship between maximum likelihood (ML), Bayes and Empirical Bayes (EB) approaches to the analysis. The EM algorithm is seen as a device for computing parameter estimates with any method of analysis, ML, Bayes, or EB, and its form in the three cases elucidates the distinction in the three approaches. We begin in Section 15.2 by considering the analysis of a very general linear growth model, then discuss a special case in Section 15.3. Methods for non-linear models are developed in the remaining section.

15.2 A Two-Stage Model for Linear Growth

The following two-stage model is often used to describe linear growth curves in a sample of individuals. Let $\mathbf{y}_i$ denote the $T_i x1$ vector of measurements for the ith individual, $\mathbf{Z}_i$ the $T_i xq$ matrix specifying the time design for the ith individual, and $\boldsymbol{\beta}_i$ the $qx1$ vector of the ith individual's growth curve parameters, for i=1,...,K. Then we write

<u>Stage 1:</u> $\mathbf{y}_i=\mathbf{Z}_i\boldsymbol{\beta}_i+\mathbf{e}_i,\ \mathbf{e}_i \sim N(0,\sigma^2 \mathbf{I}_{T_i x T_i})$ and

[1] Department of Biostatistics, Harvard School of Public Health, Boston, Massachusetts, USA

<u>Stage 2:</u> $\boldsymbol{\beta}_i=\mathbf{A}_i\alpha+\delta_i$, where $\delta_i \sim N(\mathbf{0},\mathbf{D})$ and $\mathbf{A}_i$ is a $q x p$ matrix of person-specific covariates which affect the parameters of the growth curve (such as sex, breed, etc.). Then marginally the $\mathbf{y}_i$'s are multivariate normal with mean and variance given by

$$E(\mathbf{y}_i) = \mathbf{Z}_i\mathbf{A}_i\alpha \tag{15.1}$$

$$\Sigma_i = \text{var}(\mathbf{y}_i) = \mathbf{Z}_i\mathbf{D}\mathbf{Z}_i^T+\sigma^2\mathbf{I}. \tag{15.2}$$

Although this model does not require a common design matrix nor even a common number of observations per person, it does have limitations in that no individual covariates may vary over time, and individual mean curves determine the number of covariance parameters which must be fitted. A more satisfactory model which unlinks the mean and variance is obtained by writing

$$\mathbf{y}_i = \mathbf{X}_i\alpha+\mathbf{Z}_i\delta_i+\mathbf{e}_i,$$

where α, δ_i, and $\mathbf{e}_i$ are as before, $\mathbf{X}_i$ is an arbitrary $T_i x p$ design matrix and $\mathbf{Z}_i$ is the corresponding design matrix of the random effects; it may be a subset of the columns included in the original $\mathbf{Z}_i$.

We are generally interested in inferences about three types of parameters, the fixed effects α, the random effects $\boldsymbol{\beta}_i$ (or δ_i), and the variance-covariance parameters σ^2 and $\mathbf{D}$. A classical approach based on maximum likelihood and its extensions has been developed by Harville (1976, 1977) among others. Here α, σ^2, and $\mathbf{D}$ are estimated by maximizing their likelihood based on the marginal normal distribution of the $\mathbf{y}_i$'s. The $\boldsymbol{\beta}_i$'s are estimated using an extension of the Gauss-Markov theorem for random effects (Rao 1975). We refer to this approach as the ML approach.

The general Bayes approach is to assume α, σ^2, and $\mathbf{D}$ have independent prior distributions, and base inferences about these parameters and the $\boldsymbol{\beta}_i$'s on their posterior distributions, given the observations. Because of the complexity of the computations, posterior modes are used to approximate posterior means, and posterior variances are approximated using inverse second derivatives of the log-posterior density. Such an approach is described in Fearn (1975), following a more general development in Lindley and Smith (1972). The EB approach can be viewed as a Bayes approach using non-informative priors for α, σ^2, and $\mathbf{D}$, and empirical Bayes estimates for the δ_i.

The EM algorithm (Dempster et al. 1977) is a general purpose algorithm for computing MLE's or posterior modes in incomplete data problems. In the variance component setting it can be used for the iterative computations with either the Bayes or ML analysis; we use it to illustrate the differences in the two approaches. The EM is based on identifying a set of "complete data" which, if observed, would make estimation of α, σ^2, and $\mathbf{D}$ straightforward.

To implement the EM, we first identify the two steps of each iteration. For the maximization (M) step, we identify solutions to the likelihood equations for estimating α, $\mathbf{D}$, and σ^2 from the complete data. For the expectation (E) step, we compute the expectation of the complete data log-likelihood given the observed data. If we have sufficient statistics with complete data, then the E-step reduces to finding the expectation of the sufficient statistics, given the observed data. For Bayes analyses, we replace log-likelihoods with log-posterior densities.

Following Laird (1982a), we choose the complete data vector for the ith person to be $\mathbf{u}_i^T=(\mathbf{y}_i^T,\delta_i^T,\mathbf{e}_i^T)$. With complete data, it is straightforward to see that $\mathbf{S}^T=[\Sigma\mathbf{X}_i^T(\mathbf{y}_i-\mathbf{Z}_i\delta_i),\ \Sigma\delta_i\delta_i^T,\Sigma\mathbf{e}_i^T\mathbf{e}_i]$ forms the vector of sufficient statistics for α, $\mathbf{D}$, and σ^2. Were the δ_i observed, α, $\mathbf{D}$, and σ^2 would be estimated by ML by setting $\mathbf{S}$ equal to its expectation:

$$(\Sigma\mathbf{X}_i^T\mathbf{X}_i)\hat{\alpha} = \Sigma\mathbf{X}_i^T(\mathbf{y}_i-\mathbf{Z}_i\delta_i) = \mathbf{S}_1 \tag{15.3}$$

$$K\hat{\mathbf{D}} = \Sigma\delta_i\delta_i^T = \mathbf{S}_2 \tag{15.4}$$

and

$$(\Sigma T_i)\hat{\sigma}^2 = \Sigma\mathbf{e}_i^T\mathbf{e}_i = S_3. \tag{15.5}$$

An estimate of the individual $\boldsymbol{\beta}_i$ is

$$\hat{\boldsymbol{\beta}}_i = \mathbf{A}_i\hat{\alpha}+\delta_i. \tag{15.6}$$

For the Bayes analysis, the M-step must be adjusted to maximize the joint posterior mode of the parameters. The first derivatives of the log-likelihood and the prior distributions are most conveniently expressed in terms of the natural parameters

$$\psi_1 = \alpha/\sigma^2$$
$$\psi_2 = \mathbf{D}^{-1}/2$$
$$\psi_3 = 1/2\sigma^2.$$

Following Fearn (1975), we will assume ψ_1 has a vague prior, $2\psi_2$ is Wishart $(\mathbf{L}^{-1},\nu_2)$, and $(2\lambda\psi_3)$ is $\chi^2_{\nu_3}$. With this formulation, the log of the posterior density based on complete data has exactly the same form as the log-likelihood, but with $\mathbf{S}$ replaced by $\mathbf{S}+\mathbf{T}$ where $\mathbf{T}^T=(\mathbf{0},\mathbf{L},\lambda)$, K is replaced by $K+\nu_2-q-2$, and (ΣT_i) replaced by $(\Sigma T_i)+\nu_3-2$. Thus maximizing the log posterior requires setting

$$(\Sigma\mathbf{X}_i^T\mathbf{X}_i)\tilde{\alpha} = \Sigma\mathbf{X}_i^T(\mathbf{y}_i-\mathbf{Z}_i\delta_i) \tag{15.7}$$

$$(K+\nu_2-q-2)\tilde{\mathbf{D}} = \Sigma\delta_i\delta_i^T+\mathbf{L} \tag{15.8}$$

$$(\Sigma T_i+\nu_3-2)\tilde{\sigma}^2 = \Sigma \mathbf{e}_i^T\mathbf{e}_i+\lambda. \tag{15.9}$$

Because of the way we have defined the complete data, ψ_2 and ψ_3 are a posteriori independent of each other and α, with $2(\lambda+\Sigma\mathbf{e}_i^T\mathbf{e}_i)\psi_3$ being $\chi^2_{(\nu_3+\Sigma T)}$ and $2\psi_2$ being Wishart (ν_2+K) with matrix $(\mathbf{L}+\Sigma\delta_i\delta_i^T)$. This has two implications. First, (15.8-15.9) define the marginal posterior modes of σ^2 and $\mathbf{D}$, and secondly, replacing ν_2-q-2 and ν_3-2 by ν_2 and ν_3 gives marginal posterior means. Changing (15.8-15.9) to be

$$(K+\nu_2)\tilde{\mathbf{D}} = \Sigma\delta_i\delta_i^T+\mathbf{L} \tag{15.10}$$

$$(\Sigma T_i+\nu_3)\tilde{\sigma}^2 = \Sigma \mathbf{e}_i^T\mathbf{e}+\lambda \tag{15.11}$$

gives the mean rather than the mode. Unpublished work by Credeur (1978) suggests that this device improves the mode as an approximation to the mean in the incomplete data setting, hence we use these equations instead. Note that because the posterior of α depends upon σ^2, Eq. (15.7) defines the mode (and the mean) of the posterior of α conditional on σ^2 and $\mathbf{D}$. Although a marginal mean would be preferable (integrating out σ^2 and $\mathbf{D}$), this becomes very difficult computationally when we do the incomplete data computations. Thus Eqs. (15.7, 15.10, 15.11) define the M-step for the Bayes analysis.

The E-step of the EM also has two versions, one for ML and one for posterior mode estimation. For ML, we calculate $E(\mathbf{S}|\mathbf{y},\alpha,\sigma^2,\mathbf{D})$ where $\mathbf{y}^T=(\mathbf{y}_1^T,...,\mathbf{y}_K^T)$. Because of normality it is easy to see that given $\mathbf{y}$, δ_i and $\mathbf{e}_i$ are jointly multivariate normal with means

$$\delta_i^* = \mathbf{D}\mathbf{Z}_i^T\mathbf{W}_i(\mathbf{y}_i-\mathbf{X}_i\alpha)/\sigma^2 \tag{15.12}$$

$$\mathbf{e}_i^* = \mathbf{W}_i(\mathbf{y}_i-\mathbf{X}_i\alpha) \tag{15.13}$$

and variances

$$\mathrm{var}(\delta_i|\mathbf{y}_i,\alpha,\mathbf{D},\sigma^2) = \mathbf{D}-\mathbf{D}\mathbf{Z}_i^T\mathbf{W}_i\mathbf{Z}_i\mathbf{D}/\sigma^2 \tag{15.14}$$

$$\mathrm{var}(\mathbf{e}_i|\mathbf{y}_i,\alpha,\mathbf{D},\sigma^2) = \sigma^2\mathbf{I}_{T_i\times T_i}-\sigma^2\mathbf{W}_i, \tag{15.15}$$

where

$$\mathbf{W}_i = \sigma^2\Sigma_i^{-1}. \tag{15.16}$$

Thus it follows easily that $\mathbf{S}^*=E(\mathbf{S}|\mathbf{y},\alpha,\sigma^2,\mathbf{D})$ is

$$\mathbf{S}_1^* = \Sigma \mathbf{X}_i^T(\mathbf{y}_i-\mathbf{Z}_i\delta_i^*) \tag{15.17}$$

$$\mathbf{S}_2^* = \Sigma \delta_i^* \delta_i^{*T} + \Sigma(\mathbf{D}-\mathbf{D}\mathbf{Z}_i^T\mathbf{W}_i\mathbf{Z}_i\mathbf{D}/\sigma^2) \tag{15.18}$$

and

$$\mathbf{S}_3^* = \Sigma \mathbf{e}_i^{*T}\mathbf{e}_i^* + \Sigma \mathrm{tr}(\sigma^2\mathbf{I}-\sigma^2\mathbf{W}_i). \tag{15.19}$$

Using results in Dempster et al. (1977), for ML estimation we equate the left-hand sides of Eqs. (15.3-15.5) to the corresponding right-hand sides of (15.17-15.19). After simplication, this yields:

$$(\Sigma \mathbf{X}_i^T\hat{\mathbf{W}}_i\mathbf{X}_i)\hat{\alpha} = \Sigma \mathbf{X}_i^T\hat{\mathbf{W}}_i\mathbf{y}_i \tag{15.20}$$

$$K\hat{\mathbf{D}} = \Sigma \hat{\delta}_i\hat{\delta}_i^T + \Sigma\{\hat{\mathbf{D}}-\hat{\mathbf{D}}\mathbf{Z}_i^T\hat{\mathbf{W}}_i\mathbf{Z}_i\hat{\mathbf{D}}/\hat{\sigma}^2\} \tag{15.21}$$

$$(\Sigma T_i)\hat{\sigma}^2 = \Sigma \hat{\mathbf{e}}_i^T\hat{\mathbf{e}}_i + \Sigma \mathrm{tr}(\mathbf{I}-\hat{\sigma}^2\hat{\mathbf{W}}_i), \tag{15.22}$$

where $(\hat{\delta}_i,\hat{\mathbf{e}}_i)$ is $(\delta_i^*,\mathbf{e}_i^*)$ evaluated at $(\hat{\alpha},\hat{\mathbf{D}},\hat{\sigma}^2)$ and

$$\hat{\mathbf{W}}_i = (\mathbf{I}-\mathbf{Z}_i\hat{\mathbf{D}}\mathbf{Z}_i^T\hat{\Sigma}_i^{-1}) = \hat{\sigma}^2\hat{\Sigma}_i^{-1}. \tag{15.23}$$

Since there is generally no closed form solution, some iterative method must be used. To implement the EM, we begin with initial values for $(\alpha,\sigma^2,\mathbf{D})$, evaluate $\mathbf{S}^*$ using (15.17-15.19 and 15.12-15.13), substitute $\mathbf{S}^*$ for the right-hand side of (15.3-15.5), and solve for new values of α, $\mathbf{D}$, and σ^2. These steps are repeated to convergence. Laird (1982a) shows that this algorithm is equivalent to Henderson's algorithm when we have only variance parameters in the model. The extended Gauss-Markov estimate of $\boldsymbol{\beta}_i$ is:

$$\hat{\boldsymbol{\beta}}_i = \mathbf{A}_i\hat{\alpha}+\hat{\delta}_i,$$

which can also be thought of as an empirical Bayes estimate of $\boldsymbol{\beta}_i$, since it equals $E(\boldsymbol{\beta}_i|\mathbf{y}_i,\hat{\alpha},\hat{\sigma}^2,\hat{\mathbf{D}})$.

Two adjustments need to be made to the E-step for the Bayes approach. First, we compute conditional expectations of $\mathbf{S}+\mathbf{T}$, and secondly, because we use marginal posterior modes of $\mathbf{D}$ and σ^2, in computing $E(S_2)$ and $E(S_3)$ we must also integrate out α. Because α

has a vague prior, it is easily seen that the posterior of α given $\mathbf{y}$, σ^2, and $\mathbf{D}$ (but not the δ_i's) is multivariate normal with mean

$$\tilde{\alpha} = (\Sigma \mathbf{X}_i \mathbf{W}_i \mathbf{X}_i)^{-1} \Sigma \mathbf{X}_i^T \mathbf{W}_i \mathbf{y}_i$$

and

$$\text{var}(\alpha|\mathbf{y},\sigma^2,\mathbf{D}) = \sigma^2(\Sigma \mathbf{X}_i^T \mathbf{W}_i \mathbf{X}_i)^{-1} \equiv \Gamma.$$

Thus the δ_i's and $\mathbf{e}_i$'s are marginally multivariate normal (given $\mathbf{y}_i,\sigma^2,\mathbf{D}$) with means given by (15.12) and (15.13), evaluating α at $\tilde{\alpha}$, and

$$\text{var}(\delta_i|\mathbf{y}_i,\sigma^2,\mathbf{D}) = \mathbf{D}\text{-}\mathbf{D}\mathbf{Z}_i^T\mathbf{W}_i\mathbf{Z}_i\mathbf{D}+\mathbf{D}\mathbf{Z}_i^T\mathbf{W}_i\mathbf{X}_i\Gamma\mathbf{X}_i^T\mathbf{W}_i\mathbf{Z}_i\mathbf{D}/\sigma^4 \tag{15.24}$$

$$\text{var}(\mathbf{e}_i|\mathbf{y}_i,\sigma^2,\mathbf{D}) = \sigma^2\mathbf{I}\text{-}\sigma^2\Sigma_i^{-1}+\mathbf{W}_i\mathbf{X}_i\Gamma\mathbf{X}_i^T\mathbf{W}_i. \tag{15.25}$$

So $E(\mathbf{S}_1|\mathbf{y})$ remains as before, but $E(\mathbf{S}_2|\mathbf{y})$ and $E(\mathbf{S}_3|\mathbf{y})$ have additional terms coming from their prior expectations ($\mathbf{T}$) and the addition to the variance after marginalizing on α. Computing these expectations and equating them to the left-hand sides of (15.7, 15.10-11) yields the Bayes estimates

$$(\Sigma \mathbf{X}_i^T \tilde{\mathbf{W}}_i \mathbf{X}_i)\tilde{\alpha} = \Sigma \mathbf{X}_i^T \tilde{\mathbf{W}}_i \mathbf{y}_i \tag{15.26}$$

$$(\mathbf{K}+\nu_2)\tilde{\mathbf{D}} = \Sigma\tilde{\delta}_i\tilde{\delta}_i^T+\Sigma\{\tilde{\mathbf{D}}\text{-}\tilde{\mathbf{D}}\mathbf{Z}_i^T\tilde{\mathbf{W}}_i\mathbf{Z}_i\tilde{\mathbf{D}}+\tilde{\mathbf{D}}\mathbf{Z}_i^T\tilde{\mathbf{W}}_i\mathbf{X}_i\tilde{\Gamma}\mathbf{X}_i^T\tilde{\mathbf{W}}_i\mathbf{Z}_i\tilde{\mathbf{D}}/\tilde{\sigma}^4\}+\mathbf{L} \tag{15.27}$$

$$(\Sigma \mathbf{T}_i+\nu_3)\tilde{\sigma}^2 = \Sigma\tilde{\mathbf{e}}_i^T\tilde{\mathbf{e}}_i+\Sigma\{\tilde{\sigma}^2\mathbf{T}_i\text{-}\tilde{\sigma}^2\text{tr}\Sigma_i^{-1}+\text{tr}\tilde{\mathbf{W}}_i\mathbf{X}_i\Gamma\mathbf{X}_i^T\tilde{\mathbf{W}}_i\}+\lambda, \tag{15.28}$$

where $\tilde{\delta}_i$ and $\tilde{\mathbf{e}}_i$ are δ_i^* and $\mathbf{e}_i^*$ evaluated at $\tilde{\alpha}$, $\tilde{\mathbf{D}}$, and $\tilde{\sigma}^2$. Note that (15.20) and (15.26) define identical estimators for α, apart from the fact that they use weight matrices, $\mathbf{W}_i$, evaluated at different estimates of σ^2 and $\mathbf{D}$. If we take vague priors for $\mathbf{D}$ and σ^2 as well, so $\nu_2=\nu_3=0$ and $\lambda=0$, $\mathbf{L}=0$, Eqs. (15.27-28) yield the REML estimates of σ^2 and $\mathbf{D}$ (Harville 1977; Patterson and Thompson 1971). Because the joint posterior distribution of α and the δ_i's is multivariate normal given the σ^2, $\mathbf{D}$, and $\mathbf{y}_i$'s, $\tilde{\delta}_i$ can be thought of as either a joint or marginal posterior mode or expectation, marginalizing on α, but fixing the variance components.

Thus, REML estimates of variance components can be viewed as flat prior Bayes using marginal posteriors (integrating out α and the δ_i's), while maximum likelihood estimators can be viewed as flat prior Bayes using joint modes for α, $\mathbf{D}$, and σ^2, but integrating out the δ_i's. This fact strengthens the work of O'Hagan (1976) which suggests that marginal modes are preferable to joint modal estimates.

Because $\hat{\alpha}$ and $\hat{\delta}_i$ are linear functions of $\mathbf{y}_i$, their approximate variances are given by

$$\mathrm{var}(\hat{\alpha}) = \hat{\sigma}^2(\Sigma \mathbf{X}_i^T \hat{\mathbf{W}} \mathbf{X}_i)^{-1} \tag{15.29}$$

$$\mathrm{var}(\hat{\delta}_i) = \hat{\mathbf{D}}\mathbf{Z}_i^T\{\hat{\mathbf{W}}_i - \hat{\mathbf{W}}_i\hat{\mathbf{Q}}_i\hat{\mathbf{W}}_i\}\mathbf{Z}_i\hat{\mathbf{D}}/\hat{\sigma}^2, \tag{15.30}$$

where

$$\mathbf{Q}_i = \mathbf{X}_i(\Sigma \mathbf{X}_i^T \mathbf{W}_i \mathbf{X}_i)^{-1}\mathbf{X}_i^T.$$

The marginal posterior variance of $\tilde{\alpha}$ given $\sigma^2=\tilde{\sigma}^2$ and $\mathbf{D}=\tilde{\mathbf{D}}$ is also given by (15.29) with $\hat{\mathbf{W}}_i$ replaced by $\tilde{\mathbf{W}}_i$. However, the marginal posterior variance of δ_i has an extra term:

$$\mathrm{var}(\tilde{\delta}_i) = \tilde{\mathbf{D}} + \tilde{\mathbf{D}}\mathbf{Z}_i^T\tilde{\mathbf{W}}_i\{\tilde{\mathbf{Q}}_i - \tilde{\mathbf{W}}_i^{-1}\}\tilde{\mathbf{W}}_i\mathbf{Z}_i\tilde{\mathbf{D}}/\tilde{\sigma}^2. \tag{15.31}$$

Harville (1976) notes that (15.31) can also be thought of as the sampling variance of $(\tilde{\delta}_i - \delta_i)$. Finally, an estimate of $\boldsymbol{\beta}_i$ and its variance are

$$\tilde{\boldsymbol{\beta}}_i = \mathbf{A}_i\tilde{\alpha} + \tilde{\delta}_i$$

and

$$\mathrm{var}(\tilde{\boldsymbol{\beta}}_i) = [\mathbf{A}_i - \tilde{\mathbf{D}}\mathbf{Z}_i\tilde{\mathbf{W}}_i\mathbf{X}_i/\tilde{\sigma}^2](\Sigma \mathbf{X}_i^T\tilde{\mathbf{W}}_i\mathbf{X}_i)^{-1}[\mathbf{A}_i - \tilde{\mathbf{D}}\mathbf{Z}_i^T\tilde{\mathbf{W}}_i\mathbf{X}_i/\tilde{\sigma}^2]^T + \tilde{\mathbf{D}} - \tilde{\mathbf{D}}\mathbf{Z}_i^T\tilde{\mathbf{W}}_i\mathbf{Z}_i\tilde{\mathbf{D}}/\tilde{\sigma}^2.$$

The expression for $\mathrm{var}(\hat{\alpha})$ given in (15.29) is asymptotically correct, since $\hat{\alpha}$ is ML and the asymptotic covariance between $\hat{\alpha}$ and $(\hat{\mathbf{D}},\hat{\sigma}^2)$ is zero. While expression (15.31) without the tildes is easily seen to be the appropriate variance for $\tilde{\delta}_i$ if $\mathbf{D}$ and σ^2 are known, there is no corresponding justification for it when estimates for $\mathbf{D}$ and σ^2 are substituted, since $\tilde{\delta}_i$ is not ML. This variance is likely to be seriously understated if K is small. Kackar and Harville (1984) suggest one approach to adjusting the variance of the estimated random effects for two ANOVA designs. Morris (1983a,b) takes an empirical Bayes approach to estimating the variance of the random effects, in the simple case where $q \simeq 1$ and $\delta_i \sim N(0,d)$. Laird and Louis (1986) extend Morris' approach and provide empirical Bayes confidence intervals for the random effects based on an empirical Bayes bootstrap. In principle, these EB approaches can be extended to the more general case, but further research is needed.

We remark that Eqs. (15.27, 15.28) for $\tilde{\mathbf{D}}$ and $\tilde{\sigma}^2$ differ from those given by Fearn (1975), who uses the joint posterior mode of the δ_i's, $\mathbf{D}$ and σ^2 for Bayes estimation. In the balanced data case, where $\mathbf{X}_i=\mathbf{Z}_i\equiv\mathbf{Z}$, and $T_i\equiv T$, Fearn's estimate of $\mathbf{D}$ with vague prior information reduces to

$$\hat{\mathbf{D}} = \Sigma(\delta_i^{LS})(\delta_i^{LS})^T/(K-2)$$

where

$$\delta_i^{LS} = (\mathbf{Z}^T\mathbf{Z})^{-1}\mathbf{Z}^T(\mathbf{y}_i-\bar{\mathbf{y}}).$$

Hence $\mathbf{D}$ is biased upward by failure to subtract out the sampling variance. When we add covariates (here q=p), the upward bias in Fearn's $\hat{\mathbf{D}}$ will be partially offset by the downward bias arising from not correcting for lost degrees of freedom in estimating α. In contrast, Eq. (15.27) with vague prior information yields

$$\tilde{\mathbf{D}} = \Sigma(\delta_i^{LS})(\delta_i^{LS})^T/(K-1)-\tilde{\sigma}^2(\mathbf{Z}^T\mathbf{Z})^{-1} ,$$

which is unbiased for $\mathbf{D}$. Lange and Laird (1986) show how to adjust this estimate if the expression for $\tilde{\mathbf{D}}$ is not non-negative definite.

15.3 Two-Step Methods for Linear Models

In this section we make the simplifying assumption that the two-stage model holds, so that $\mathbf{X}_i=\mathbf{Z}_i\mathbf{A}_i$ for i=1,...,K. When this model holds, an attractive approach is to first obtain the least-squares estimate of each individual's growth curve, $\bar{\boldsymbol{\beta}}_i$, and then analyze the least-squares coefficients. Such simplified approaches have intuitive appeal to investigators. It is straightforward to derive a full ML, REML, or Bayes analysis for this situation, either directly or using the equations in Section 15.2.

The direct approach is to assume that $\bar{\boldsymbol{\beta}}_i$ is $N[\boldsymbol{\beta}_i,\sigma^2(\mathbf{Z}_i^T\mathbf{Z}_i)^{-1}]$,

$$s_i^2 = (\mathbf{y}_i-\mathbf{Z}_i\bar{\boldsymbol{\beta}}_i)^T(\mathbf{y}_i-\mathbf{Z}_i\bar{\boldsymbol{\beta}}_i)/\sigma^2$$

is $\chi^2_{(T_i-q)}$, and $\bar{\boldsymbol{\beta}}_i$ and s_i^2 are independent. Hence, marginally the $\bar{\boldsymbol{\beta}}_i$ are $N[\mathbf{A}_i\alpha,\sigma^2(\mathbf{Z}_i^T\mathbf{Z}_i)^{-1}+\mathbf{D}]$, still independent of the s_i^2's, whose distribution does not depend upon $\boldsymbol{\beta}_i$. Deriving ML or REML equations for this model is straightforward, and yields (for REML):

$$\hat{\alpha} = (\Sigma\mathbf{A}_i^T\hat{\mathbf{U}}_i\mathbf{A}_i)^{-1}\Sigma\mathbf{A}_i^T\hat{\mathbf{U}}_i\bar{\boldsymbol{\beta}}_i \qquad (15.32)$$

$$\hat{\sigma}^2 = \{\Sigma(s_i^2+\hat{S}_i^2)+\hat{\sigma}^2\mathrm{tr}[(\mathbf{Z}_i^T\mathbf{Z}_i)^{-1}\hat{\mathbf{P}}_i]\}/(\Sigma T_i) \tag{15.33}$$

$$\hat{\mathbf{D}} = \Sigma\{\hat{\mathbf{D}}\hat{\mathbf{U}}_i(\bar{\boldsymbol{\beta}}_i-\mathbf{A}_i\hat{\alpha})(\bar{\boldsymbol{\beta}}_i-\mathbf{A}_i\hat{\alpha})^T\hat{\mathbf{U}}_i^T\hat{\mathbf{D}}+\hat{\mathbf{D}}[\mathbf{I}-\hat{\mathbf{P}}_i\hat{\mathbf{D}}]\}/K, \tag{15.34}$$

where

$$\mathbf{U}_i = [\sigma^2(\mathbf{Z}_i^T\mathbf{Z}_i)^{-1}+\mathbf{D}]^{-1}$$

$$\mathbf{P}_i = \mathbf{U}_i[\mathbf{I}-\mathbf{A}_i(\Sigma\mathbf{A}_i^T\mathbf{U}_i\mathbf{A}_i)^{-1}\mathbf{A}_i\mathbf{U}_i]$$

and

$$\hat{S}_i^2 = \Sigma(\bar{\boldsymbol{\beta}}_i-\hat{\boldsymbol{\beta}}_i)^T\mathbf{Z}_i^T\mathbf{Z}_i(\bar{\boldsymbol{\beta}}_i-\hat{\boldsymbol{\beta}}_i),$$

where

$$\hat{\boldsymbol{\beta}}_i = \mathbf{A}_i\hat{\alpha}+\hat{\mathbf{D}}\hat{\mathbf{U}}_i(\bar{\boldsymbol{\beta}}_i-\mathbf{A}_i\hat{\alpha})$$

is the empirical Bayes estimate of $\boldsymbol{\beta}_i$. Iterative EM equations set the left-hand sides of (15.32-15.34) to $\hat{\alpha}^{(p+1)}$, $\hat{\mathbf{D}}^{(p+1)}$, and $\hat{\sigma}^{2(p+1)}$ and evaluate the right-hand sides at the values of the parameters at the pth iteration.

This formulation has the drawback of requiring that each $\mathbf{Z}_i$ has full rank. However, it is easily seen that this is not necessary for the above equations to be valid whenever $\bar{\boldsymbol{\beta}}_i$ satisfies

$$(\mathbf{Z}_i^T\mathbf{Z}_i)\bar{\boldsymbol{\beta}}_i = \mathbf{Z}_i^T\mathbf{y}_i. \tag{15.35}$$

This follows from noting that

$$\mathbf{U}_i = \mathbf{D}^{-1}(\sigma^2\mathbf{D}^{-1}+\mathbf{Z}_i^T\mathbf{Z}_i)^{-1}\mathbf{Z}_i^T\mathbf{Z}_i,$$

so that $\mathbf{U}_i$ exists even when $(\mathbf{Z}_i^T\mathbf{Z}_i)^{-1}$ does not, and thus $\mathbf{U}_i\bar{\boldsymbol{\beta}}_i$ is unique for any $\bar{\boldsymbol{\beta}}_i$ satisfying (15.35). In addition, since $\mathbf{U}_i$ is symmetric, $(\mathbf{Z}_i^T\mathbf{Z}_i)^{-1}\mathbf{U}_i=(\sigma^2\mathbf{D}^{-1}+\mathbf{Z}_i^T\mathbf{Z}_i)^{-1}\mathbf{D}^{-1}$, and

$$\hat{S}_i^2+s_i^2 \equiv (\mathbf{y}_i-\mathbf{Z}_i\hat{\boldsymbol{\beta}}_i)^T(\mathbf{y}_i-\mathbf{Z}_i\hat{\boldsymbol{\beta}}_i)$$

is unique for any $\bar{\boldsymbol{\beta}}_i$ satisfying (15.35), since $\hat{\boldsymbol{\beta}}_i$ is unique.

Note that it is unnecessary to assume $\mathbf{D}$ is arbitrary; in fact, we may prefer to specify a model which assumes a block diagonal subset of the elements is non-zero. The analysis under this model is easily implemented using the analysis for the full model, but zeroing out the appropriate elements of $\mathbf{D}$ at the end of each iteration. In this case, $\mathbf{D}^{-1}$ does not exist, thus we must have $\mathbf{Z}_i$ at full rank for a fully efficient two-stage analysis.

15.4 Methods for Non-Linear Growth Curves

Although linear growth models have sufficient flexibility for most purposes, there are several areas where they fail, most notably in modeling growth over any appreciable range or in the analysis of pharmacokinetic data. This section reviews methods of analysis using non-linear models. Specifically we assume that the data on the ith individual may be written as

$$\mathbf{y}_i = \mathbf{g}_i + \mathbf{e}_i, \tag{15.36}$$

where $\mathbf{g}_i \equiv \mathbf{g}(\mathbf{Z}_i, \boldsymbol{\beta}_i)$ defines some non-linear function of $\boldsymbol{\beta}_i$, with $\mathbf{e}_i \sim N(\mathbf{0}, \sigma^2\mathbf{I})$ and the $\boldsymbol{\beta}_i$'s are $N(\mathbf{A}_i\alpha, \mathbf{D})$. Several new issues immediately arise with non-linear models. First is the interpretability of the parameters, $\boldsymbol{\beta}_i$, and the assumption of multivariate normality for the $\boldsymbol{\beta}_i$'s. Many non-linear models require constrained parameters; hence, transformations on the original parameters may be necessary to make the normality assumption tenable. In contrast to the linear models setting, the full maximum likelihood analysis for estimating α, $\mathbf{D}$, and σ^2 is generally quite unfeasible, since closed form expressions may not exist for the distribution of $\mathbf{y}_i$. It is, however, instructive to see what form the EM equations would take for a full ML (or REML) analysis. To do this, we again let the complete data consist of $(\mathbf{y}_i, \delta_i, \mathbf{e}_i)$, where

$$\delta_i = \boldsymbol{\beta}_i - \mathbf{A}_i\alpha.$$

Implementation of the EM is not straightforward since, because of the non-linearity, sufficient statistics do not in general exist for α, even with complete data. Hence, the M-step must be replaced by maximizing the expected log-likelihood of the complete data, given the observed data. For σ^2 and $\mathbf{D}$ it is easily seen that both the M-step and the E-steps remain the same, requiring:

$$K\hat{\mathbf{D}} = \Sigma E(\delta_i\delta_i^T|\mathbf{y}) \tag{15.37}$$

$$(\Sigma T_i)\hat{\sigma}^2 = \Sigma E(\mathbf{e}_i^T\mathbf{e}_i|\mathbf{y}_i), \tag{15.38}$$

although the δ_i's and $\mathbf{e}_i$'s are no longer multivariate normal, given the $\mathbf{y}_i$'s. For α with complete data, we must minimize $\Sigma(\mathbf{y}_i-\mathbf{g}_i)^T(\mathbf{y}_i-\mathbf{g}_i)$ where we let $\boldsymbol{\beta}_i=\mathbf{A}_i\alpha+\delta_i$; equivalently we set $\Sigma(\mathbf{y}_i-\mathbf{g}_i)^T\partial\mathbf{g}_i/\partial\alpha=\mathbf{0}$. With incomplete data we must minimize

$$\Sigma[\mu_i(\alpha)^T\mu_i(\alpha)+\mathrm{tr}\mathbf{V}_i(\alpha)] \tag{15.39}$$

where

$$\mu_i(\alpha) = \mathbf{y}_i-\mathrm{E}(\mathbf{g}_i|\mathbf{y}_i,\alpha,\sigma^2,\mathbf{D}), \tag{15.40}$$

and $\mathbf{V}_i(\alpha)$ is the conditional variance of $\mathbf{g}_i$ given $\mathbf{y}_i$. In general, both $\mu_i(\alpha)$ and $\mathbf{V}_i(\alpha)$ will also be functions of $\mathbf{D}$ and σ^2, but these will be evaluated at the solution of (15.37-15.38) in computing the solution to (15.39). Closed forms for $\mu(\alpha)$, $\mathbf{V}(\alpha)$, and $\mathrm{E}(\delta_i\delta_i^T|\mathbf{y})$ and $\mathrm{E}(\mathbf{e}_i^T\mathbf{e}_i|\mathbf{y})$ do not usually exist and these expectations must be approximated.

One approximate method draws on work of Laird (1978a), and is also used in Harville and Mee (1984). The idea is to estimate α and the δ_i's from their joint posterior, given $\mathbf{D}$ and σ^2, then estimate σ^2 and $\mathbf{D}$ from their marginal likelihood, integrating out the δ_i's and α. Since in this setting, the posterior of α and the δ's is not readily computed, we approximate the posterior mean by the posterior mode and the posterior variance by the negative inverse of the second derivative matrix of the log-posterior density. Thus, $\hat{\alpha}$ and $\hat{\delta}_i$ are chosen to minimize

$$-2\ell(\alpha,\delta) = \Sigma(\mathbf{y}_i-\mathbf{g}_i)^T(\mathbf{y}_i-\mathbf{g}_i)/\sigma^2+\delta_i^T\mathbf{D}^{-1}\delta_i, \tag{15.41}$$

and their variance is approximated by

$$\left[-\frac{\partial^2\ell(\alpha,\delta)}{[\partial(\alpha,\delta)][\partial(\alpha,\delta)]^T}\right]^{-1}. \tag{15.42}$$

These computations should not be especially difficult, although iterative. We note that this approach to estimating individual growth curves $[\hat{\mathbf{g}}_i=\mathbf{g}_i(\mathbf{Z}_i,\mathbf{A}_i\hat{\alpha}+\hat{\delta}_i)]$ has proven to be very satisfactory (Berkey 1982; Bock and Thissen 1980).

To estimate $\mathbf{D}$ and σ^2 from their marginal posterior, we again use Eqs. (15.37-15.38), but we assume that given $\mathbf{y}_i$, δ_i is approximately normal with a mean equal to $\hat{\delta}_i$ and variance given by the appropriate submatrix of (15.42). Since

$$\mathbf{e}_i = \mathbf{y}_i-\mathbf{g}_i,$$

the posterior mean and variance of $\mathbf{e}_i$ will be approximated by

$$E(\mathbf{e}_i) \doteq \mathbf{y}_i - \mathbf{g}_i(\mathbf{Z}_i, \mathbf{A}_i\alpha + \delta_i)$$

and

$$\mathrm{var}(\mathbf{e}_i) \doteq (\partial \mathbf{g}_i/\partial \delta_i)(\mathrm{var}\ \delta_i)(\partial \mathbf{g}_i/\partial \delta_i)^T,$$

or by

$$E(\mathbf{e}_i) = \mathbf{y}_i - E(\mathbf{g}_i)$$

$$\mathrm{var}(\mathbf{e}_i) = \mathrm{var}(\mathbf{g}_i),$$

if these last two expressions can be calculated, assuming approximate normality for δ_i. If $\mathbf{g}_i$ is a sum of exponentials, this may be feasible. Although this method simplifies calculation of $\hat{\alpha}$ and $\hat{\delta}_i$, it requires iteration at every E-step to compute the posterior moment approximations of δ_i. These approximations will not be very good, unless each T_i is large.

If each T_i is large, a simpler method, used by Berkey and Laird (1986), and in a Bayes analysis by Racine-Poon (1985), works well for estimating α and $\mathbf{D}$. This builds on the two-step approach for linear models. First estimate $\boldsymbol{\beta}_i$ and its approximate variance-covariance matrix $\mathbf{C}_i$ separately for each person using non-linear least-squares. Then assuming each $\bar{\boldsymbol{\beta}}_i$ is approximately $N(\boldsymbol{\beta}_i, \mathbf{C}_i)$ and $\boldsymbol{\beta}_i$ is $N(\mathbf{A}_i\alpha, \mathbf{D})$, ML or REML estimation of α and $\mathbf{D}$ is straightforward. As Berkey and Laird (1986) show, the ML equations for α and $\mathbf{D}$ satisfy

$$\hat{\alpha} = \Sigma(\mathbf{A}_i^T\hat{\mathbf{Q}}_i\mathbf{A}_i)\Sigma\mathbf{A}_i^T\hat{\mathbf{Q}}_i\bar{\boldsymbol{\beta}}_i$$

$$K\hat{\mathbf{D}} = \Sigma\{\hat{\mathbf{D}}\hat{\mathbf{Q}}_i(\bar{\boldsymbol{\beta}}_i - \mathbf{A}_i\hat{\alpha})(\bar{\boldsymbol{\beta}}_i - \mathbf{A}_i\hat{\alpha})^T\hat{\mathbf{Q}}_i\hat{\mathbf{D}} + \hat{\mathbf{D}}[\mathbf{I} - \hat{\mathbf{Q}}_i\hat{\mathbf{D}}]\},$$

where

$$\hat{\mathbf{Q}}_i = (\hat{\mathbf{D}} + \mathbf{C}_i)^{-1}$$

and $\bar{\boldsymbol{\beta}}_i$ is the least-squares estimate of $\boldsymbol{\beta}_i$. The expression for $\hat{\mathbf{D}}$ is easily modified to give the REML estimate:

$$K\tilde{\mathbf{D}} = \Sigma\{\tilde{\mathbf{D}}\tilde{\mathbf{Q}}_i(\bar{\boldsymbol{\beta}}_i - \mathbf{A}_i\tilde{\alpha})(\bar{\boldsymbol{\beta}}_i - \mathbf{A}_i\tilde{\alpha})^T\tilde{\mathbf{Q}}_i\tilde{\mathbf{D}} + \tilde{\mathbf{D}}[\mathbf{I} - \tilde{\mathbf{P}}_i\tilde{\mathbf{D}}]\},$$

where

$$\tilde{P}_i = \tilde{Q}_i[I - A_i(\Sigma A_i^T \tilde{Q}_i A_i)^{-1} A_i^T \tilde{Q}_i] \; .$$

Racine-Poon (1985) gives similar equations for estimating α and $\mathbf{D}$ using a Bayes approach outlined by Fearn (1975). The ML approach used by Berkey and Laird (1986) was found to give good results for estimating population parameters; however, the EB estimates of the growth curves:

$$\tilde{g}_i = g_i(\mathbf{Z}_i, \mathbf{A}_i \tilde{\alpha}_i + \tilde{\delta}_i)$$

for

$$\tilde{\delta}_i = \tilde{\mathbf{D}}\tilde{\mathbf{Q}}_i(\bar{\boldsymbol{\beta}}_i - \mathbf{A}_i \tilde{\alpha})$$

were found to be poor for estimating the individual growth curves. By contrast, the posterior modes obtained by minimizing (15.41) gave good results in this setting.

An alternative approach to this problem is given by Mallet (1986), who extends the work of Laird (1978b) and Lindsay (1983) on non-parametric estimation. The general approach, for the special case of no covariates ($\mathbf{A}_i = \mathbf{I}$), is to estimate the distribution of the $\boldsymbol{\beta}_i$'s non-parametrically, by maximizing the marginal likelihood:

$$L(G, \sigma^2) = \Pi \int_{R^q} \sigma^{-1} e^{-(y_i - g_i)^T (y_i - g_i)/2\sigma^2} dG(\boldsymbol{\beta}_i)$$

as function of σ^2 and G, where G is any distribution in R^q. Empirical Bayes estimates of the individual $\boldsymbol{\beta}_i$'s can be obtained using the approach outlined in Laird (1982b). Although the computations for this approach are relatively straightforward, there is no obvious way to introduce covariates into the model.

Acknowledgements. Dr. Laird was supported by grant GM29745 from the National Institutes of Health.

References

Berkey CS (1982) Bayesian approach for a non-linear growth model. Biometrics 38:953-961

Berkey CS, Laird NM (1986) Non-linear growth curve analyses: estimating the population parameters. Ann Hum Biol 13:111-128

Bock RD, Thissen D (1980) Statistical problems of fitting individual growth curves. In: Johnstone FE, Roche AF, Susanne C (eds) Human physical growth and maturation. Plenum Press, New York

Credeur KR (1980) Estimation from incomplete multinomial data. PhD Thesis, Harvard Univ

Dempster AP, Laird NM, Rubin DB (1977) Maximum likelihood from incomplete data via the EM algorithm. J R Stat Soc Ser B 39:1-38

Fearn T (1975) A Bayesian approach to growth curves. Biometrika 62:89-100

Harville DA (1976) Extension of the Gauss-Markov theorem to include the estimation of random effects. Ann Stat 4:384-395

Harville DA (1977) Maximum likelihood approaches to variance component estimation and to related problems. J Am Stat Assoc 72:320-340

Harville DA and Mee RW (1984) A mixed-model procedure for analyzing ordered categorical data. Biometrics 40:393-408

Kackar RN, Harville DA (1984) Approximations for standard errors of estimates of fixed and random effects in mixed linear models. J Am Stat Assoc 79:853-862

Laird NM (1978a) Bayesian estimation methods for two-way contingency tables. Biometrika 65:581-590

Laird NM (1978b) Non-parametric maximum-likelihood estimation of a mixing distribution. J Am Stat Assoc 73:805-813

Laird NM (1982a) Computation of variance components using the E-M algorithm. J Stat Comput Simul 14:295-303

Laird NM (1982b) Empirical Bayes estimates using the non-parametric maximum likelihood estimate of the prior. J Stat Comput Simul 1:211-220

Laird NM, Louis TA (1986) Empirical Bayes confidence intervals based on bootstrap samples. Tech Rep, Dep Biostatistics, Harvard School of Public Health, Boston

Lange N, Laird NM (1986) Random-effects and growth curve modeling for balanced and complete longitudinal data. Tech Rep, Dep Biostatistics, Harvard School of Public Health, Boston

Lindley DV, Smith AFM (1972) Bayes estimates for the linear model. J R Stat Soc Ser B 34:1-41

Lindsay BG (1983) The geometry of mixture likelihoods: a general theory. Ann Stat 11:86-94

Mallet A (1986) A maximum likelihood estimation method for random coefficient regression models. Biometrika 73:645-656

Morris CN (1983a) Parametric empirical Bayes inference: theory and applications. J Am Stat Assoc 78:47-59

Morris CN (1983b) Parametric empirical bayes confidence intervals. In: Box GEP, Leonard T, Wu CF (eds) Scientific inference, data analysis, and robustness. Academic Press, New York pp 25-50

O'Hagan A (1976) On posterior joint and marginal models. Biometrika 63:329-333

Patterson HD, Thompson R (1971) Recovery of interblock information when block sizes are unequal. Biometrika 58:545-554

Racine-Poon A (1985) A Bayesian approach to non-linear random effects models. Biometrics 41:1015-1024

Rao CR (1975) Simultaneous estimation of parameters in different linear models and applications to biometric problems. Biometrics 31:545-554

16 Survival, Endurance and Censored Observations in Animal Breeding

S.P. Smith[1]

Longevity and other survival measures are important traits and these are usually subjected to censoring. Methods in failure time analysis, useful for the censored data encountered in animal breeding, are described. The value of the Cox model and of other rank regression models is highlighted. These methods incorporate an unknown transformation in the model, solving the scale problems frequently confronted in genetic analysis. Bayesian approaches allow models to accommodate random factors such as additive genetic values. Effects are estimated by solving, perhaps several times, a linear system of equations that resembles Henderson's mixed model equations. Proposed numerical techniques are tested on an actual data set.

16.1 Introduction

In dairy cattle, productive life span is subjected to censoring as are most survival measures. Censoring occurs because of time constraints and it complicates the statistical analysis. Everett et al. (1976) accommodated censoring by performing an analysis conditional on the opportunity to survive. This was a linear analysis of zero-one survival codes and, consequently, improvements can be made (De Lorenzo and Everett 1986).

Productive life span is regarded as an economically important trait in dairy cows (Burnside et al. 1984). Thus, there is justification in attempting to improve statistical methods, e.g., techniques for evaluating sires on daughter survival. Failure time analysis is appropriate for censored data and, therefore, it should be considered in any survival analysis.

It has been suggested that failure time analysis has application in animal breeding (Famula 1981; Wolynetz and Binns 1983; Smith and Allaire 1986). Recently, Smith and Quaas (1984) conducted a failure time analysis on dairy cow survival. This study involved 587 013 animals and it demonstrated that a large-scale failure time analysis in animal breeding is feasible.

---

1 Animal Genetics and Breeding Unit, University of New England, Armidale, Australia

The purpose of this paper is to present refinements to the statistical theory that encompasses the genetic analysis of survival and endurance. A brief discussion pertaining to the characterization of survival times and endurance measures follows, and examples are given of these measures in animal breeding. Methodology appropriate for analysis is described in a framework consistent with quantitative genetics. Numerical considerations are emphasized, and application of these methods to a real data set is illustrated.

16.2 Characterization of Survival Times and Endurance Measures

16.2.1 Properties

Survival times generally possess the following properties:

1. The measures are usually related to time. Like time, the units of measure generally accumulate and are typically positive. The survival period is initiated with a random or planned event, e.g., birth, first calving, start of a clinical trial. The end of the survival period is defined when an event such as failure occurs.
2. Because survival measures are so closely related to time, censoring is prevalent. Censoring occurs when any particular survival time is only known to be above or below some value. For example, any observation on an item that has not failed at the end of an experiment is censored.

Censoring is perhaps the most important feature of survival measures. Indeed, failure time analysis is generally synonymous with the analysis of censored data. Property 1 above is also important as it allows the risk of failure to be related to time or to the length of time at risk. For example, the stochastic behavior of the survival time may be determined in part by aging. Direct influences on survival may be time dependent effects that can be modelled accordingly (e.g., Kalbfleisch and Prentice 1980).

Examples of survival traits in animal breeding are: longevity, productive life span, survival within first lactation, survival from birth to breeding, prenatal and postnatal survival. These traits may be measured on continuous or discrete scales. For example, Everett et al. (1976) used a discrete measure of dairy cow survival termed stayability while Smith and Quaas (1984) worked with the continuous trait, days of productive life.

Not all measurements subject to censoring are survival measures. For example, Iniguez et al. (1984) considered the number of mating attempts until conception in ewes, and this trait is certainly subject to censoring. Methods designed for survival analysis can be applied to these cases as well.

Many endurance measures are neither survival measures nor subject to censoring. For these cases, ordinary methods designed for uncensored data are usually adequate for analysis. Nevertheless, methods described herein can also find application with endurance measures like racing time in horses.

16.2.2 Censoring

There are several possible censoring mechanisms (Lagakos 1979). However, we will only consider independent censoring, and this is frequently assumed for field data.

Observations consist of (t_i,d_i), $i=1,2,...,n$, where

$$(t_i,d_i) = \begin{cases} (z_i, 1) & \text{if } z_i \leq c_i \\ (c_i, 0) & \text{if } c_i < z_i \end{cases}$$

such that z_i is an underlying failure time, and c_i is a random interruption time. We see that when the i-th observation (t_i) is censored $d_i=0$, but $d_i=1$ when t_i is uncensored. Under independent censoring, z_i and c_i are statistically independent.

Common causes of censoring found with productive life span measures in dairy cows are: herd goes off test; cow sold for breeding or dairy purposes; cow still alive; and accidental death or sold due to injury. The assumption of independent censoring seems reasonable as the above causes appear unrelated to failure caused, for example, by functional "breakdown".

16.3 Models

16.3.1 Parametric Models

To specify a parametric model, we associate the stochastic behavior of the underlying failure times (z_i) with a suitable density function. This function depends on unknown parameters (**b**) that can be regarded as effects in the model. These effects can be estimated by maximum likelihood using an appropriate algorithm.

Under conditions of independent censoring and when observations are independent the likelihood is proportional to

$$L(\mathbf{b}) = \prod_{i=1}^{n} f_i(t_i)^{d_i} S_i(t_i)^{1-d_i} , \tag{16.1}$$

where f_i and S_i are the probability density function and the survival function for the i-th observation, respectively. The survival function is defined as $S_i(t_i)=Pr[z_i>t_i]$, where Pr[A] is the probability of event A.

The linear model has been used extensively in animal breeding. With normal residuals the least-squares equations can be derived from the maximum likelihood approach. We can generalize the linear model for censored data using (16.1). For example, we might take z_i as a log-normal random variable, as was suggested by Wolynetz and Binns (1983). With censoring, the equations that result from maximizing (16.1) will be non-linear.

We may wish to model the risk of failure as a function of time. For this situation we describe a model that depicts the so-called hazard function, denoted by g(t). Simple examples of such models can be found in Gross and Clark (1975), e.g., the Weibull $\{g(t)=at^b;a>0,b>-1\}$ and Rayleigh $\{g(t)=a+bt;a>0,b>0\}$ distributions. Other models, including examples that mimic regression analysis, can be found in Elandt-Johnson and Johnson (1980).

Mathematically, the hazard function is defined as the ratio f(t)/S(t). Thus, g(t) is the conditional density function for failure at t, given non-failure to t. The g(t) can simply be thought of as a failure rate at t, where g(t) can assume any positive value.

Despite the fact that g(t) is defined in terms of f(t) and S(t), given g(t) we have

$$f(t) = g(t)\exp\{-\int_0^t g(u)du\}$$

$$S(t) = \exp\{-\int_0^t g(u)du\} .$$

That is, f(t) and S(t) are defined uniquely in terms of g(t). Consequently, (16.1) can be written as

$$L(\mathbf{b}) = \prod_{i=1}^{n} g_i(t)^{d_i}\exp\{-G_i(t)\}, \tag{16.2}$$

where $g_i(t)$ is the hazard function for the i-th observation, and $G_i(t)$ is called the cumulative hazard which equals

$$\int_0^t g_i(u)du.$$

When models are used to describe g(t), it is convenient to work with (16.2) rather than (16.1) for maximum likelihood estimation.

16.3.2 Semi-Parametric Models

To define a semi-parametric model, we represent the random behavior of the failure times by an appropriate class of density functions. As with parametric models, any density function within the class depends on the unknown parameters, **b**. The parameters can be estimated by maximizing a marginal likelihood or some approximate likelihood.

We will not describe non-parametric or semi-parametric models appropriate for categorical observations like stayability. Techniques designed for categorical data (e.g., Gianola and Foulley 1983; Harville and Mee 1984) can sometimes be used to analyze discrete failure times. DeLorenzo and Everett (1986) used a logistic model to analyze dichotomous survival measures in dairy cows. Kalbfleisch and Prentice (1980) presented a discrete proportional hazards model that resembles the model described next: the Cox regression model.

The Cox Regression Model. Cox's (1972) regression model is specified by representing the hazard function as

$$g_i(t) = g_o(t)\exp\{\mathbf{x}_i'\mathbf{b}\}, \tag{16.3}$$

where $g_o(t)$ is an unknown positive function, $\mathbf{x}_i$ is a column vector of known covariates or indicator variables for the i-th observation and **b** is the parameter vector. The function $g_o(t)$ is called the baseline hazard function and in (16.3) it is not a function of i. Because $g_i(t)$ equals $g_o(t)$ times the factor $\exp\{\mathbf{x}_i'\mathbf{b}\}$, (16.3) is called a proportional hazards model.

Model (16.3) is popular because **b** can be estimated without ever knowing or estimating $g_o(t)$. This is due to the nature of the information that is used in estimation, i.e., the marginal information. The likelihood constructed from the marginal information is invariant to $g_o(t)$ (Kalbfleisch and Prentice 1973) and **b** can be estimated by maximizing this likelihood, the marginal likelihood.

When there is no censoring, the marginal information is equivalent to the rank information. For example, if we observe the uncensored records t_1=3, t_2=7, and t_3=8, the marginal information is $z_1<z_2<z_3$. With censoring, the marginal information corresponds to what is known about the rank information of underlying failure times. If t_1 is censored, the

marginal information is $z_2<z_3$, that is, we lose the first observation because the underlying failure time, z_1, may have any rank order. Alternatively, if t_3 is censored the marginal information is still $z_1<z_2<z_3$. If t_2 is censored the marginal information is $z_1<z_2$ and $z_1<z_3$.

In a continuous model, failure times will be distinct. However, because measurements are not perfect, failure times will occasionally be tied and our knowledge about the correct rank order will be incomplete. Marginal information can also be used to represent ties. If we observe the uncensored records $t_1=3$, $t_2=7$, and $t_3=7$, the marginal information is the union of $z_1<z_2<z_3$ and $z_1<z_3<z_2$. This information can be incorporated into the likelihood. However, ties will cause numerical calculation to be cumbersome. The interested reader is referred to Kalbfleisch and Prentice (1980).

In the most general setting the marginal likelihood is Pr[M] where M represents the marginal information. Kalbfleisch and Prentice (1980) justify the use of the marginal likelihood under conditions of independent censoring. When there are no ties and when observations are independent, Pr[M] calculated under (16.3) is

$$L(\mathbf{b}) = \prod_{j=1}^{m} \exp\{\mathbf{x}_j'\mathbf{b}\}/[\sum_{i\varepsilon R(t_j)} \exp\{\mathbf{x}_i'\mathbf{b}\}], \tag{16.4}$$

where $t_1,t_2,\ldots,t_m$ ($m\leq n$) are assumed to be the uncensored observations, and $R(t_j)$ is the index set of all subjects at risk prior to t_j. An observation is at risk prior to t_j if it survived beyond t_j or failed at t_j.

Because the marginal likelihood is difficult to evaluate when there are ties, Peto (1972) suggests the following approximation

$$\prod_{i=1}^{r} \exp\{\mathbf{w}_i'\mathbf{b}\}(\sum_{j\varepsilon R(v_i)} \exp\{\mathbf{x}_j'\mathbf{b}\})^{-D_i} \tag{16.5}$$

where $v_1,v_2,\ldots,v_r$ are distinct uncensored failure times, D_i is the number of failures at v_i and

$$\mathbf{w}_i = \sum_{j\varepsilon F(v_i)} \mathbf{x}_j,$$

where $F(v_i)$ is the index set of all subjects that failed at v_i. With no censoring (16.5) is identical to (16.4).

An important generalization of (16.3) is the incorporation of strata (Kalbfleisch and Prentice 1980). If observation i is in stratum s(i), then the hazard function is

$$g_i(t) = g_{s(i)}(t)\exp\{\mathbf{x}_i'\mathbf{b}\}, \tag{16.6}$$

where $g_{s(i)}(t)$ is an unknown positive function defined for stratum s(i). Model (16.6) allows stratum effects to operate on the hazard function, as non-multiplicative influences. The likelihood appropriate for analysis is

$$L(\mathbf{b}) = \prod_k \Pr[M_k], \tag{16.7}$$

where M_k is the marginal information from stratum k and, $\Pr[M_k]$ can be evaluated using (16.4), or approximated by (16.5) when there are ties.

Model (16.6) has been used in a dairy cattle study with strata representing years of birth (Smith and Quaas 1984). Contemporary groups are frequently defined in animal breeding and these can also be taken as strata in model (16.6). Such a model would be very flexible with respect to the variety of ways contemporary groups could affect survival. For example, a model that accommodates strata may be necessary as the criterion for failure (e.g., culling criterion) can be different for each group.

Rank Regression Models. Rank regression models are represented by

$$h(t_i) = \mathbf{x}_i'\mathbf{b}+e_i, \tag{16.8}$$

where $h(\cdot)$ is a monotone increasing function that is assumed unknown, and e_i is an independent residual. It is possible to estimate **b** without ever knowing or estimating $h(\cdot)$. This is done by maximizing a likelihood based on rank information. This likelihood is invariant to $h(\cdot)$ because monotone transformations do not alter rank order.

To construct the likelihood, the residual in (16.8) is assumed to have a particular probability density function. Several possible distributions are given in Pettitt (1983), and Smith and Hammond (1988) describe a model where the residual is a log-gamma variable that approaches normality. Clayton and Cuzick (1985) describe methods appropriate for the Pareto family of distributions, of which the logistic distribution is an example. Once a distribution is specified the likelihood based on rank information is

$$L(\mathbf{b}) = \Pr[R], \tag{16.9}$$

where R is the rank information and **b** is taken as fixed. When some of the observations are censored we may use the marginal likelihood, Pr[M], which is evaluated assuming model (16.8). This likelihood is also invariant to h(·).

The Cox regression model is in fact a rank regression model where the residual is log-exponential. The log-exponential distribution is a non-symmetric unimodal distribution that is continuous on the real line. In the Cox regression model:

$$h(t) = \log[G_0(t)],$$

where $G_0(t)=\int_0^t g_0(u)du$. To make the correspondence complete, **b** in (16.8) must be equated to -**b** in (16.3). Assuming $g_0(\cdot)$ in (16.3) is arbitrary is tantamount to assuming h(·) in (16.8) is arbitrary.

With the exception of the log-exponential and log-gamma distributions, evaluating (16.9), let alone maximizing it, can be very difficult if not impossible. Nevertheless, Pettitt (1982, 1983) suggests that the log of (16.9) can be approximated adequately by a Taylor series expansion around **b**=**0**. This approximation reduces the computational demand considerably and results in a system of linear equations that must be solved to obtain estimates.

The use of ranks might be criticized due to the loss of information accruing when changing continuous measures into ranks. This loss is the price paid for robustness. However, it is surprising how little information is actually lost. Less information is lost as more observations are used, and for a particular case involving the exponential distribution, the relative efficiency grows at a fast rate, e.g., 82, 94, and 99% for 5, 20, and 100 observations, respectively (Kalbfleisch and Prentice 1980).

The ramifications of rank regression methodology in quantitative genetics are interesting. Mather (1949) and Wright (1952) both stressed the importance of an adequate scale of measure when studying continuous variation. Ideally, a scale or transformation is needed so as to minimize or remove epistasis and genotype by environment interactions. Whereas these interactions may exist on the original scale we can rarely accommodate them in a linear analysis of untransformed data. The argument that interactions should best be left as interactions (e.g., Falconer 1981) can be impractical. Rank regression methodology offers an appealing solution to problems of scale. While the need to transform or not to transform must be determined empirically (Mather 1949), we need not labor over the choice of transformation because this is assumed unknown with rank regression techniques. Furthermore, an analysis of untransformed data generally involves more assumptions.

A common view is that transformation complicates the interpretation of effects in the model. However, because $h(\cdot)$ is a monotone increasing function, large effects on the underlying scale translate into large effects on the observed scale. This relationship can be formalized using invariance properties that involve medians and probabilities (Pettitt 1983).

In terms of additive genetic effects we have a stronger relationship. Let A and A_h represent correct additive genetic effects for the observed and the transformed scale, respectively. Goddard (1985) suggested that the correlation between the ideal measures, A and A_h, is essentially one. This result is only approximate and it holds either when the additive contribution from individual genes are similar in magnitude (Goddard 1986, personal communication), or when there are a large number of additive and unlinked loci each contributing a small effect. Consequently, when we invoke normality as is frequently done when using linear mixed model methodology (Henderson 1973), we must admit that the correlation between A and A_h is one. Monotone transformations do not corrupt our interpretation of the additive genetic scale.

Like the Cox model, rank regression techniques can be generalized to accommodate strata (Pettit 1984). The following is a stratified rank regression model

$$h_{s(i)}(t_i) = \mathbf{x}_i'\mathbf{b}+e_i, \tag{16.10}$$

where observation i belongs to stratum s(i), and $h_{s(i)}(\cdot)$ is an unknown transformation defined for stratum s(i). To estimate **b** we maximize

$$L(\mathbf{b}) = \prod_k \Pr[R_k], \tag{16.11}$$

where R_k is the rank information from stratum k. If some of the observations are censored, **b** can be estimated by maximizing (16.7) where $\Pr[M_k]$ is computed under (16.10). To estimate **b**, the approximation of Pettitt (1983) can also be extended to accommodate strata.

The information lost by changing continuous measures into within-stratum rankings is greater than when an overall ranking is made. The relative efficiency is a function of the number of observations per stratum. For the case of the exponential distribution described earlier, there is still 82% efficiency when there are only five observations per stratum.

Stratification would be required in field data where there is little control over environmental influences. Different unknown transformations may be required for each contemporary group, particularly if heteroscedasticity is a problem. Model (16.10) can be used to avoid the problems recognized by Everett et al. (1982) and Brotherstone and Hill

(1986) regarding within-herd heterogeneity of residual variances. The model (16.10) could also be used, for example, to perform a genetic analysis of racing time in horses where an unknown transformation is defined for each horse race.

16.4 Maximum a Posteriori

Until now we have treated **b** as fixed parameters. In this section methodology is described that allows some of the elements in **b** to be random. This generalization will allow techniques described herein to compete with Henderson's mixed linear model methods. For example, we would like to estimate additive genetic effects, and these are random.

Consider the case where **b** is a vector of location parameters, i.e., ignore shape parameters, variance components, and other parameters that are not analogous to effects in a linear model. Let $\mathbf{b}_f$ and $\mathbf{b}_r$ denote the fixed and random effects in **b**, respectively. To complete model specification, prior distributions are assigned to $\mathbf{b}_f$ and $\mathbf{b}_r$ in accordance to the Bayesian approach (e.g., Gianola and Fernando 1986). The prior given to $\mathbf{b}_r$, e.g. $n(\mathbf{b}_r)$, is assumed to be multivariate normal with mean null and known covariance matrix **G**. The prior given to $\mathbf{b}_f$, e.g. $m(\mathbf{b}_f)$, is assumed to assign equal likelihood to all possible realizations of $\mathbf{b}_f$, i.e., $m(\mathbf{b}_f)=1$. Fixed and random effects are assumed to be independent and, hence, the prior for **b**, e.g. $p(\mathbf{b})$, is

$$p(\mathbf{b}) \propto n(\mathbf{b}_r). \tag{16.12}$$

The vector **b** contains only location effects for the semi-parametric models described in this paper. However, other parameters may be needed for the parametric models, and some prior other than (16.12) may be appropriate. For example, a non-informative prior for a variance component might be used and this prior need not be flat (Berger 1980). It may also be desired to assign priors to the parameters that determine **G** (Gianola and Fernando 1986). This issue will not be considered here.

The posterior distribution of **b**, e.g. $c(\mathbf{b})$, given data, rank information, or marginal information, is proportional to the joint density

$$L(\mathbf{b})\cdot p(\mathbf{b}), \tag{16.13}$$

where $L(\mathbf{b})$ is one of the likelihoods described earlier. To predict **b**, the maximum *a posteriori* estimate (Melsa and Cohn 1978) is proposed. This estimate is found by maximizing $c(\mathbf{b})$ or equivalently (16.13). The maximum *a posteriori* approach has been proposed and justified for various non-linear models found in animal breeding (Gianola and Foulley 1983; Harville and Mee 1984; Smith and Allaire 1986). However, these authors have used different terminology.

Random effects estimated by maximizing (16.13) share similarities with random effects estimated by solving Henderson's mixed model equations; the effects are regressed back to the mean of the prior distribution particularly when they are associated with little information. Indeed, the equations that result from maximizing (16.13) with respect to **b** generally resemble the mixed model equations. Similar to solutions to Henderson's equations, the value of **b** that maximizes (16.13) depends on **G** and hence, on unknown variance components. Methods for estimating **G** will not be described in this paper. However, useful suggestions can be found in Harville and Mee (1984), Foulley et al. (1983), and Smith, Chapter 10, this Volume.

16.5 Numerical Methods

Let us consider the case where **b** is a vector of location parameters such that L(**b**) is a function of **b** only through

$$\mathbf{u} = \mathbf{X}\mathbf{b}_f + \mathbf{Z}\mathbf{b}_r \,.$$

In most situations elements in **u** correspond to observations. This correspondence is not a technical requirement and indeed it may even be false for some non-linear models.

To estimate **b** by maximizing (16.13) we may iterate on

$$\begin{bmatrix} \mathbf{X}'\mathbf{W}^k\mathbf{X} & \mathbf{X}'\mathbf{W}^k\mathbf{Z} \\ \mathbf{Z}'\mathbf{W}^k\mathbf{X} & \mathbf{Z}'\mathbf{W}^k\mathbf{Z}+\mathbf{G}^{-1} \end{bmatrix} \begin{bmatrix} \mathbf{b}_f^{k+1} \\ \mathbf{b}_r^{k+1} \end{bmatrix} = \begin{bmatrix} \mathbf{X}'(\mathbf{W}^k\mathbf{u}^k+\mathbf{q}^k) \\ \mathbf{Z}'(\mathbf{W}^k\mathbf{u}^k+\mathbf{q}^k) \end{bmatrix}, \tag{16.14}$$

where

$$\mathbf{u}^k = \mathbf{X}\mathbf{b}_f^k + \mathbf{Z}\mathbf{b}_r^k,$$

$$\mathbf{q}^k = \partial\log(L)/\partial\mathbf{u}|_{\mathbf{u}=\mathbf{u}^k}$$

and

$$\mathbf{W}^k = -\partial^2\log(L)/(\partial\mathbf{u}\partial\mathbf{u}')|_{\mathbf{u}=\mathbf{u}^k}.$$

The matrix $\mathbf{W}^k$ is assumed to be non-negative definite as is usually the case. Scheme (16.14) was derived from Newton's method.

To estimate $\mathbf{b}$ by maximizing L we may use (16.14) after deleting $\mathbf{G}^{-1}$. However, maximizing (16.13) is more appropriate when the prior is reasonable.

The right side of (16.14) can be written as

$$\begin{bmatrix} \mathbf{X}' & \mathbf{W}^k & \mathbf{y}^k \\ \mathbf{Z}' & \mathbf{W}^k & \mathbf{y}^k \end{bmatrix}, \qquad (16.15)$$

where $\mathbf{y}^k=\mathbf{u}^k+(\mathbf{W}^k)^{-1}\mathbf{q}^k$. With this modification,the system in (16.14) is directly analogous to Henderson's mixed model equations and this correspondence has been pointed out by some authors working with various non-linear models (e.g., Foulley and Gianola 1984). Calculating the "working" variate, $\mathbf{y}^k$, may be prohibitive because a matrix inverse is involved. For rank regression models the inverse matrix does not even exist. Luckily, (16.15) is redundant given the right side of (16.14).

At any iterate, solving (16.14) may require a procedure like the Gauss-Seidel method (Stoer and Bulirsch 1980), and this will generally be the case when $\mathbf{b}$ is large. Fortunately, Newton's method typically converges fast, if it converges at all, and, consequently, system (16.14) may need to be solved only a few times. Provided $\mathbf{W}^k$ and $\mathbf{q}^k$ can be easily evaluated, fitting a non-linear model is only J times "harder" than fitting a linear model, where J is the number of Newton's steps. We will see that J can be as small as one for some cases.

Formulae for $\mathbf{W}^k$ and $\mathbf{q}^k$ depend on the model and can be derived by application of calculus. For likelihood (16.5) the i-th element of $\mathbf{q}^k$ is

$$d_i-\exp\{u_i\}\sum_{j=1}^{r} I(i,j)D_j/[\sum_{k\varepsilon R(v_j)} \exp\{u_k\}],$$

$$I(i,j) = \begin{cases} 1 & \text{if } i \ \varepsilon\ R(v_j) \\ 0 & \text{otherwise} \end{cases},$$

It is understood that the i-th element of $\mathbf{q}^k$ corresponds to the i-th observation and i-th row of $\mathbf{X}$ or $\mathbf{Z}$. The ij-th element of $\mathbf{W}_k$ is

$$H(i,j)-\exp\{u_i+u_j\}\sum_{a=1}^{r} I(i^*,a)D_a/[\sum_{k\varepsilon R(V_a)} \exp\{u_k\}]^2,$$

where $i^*=i$ if $t_i \leq t_j$ and $i^*=j$ otherwise, $H(i,j)=0$ if $i \neq j$, and $H(i,i)$ equals

$$\exp\{u_i\} \sum_{a=1}^{r} I(i,a)D_a / [\sum_{k \varepsilon R(v_a)} \exp\{u_k\}].$$

For the Cox model, evaluating $\mathbf{W}^k$ and $\mathbf{q}^k$ is relatively easy. For some rank regression models evaluating $\mathbf{W}^k$ and $\mathbf{q}^k$ can be very difficult and this is the case for the log-gamma model described by Smith and Hammond (1988). Pettitt's (1982, 1983) approximation is equivalent to evaluating $\mathbf{u}^k$, $\mathbf{W}^k$, and $\mathbf{q}^k$ at $\mathbf{b}=\mathbf{0}$. The quantities can be used in (16.14) and estimates are obtained by solving (16.14) once. Pettitt's procedure can be used where other likelihood-based methods are currently intractable. In some cases it may be better to abandon scheme (16.14) in preference to an approximate (Clayton and Cuzick 1985) or an exact (Smith and Hammond 1988) EM algorithm.

Even if $\mathbf{W}^k$ can be evaluated or approximated we are left with a serious problem: for an analysis of ranks $\mathbf{W}^k$ will be non-sparse, and consequently, the coefficient matrix in (16.14) will not be sparse. When $\mathbf{b}$ is large, Newton's method may not be practical, as noted by Smith and Allaire (1986) for the Cox model. They give an alternative to Newton's method that can be used with large Cox models.

Fortunately, the above problem does not arise in some stratified Cox or rank models. If strata are defined by contemporary groups and if each group has only a few observations, then $\mathbf{W}_k$ will be block diagonal and sparse. The coefficient matrix of (16.14) will be sparse, and the number of non-zero elements will equal the number of non-zero elements in an analogous matrix constructed for a linear model (diagonal residual variance structure) where effects representing contemporary groups are absorbed out. Stratified Cox or rank models should be preferred in the analysis of field data. Hence, the problem posed by Newton's method does not exist for many animal breeding problems.

16.6 A Preliminary Investigation

For categorical data, comparisons between best linear unbiased prediction (BLUP) and non-linear prediction have sometimes shown no difference between the two approaches (e.g., Meijering 1985). It is not technically correct to call the linear approach BLUP in these cases because necessary assumptions are hard to justify with categorical data. Consequently, it is understood that BLUP refers to a "quasi-method" (Hoschele 1986).

The fact that BLUP can sometimes compete with non-linear prediction when a non-linear model is appropriate suggests more than the power of BLUP. It also implies that a linear model sometimes approximates well a non-linear model and, consequently, a non-linear analysis can be as simple as a linear analysis; for example, one could iterate on (16.14) only once with user-defined starting values. However, it is important to differentiate between applying BLUP techniques and finding a linear approximation to a non-linear model. In particular, finding a suitable linear approximation implies that an effort is made to accommodate assumptions of the non-linear model. For example, using system (16.14) involves not just evaluating the weight matrix $\mathbf{W}^k$ but also constructing the variates $\mathbf{y}^k$. Evaluating $\mathbf{y}^k$ can be viewed as transforming data and this may be ignored in a BLUP analysis. When BLUP techniques are used, it is advisable to search for a suitable linear model that approximates a non-linear model, although this may not always make a difference.

I do not recommend the blind use of (16.14) as a method to linearize a non-linear model. The practitioner must first define a sensible model. Failure to do so can cause problems. For example, treating binomial probabilities as location parameters will cause the weight matrix, $\mathbf{W}^k$ (or its expectation), to emphasize information that should not be emphasized (Meijering and Gianola 1985). The problem is circumvented when the logistic function is used to describe the binomial probabilities.

The approximation of Pettitt (1983) is a linear model that approximates the rank regression model. The method involves solving (16.14) once with necessary quantities evaluated at $\mathbf{b}=\mathbf{0}$. This technique was investigated for a Cox regression model using field records. The edited data consisted of 12 705 survival records of Holstein dairy cows born after May 1976 on dairy farms in New South Wales, Australia. The analysis included only records from cows that were on test during their first lactation. Only 26% of the records where found to have "failed" (death or sold for beef). This figure is low and probably constitutes a gross underestimate, as a result of inefficiencies in the recording collection system.

The hazard function for the i-th observation was assumed to be

$$g_{s(i)}(t)\exp\{b_o a_i + b_1 a_i^2 + s_{j(i)}\},$$

where s(i) is the herd-year group to which observation i belongs, $g_{s(i)}(t)$ is an arbitrary function defined for herd-year s(i), a_i is age in years at first calving for observation i, and $s_{j(i)}$ is the j(i)-th sire effect. Herd-year groups were defined by herd and year of first calving. There were 1275 herd-year groups and 395 random sires. Sires were assumed to be random and unrelated, and the sire variance was calculated from a heritability of 0.10. Survival in daughters was the length of time in days between first freshening and failure.

Pettitt's approximation was used, and system (16.14) was iterated seven times to obtain the maximum *a posteriori* estimates. The correlation between 395 sire effects computed the two different ways was 0.99990. The rank correlation was 0.99995. These results imply that the short-cut procedure of Pettitt (1983) can be used to obtain satisfactory estimates in Cox models. If this is true a survival analysis need not require any more work than a single linear analysis. Pettitt's approximation needs to be investigated for other data sets.

16.7 Conclusion

When researchers insist on conducting a linear analysis they may iterate on (16.14) once. However, software should allow for multiple rounds of iteration because this option usually involves little effort in coding. Researchers can then iterate several times if the need arises, in which case the non-linear analysis can be computed.

Whereas emphasis in this paper was on the analysis of censored data, rank regression methods show significant promise as a tool in genetic analysis of uncensored data. Rank regression techniques solve the problem of adequacy of scale and, consequently, such methods may eventually replace the linear model. Before this transition can occur, rank regression methods must be extended to the multivariate case. This extension will complicate prediction of breeding values and estimation of genetic parameters. Perhaps the approximation of Pettitt (1982, 1983) can be extended to the multivariate case. The multivariate generalization of Clayton and Cuzick (1985) warrants further study.

References

Berger JO (1980) Statistical decision theory. Springer-Verlag, Berlin Heidelberg New York

Brotherstone S, Hill WG (1986) Heterogeneity of variance amongst herds for milk production. Anim Prod 42:297-303

Burnside EB, McClintock AE, Hammond K (1984) Type, production and longevity in dairy cattle: a review. Anim Breed Abstr 52:711-719

Cox DR (1972) Regression models and life tables. J R Stat Soc Ser B 34:187-202

Clayton D, Cuzick J (1985) Multivariate generalizations of the proportional hazards models. J R Stat Soc Ser A 148:82-117

De Lorenzo MA, Everett RW (1986) Prediction of sires effects for probability of survival to fixed ages with a logistic linear model. J Dairy Sci 69:501-509

Elandt-Johnson RC, Johnson NL (1980) Survival models and data analysis. Wiley, New York

Everett RW, Keown JF, Clapp EE (1976) Production and stayability trends in dairy cattle. J Dairy Sci 59:1532-1539

Everett RW, Keown JF, Taylor JF (1982) The problem of heterogeneous within herd error variances when identifying elite cows. J Dairy Sci 65 suppl 1:100

Falconer DS (1981) Introduction to quantitative genetics. 2nd edn. Longman, New York

Famula TR (1981) Exponential stayability model with censoring and covariates. J Dairy Sci 64:538-547

Foulley JL, Gianola D (1984) Estimation of genetic merit from bivariate "all or none" responses. Genet Sel Evol 16:285-306

Foulley JL, Gianola D, Thompson R (1983) Prediction of genetic merit from data on binary and quantitative variates with an application to calving difficulty, birth weight and pelvic opening. Genet Sel Evol 15:401-423

Gianola D, Fernando RL (1986) Bayesian methods in animal breeding theory. J Anim Sci 63:217-244

Gianola D, Foulley JL (1983) Sire evaluation for ordered categorical data with a threshold model. Genet Sel Evol 15:201-223

Goddard ME (1985) Genotype X environment interactions in dairy cattle. In: Proc 5th Conf Aust Assoc Anim Breed Genet, Adelaide, pp 52-56

Gross AJ, Clark VA (1975) Survival distributions: reliability applications in the biomedical sciences. Wiley, New York

Harville DA, Mee RW (1984) A mixed model procedure for analyzing ordered categorical data. Biometrics 40:393-408

Henderson CR (1973) Sire evaluation and genetic trends. In: Proc Anim Breed Genet Symp in Honor of Dr. J. L. Lush. ASAS and ADSA, Champaign, Illinois, pp 10-41

Hoschele I (1986) Estimation of breeding values and variance components with quasi-continuous data. PhD Diss, Hohenheim Univ, Stuttgart

Iniguez L, Quaas RL, Robson DS, LD (1984) Modelling and non-linear methodology in the evaluation of conception in sheep under accelerated lambing systems (A.L.S.). J Anim Sci 59 suppl 1:154

Kalbfleisch JD, Prentice RL (1973) Marginal likelihood based on Cox's regression and life model. Biometrika. 60:267-278

Kalbfleisch JD, Prentice RL (1980) The statistical analysis of failure time data. Wiley, New York

Lagakos SW (1979) General right censoring and its impact on the analysis of survival data. Biometrics 35:139-156

Mather K (1949) The genetical theory of continuous variation. Hereditas suppl:376-401

Meijering A (1985) Sire evaluation for calving traits by best linear unbiased prediction and non-linear methodology. Z Tierz Züchtungsbiol 102:95-105

Meijering A, Gianola D (1985) Observations on sire evaluation with categorical data using heteroscedastic mixed linear models. J Dairy Sci 68:1226-1232

Melsa JL, Cohn DL (1978) Decision and estimation theory. McGraw-Hill, New York

Peto R (1972) Contribution to the discussion on the paper of DR Cox. J R Stat Soc Ser B 34:205-207

Pettitt AN (1982) Inferences for the linear model using a likelihood based on ranks. J R Stat Soc Ser B 44:234-243

Pettitt AN (1983) Approximate methods using ranks for regression with censored data. Biometrika 70:121-132

Pettitt AN (1984) Fitting a sinusoid to biological rhythm data using ranks. Biometrics 40:295-300

Smith SP, Allaire FR (1986) Analysis of failure times measured on dairy cows: theoretical considerations in animal breeding. J Dairy Sci 69:217-227

Smith SP, Hammond K (1988) Rank regression with log-gamma residuals. Biometrika 75:741-751

Smith SP, Quaas RL (1984) Productive life span of bull progeny groups: failure time analysis. J Dairy Sci 67:2999-3007

Stoer J, Bulirsch R (1980) Introduction to numerical analysis. Springer-Verlag, Berlin Heidelberg New York

Wolynetz MS, Binns MR (1983) Stayability of dairy cattle: models with censoring and covariates. J Dairy Sci 66:935-942

Wright S (1952) The genetics of quantitative variability. In: Reeve ECR (ed) Quantitative inheritance. Her Majesty's Stationery Office, London pp 5-41

17 Genetic Evaluation for Discrete Polygenic Traits in Animal Breeding

J.L. Foulley[1], D. Gianola[2], and S. Im[3]

Linear and non-linear models for the analysis of categorical data in animal breeding are reviewed and discussed on account of recent research made in this area. Only non-linear methods based on the threshold-liability concept introduced by Wright are described. Emphasis is on describing statistical techniques for estimating genetic merit and parameters of genetic and phenotypic variation. For each kind of methodology, the simple case of dichotomous responses is discussed in more detail as it serves as a basis for the presentation. Special consideration also is given to mixed model structures of data involving genetic effects and nuisance environmental parameters as fixed effects, as well as sire transmitting abilities, breeding values or producing abilities as random effects. A linear mixed model approach developed recently is examined in detail and extended to more general situations. For the non-linear threshold model, it is shown how Bayesian methodology is particularly well suited for estimating location and dispersion parameters in the underlying scale under mixed sources of variation. The generality of the approach is illustrated through a discussion of extensions of the procedure.

17.1 Introduction

Categorical variables are ubiquitous in animal breeding. They arise in many type trait scoring systems, and most components of numerical productivity such as ovulation rate, fertility, pre- and postnatal survival and birth problems are scored categorically. From an economic point of view these traits are extremely important because reproductive performance is often a factor limiting the efficiency of many animal production systems. However, most applications of quantitative genetics theory have been devoted to traits exhibiting a continuous distribution. Moreover, the normality assumption legitimates the use of linear models and linear predictors of breeding values. A very large amount of work has been done in that area resulting in the

[1] INRA, Station de Génétique Quantitative et Appliquée, Jouy-en-Josas, France
[2] Department of Animal Sciences, University of Illinois, Urbana, Illinois, USA
[3] INRA, Laboratoire de Biometrie, 31326 Castanet-Tolosan, France

development of best linear unbiased prediction (BLUP) and mixed model methodology for estimating location and dispersion parameters (Henderson 1973).

Because the genetic determinism of such traits is mainly polygenic, this approach to the genetic analysis of continuous traits is well justified. On the other hand, many traits with a discontinuous phenotypic distribution have been related to a genetic determinism involving a limited number of genes. However, in many instances such as with fertility and survival records, traits showing a discrete distribution cannot be readily analyzed as Mendelian characters, and might be explained with a polygenic inheritance model. Due to Wright's (1934a,b) pioneering work on the variability of number of digits in lines of guinea pigs, a conceptual model is now available to deal with such traits. The threshold and liability concepts involving a hypothetical underlying distribution of phenotypes have been widely used in further theoretical developments pertaining to these so-called quasi-continuous traits (Robertson and Lerner 1949; Dempster and Lerner 1950), especially in the field of human genetics and susceptibility to disease (Falconer 1965; Morton and McLean 1974). Despite the existence of other alternative concepts, e.g., the "risk function" (Edwards 1969; Smith 1970; Curnow 1972) and the "extremal concept" (Bock 1975), the abrupt threshold model remains the cornerstone of quantitative genetics theory of discrete traits with ordinal responses showing a (conditional) multinomial distribution. This is also true under complex segregation patterns and mixed models of inheritance (Lalouel et al. 1983).

From a statistical point of view, there is a large amount of literature dealing with the analysis of discrete data as reflected by the large number of books published on the topic during the last decade; see, for instance, Haberman (1978, 1979), Bishop et al. (1975), Plackett (1981), Agresti (1984). Furthermore, very interesting work has also been done by the "data analysis" school (Tukey 1962; Benzecri 1973) and in other areas of applications such as, for instance, industrial quality control (Nair 1986) and econometrics (Judge et al. 1985). Within the animal breeding domain, several comprehensive reviews have been published such as those by Gianola (1982), Danell (1985), Meijering (1985a), and Hoschele (1986), among the most recent ones.

The purpose of this paper is deliberately not to repeat or summarize all this material, but to give a view of recent procedures applicable to animal breeding. The threshold-liability model has been known for a long time. Surprisingly, however, inference about genetic and phenotypic parameters defined on the "observed" scale is often carried out in practice via estimators derived from statistical procedures that do not hypothesize an underlying normal distribution with thresholds. These procedures include linear models derived for continuous variates and these, in particular, have been applied frequently for predicting breeding values for discrete traits. As pointed out by Gianola (1982), this ambivalent attitude can lead to serious difficulties because: (1) linear models used are often inadequate to be fitted to discrete distributions, and (2) assumptions underlying the linear model are hardly compatible with those of the threshold model. On the other hand, it seems inappropriate to dismiss a priori linear models for analyzing discrete data just because they are not consistent with the threshold-liability model, however sensible the latter is. Therefore, each of these models will be discussed separately, with special emphasis on the assumptions involved and on their consequences on the statistical procedures derived. For each model, primary interest will center on the description of statistical procedures suitable for estimating genetic merit of candidates for selection and for estimating parameters of genetic and phenotypic variability.

This discussion will be necessarily incomplete. In particular, nothing will be said about the analysis of contingency tables with log-linear models or the analysis of survival and failure time. Readers interested in these areas, especially in applications to animal science and breeding, are referred to expository papers such as those by Rutledge and Gunsett (1982) and Gill (1986) for the first subject, and by Smith and Allaire (1986) for the second one.

17.2 Analysis of the Discontinuous Scale with Linear Models

17.2.1 Single Population Analysis.

Model. Linear models for analyzing dichotomous responses are very well developed for the case of one population and one random factor (Guiard et al. 1985). The corresponding model can be written as:

$$y_{ij} = \pi_i + e_{ij} \tag{17.1a}$$

$$\pi_i = \pi + u_i \quad \text{for } i = 1,2,...,m \qquad j = 1,2,...,n_i, \tag{17.1b}$$

where y_{ij} is the jth observation for the ith level of the random classification u; y_{ij} takes the values 1 and 0 with conditional probabilities π_i and $1-\pi_i$, respectively. In (17.1a), π_i is decomposed linearly into π, a general population mean, plus u_i, the effect of the ith level of the random factor considered (e.g., transmitting abilities of sires), and e_{ij}, a residual term.

Setting the linear decomposition in (17.1b) and assuming that conditionally on u_i, the data y_{ij} are independent Bernoulli variables:

$$y_{ij}|u_i \sim B(1,\pi_i), \tag{17.2}$$

one can develop a model similar to the conventional linear model for normal data. However, two important features deserve special attention. First, the assumption of independence between main effects and residuals must be relaxed in favor of the assumption of lack of correlation. Secondly, as $y_{ij}=y_{ij}^2$, the total variance

$$\text{var}(y_{ij}) = \pi(1-\pi) \tag{17.3}$$

is a function of the location parameter π.

Variance Component Estimation. As the basic assumptions of the analysis of variance hold, except for normality, this standard technique can be applied to estimate the variance components σ_u^2, σ_e^2, and the intra-class correlation ρ using the "between" (MS_u) and "within" (MS_e) mean squares. Letting

$$p_i = (\sum_j y_{ij})/n_i \text{ and } \bar{p} = (\sum_{ij} y_{ij})/N,$$

with $N = \sum_i n_i$, these statistics are:

$$MS_u = (\sum_i n_i p_i^2 - N\bar{p}^2)/(m-1) \tag{17.4a}$$

$$MS_e = [\sum_i n_i p_i(1-p_i)]/(N-m)$$

$$= (N\bar{p} - \sum_i n_i p_i^2)/(N-m). \tag{17.4b}$$

This procedure has been used for a long time to analyze all-or-none traits in animal breeding, especially to estimate the genetic variation. For instance, see the early work by Lush et al. (1948) about the inheritance of resistance to disease in poultry, and also articles by Razungles (1977), Elston (1977), and Hill and Smith (1977).

Robertson and Lerner (1949) suggested a slightly different approach for estimating heritability. They took advantage of expression (17.3) for the total variance and estimated it as $\bar{p}(1-\bar{p})$. After equating the between (family) mean square to its expectation, one solves for σ_u^2 as a function of $E(MS_u)$ and σ_y^2, the total variance:

$$\sigma_u^2 = \sigma_y^2 \{[E(MS_u)/\sigma_y^2]-1\}/(k-1), \tag{17.5}$$

where

$$k = [N-(\sum_i n_i^2)/N]/(m-1).$$

By multiplying the numerator and denominator in (17.5) by m-1, and replacing $E(MS_u)$ and σ_y^2 by their estimated values, one obtains the following estimator ($\hat{\rho}$) of the intra-class correlation coefficient:

$$\hat{\rho} = \{[(\sum_i n_i p_i^2 - N\bar{p}^2)/\bar{p}(1-\bar{p})] - (m-1)\}/n_o \qquad (17.6)$$

with, $n_o=(m-1)(k-1)$. Letting $n_{i1}=n_i p_i$, the first term in the numerator is equal to

$$\left[\sum_i (n_{i1}-n_i\bar{p})^2/n_i\bar{p}(1-\bar{p})\right],$$

which is the heterogeneity χ^2 in an m x 2 contingency table. Therefore, the numerator of $\hat{\rho}$ in (17.6) can be viewed as the excess of the observed χ^2 over its expected value (m-1). This property allows one to derive an expression for the sampling variance of this estimator. With large n_i's

$$\text{var}(\hat{\rho}) = 2(m-1)/n_o^2.$$

For more details about criteria proposed for testing the equality of proportions π_i, and asymptotic theory, see Brier (1980) and Shah and Claypool (1984).

Genetic Evaluation. Consider the case where u_i in model (17.1b) is the transmitting ability of sire i. The joint distribution of u_i, e_{ij}, and y_{ij} has the following expectation vector and variance matrix:

$$E\begin{bmatrix} u_i \\ e_{ij} \\ y_{ij} \end{bmatrix} = \begin{bmatrix} 0 \\ 0 \\ \pi \end{bmatrix}; \text{Var}\begin{bmatrix} u_i \\ e_{ij} \\ y_{ij} \end{bmatrix} = \begin{bmatrix} \sigma_u^2 & 0 & \sigma_u^2 \\ 0 & \sigma_e^2 & 0 \\ \sigma_u^2 & 0 & \pi(1-\pi) \end{bmatrix}.$$

When the mean π and the variance components σ_u^2, σ_e^2 are known, a selection index for predicting the u's may be constructed whatever the distribution involved (Henderson 1973). The corresponding predictor may be expressed as

$$\pi + u_i = (n_i p_i + \lambda\pi)/(n_i + \lambda), \qquad (17.7)$$

where $\lambda=\sigma_e^2/\sigma_u^2$ is the error to sire variance ratio; note that (17.7) is a weighted average of p_i and π, the weights being n_i and λ, respectively. When the mean π is unknown, BLUP replaces the selection index; the formula is as in (17.7) but with π replaced by its generalized least-squares estimator

$$\hat{\pi} = \left[\sum_i n_i p_i/(n_i+\lambda)\right] / \sum_i n_i/(n_i+\lambda). \tag{17.8}$$

A further step would be to replace the variance component by estimated values (Gianola et al. 1986).

It is worth noting that the predictor in (17.7) is also a Bayesian estimator of π_i (more specifically the posterior mean) when a beta distribution is used as prior. Assuming that a priori the π_i's are independent and identically distributed as beta (α, β), the posterior distribution remains in the same family because in this case the prior and the likelihood (product binomial) are conjugate (Cox and Hinkley 1974). Letting

$$y_{i.} = \sum_j y_{ij} \, ,$$

this posterior distribution can be written as:

$$f(\pi_1,..,\pi_i,..,\pi_m | y_{i.};\ i=1,2,..,m)$$

$$\propto \prod_i \pi^{y_{i.}+\alpha-1} (1-\pi_i)^{n_i-y_{i.}+\beta-1} \tag{17.9}$$

with the normalizing constant

$$\left[\prod_i B(y_{i.}+\alpha;\ n_i-y_{i.}+\beta)\right]^{-1},$$

B(x,y) designating the beta function. The posterior mean of π_i is then:

$$E(\pi_i | y_{i.};\ i=1,2,..,m)$$

$$= (y_{i.}+\alpha)/(n_i+\alpha+\beta)\,. \qquad (17.10)$$

Using the same reparameterization as in Im (1982) with $\pi=\alpha/(\alpha+\beta)$ as the prior mean of π_i and $\gamma=(\alpha+\beta+1)^{-1}$, (17.10) becomes

$$E(\pi_i|y_{i.};\ i=1,2,..,m)$$

$$= [n_ip_i+(\gamma^{-1}-1)\pi]/[n_i+(\gamma^{-1}-1)]. \qquad (17.11)$$

The parameter γ is interpretable as the intra-class correlation (Im 1982), and γ^{-1}-1 is equivalent to $\lambda=\sigma_e^2/\sigma_u^2$ in the linear model. This makes very clear the equivalence between the Bayesian estimator in (17.11) and the selection index in (17.7). However, BLUP and its extension with σ_e^2 and σ_u^2 replaced by their estimates derived from (17.4a,b), or with λ replaced by $(\hat{\rho}^{-1}-1)$ from (17.6), will differ from the empirical Bayesian estimator computed with the parameters π and γ estimated by maximum likelihood (ML) from the marginal distribution of the data. Griffiths (1973) and Williams (1975) describe this ML procedure.

Extension to Polychotomous Responses. The approach employed for all-or-none traits can also be used for polychotomous traits by extending model (17.1a,b) to the multivariate domain. Each of the (c+1) categories is then viewed as a different trait. In fact, one category is deleted, for instance the last one, because probabilities add up to 1. The model may be written as in (17.1a) and (17.1b):

$$y_{ijk} = \pi_{ik}+e_{ijk} = \pi_k+u_{ik}+e_{ijk} \quad k = 1,2,...,c,$$

where y_{ijk} is a 0-1 variable indicating whether the response falls in category k (1) or not (0) for observation j (j=1,...,n_i) in treatment u_i (i=1,...,m). In matrix notation:

$$\mathbf{y}_{ij} = \pi_i+\mathbf{e}_{ij} = \pi+\mathbf{u}_i+\mathbf{e}_{ij} \qquad (17.12)$$

with

$$\mathbf{y}_{ij} = \{y_{ijk}\},\ \pi_i = \{\pi_{ik}\},\ \pi = \{\pi_k\},$$

$$\mathbf{u}_i = \{u_{ik}\},\ \mathbf{e}_{ij} = \{e_{ijk}\}$$

being column vectors with c elements. As in the univariate case it is assumed that the $\mathbf{u}_i$'s are u.i.d. $(\mathbf{0},\mathbf{G}_o)$, the $\mathbf{e}_{ij}$'s are u.i.d. $(\mathbf{0}, \mathbf{R}_o)$, and the $\mathbf{u}_i$ and $\mathbf{e}_{ij}$'s are uncorrelated where u.i.d.

(**0**, **V**) stands for uncorrelated and identically distributed with mean **0** and variance covariance matrix **V**. In this model,

$$E(\mathbf{y}_{ij}) = \pi$$

$$Var(\mathbf{y}_{ij}) = \mathbf{V}_o = Diag(\pi) - \pi\,\pi', \tag{17.13}$$

where $\mathbf{V}_o = \mathbf{G}_o + \mathbf{R}_o$.

This model was proposed by Landis and Koch (1977) to estimate variance-covariance components, and intra-class (k=k') and inter-class (k≠k') correlation coefficients using MANOVA procedures. It was utilized later by Quaas and Van Vleck (1980) to derive a multiple trait BLUP procedure of genetic evaluation for categorical data. This procedure was applied to sire evaluation of Brown Swiss bulls for type traits by Van Vleck and Karner (1980) and to the evaluation of dairy bulls in Ontario for dystocia and mortality by Cady and Burnside (1982). Im (1982) considered model (17.12) for genetic evaluation purposes, but specified a Dirichlet distribution for the π_i's.

The formula for variance-covariance estimators and BLUP will not be derived again here. Although awkward, the algebra is simple in principle. More attention will be given to similarities with results in genetic evaluation obtained in the univariate case. After some manipulation of the mixed model equations, it can be shown that the BLUP of $\pi + \mathbf{u}_i$ is given by:

$$\hat{\pi} + \hat{\mathbf{u}}_i = (n_i \mathbf{I}_c + \mathbf{R}_o \mathbf{G}_o^{-1})^{-1}(n_i \mathbf{p}_i + \mathbf{R}_o \mathbf{G}_o^{-1}\,\hat{\pi}) \tag{17.14}$$

with

$$\hat{\pi} = \left[\sum_{i=1}^{m} n_i (n_i \mathbf{I}_c + \mathbf{R}_o \mathbf{G}_o^{-1})^{-1}\right]^{-1} \cdot \sum_{i=1}^{m} n_i\,(n_i \mathbf{I}_c + \mathbf{R}_o \mathbf{G}_o^{-1})^{-1} \mathbf{p}_i. \tag{17.15}$$

Formulae (17.14) and (17.15) appear to be exactly the multivariate counterparts of (17.7) and (17.8) derived in the univariate case, with $\mathbf{R}_o \mathbf{G}_o^{-1}$ replacing the ratio σ_e^2/σ_u^2. Selection index corresponds to formula (17.14) with the true mean π instead of the estimate. This highlights the nature of the linear predictor of $\pi + \mathbf{u}_i$, which may be viewed as a weighted average of π and of the marginal frequency $\mathbf{p}_i$ of responses in family i, suggesting a Bayesian interpretation. In fact, if the π_i's are assumed to be i.i.d. (independently identically distributed) random variables following a Dirichlet distribution $D_c\,(\alpha_1, \alpha_2, \ldots, \alpha_k, \ldots, \alpha_{c+1})$ (Brier 1980 and Im 1982), it can be shown that the posterior mean of π_{ik} (probability of response in category k of a progeny of sire i) can be expressed as:

$$E(\pi_{ik}|\text{data}) = [n_i p_{ik}+(\rho^{-1}-1)\pi_k]/[n_i+(\rho^{-1}-1)] \tag{17.16}$$

where

$$\pi_k = \alpha_k / \sum_{\ell=1}^{c+1} \alpha_\ell$$

is the prior mean of π_{ik}, and

$$\rho = (\sum_{\ell=1}^{c+1} \alpha_\ell + 1)^{-1}$$

is a parameter interpretable as an overall intra-class correlation. Formula (17.16) provides the same expression as that arising for the selection index (with π instead of $\hat{\pi}$) provided $\mathbf{G}_o=\rho\mathbf{V}_o$ so that $\mathbf{R}_o\mathbf{G}_o=(\rho^{-1}-1)\mathbf{I}_c$. This was already mentioned by Quaas and Van Vleck (1980) who, however, pointed out that this would reduce the flexibility of the model. Indeed, the condition above implies not only the same heritability and repeatability for all categories ($\sigma^2_{ek}/\sigma^2_{uk}$=constant) but also the equality of genetic and residual correlations ($r_{e,kk'}=r_{G,kk'}$, r_e and r_G designating elements of the correlation matrices corresponding to $\mathbf{R}_o$ and $\mathbf{G}_o$, respectively). It is difficult to assess whether or not this assumption is practically tenable. Landis and Koch (1977) were probably concerned with a similar simplification when introducing a pooled intra-class correlation. This clearly highlights a critical aspect of the multidimensional model, i.e., its high degree of parameterization for at least some situations encountered with polychotomous responses. Another important limitation is its inability to take into account ordered responses.

Scoring Procedures. It is common and sometimes reasonable to assign scores to the different categories; see, for instance, Schaeffer and Wilton (1976) and Pollak and Freeman (1976) for an application of this approach to dystocia. The procedure requires one to choose a vector of weights $\eta'=[\eta_1,\eta_2,...,\eta_k,...,\eta_{c+1}]$ for the different categories. The new variate $w_{ij}=\eta' y_{ij}$, although still discrete (i.e., showing a finite number of values) is often analyzed using linear procedures available for continuous data (Quaas and Van Vleck 1980). However, these authors suggested that it would be more sensible to carry out the analysis as described before and apply the scoring procedure to the final estimates, so that, if $\eta^{*\prime}=[\eta_1,\eta_2,...,\eta_c]/\eta_{c+1}$, the prediction is $\eta^{*\prime}(\hat{\pi}+\hat{\mathbf{u}}_i)$. On the other hand, Gianola (1982) proposed an extension of the procedures of Robertson and Lerner (1949), for estimating intra-class correlation when applied to scores. The model may now be written as $w_{ij}=\eta'y_{ij}=\mu+a_i+r_{ij}$. Letting SS_{wa} be the between-family adjusted sum of squares and knowing that $\sigma_r^2=\eta'Var(y_{ij})\eta-\sigma_a^2$, the estimator of the intra-class correlation is given by

$$\rho_w = [\{SS_{wa}/[\eta'var(y_{ij})\eta]\}-(m-1)]/n_o, \tag{17.17}$$

where $var(y_{ij})$ is computed from (17.13) with π including category (c+1).

17.2.2 Multipopulation Analysis

The methodology for a single population is very limited for application to sire evaluation due to its inability to deal with mixed model data structures. Some attempts have been made to extend the mixed model procedures developed by Henderson for continuous variables, to categorical data. Schaeffer and Wilton (1976) presented a mixed model version of the method of Grizzle et al. (1969) applied to a linear model of cell probabilities for calving ease. Grizzle's method involves a weighted least-squares approach for estimating the effects of factors which are all considered as fixed. Schaeffer and Wilton (1976) suggested to replace the equations arising in Grizzle's approach by mixed model equations; the rationale for this was provided by the least-squares coefficient matrix of Grizzle's equations. However, in Grizzle's method, the weights are functions of the incidence of categories in each subclass and, consequently, vary from one subclass to another. Schaeffer and Wilton (1976) computed these weights from an overall estimate of each category probability, i.e., as if each subclass were a random sample from a single population. Whether this approximation is critical or not from a practical point of view is difficult to assess in general. This procedure, however, is inconsistent with the modelling of each subclass probability as a specific combination of fixed effects.

Pollak and Freeman (1976) applied conventional mixed model methodology to calving scores, and so did Berger and Freeman (1978) although these authors allowed the residual variances to vary with parity of dam. More recently, Van Vleck and Edlin (1984) developed a multiple trait random model to evaluate sires on calving scores of their calves. They regarded calving scores expressed in four combinations of parity of dam x sex of calf as four different traits. As each calf has a measurement in only one combination, it provides information on one trait only and the variance covariance matrix of residuals becomes diagonal (when the cow effect is absent or ignored). In that way, this procedure adjusts automatically for heterogeneity in means and heteroscedasticity in sire and residual variances among parity x sex subclasses. However, fixed factors are not included in the model so that no adjustment for different variance among combinations of other nuisance factors (e.g., herd x year x season) is done. Finally, in all these three studies, calving scores were treated as a continuous variable ignoring the discrete structure (five categories).

Although these procedures might be statistically robust and practically efficient for selection purposes, it seems reasonable to investigate whether more consistent and appealing approaches of genetic analysis (especially genetic evaluation) for discrete traits by linear models might be developed in a mixed model structure. A recent paper published by Beitler and Landis (1985) gives us the opportunity to address this problem in the simplest case of a dichotomous trait, with several populations modelled via levels of a single fixed factor.

A Two-Way Mixed Model for Binary Traits. Let the model be

$$y_{ijk} = \pi+u_i+\beta_j+e_{ijk}, \tag{17.18}$$

where y_{ijk} is the observation on the kth ($k=1,2,...,n_{ij}$) subject assigned to the jth treatment ($j=1,2,...,p$) in the ith level ($i=1, 2,...,m$) of the random classification u (usually families in animal breeding). The random variable y_{ijk} takes the value 1 if the response occurs and 0 otherwise, with conditional probabilities π_{ij} and $1-\pi_{ij}$, respectively, where π_{ij}, the probability of "success", is written as:

$$\pi_{ij} = \pi_j+u_i = \pi+\beta_j+u_i. \tag{17.19}$$

Beitler and Landis (1985) postulated that the $\{u_i\}$ are i.i.d. $(0,\sigma_u^2)$, the $\{e_{ijk}\}$ are i.i.d. $(0,\sigma_{ej}^2)$ and that these two sets are mutually independent. As shown previously for the one-way random model, although this last assumption cannot be satisfied with the form in (17.18) (see also, Im 1986), the two sets remain uncorrelated. Furthermore, the assumption of independence among the u_i's, on the one hand, and among the e_{ijk}'s, on the other, does not appear necessary to develop the model, the assumption of no correlation being sufficient. The basic assumptions needed in this model are the following: (1) π_{ij} is decomposed as in (17.19) into the sum of a population mean π, a fixed treatment effect β_j, and a random component u_i. For simplicity, these factors will be considered here to be additive. An interaction component was incorporated by Beitler and Landis (1985) as well as by Lavergne (1984), but within the framework of a pure random model in the case of the last author. (2) Conditionally to π_{ij} (i.e., within the ij^{th} subclass) the y_{ijk} are independently distributed as Bernoulli variates with parameter π_{ij}:

$$y_{ijk}|\pi_{ij} \sim \text{i.d. } B(\pi_{ij}). \tag{17.20}$$

Conversely, the residuals $e_{ijk}=y_{ijk}-\pi_{ij}$ are conditionally independent with

$$E(e_{ijk}|\pi_{ij}) = 0 \tag{17.21a}$$

$$\text{var}(e_{ijk}|\pi_{ij}) = \pi_{ij}(1-\pi_{ij}). \tag{17.21b}$$

The residual takes the values $1-\pi_{ij}$ and $-\pi_{ij}$ with conditional probabilities π_{ij} and $1-\pi_{ij}$, respectively. This shows very clearly that e_{ijk} cannot be independent of u_i involved in the decomposition of π_{ij}. However, these two variates are uncorrelated because, under the assumption that $E(u_i)=0$

$$\mathrm{Cov}(u_i,e_{ijk}) = E_{\pi_{ij}}[E(u_i e_{ijk}|\pi_{ij})]$$

$$= E[u_i E(e_{ijk}|\pi_{ij})] ,$$

which, from (17.21a) reduces to

$$\mathrm{Cov}(u_i,e_{ijk}) = 0. \qquad (17.22)$$

Given the expressions (17.21a,b) for the conditional mean and variance, those for the unconditional distribution follow immediately:

$$E(e_{ijk}) = 0 \qquad (17.23a)$$

$$\mathrm{var}(e_{ijk}) = E_{\pi_{ij}}[\pi_{ij}(1-\pi_{ij})].$$

Referring to the decomposition in (17.19), this expectation is computed as:

$$\mathrm{var}(e_{ijk}) = \pi_j(1-\pi_j)+E[u_i(1-u_i)].$$

If at this stage, we assume that the $\{u_i\}$ are identically distributed with mean zero and variance σ_u^2, this variance becomes (using Beitler and Landis' notation)

$$\sigma_{e_j}^2 = \pi_j(1-\pi_j)-\sigma_u^2 . \qquad (17.23b)$$

From (17.23a) it turns out that

$$E(y_{ijk}) = \pi_j, \qquad (17.24a)$$

and, knowing that $y_{ijk}=y_{ijk}^2$

$$\mathrm{var}(y_{ijk}) = \pi_j(1-\pi_j). \qquad (17.24b)$$

Hence (17.23b) can be rewritten as

$$\mathrm{var}(y_{ijk}) = \sigma_u^2+\sigma_{e_j}^2, \qquad (17.24c)$$

which is consistent with the property of no correlation between u_i and e_{ijk} derived from the basic assumptions underlined in (1) and (2) as given earlier.

From (17.24b) and (17.24c), it can be inferred that

$$\sigma_u^2 \leq \pi_j(1-\pi_j) \text{ for any } j=1,2,...,p. \tag{17.25}$$

Although very important, this constraint was not mentioned by Beitler and Landis (1985). This condition indicates, contrarily to what they assumed, that the distribution of u_i is not entirely free of the incidence of the attribute in β_j subclasses.

Variance Component Estimation under the Beitler and Landis Model. Due to the assumptions of this model, where no specific distribution is ascribed to π_{ij}, it is not possible to apply maximum likelihood procedures for estimating variance components. However, quadratic unbiased estimators can be derived; in particular, Henderson's Method 3 can be utilized for estimating σ_u^2 and $\sigma_{e_j}^2$ (j=1,2,..,p) considered, as in Beitler and Landis, to be the unknown parameters.

Letting $\boldsymbol{\beta}=\{\beta_j\}$, $\mathbf{u}=\{u_i\}$ and using the R notation for reduction in sums of squares, the procedure for the two-way additive model is based on the quadratic:

$$R(\mathbf{u}|\pi,\boldsymbol{\beta}) = R(\pi,\boldsymbol{\beta},\mathbf{u})-R(\pi,\boldsymbol{\beta}),$$

which may be computed from the normal equations after absorbing the $\boldsymbol{\beta}$ equations (Searle 1971; Searle et al. 1981)

$$R(\mathbf{u}|\pi,\boldsymbol{\beta}) = \mathbf{r}'\mathbf{C}^-\mathbf{r}\ , \tag{17.26}$$

where $\mathbf{C}=\{C_{ii'}\}$ and $\mathbf{r}=\{r_i\}$ for i=1,2,..,p, have elements

$$\left.\begin{aligned} C_{ii} &= n_{i.}-\sum_j n_{ij}^2/n_{.j} \\ C_{ii'} &= -\sum_j n_{ij}n_{i'j}/n_{.j} \text{ for } i\neq i' \end{aligned}\right\} \tag{17.27}$$

$$r_i = y_{i..}-\sum_j n_{ij}y_{.j.}/n_{.j}\ . \tag{17.28}$$

The dot symbol indicates summation over the corresponding subscript and $\mathbf{C}^-$ is any generalized inverse of $\mathbf{C}$. Consider $\mathbf{y}_j$, the vector of observations within the β_j subclass, which can be written as $\mathbf{y}_j = \mathbf{X}_j b_j + \mathbf{Z}_j \mathbf{u} + \mathbf{e}_j$, where $b_j = \pi + \beta_j$, and $\mathbf{X}_j$, $\mathbf{Z}_j$ are the appropriate ($n_{.j}$ x 1), ($n_{.j}$ x m) incidence matrices for b_j and $\mathbf{u}$ effects, respectively. Moreover, letting $\mathbf{C}_j = \mathbf{Z}_j' \mathbf{P}_j \mathbf{Z}_j$ such that $\mathbf{C} = \sum_j \mathbf{C}_j$, with $\mathbf{P}_j$ being the projector matrix $\mathbf{P}_j = \mathbf{I} - \mathbf{X}_j(\mathbf{X}_j'\mathbf{X}_j)^{-1}\mathbf{X}_j'$, it can be shown that the quadratic in (17.26) has the expectation:

$$E[R(\mathbf{u}|\pi,\boldsymbol{\beta})] = \mathrm{tr}(\mathbf{AC})\sigma_u^2 + \sum_j \mathrm{tr}(\mathbf{C}^-\mathbf{C}_j)\sigma_{e_j}^2 \tag{17.29}$$

with $\mathbf{A}$ being a known symmetric matrix, taken to be the identity matrix in Beitler and Landis, although more generally one can write:

$$\mathrm{var}(\mathbf{u}) = \mathbf{A}\sigma_u^2.$$

A simple estimator of $\sigma_{e_j}^2$, the residual variance within treatment j, can be obtained from the quadratic $\mathbf{y}_j'\mathbf{y}_j - R(\pi, \boldsymbol{\beta}_j, \mathbf{u})$, having the expectation $(n_{.j} - m)\sigma_{e_j}^2$. The corresponding estimator

$$\hat{\sigma}_{e_j}^2 = [\sum_i n_{ij} p_{ij}(1 - p_{ij})]/(n_{.j} - m) \tag{17.30}$$

is the one-way ANOVA estimator of (17.4b) restricted to data in treatment j; $p_{ij} = y_{ij.}/n_{ij}$ designates the frequency of the attribute under study in subclass ij.

The σ_u^2-solution is obtained solving (17.29) for σ_u^2 after replacing the expected by the observed ($\mathbf{r}'\mathbf{C}^-\mathbf{r}$) reduction, and $\sigma_{e_j}^2$ by its estimate from (17.30). In the case where $\mathbf{A}=\mathbf{I}$, the coefficient of σ_u^2 in (17.29) can be easily computed and is equal to $N - \sum_{ij} n_{ij}^2/n_{.j}$, where N is the total number of observations. Beitler and Landis suggested for the "main-effects" model another estimator of $\sigma_{e_j}^2$:

$$\hat{\sigma}_{e_j}^2 = \left[\sum_{ik} y_{jk}^2 - (y_{.j.}^2/n_{.j}) - n_{.j}\mathbf{r}'\mathbf{C}^-\mathbf{r}/N\right] \Big/ \left[n_{.j} - 1 - \mathrm{tr}(\mathbf{C}^-\mathbf{C}_j)\right]. \tag{17.31}$$

As mentioned by Beitler (1981, unpublished PhD thesis, Univ. of Michigan) this estimator is biased and so is the estimator of σ_u^2 obtained using $\hat{\sigma}_{e_j}^2$ from (17.31) in lieu of its true value in (17.29). A simple alternative is to compute p quadratic forms (algebraically independent) in addition to $\mathbf{r'C^-r}$, equate them to their theoretical expectations, and solve the resulting system of (p+1) equations with p+1 unknowns $(\sigma_u^2,\sigma_{e_1}^2,\ldots,\sigma_{e_j}^2,\ldots,\sigma_{e_p}^2)$. One possible choice is simply:

$$Q_j = \sum_{ik} y_{ijk}^2 - (y_{.j.}^2/n_{.j}) \text{ for } j=1,2,\ldots,p,$$

which under the condition **A=I** has the expectation

$$E(Q_j) = tr(C_j)\sigma_u^2+(n_{.j}-1)\sigma_{e_j}^2.$$

Finally, remembering that $\sigma_{e_j}^2=\pi_j(1-\pi_j)-\sigma_u^2$, one may argue that the elementary unknowns in this model are the location parameters π_j and the variance of the random factor, σ_u^2. Let $\hat{\pi}_j$ be an estimate of π_j; then, in the main effects model, σ_u^2 can be estimated from (17.29) as:

$$\hat{\sigma}_u^2 = \left[R(\mathbf{u}|\pi,\boldsymbol{\beta})-\sum_j \hat{\pi}_j (1-\hat{\pi}_j)tr(\mathbf{C}^-\mathbf{C}_j)\right]/[tr(\mathbf{AC})-m+1]. \tag{17.32}$$

This leads immediately to an extension of Robertson and Lerner's formula in (17.6) for the estimation of the intra-class correlation ρ_j, which now varies across β_j subclasses:

$$\hat{\rho}_j = \left[\frac{R(\mathbf{u}|\pi,\boldsymbol{\beta})}{\hat{\pi}_j(1-\hat{\pi}_j)} - tr(\mathbf{C}^-\mathbf{C}_j) - \frac{\sum_{k\neq j}\hat{\pi}_k(1-\hat{\pi}_k)tr(\mathbf{C}^-\mathbf{C}_k)}{\hat{\pi}_j(1-\hat{\pi}_j)}\right]/[tr(\mathbf{AC})-m+1]\,. \tag{17.33}$$

Different approaches may be pursued for estimating π_j. One may simply take the observed frequency, $\hat{\pi}_j=y_{.j.}/n_{.j}$. A more appealing estimator might be OLS (Daudin and Meyniel 1980; Harvey 1982) or, better, the GLS solution of the linear probability model taking into account the correlation between residuals due to the mixed model structure. This is a generalization to the mixed model situation of the approach of Grizzle et al. (1969) when applied to a linear function of probabilities, and also of the procedure described by Judge et al.

(1985). As the matrix of variances and covariances between frequencies of success is unknown, the usual procedure is to employ what Schmidt (1976) called a "feasible" GLS estimator, and Judge et al. (1985) an "estimated" GLS, using estimates instead of true parameter values.

Genetic Evaluation. Since BLUP involves only linear combinations of elementary subclass totals or means, advantage can be taken of this, at least from an expository point of view, for deriving BLUP of **u** in a frequency model. Let $p_{ij}=\sum_k y_{ijk}/n_{ij}$ be the observed proportion of successes for "treatment" j in the ith level of the random factor u. These proportions can be expressed in vector notation writing:

$$\mathbf{p}_i' = (p_{i1}, p_{i2}, \ldots, p_{ij}, \ldots, p_{ip_i})$$

and across the m levels of u by

$$\mathbf{p}' = (\mathbf{p}_1', \mathbf{p}_2', \ldots, \mathbf{p}_i', \ldots, \mathbf{p}_m').$$

In the same way, one can define a vector of residuals

$$\mathbf{e}. = (\mathbf{e}_1', \mathbf{e}_2', \ldots, \mathbf{e}_i', \ldots, \mathbf{e}_m') \text{ with } \mathbf{e}_i' = \{\sum_k e_{ijk}/n_{ij}\}.$$

Using frequencies, the model in (17.18) reduces to:

$$\mathbf{p} = \mathbf{Xb} + \mathbf{Zu} + \mathbf{e}, \tag{17.34}$$

where $\mathbf{b}=\{\pi+\beta_j\}$ and $\mathbf{u}=\{u_i\}$, and **X**, **Z** are now $\sum_i p_i$ x p, $\sum_i p_i$ x q incidence matrices. The form of these matrices is:

$$\mathbf{X}' = [\mathbf{I}_1', \mathbf{I}_2', \ldots, \mathbf{I}_i', \ldots, \mathbf{I}_m'], \tag{17.35}$$

where $\mathbf{I}_i$ is a p_i x p matrix formed from a p x p identity matrix where only the rows pertaining to the p_i levels of β_j represented within the ith level of u are kept. Further

$$\mathbf{Z} = \bigoplus_{i=1}^{m} \mathbf{1}_{p_i} \tag{17.36}$$

with $\mathbf{1}_{p_i}$ being a (p_i x 1) column vector of ones, and $\oplus$ is the direct sum operator.

Random variables in (17.34) have the following expectations and variance covariance matrix:

$$E\begin{bmatrix}\mathbf{u}\\ \mathbf{e.}\\ \mathbf{p}\end{bmatrix}=\begin{bmatrix}\mathbf{0}\\ \mathbf{0}\\ \mathbf{Xb}\end{bmatrix};\ \mathrm{Var}\begin{bmatrix}\mathbf{u}\\ \mathbf{e.}\\ \mathbf{p}\end{bmatrix}=\begin{bmatrix}\Sigma_u & \mathbf{0} & \Sigma_u\mathbf{Z}'\\ \mathbf{0} & \Sigma_e & \mathbf{0}\\ \mathbf{Z}\Sigma_u & \mathbf{0} & \Sigma_p\end{bmatrix} \tag{17.37}$$

with:

$$\Sigma_u=\mathbf{A}\sigma_u^2$$

A being the relationship matrix, (17.38a)

$$\Sigma_e=\bigoplus_{i=1}^{m}\mathrm{Diag}\left\{\sigma_{e_1}^2 n_{i1}^{-1}\ldots,\sigma_{e_j}^2 n_{ij}^{-1},\ldots,\sigma_{e_{p_i}}^2 n_{ip_i}^{-1}\right\} \tag{17.38b}$$

$$\Sigma_p=\mathbf{ZAZ}'\sigma_u^2+\Sigma_e. \tag{17.38c}$$

Assuming σ_u^2 and $\sigma_{e_j}^2$ (j=1,2,...,p) are known, BLUP of $\mathbf{k}'\mathbf{b}+\boldsymbol{\ell}'\mathbf{u}$ can be obtained solving Henderson's mixed model equations:

$$\begin{bmatrix}\mathbf{X}'\Sigma_e^{-1}\mathbf{X} & \mathbf{X}'\Sigma_e^{-1}\mathbf{Z}\\ \mathbf{Z}'\Sigma_e^{-1}\mathbf{X} & \mathbf{Z}'\Sigma_e^{-1}\mathbf{Z}+\mathbf{A}^{-1}/\sigma_u^2\end{bmatrix}\begin{bmatrix}\hat{\mathbf{b}}\\ \hat{\mathbf{u}}\end{bmatrix}=\begin{bmatrix}\mathbf{X}'\Sigma_e^{-1}\mathbf{p}\\ \mathbf{Z}'\Sigma_e^{-1}\mathbf{p}\end{bmatrix}, \tag{17.39}$$

where $\hat{b}$ gives a GLS estimation (Henderson et al. 1959) which, as mentioned earlier, can be used in (17.33).

Using (17.35-38c) in (17.39) and now assuming that **A**=**I**, the solutions in (17.39) can be written as:

$$\hat{u}_i = \left[\sum_j n^*_{ij}(p_{ij} - \hat{\pi}_j) \right] / (n^*_{i.} + \sigma_u^{-2}), \tag{17.40}$$

where $n^*_{ij} = n_{ij}\sigma_{ej}^{-2}$; $n^*_{i.} = \sum_j n^*_{ij}$, and $\hat{\pi}_j$ (the GLS estimator of π_j) is the solution of the system:

$$\left[(n^*_{.j} - \sum_i n^{*}_{ij}/(n^*_{i.} + \sigma_u^{-2})) \right] \hat{\pi}_j - \sum_{j \neq j'} \left[\sum_i n^*_{ij} n^*_{ij'}/(n^*_{i.} + \sigma_u^{-2}) \right] \hat{\pi}_{j'}$$

$$= \sum_i n^*_{ij} \left[p_{ij} - \sum_j n^*_{ij} p_{ij}/(n^*_{i.} + \sigma_u^{-2}) \right] \text{ for } j = 1, 2, \ldots, p. \tag{17.41}$$

BLUP of the u_i's requires that σ_u^2 and $\sigma_{e_j}^2$ be known, in the same way that GLS of the π_j's does. Therefore, a "feasible" BLUP requires that these parameters be replaced by some estimates. One may use method 3 (see 17.29, 17.30, 17.31) quadratic unbiased estimators, or because $\sigma_{e_j}^2 = \pi_j(1-\pi_j) - \sigma_u^2$, employ an estimate of π_j and $\hat{\sigma}_u^2$ of (17.32). It is also tempting to iterate with these expressions.

This procedure can be extended (Appendix 17.A) to more than one fixed factor and to several categories using the multivariate approach of Quaas and Van Vleck (1980). The mixed model equations in (17.39) are closely related to the procedure of sire evaluation for calving difficulty developed by Berger and Freeman (1978) except that the trait analyzed by these authors was not a binary variate but a score ranging from 1 to 5. They considered a constant sire variance and residual variances varying across parity subclasses. Formula (17.40), as in Gianola (1986), indicates that the smaller the residual variance in subclass j (i.e., the lower the incidence π_j in that subclass), the larger is the weight applied to the data in that particular subclass for BLUP of u_i. Meijering and Gianola (1985b), in the context of sire evaluation for calving difficulty as a dichotomous or tetrachotomous trait, observed that this is exactly the opposite of what would occur in a threshold-liability model.

17.3 Models Postulating an Underlying Scale

New procedures of genetic evaluation for discrete traits in animal breeding have been introduced recently by several authors (Gianola and Foulley 1982, 1983a,b; Foulley et al. 1983, 1987a,b; Harville and Mee 1984; Gilmour 1983; Gilmour et al. 1985; Hoschele et al. 1987). Whatever the situation addressed (dichotomous or ordered polychotomous responses; univariate or multivariate analysis), all these procedures are based on Wright's model postulating an underlying normal distribution rendered discrete via thresholds.

It is important to bear in mind the basic assumptions of this model. The definition of "liability" as a normally distributed variable (x) is valid only if polygenic inheritance holds (Goux 1970) with a large number of loci with linkage equilibrium and no epistasis (Bulmer 1980). Consider an all-or-none trait and a purely additive genetic model for the underlying liability:

$$x_i = \mu + a_i + e_i,$$

where μ, a_i, and e_i are the population mean, the (additive) genetic value, and an environmental deviation, respectively, for individual i. The genotypic value w_i in the "outward" scale is defined to be

$$w_i = \Phi[(\tau - \mu - a_i)/\sigma_e],$$

where τ is a threshold, σ_e is the residual standard deviation on the continuous scale and $\Phi(.)$ is the normal cumulative distribution function. Clearly, w_i is a non-linear function of a_i, which implies that this model generates non-additive variance in the outward scale due to non-linearity (Dempster and Lerner 1950). The amount of this type of variance depends on both μ and $h_c^2 = \sigma_a^2/(\sigma_a^2 + \sigma_e^2)$, the heritability in the continuous scale. This is so because

$$\partial w_i/\partial a_i = -\,\phi[(\tau - \mu - a_i)/\sigma_e]\sigma_e^{-1} \tag{17.42}$$

is a function of μ and a_i, ϕ being the standard normal density function. Dempster and Lerner (1950) noted the non-linearity of (17.42) and introduced the additive genetic variance in the outward scale $\sigma^2_{w_a}$ using a first-order approximation of $\mathrm{var}(w_i)$, i.e., $\sigma^2_{w_a} = [E(\partial w_i/\partial a_i)]^2 \mathrm{var}(a_i)$. Hence, letting σ_p be the phenotypic standard deviation and integrating out a_i

$$\sigma^2_{w_a} = \phi^2[(\tau - \mu)/\sigma_p]h_c^2. \tag{17.43}$$

The difference between the exact genetic variance and (17.43) gives the non-additive variance in the outward scale, which increases as h_c^2 increases and as the prevalence of the trait deviates from 1/2. Thus, the threshold-liability concept can hardly legitimate an additive model in the outward scale except as an approximation and, in addition, provided the model describes a single population only. Therefore, formula (17.43) precludes the use of mixed linear models with constant "genetic variance" as shown in the first part of this paper.

The purpose of this section is to review alternative statistical procedures based on the threshold-liability model. For the sake of simplicity and clarity, most of the discussion will be devoted to traits having a "threshold-dichotomy distribution" (Wright 1968) even though other distributions can be analyzed with threshold models.

17.3.1 Binary Responses

The data can be arranged as an s x 2 contingency table having subpopulations in rows (k=1,2,...,s) and responses in columns such that the data:

$$\mathbf{y}' = [\mathbf{y}_1, \mathbf{y}_2, \ldots, \mathbf{y}_k, \ldots, \mathbf{y}_s],$$

where $\mathbf{y}_k$ is the 2 x 1 vector

$$y_k = \sum_{r=1}^{n_{k+}} y_{kr} \,.$$

y_{kr} being a (2 x 1) vector having a 1 in the position of the category of response and a 0 elsewhere. The two categories of response can be coded as [0] and [1], with n_{ko} and n_{k1} being the number of responses in categories [0] and [1] for subpopulation k, respectively, the total $n_{k+}=n_{ko}+n_{k1}$ being assumed fixed by the sampling procedure (Fienberg 1980).

Model. The conceptual variable associated with the rth observation in subpopulation k is modelled as:

$$\ell_{kr} = \mu_k + e_{kr}, \tag{17.44}$$

where μ_k is a location parameter for population k and e_{kr} is a residual. With polygenic inheritance and small environmental influences, it can be assumed that the e_{kr}'s are independent variables following the normal distribution $N(0,\sigma_k^2)$.

Conditionally to μ_k, the probability that an observation from subpopulation k falls into a given category of response, e.g., [0], is written as

$$\pi_{k,o} = \Phi[(\tau-\mu_k)/\sigma_k], \tag{17.45}$$

where $(\tau-\mu_k)/\sigma_k$ is the standardized distance between the population mean (μ) and the threshold (τ), and $\Phi(.)$ is the CDF of a standard normal distribution. When σ_k is the same for different subpopulations (Wright 1968), (17.45) permits one to compare populations for dichotomous responses from their means. Taking a common σ (σ=1, e.g.) as unit of measurement and the origin at the threshold (τ=0), then

$$\pi_{k,o} = 1-\Phi(\mu_k). \tag{17.46}$$

Because of the normality assumption, it is logical to adopt a linear model to describe the effects of factors and/or covariates on μ_k. Letting these effects be designated by the vector θ, we write:

$$\mu_k = \mathbf{w}_k'\theta, \tag{17.47}$$

where $\mathbf{w}_k'$ is an incidence row vector.

The assumption of constant variance on the underlying continuum can be relaxed. In some cases the adjustment can be readily done if the prior values of residual variances in the different subpopulations are known. For instance, in a sire and maternal grandsire model, the ratio between residual variances of records of animals having both ancestors identified and those with only sires identified is $[1-(5/16)h_c^2]$ / $[1-(h_c^2/4)]$, h_c^2 being heritability in the underlying scale. Alternatively, heterogeneity of residual variances can be also assessed from the data. For example, McCullagh (1980, 1986) proposed using the model:

$$\pi_{k,o} = 1-\Phi(\mathbf{w}_k'\theta/\sigma_k)$$

with

$$\log(\sigma_k) = \mathbf{h}_k'\psi,$$

where ψ is a vector of unknown effects (possibly a subset of the θ's), and $\mathbf{h}_k'$ is a row incidence vector. From now on, we will restrict attention to models with constant variance.

As usually done in animal breeding, θ is partitioned into two sets of effects, $\boldsymbol{\beta}$ and $\mathbf{u}$, with incidence vectors $\mathbf{x}_k'$ and $\mathbf{z}_k'$, respectively. The p x 1 vector $\boldsymbol{\beta}$ represents nuisance environmental parameters (e.g., herd x year x season, age of dam), or effects of genetic groups. The m x 1 vector $\mathbf{u}$ can contain other environmental effects or breeding values, producing abilities or, typically, as considered here, transmitting abilities of sires (Foulley and Elsen 1987). While the presentation is restricted for simplicity to a single factor in **u**, the approach is more general. Conditionally to θ, the $\mathbf{y}_k$ are independent following a binomial distribution with parameters n_{k+} and $\Phi(\mathbf{w}_k'\theta)$ so that

$$p(\mathbf{y}|\theta) \propto \prod_k \left[\Phi(\mathbf{w}_k'\theta)\right]^{n_{k1}} \left[1-\Phi(\mathbf{w}_k'\theta)\right]^{n_{k+}-n_{k1}} . \tag{17.48}$$

Point Estimation of Genetic Merit. As in other genetic evaluation problems (Ronningen 1971; Dempfle 1977; Lefort 1980; Gianola and Fernando 1986), it is convenient here to employ a Bayesian procedure. This is the approach followed by Gianola and Foulley (1982; 1983a,b), Foulley and Elsen (1987), Foulley et al. (1983; 1987a,b) and also by Stiratelli et al. (1984) and Knuiman and Laird, Chapter 9, this Volume. On the other hand, maximum likelihood approaches for estimating the location parameters with binary data were presented by Kolakowski and Bock (1981) and McCullagh and Nelder (1983). A "natural" choice for the prior distribution of θ is the conjugate distribution of the ℓ_{kr}'s

$$\theta|\Sigma \sim N(\alpha,\Sigma), \tag{17.49}$$

where

$$\Sigma = \begin{bmatrix} \Sigma_{\beta\beta} & \Sigma_{\beta u} \\ \Sigma_{u\beta} & \Sigma_{uu} \end{bmatrix}.$$

Furthermore, it will be assumed as, for instance, in Stiratelli et al. (1984) that prior knowledge about β is vague (i.e., $\text{diag}\{\Sigma_{\beta\beta}\}\to\infty$) so as to mimic the traditional mixed model analysis with β fixed and **u** random. With these conditions, the conditional distribution of β given **u** tends to be uniform and the prior density of θ is proportional to the prior marginal density of **u** so:

$$p(\theta|\Sigma) \propto p(\mathbf{u}|\Sigma_{uu}). \tag{17.50}$$

When the **u**'s are transmitting abilities, one can write $\Sigma_{uu}= \mathbf{A}\sigma_u^2$, where **A** is a qxq symmetric matrix with elements equal to twice Malecot's coefficients of coancestry between the q sires, and σ_u^2 is the sire component of variance, equal to one-quarter of the additive genetic variance. Hence

$$p(\mathbf{u}|\sigma_u^2) \propto \exp(-\mathbf{u}'\mathbf{A}^{-1}\mathbf{u}/2\sigma_u^2) \tag{17.51}$$

and

$$p(\boldsymbol{\beta},\mathbf{u}|\mathbf{y},\sigma_u^2) \propto (17.48)\text{x}(17.51). \tag{17.52}$$

The best selection rule in the sense of maximizing the expected transmitting ability of a fixed number of sires is the mean of the posterior distribution of **u** (Goffinet and Elsen 1984; Fernando and Gianola 1984). As its calculation is difficult for the model discussed here, we consider modal estimators of θ, and regard the **u**-component of the posterior mode or MAP (maximum a posteriori) of θ as an approximation to the posterior mean. This is equivalent to the maximization of a joint probability mass density function as done by Harville and Mee (1984).

Letting $L(\boldsymbol{\beta},\mathbf{u}|\sigma_u^2)$ be the log-posterior density derived from (17.52):

$$L(\boldsymbol{\beta},\mathbf{u}|\sigma_u^2)=\sum_{k=1}^{s}\left\{n_{k1}\ln\left[\Phi(\mathbf{x}_k'\boldsymbol{\beta}+\mathbf{z}_k'\mathbf{u})\right]\right.$$

$$\left.+(n_{k+}-n_{k1})\ln\left[1-\Phi(\mathbf{x}_k'\boldsymbol{\beta}+\mathbf{z}_k'\mathbf{u})\right]\right\} -(\mathbf{u}'\mathbf{A}^{-1}\mathbf{u}/2\sigma_u^2)+\text{constant}. \tag{17.53}$$

Finding the maximum of (17.53) with respect to $\boldsymbol{\beta}$ and **u** leads to a non-linear system (Gianola and Foulley 1983a) which can be solved iteratively using a second-order algorithm such as Newton-Raphson or Fisher's scoring procedure. These consist in iterating from round t to t+1 with:

$$\begin{bmatrix} \mathbf{X}'\mathbf{Q}^{[t]}\mathbf{X} & \mathbf{X}'\mathbf{Q}^{[t]}\mathbf{Z} \\ \mathbf{Z}'\mathbf{Q}^{[t]}\mathbf{X} & \mathbf{Z}'\mathbf{Q}^{[t]}\mathbf{Z}+\lambda\mathbf{A}^{-1} \end{bmatrix}\begin{bmatrix} \boldsymbol{\beta}^{[t+1]} \\ \mathbf{u}^{[t+1]} \end{bmatrix}=\begin{bmatrix} \mathbf{X}'\mathbf{Q}^{[t]}\mathbf{y}^{[t]} \\ \mathbf{Z}'\mathbf{Q}^{[t]}\mathbf{y}^{[t]} \end{bmatrix}, \tag{17.54}$$

where:

$$\mathbf{y}^{[t]} = \mathbf{X}\boldsymbol{\beta}^{[t]}+\mathbf{Z}\mathbf{u}^{[t]}+[\mathbf{Q}^{-1}\mathbf{v}]^{[t]} , \tag{17.55}$$

$\lambda=1/\sigma_u^2$ equals $(4-h_c^2)/h_c^2$ in a sire model,

$\mathbf{v}^{[t]}=\{v_k\}^{[t]}$ is an s x 1 vector having elements

$$v_k = \phi(\mathbf{w}_k'\theta)\left[\frac{n_{k1}-n_{k+}\Phi(\mathbf{w}_k'\theta)}{\Phi(\mathbf{w}_k'\theta)[1-\Phi(\mathbf{w}_k'\theta)]}\right], \tag{17.56}$$

$\mathbf{Q}$ is an s x s diagonal matrix which can be expressed as:

1. with Newton-Raphson

$$\mathbf{Q} = \text{Diag}\{n_{ko}I_{ko}(I_{ko}+\mathbf{w}_k'\theta)+n_{k1}I_{k1}(I_{k1}+\mathbf{w}_k'\theta)\}, \tag{17.57}$$

$$I_{ko} = -\phi(\mathbf{w}_k'\theta)/[1-\Phi(\mathbf{w}_k'\theta)], \tag{17.58a}$$

$$I_{k1} = \phi(\mathbf{w}_k'\theta)/\Phi(\mathbf{w}_k'\theta), \tag{17.58b}$$

2. with scoring

$$\mathbf{Q} = \text{Diag}\left\{\frac{n_{k+}\phi^2(\mathbf{w}_k'\theta)}{\Phi(\mathbf{w}_k'\theta)[1-\Phi(\mathbf{w}_k'\theta)]}\right\} \quad \text{for } k=1,2,\ldots,s. \tag{17.59}$$

When subpopulations are defined at an elementary level with a single binary observation per subclass, (17.54-59) hold putting $n_{k+}=1$, and $n_{k1}=0$ or 1. In particular one obtains $v_k=I_{ko}$ if the response is in category [o] and $v_k=I_{k1}$ if in category [1]. Expressions (17.57), (17.58a), and (17.58b) suggest a relationship between MAP and the "normal score" procedure (Kendall and Stuart 1961; Millier 1977) as discussed in Gianola and Foulley (1983a). Numerical issues raised when solving the system in (17.54) especially with large data sets, such as comparisons between Newton-Raphson, scoring, and other algorithms, as well as convergence properties, were discussed by Foulley et al. (1983, 1987b), Harville and Mee (1984); Meijering (1985b), and Hoschele (1986).

Regarding inferences, the posterior density (17.52) does not fall into "typical" forms of distribution. However, this posterior is asymptotically normal (Cox and Hinkley 1974), and a possible approximation to (17.52) (Berger 1985) is:

$$\theta|\mathbf{y}, \sigma_u^2 \sim N\,[\theta^*(\sigma_u^2), C(\sigma_u^2)], \tag{17.60}$$

where $\theta^*(\sigma_u^2)$ is the MAP solution from (17.54) and

$$C(\sigma_u^2) = \left[- \frac{\partial^2 L\ (\theta;\sigma_u^2)}{\partial\theta\partial\theta'} \right]^{-1}_{\theta=\theta^*(\sigma_u^2)} . \tag{17.61}$$

This permits making approximate posterior probability statements about θ (Gianola and Foulley 1983a). Zellner and Rossi (1984) have started to investigate using Monte Carlo numerical procedures how these normal approximations work, with emphasis on a two-parameter logit model.

Joint Estimation of Breeding Values and Dispersion Parameters. The point estimator described above is the mode of the posterior distribution of θ conditionally on the sire variance σ_u^2. When this variance is unknown, inferences on θ can be based on $p(\theta|\mathbf{y},\sigma_u^2=\sigma_u^{2*})$ where σ_u^2 is replaced by the mode of its marginal posterior distribution (Box and Tiao 1973; O'Hagan 1980). When a uniform prior distribution is chosen for σ_u^2, the modal value σ_u^{2*} is the "marginal" maximum likelihood estimator (Malecot 1947) obtained from $p(\mathbf{y}|\sigma_u^2)$, and is equivalent to REML under multivariate normality (Harville 1974). The point estimator of θ so obtained, i.e., $\theta^*(\sigma_u^{2*})$, can be viewed as an empirical Bayes estimator.

This approach was suggested by Gianola et al. (1986) in the context of optimum prediction of breeding values, and applied by Foulley et al. (1987a,b) to genetic evaluation for binary responses in situations of uncertain paternity and multiple trait analysis, respectively. These authors have shown that maximization of $p(\sigma_u^2|\mathbf{y})$ with respect to σ_u^2 with a flat prior distribution for σ_u^2 leads to the equation:

$$\mathbf{E}_c \left[\frac{\partial}{\partial\sigma_u^2} \ln\ p(\mathbf{u}|\sigma_u^2) \right] = 0\ , \tag{17.62}$$

where E_c indicates expectation with respect to the distribution $p(\mathbf{u}|\mathbf{y},\sigma_u^2)$. The derivation is based on the decomposition of $p(\boldsymbol{\beta},\mathbf{u},\sigma_u^2|\mathbf{y})$ into $p(\mathbf{y}|\boldsymbol{\beta},\mathbf{u},\sigma_u^2).p(\mathbf{u}|\sigma_u^2).\ p(\boldsymbol{\beta}).p(\sigma_u^2)$. An important simplification arises when solving (17.62) iteratively. If at iteration t+1, the expectation in (17.62) is evaluated at $\sigma_u^{2[t]}$, the value of σ_u^2 at the previous iteration, maximization of $\ln p(\sigma_u^2|\mathbf{y})$ can be done by taking at each iterate:

$$\underset{\sigma_u^2}{\mathrm{Max}}\ E_c^{[t]}\left[\ \ell n\ p(\mathbf{u}|\sigma_u^2)\ \right].\tag{17.63}$$

Now, as $p(\mathbf{u}|\sigma_u^2)$ is a normal density function, the expectation of its logarithm is equal to

$$\text{constant} -(q/2)\ \ell n(\sigma_u^2)-(1/2)\sigma_u^{-2}E_c(\mathbf{u}'\mathbf{A}^{-1}\mathbf{u}),$$

which is maximum with respect to σ_u^2 (Anderson 1984) when:

$$\sigma_u^{2[t+1]} = E_c^{[t]}(\mathbf{u}'\mathbf{A}^{-1}\mathbf{u})/m.\tag{17.64}$$

Assuming as others (Harville and Mee 1984; Stiratelli et al. 1984) that the expectation and variance of $\mathbf{u}|\mathbf{y},\sigma_u^2$ can be approximated by the parameters in (17.60), one obtains in (17.64):

$$\sigma_u^{2[t+1]} = \left[\mathbf{u}^{*\prime}\mathbf{A}^{-1}\mathbf{u}^{*}+\mathrm{tr}(\mathbf{A}^{-1}\mathbf{C}_{uu}^{*})\right]^{[t]}/m,\tag{17.65}$$

where $\mathbf{u}^*$ is the u-solution to (17.54), and $\mathbf{C}_{uu}^*$ is the u-part of the inverse of the (Newton-Raphson) coefficient matrix in (17.54) evaluated at $\sigma_u^2=\sigma_u^{2[t]}$.

This expression was also derived by Harville and Mee (1984) and it appears in the EM algorithm (Dempster et al. 1977) for estimation of σ_u^2 by REML under normality. Since the likelihood $p(\mathbf{y}|\boldsymbol{\beta},\mathbf{u},\sigma_u^2)$ does not depend on σ_u^2, (17.62) is very general and can be applied to more complex situations, e.g., uncertain paternity (Foulley et al. 1987a), multivariate analysis (Foulley et al. 1987b), and Poisson-binomial traits (Foulley and Elsen 1987). This procedure can also be extended to several random factors (Foulley and Elsen 1987; Hoschele et al. 1987). In any case, successive solutions calculated with (17.65) are strictly positive (or more generally positive-definite) provided starting values are strictly positive (or positive-definite). Because this algorithm often converges slowly, alternative techniques have been proposed (Schaeffer 1979; Smith and Graser 1986; Thompson and Cameron 1986; Van Raden and Freeman 1986; Hoschele et al. 1987; Laird et al. 1987). As shown in a simulation study, the marginal ML estimator of σ_u^2 leads to smaller mean squared errors than conventional REML applied directly to data or transformed scores except with very small subclass size (Hoschele et al. 1987). The use of informative priors for variance components has also started to be investigated (Foulley et al. 1987b; Hoschele et al. 1987).

Gilmour's Approach. The methods proposed by Gianola and Foulley (1982; 1983a,b), Foulley et al. (1983; 1987a,b), Foulley and Gianola (1984, 1986), Stiratelli et al. (1984), and Harville and Mee (1984) are essentially the same and lead to the equations anticipated by Thompson (1979). An alternative approach was developed by Gilmour (1983) and

Gilmour et al. (1985) which is based on generalized linear models (McCullagh and Nelder 1983). The starting point is the generalized linear model:

$$\ell_k \doteq \mathbf{x}_k'\boldsymbol{\beta}+\left[\frac{\partial\Phi^{-1}(p)}{\partial p}\right]_{p=\pi_k}(p_k-\pi_k)\ , \tag{17.66}$$

where p_k and π_k are the observed frequency and true probability of response ("success") for an observation in the kth subpopulation. Φ^{-1} designates the probit "link" function. $\mathbf{x}_k'\boldsymbol{\beta}$ is the expectation of the working variable ℓ_k that, by definition, is a linear function of the unknown fixed effects $\boldsymbol{\beta}$.

The maximum quasi-likelihood equations for $\boldsymbol{\beta}$ are

$$\mathbf{X'W^*X}\tilde{\boldsymbol{\beta}} = \mathbf{X'W^*}\boldsymbol{\ell}, \tag{17.67}$$

where $\mathbf{X}=[\mathbf{x}_1,\mathbf{x}_2,...,\mathbf{x}_k,...,\mathbf{x}_s]'$, $\boldsymbol{\ell}=\{\ell_k\}$ and $\mathbf{W}^*$ is the s x s symmetric matrix of weights given by

$$\mathbf{W}^* = \mathbf{D}\mathbf{V}^{*-1}\mathbf{D} \tag{17.68a}$$

with $\mathbf{D}^{-1}=\text{Diag}\{[\partial\Phi^{-1}(p)/\partial p]\}_{p=\pi_k}$ (k=1,2,...,s). Also letting $\mathbf{p}=\{p_k\}$

$$\mathbf{V}^* = \text{var}(\mathbf{p}). \tag{17.68b}$$

One critical issue is the specification of the elements of $\mathbf{V}^*$. Let, as previously, y_{kr} be an indicator variable taking the value 1 if a response occurs in observation r ($r=1,2,...,n_{k+}$) and population k, and **0** otherwise. Then, $p_k=(\sum_r y_{kr})/n_{k+}$, and $\sum_r y_{kr}$ has a (conditional) binomial distribution with parameters n_{k+} and $E(p_k|\theta)$; the expectation of p_k can be expressed as a function of parameters $\theta'=[\boldsymbol{\beta}',\mathbf{u}']$ containing levels of factors involved in the definition of populations 1,...,k. The conditional variance of p_k given θ is

$$\text{var}(p_k|\theta) = \{E(p_k|\theta)[1-E(p_k|\theta)]\}/n_{k+} \tag{17.69a}$$

and the unconditional variance can be obtained as:

$$\text{var}(p_k) = E_\theta[\text{var}(p_k|\theta)]+\text{var}_\theta[E(p_k|\theta)]. \tag{17.69b}$$

One possible way to write $E(p_k|\theta)$ is to make a first-order expansion of p_k at $\mathbf{x}_k'\boldsymbol{\beta}$ using the inverse link function in (17.66); then:

$$p_k \doteq \Phi(\mathbf{x}_k'\boldsymbol{\beta}) + \left[\frac{\partial\Phi(\ell)}{\partial\ell}\right]_{\ell=\mathbf{x}_k'\boldsymbol{\beta}} (\ell_k - \mathbf{x}_k'\boldsymbol{\beta}). \tag{17.70}$$

Moreover, one can write:

$$\ell_k = (\sum_r \ell_{kr})/n_{k+}, \tag{17.71a}$$

where

$$\ell_{kr} = \mathbf{x}_k'\boldsymbol{\beta} + \mathbf{z}_k'\mathbf{u} + e_{kr}, \tag{17.71b}$$

where $\mathbf{x}_k'$ and $\mathbf{z}_k'$ are incidence row vectors, and the e_{kr}'s are independent standardized residual components, also uncorrelated with $\mathbf{u}$. Provided $\boldsymbol{\beta}$ and $\mathbf{u}$ are in units of the standard deviation of ℓ_{kr} in (17.71b), i.e., $\sigma_\ell=(1+\sigma_u^2)^{1/2}$, with $\sigma_e^2=1$, and letting $\pi_k=\Phi(\mathbf{x}_k'\boldsymbol{\beta})$, (17.70) reduces to

$$p_k \doteq \pi_k + \phi(\mathbf{x}_k'\boldsymbol{\beta})(\mathbf{z}_k'\mathbf{u} + e_k). \tag{17.72}$$

Hence

$$E(p_k|\theta) \doteq \pi_k + \phi(\mathbf{x}_k'\boldsymbol{\beta})\mathbf{z}_k'\mathbf{u} \tag{17.73a}$$

and, since the $\mathbf{u}$'s are the only random effects in θ,

$$E_u[\mathrm{var}(p_k|\theta)] \doteq [\pi_k(1-\pi_k) - \phi^2(\mathbf{x}_k'\boldsymbol{\beta})t]/n_{k+} \tag{17.73b}$$

because $\mathbf{z}_k'\mathbf{z}_k=1$. Also

$$\mathrm{var}_{\mathbf{u}}[E(p_k|\theta)] \doteq \phi^2(\mathbf{x}_k'\boldsymbol{\beta})t. \tag{17.73c}$$

Above, $t=\sigma_u^2/(1+\sigma_u^2)$ is the intra-class correlation when u_i is expressed in units of $(1+\sigma_u^2)^{1/2}$.

In addition, Gilmour et al. (1985) used (17.72) to approximate the (k,ℓ) element of V^* in (17.68b) as

$$\mathrm{Cov}(p_k, p_\ell) \doteq \phi(\mathbf{x}_k'\boldsymbol{\beta})\mathbf{z}_k'\mathbf{G}\mathbf{z}_\ell\phi(\mathbf{x}_\ell'\boldsymbol{\beta}) \tag{17.74}$$

with **G**=**A**t, **A** being the additive genetic relationship matrix. Collecting (17.73b), (17.73c), and (17.74), and using these expressions in (17.68b), one obtains

$$\mathbf{V}^* = \mathbf{V}+\mathbf{DZGZ'D} \tag{17.75}$$

with **V** defined by Gilmour et al. (1985) as:

$$\mathbf{V} = \mathrm{Diag}\left\{[\pi_k(1-\pi_k)-\phi^2(\mathbf{x}_k'\boldsymbol{\beta})t]/n_{k+}\right\} \tag{17.76a}$$

and

$$\mathbf{D} = \mathrm{Diag}\left\{\phi(\mathbf{x}_k'\boldsymbol{\beta})\right\}. \tag{17.76b}$$

Finally, the matrix of weights $\mathbf{W}^*$ in (17.68a) becomes

$$\mathbf{W}^* = (\mathbf{D}^{-1}\mathbf{V}\mathbf{D}^{-1}+\mathbf{ZGZ'})^{-1}. \tag{17.77}$$

If the data vector $\boldsymbol{\ell}=\{\ell_k\}$ in (17.71a) had expectation $E(\boldsymbol{\ell})=\mathbf{X}\boldsymbol{\beta}$ and $\mathrm{var}(\boldsymbol{\ell})=\mathbf{R}+\mathbf{ZGZ'}$, and if it were observable, the BLUP of **u** would be $\hat{\mathbf{u}}=\mathrm{Cov}(\mathbf{u},\boldsymbol{\ell}')[\mathrm{var}(\boldsymbol{\ell})]^{-1}(\boldsymbol{\ell}-\mathbf{X}\tilde{\boldsymbol{\beta}})$ where $\tilde{\boldsymbol{\beta}}$ is the GLS estimator of $\boldsymbol{\beta}$ based on $\boldsymbol{\ell}$. In Gilmour et al.(1985), $\boldsymbol{\ell}-\mathbf{X}\tilde{\boldsymbol{\beta}}$ is replaced by $[\mathbf{D}^{-1}(\mathbf{p}-\boldsymbol{\pi})]$ evaluated at $\boldsymbol{\beta}=\tilde{\boldsymbol{\beta}}$. This is the difference between the working variable (17.66) evaluated at $\boldsymbol{\beta}=\tilde{\boldsymbol{\beta}}$, which is the estimator of $\boldsymbol{\beta}$ from (17.67), and $\mathbf{X}\tilde{\boldsymbol{\beta}}$. Taking advantage of the fact that $\mathbf{W}^{*-1}$ (the variance of the working variable) has the same structure (see 17.77) as the variance covariance matrix of data following the usual mixed linear model, Gilmour et al. (1985) suggested that the mixed model equations could be used to compute $\tilde{\boldsymbol{\beta}}$ and $\tilde{\mathbf{u}}$, an analog of $\hat{\mathbf{u}}$. Letting $\mathbf{S}^{-1}=\mathbf{D}^{-1}\mathbf{V}\mathbf{D}^{-1}$, the system of equations to solve may be written as:

$$\begin{bmatrix} \mathbf{X'S}^{[t]}\mathbf{X} & \mathbf{X'S}^{[t]}\mathbf{Z} \\ \mathbf{Z'S}^{[t]}\mathbf{X} & \mathbf{Z'S}^{[t]}\mathbf{Z}+(\mathbf{A}^{-1}/t) \end{bmatrix} \begin{bmatrix} \tilde{\boldsymbol{\beta}}^{[t+1]} \\ \tilde{\mathbf{u}}^{[t+1]} \end{bmatrix} = \begin{bmatrix} \mathbf{X'S}^{[t]}\boldsymbol{\ell}^{[t]} \\ \mathbf{Z'S}^{[t]}\boldsymbol{\ell}^{[t]} \end{bmatrix}, \tag{17.78}$$

where

$$\ell^{[t]} = \mathbf{X}\tilde{\boldsymbol{\beta}}^{[t]} + \mathbf{S}^{-1[t]}\mathbf{v}^{[t]} \tag{17.79}$$

$$\mathbf{S} = \mathrm{Diag}\left\{ \frac{n_{k+}\phi^2(\mathbf{x}_k'\boldsymbol{\beta})}{\Phi(\mathbf{x}_k'\boldsymbol{\beta})[1-\Phi(\mathbf{x}_k'\boldsymbol{\beta})]-\phi^2(\mathbf{x}_k'\boldsymbol{\beta})t} \right\} \tag{17.80}$$

$$\mathbf{v} = \left\{ \frac{n_{k1}-n_{k+}\Phi(\mathbf{x}_k'\boldsymbol{\beta})}{\Phi(\mathbf{x}_k'\boldsymbol{\beta})\,[1-\Phi(\mathbf{x}_k'\boldsymbol{\beta})]-\phi^2(\mathbf{x}_k'\boldsymbol{\beta})t}\ \phi(\mathbf{x}_k'\boldsymbol{\beta}) \right\}. \tag{17.81}$$

The denominator of elements in (17.80) and (17.81) is an approximation to

$$E_u\left\{ \Phi(\mathbf{x}_k'\boldsymbol{\beta}+\mathbf{z}_k'\mathbf{u})[1-\Phi(\mathbf{x}_k'\boldsymbol{\beta}+\mathbf{z}_k'\mathbf{u})] \right\}.$$

The similarities between **S** in (17.80) and **Q** in (17.59), and between (17.81) and (17.56) are clear. Therefore, the systems of equations to be solved in the Gilmour et al. (1985) and in the "joint maximization" procedures are similar. It can be shown that the exact expressions for the elements of **V*** in (17.75) are:

$$V^*_{kk} = \{\pi_k(1-\pi_k)+(n_{k+}-1)[\Phi_2(\mathbf{x}_k'\boldsymbol{\beta},\mathbf{x}_k'\boldsymbol{\beta};t)-\pi_k^2]\}/n_{k+}$$

$$V^*_{kk'} = \Phi_2(\mathbf{x}_k'\boldsymbol{\beta},\mathbf{x}_{k'}'\boldsymbol{\beta};\mathbf{z}_k'\mathbf{G}\mathbf{z}_{k'}t)-\pi_k\pi_{k'},\ \text{for } k\neq k',$$

where $\Phi_2(x,y;\rho)$ is the standardized bivariate cumulative distribution function with mean 0 and correlation ρ. Note that this complicates computations somewhat because bivariate volumes need to be calculated for each unique $\mathbf{x}_k'\boldsymbol{\beta}$ at every iterate (Ducrocq and Colleau 1986). Note also that with this expression of $\mathbf{V}^*$, $\mathbf{W}^*$ has no longer the same structure as (17.77). This precludes using (17.78) to compute $\tilde{\boldsymbol{\beta}}$ and one has to solve the original system (17.67).

The meaning of the $\tilde{\boldsymbol{\beta}}$ solution in (17.78) is clear because of the equivalence with the solution to (17.67) due to the relationship:

$$\mathbf{W} = \mathbf{S}\text{-}\mathbf{S}'\mathbf{Z}(\mathbf{Z}'\mathbf{S}\mathbf{Z}+\mathbf{G}^{-1})^{-1}\mathbf{Z}'\mathbf{S}. \tag{17.82}$$

However, as pointed out by Knuiman and Laird, Chapter 9, this Volume, the interpretation and properties of $\tilde{\mathbf{u}}$ in (17.78) need to be clarified.

Thompson et al. (1985) have also proposed to use generalized linear model techniques to estimate heritability of liability from parent-offspring relationships (Falconer 1965; James and McGuirk 1982) when both generation means are affected by fixed effects.

17.3.2 Extension to Other Situations

Ordered Polychotomies. The data $\mathbf{y}$ now form an s x (c+1) contingency table with the typical element n_{kq}, the number of responses in the qth category of response (q=1,2,...,c+1) for the kth subpopulation (k=1,2,...,s). As before, an underlying normal distribution resulting from polygenic and random environmental effects is assumed. The underlying scale is rendered discrete through a set of c fixed ordered thresholds on the conceptual scale $(\tau_1,\tau_2,...,\tau_q,...,\tau_c)$, with $\tau_0=-\infty$, $\tau_{c+1}=+\infty$. The rth experimental unit in "population" k is characterized by the underlying variable ℓ_{kr} with location parameter η_k:

$$\ell_{kr} = \eta_k+e_{kr}.$$

Given η_k, the probability of response in category q is:

$$\pi_{kq} = \text{Prob}(\tau_{q-1}<\ell\leq\tau_q|\eta_k)$$

$$= \Phi[(\tau_q-\eta_k)/\sigma]-\Phi[(\tau_{q-1}-\eta_k)/\sigma], \tag{17.83}$$

where Φ is the normal cumulative distribution function.

As previously, we take σ=1 and decompose η_k as a linear combination of fixed (α) and random ($\mathbf{u}$) effects so that:

$$\eta_k = \mathbf{h}_k'\alpha+\mathbf{z}_k'\mathbf{u}$$

or, in matrix notation:

$$\eta = \mathbf{H}\alpha+\mathbf{Z}\mathbf{u} \tag{17.84}$$

with $\mathbf{H}$ having full column rank p.

Write now, $\alpha'=(\alpha_1,\boldsymbol{\beta}')$, and $\mathbf{H}=[\mathbf{1},\mathbf{X}]$ with rank $(\mathbf{X})$=p-1; then (17.84) becomes:

$$\eta = \mathbf{1}\alpha_1+\mathbf{X}\boldsymbol{\beta}+\mathbf{Z}\mathbf{u}. \tag{17.85}$$

Because $\tau_q-\eta_k$ in (17.83) is then equal to $(\tau_q-\alpha_1)-(\mathbf{x}_k'\boldsymbol{\beta}+\mathbf{z}_k'\mathbf{u})$, the probability π_{kq} can be parameterized as a function of

i) $\mathbf{t}'=\{t_q\}$, with $t_q = \tau_q-\alpha_1$,

a c x 1 vector placing a reference "population" parameter α_1 (e.g., male calves born out of heifer matings) in relation to the c thresholds, and

ii) $\mu = \mathbf{X}\boldsymbol{\beta}+\mathbf{Z}\mathbf{u}$,

an s x 1 vector of location parameters describing the s population means. Harville and Mee (1984) give an equivalent parameterization, setting the origin at a given threshold (say $\tau_1=0$) and parameterizing t in (i) in terms of α_1 and (c-1) differences ($\tau_q-\tau_1$; q=2, ...,c) between thresholds.

Obtaining the point estimators requires iterating a system similar to (17.54), but augmented with a sector pertaining to the vector **t** (Gianola and Foulley 1983a; Harville and Mee 1984). The marginal ML estimator of σ_u^2 can also be computed using the algorithm in (17.65). Gilmour et al. (1987) describe a multiple threshold version of their procedure.

Multiple Binary Responses. This problem was addressed by Foulley and Gianola (1984, 1986) for two traits, and extended to n traits by Hoschele et al. (1986) and Foulley et al. (1987b). The contingency table has the same form as before, with n_{kq} responses in population k(k=1,2,...,s) falling in category q(q=1,2,...,2^n). The qth category can be designated by an n-bit digit $(\Omega_{1q},\Omega_{2q},...,\Omega_{jq}, ...,\Omega_{nq})$ with Ω_{jq}=0 or 1 for attributes coded [0] and [1], respectively, in trait i.

The model used is an extension of the threshold-liability concept to the multidimensional domain. The conditional probability of response in category q is related to the values of n jointly normal continuous underlying variables (ℓ_j;j=1,2,...,n), with thresholds (τ_j;j=1,2,...,n). For the rth experimental unit, one uses the model:

$$\ell_{jkr} = \eta_{jk}+e_{jkr} \tag{17.86}$$

with $\mathrm{var}(e_{jkr})=\sigma_{e_j}^2$

$$\mathrm{Cov}(e_{jkr},e_{j'k'r'}) = [\delta_{kk'rr'}]\rho_{jj'}\sigma_{e_j}\sigma_{e_{j'}},$$

where $\delta_{kk'rr'}=1$ if k=k' and r=r', and 0 otherwise; $\rho_{jj'}$ is the residual correlation between traits j and j', and σ_{e_j} is the residual standard deviation, assumed to be the same for all traits and taken as unit of measurement ($\sigma_{e_j}=1$). Hoschele et al. (1987) wrote the probability of response in category q in terms of the n-normal probability integral (Φ_n):

$$\pi_{kq} = \Phi_n[\mu_k^{(q)};\mathbf{R}^{(q)}], \tag{17.87}$$

where

$$\mu_k^{(q)} = \{\omega_{jq}\mu_{jk}\};\ j=1,2,...,n \tag{17.88}$$

$$\omega_{jq} = (-1)^{\Omega_{jq}} \tag{17.89a}$$

$$\mu_{jk} = (\tau_j-\eta_{jk})/\sigma_{e_j} \tag{17.89b}$$

and $\mathbf{R}^{(q)}$ is a symmetric (n x n) matrix having elements

$$r_{jj',q} = \omega_{jq}\omega_{j'q}\rho_{jj'}. \tag{17.90}$$

Letting $\mu_j=\{\mu_{jk}\}$ we adopt the same decomposition as in (17.47) for each trait j

$$\mu_j = \mathbf{X}_j\boldsymbol{\beta}_j+\mathbf{Z}_j\mathbf{u}_j \tag{17.91a}$$

or, after concatenating vectors and matrices, trait by trait:

$$\mu = \mathbf{X}\boldsymbol{\beta}+\mathbf{Z}\mathbf{u} = \mathbf{W}\theta. \tag{17.91b}$$

It is assumed a priori that $p(\theta|\Sigma)$ is proportional to $p(\mathbf{u}|\Sigma_u)$, with **u** following the multivariate normal distribution $N(\mathbf{0},\Sigma_u)$ such that for one random factor, $\Sigma_u=\mathbf{G}*\mathbf{A}$, **G** being an (n x n) matrix of "u" variance-covariance components, usually a fraction of the matrix of additive genetic variances and covariances. The modal estimators of $\boldsymbol{\beta}$ and **u**, given **G** and **R**, can be obtained solving a system formally identical to (17.54), but in the multidimensional domain. This system is, in a condensed form:

$$(\mathbf{W}'\mathbf{S}^{[t]}\mathbf{W}+\Sigma^{-1})\theta^{[t+1]} = \mathbf{W}'\mathbf{S}^{[t]}\mathbf{y}^{[t]} \tag{17.92}$$

with $\mathbf{y}^{[t]}=\mathbf{W}\theta^{[t]}+\mathbf{S}^{-1[t]}\mathbf{v}^{[t]}$. Elements of **v** and **S** have been given by Foulley and Gianola (1984) and Hoschele et al. (1987).

The marginal ML estimator of **G** can be approximated iterating with the EM-type expression:

$$g_{jj'}^{*[t+1]} = E_c^{[t]}(\mathbf{u}_j'\mathbf{A}^{-1}\mathbf{u}_{j'})/m, \tag{17.93}$$

where $E_c^{[t]}$ denotes expectation with respect to $\mathbf{u}|\mathbf{G}^{[t]},\mathbf{R},\mathbf{y}$. Foulley et al. (1987b) suggested an ML type estimator of **R** which is closely related to results of Tallis (1962) and Thompson (1972).

This procedure has also been extended to accommodate missing information on some traits (Foulley and Gianola 1986). In this case, the likelihood function can be written as a product of several conditionally independent multinomial distributions, each corresponding to a particular class of information. This procedure might be useful to adjust for bias due to sequential selection on categorical traits. A similar problem was addressed by Basu and Pereira (1982) using Dirichlet or Dirichlet-multinomial prior distributions. Other formulations for incomplete data were given by Koch et al. (1972) and Woolson and Clarke (1984), based on a modification of the method of Grizzle et al. (1969).

Mixture of Binary and Normal Traits. In many animal breeding applications, data sets comprise observations on categorical and quantitative variates. A typical example is dystocia, appraised subjectively, and related records on birth weight, gestation length of the calf, and possibly pelvic opening of the dam. For simplicity, consider a continuous variable $\mathbf{y}_1$ which depends on parameters $\theta_1'=[\boldsymbol{\beta}_1',\mathbf{u}_1']$, and a binary variable $\mathbf{y}_2$ with parameters $\theta_2'=[\boldsymbol{\beta}_2',\mathbf{u}_2']$. As usual, the conditional distribution of $\mathbf{y}_1$ given θ_1 is taken as:

$$\mathbf{y}_1|\theta_1 \sim N(\mathbf{W}_1\theta_1,\sigma_{e_1}^2\mathbf{I}). \tag{17.94}$$

Furthermore, it will be assumed that the conditional probability of response for individual k given the information on $\mathbf{y}_1$ can be modelled as:

$$\pi_k = \Phi(\mu_{2k}^*) \tag{17.95a}$$

with

$$\mu_{2k}^* = \mathbf{w}_{2k}'\theta_2+b(y_{1k}-\mathbf{w}_{1k}'\theta_1), \tag{17.95b}$$

where $\mu_{2k}^*=E(\ell_{2k}|y_{1k},\theta_1,\theta_2)$ is the conditional mean of the underlying normal variable ℓ_{2k} given y_{1k} (and θ_1,θ_2), b being the residual regression coefficient.

The posterior distribution of θ_1 and θ_2, assuming that b and Σ_u, the dispersion matrix of $\mathbf{u}=[\mathbf{u}_1',\mathbf{u}_2']'$, are known, can be written as:

$$p(\theta_1,\theta_2|\mathbf{y}_1,\mathbf{y}_2,\mathrm{b},\Sigma_u) \propto$$

$$p(\mathbf{y}_2|\mathbf{y}_1,\theta_1,\theta_2,\mathrm{b})p(\mathbf{y}_1|\theta_1)p(\mathbf{u}|\Sigma_u), \tag{17.96}$$

where:

$$p(\mathbf{u}|\Sigma_u) \propto \exp(-1/2\ \mathbf{u}'\Sigma_u^{-1}\mathbf{u}) \tag{17.97a}$$

$$p(\mathbf{y}_1|\theta_1) \propto \exp[-(\mathbf{y}_1-\mathbf{w}_1'\theta_1)'(\mathbf{y}_1-\mathbf{w}_1'\theta_1)/2\sigma_{e_1}^2] \tag{17.97b}$$

$$p(\mathbf{y}_2|\mathbf{y}_1,\theta_1,\theta_2,\mathrm{b}) \propto \prod_k[\Phi(\mu_{2k}^*)]^{y_{2k}}[1-\Phi(\mu_{2k}^*)]^{1-y_{2k}}, \tag{17.97c}$$

where $y_{2k}=1$ if a response occurs, and 0 otherwise. The mode of the posterior density (17.96) can be obtained upon iteration with:

$$(\mathbf{W}'\mathbf{Q}^{[t]}\mathbf{W}+\Sigma^{-1})\theta^{[t+1]} = \mathbf{W}'\mathbf{Q}^{[t]}\mathbf{y}^{[t]}, \tag{17.98}$$

where

$$\mathbf{W} = [\mathbf{W}_1,\mathbf{W}_2];\ \Sigma = \mathrm{var}(\theta)$$

$$\mathbf{Q} = \begin{bmatrix} \mathbf{Q}_{11} & \mathbf{Q}_{12} \\ \mathbf{Q}_{12}' & \mathbf{Q}_{22} \end{bmatrix}$$

$$\mathbf{y}^{[t]} = \begin{bmatrix} y_1^{[t]} \\ y_2^{[t]} \end{bmatrix} = \begin{bmatrix} \mathbf{W}_1\theta_1^{[t]} \\ \mathbf{W}_2\theta_2^{[t]} \end{bmatrix} + \begin{bmatrix} \mathbf{Q}_{11}^{[t]} & \mathbf{Q}_{12}^{[t]} \\ \mathbf{Q}_{12}'^{[t]} & \mathbf{Q}_{22}^{[t]} \end{bmatrix}^{-1} \begin{bmatrix} \mathbf{v}_1^{[t]} \\ \mathbf{v}_2^{[t]} \end{bmatrix}.$$

$\mathbf{v}_1$, $\mathbf{v}_2$, and partitions in $\mathbf{Q}$ are computed as follows:

$$\mathbf{v}_1 = (\mathbf{y}_1-\mathbf{W}_1\theta_1)\sigma_{e_1}^{-2}-\mathrm{b}\{\mathrm{v}_2\}$$

$$\mathbf{v}_2 = \{\phi(\mu_{2k}^*)[y_{2k}-\Phi(\mu_{2k}^*)]/\Phi(\mu_{2k}^*)[1-\Phi(\mu_{2k}^*]\}$$

$$Q_{11} = \mathrm{Diag}\{\sigma_{e_1}^{-2} + b^2 v_{2k}(v_{2k} + \mu_{2k}^*)\}$$

$$Q_{12} = \mathrm{Diag}\{-bv_{2k}(v_{2k} + \mu_{2k}^*)\}$$

$$Q_{22} = \mathrm{Diag}\{v_{2k}(v_{2k} + \mu_{2k}^*)\}.$$

Simplifications arise when the columns of $\mathbf{X}_1$, $\mathbf{Z}_1$, are subsets of columns of $\mathbf{X}_2$ and $\mathbf{Z}_2$, respectively, as discussed by Foulley et al. (1983). Estimators of b and Σ_u can also be derived along the lines of Foulley et al. (1987b). An alternative procedure based on either the conditional distribution of the binary variable given the continuous variate, or vice versa, was presented by Im (1985).

17.4 Discussion and Conclusion

The Bayesian paradigm provides a very powerful tool for solving problems in the estimation of breeding values and genetic parameters for categorical traits. Most of this paper was devoted to traits having (conditional) binomial or multinomial distributions. However, traits such as ovulation rate and litter size, which are count variates, might be better described with a Poisson distribution. An illustration of Bayesian inference applied to reproductive data with a mixture of Poisson and binomial distributions is in Foulley et al. (1987c).

Categorical traits were assumed in this paper to have a multifactorial genetic determinism. However, the approach can be extended to mixed models of inheritance involving a major gene and polygenic effects (Chap. 3, this Vol.). An application to the problem of assigning a sire's genotype using progeny data having a discrete distribution has been considered by Foulley and Elsen (1987).

Because of technical difficulties in evaluating analytically the distributions involved, several approximations need to be made, and this may limit the efficiency of the Bayesian techniques. The genetic evaluation procedure described in this paper consists of basing inferences on θ on the conditional distribution $p(\theta|\mathbf{Y},\gamma)$ where γ, the dispersion parameters, are assumed to be known or estimated via their marginal ML estimator. From a theoretical point of view, it would be desirable: (1) to take the expectation rather than the mode of the posterior distribution of genetic merit, (2) if modal approximations are used, it would be interesting to compute these from the completely marginalized posterior distribution of $\mathbf{u}$ after integrating out all "nuisance" parameters (Box and Tiao 1973), i.e., the fixed effects $\boldsymbol{\beta}$ and the dispersion parameters γ. Moreover, modal approximations for $\mathbf{u}$ could be especially critical when estimating $\mathbf{G}$ as done by Foulley et al. (1987b) and Hoschele et al. (1987).

With respect to aspect (2), the approach of Gilmour et al. (1985) can be interpreted as an attempt to integrate out **u** for estimating the fixed effects $\boldsymbol{\beta}$. This may explain why their procedure may give "better" (in the sense of smaller bias) estimates of variance components than joint maximization (Harville and Mee 1984), when subclasses defined as combinations of all factors involved have few observations (Gilmour et al. 1985). It is necessary to investigate more deeply the relative merits of these two procedures for estimating genetic merit and dispersion parameters. Further research is also needed to assess the adequacy of modal approximations using exact posterior density functions with, for instance, numerical integration techniques as in Zellner and Rossi (1984). Progress would also be desirable in the area of estimating directly elementary probabilities needed in the evaluation of candidates. This is so because modal estimators are not functionally invariant, and a simple back transformation to the probability scale may not be satisfactory.

Practically, the emergence of Bayesian techniques for the threshold model has led to comparisons between this technique and "BLUP" on 0,1 data or raw scores. These comparisons were primarily motivated by the apparent complexity and higher computing costs of the new techniques over those, now common, of the mixed linear model. A first set of studies involved real data sets and comparisons between the two procedures for single trait sire evaluation in cattle. Traits analyzed were: survival to 48 months of age and 56-day non-return rate (Hoschele 1986), stillbirth (Meijering 1985b), calving difficulty (Meijering 1985b; Djemali 1985; Djemali et al. 1986), and type traits (Jensen 1986). Product moment or rank correlations, not surprisingly, were very high (r=0.94 to 0.999), except in one study (r= 0.79, Djemali 1985). This is not a compelling argument to dismiss the new techniques, especially because the extra computing effort required (1.3 to 2.8 times in CPU units; Hoschele 1986) should not be overemphasized in view of the rapid developments in computing technology. More pertinent are comparisons based on simulated data (Meijering and Gianola 1985a,b; Hoschele 1986). These studies indicate that the superiority of non-linear estimators over BLUP, in terms of genetic response to selection, is larger as: (1) the number of categories decreases, (2) the overall incidences in categories are more dissimilar from each other, (3) the higher the differences in incidence among sub-classes, (4) the more unbalanced the layout, and (5) the higher the underlying heritability. For binary traits, one may expect an increase in response of about 4 to 12% with 1 to 5% incidence levels and h^2 ranging from 0.20 to 0.50. The relative efficiency of modal estimators based on the threshold concept over BLUP would even be greater if mixed model equations were amended for heteroscedasticity in residual variances, as shown by Meijering and Gianola (1985b) and explained earlier. This highlights a critical difference between the linear and the threshold model. The variance of true transmitting abilities on the observed scale is proportional to $\phi^2(\mathbf{x}_k'\boldsymbol{\beta})\sigma_u^2$ (given the threshold model) and is consequently frequency-dependent. However, this variance is assumed to be constant in the linear model. This is the reason why information is weighted in opposite ways in these two models. However, this peculiarity does not validate per se the threshold model. It is important to bear in mind that all simulation studies have employed threshold models to generate the data, thus assuming implicitly that this was the "true" model.

As seen earlier, it is awkward that σ_u^2 in the linear model must be smaller than or equal to $\pi_j(1-\pi_j)$ for any j level of fixed effects. On the other hand, the threshold-liability model has many interesting features. It permits taking into account the ordinal nature of data in a simple way, whereas procedures presented in the first part of this paper treat variables as nominal. As it can be observed clearly in multiple trait situations, this model is undoubtedly more parsimonious than the linear model, and this is a very desirable feature (McCullagh and Nelder 1983). In particular, as pointed out by Gianola and Foulley (1983a,b) and shown very clearly by Quaas et al. (1988) in an application to calving ease scores in Simmental cattle, the threshold model can adjust automatically for some scale interactions when analyses with linear models would indicate such interactions without real biological meaning.

Finally, ways of "challenging" this or other models with real data would be worth investigating. In genetic evaluation, one is primarily interested in testing the predictive ability of the model with a future data set $\mathbf{y}_2$ rather than with the observed data $\mathbf{y}_1$. In a Bayesian context, such a procedure would use the predictive distribution $p(\mathbf{y}_2|\mathbf{y}_1)$ (Berger 1985; Judge et al. 1985)

$$p(\mathbf{y}_2|\mathbf{y}_1) = \int_{R_{\theta_1}} \int_{R_{\theta_2}} p(\mathbf{y}_2|\theta_2,\theta_1)p(\theta_2,\theta_1|\mathbf{y}_1)d\theta_2$$

$$= \int_{R_{\theta_2}} p(\mathbf{y}_2|\theta_2)p(\theta_2|\mathbf{y}_1)d\theta_2, \qquad (17.99)$$

where $\theta_1=[\boldsymbol{\beta}_1',\mathbf{u}_1']'$ and $\theta_2=[\boldsymbol{\beta}_2',\mathbf{u}_2']'$ are parameters used in the models describing $\mathbf{y}_1$ and $\mathbf{y}_2$, respectively. Note that $\mathbf{p}(\mathbf{y}_2|\theta_2,\theta_1)$ reduces to $\mathbf{p}(\mathbf{y}_2|\theta_2)$ because as soon as the transmitting abilities $\mathbf{u}_2$ are known, the conditional distribution of $\mathbf{y}_2$ does not depend on $\mathbf{u}_1$. When $\boldsymbol{\beta}_1$ and $\boldsymbol{\beta}_2$ are treated as "fixed", one needs to challenge the model using a data set $\mathbf{y}_2$ with the same effects $\boldsymbol{\beta}_2 \equiv \boldsymbol{\beta}_1$ as in $\mathbf{y}_1$. This is so because $\mathbf{y}_1$ does not convey any information on effects in $\boldsymbol{\beta}_2$ which are not in $\boldsymbol{\beta}_1$. Moreover, the predictive distribution (17.99) implicitly assumes that the dispersion parameters are integrated out. Deriving analytically the exact form of this predictive distribution can be extremely difficult even in the case of a single binary trait.

Some approximations might be considered. For instance, Beitler and Landis (1985) suggested a simple chi-square test for lack of fit of the linear model. When applied here this would give the quadratic "loss function":

$$(\mathbf{y}_2-\hat{\mathbf{y}}_2)'[\text{var}(\mathbf{y}_2-\hat{\mathbf{y}}_2)]^{-1}(\mathbf{y}_2-\hat{\mathbf{y}}_2), \qquad (17.100)$$

where $\hat{y}_2$ and var(y_2-$\hat{y}_2$) are obtained when predicting θ_1, θ_2 from $\mathbf{y}_1$, using mixed model equations similar to (17.39) and (17.A12) in the appendix:

$$\hat{y}_2 = \mathbf{W}_2\hat{\theta}_2(y_1)$$

$$\text{var}(y_2\text{-}\hat{y}_2) = \mathbf{W}_2\mathbf{C}_{22}\mathbf{W}_2',$$

where $\mathbf{W}_2$ is the incidence matrix relating $\mathbf{y}_2$ to θ_2, and $\mathbf{C}_{22}$ is the θ_2 part in the inverse of the coefficient matrix.

A similar criterion could be employed as a first attempt to assess the adequacy of the threshold model using the following asymptotic approximations (Zellner and Rossi 1984)

$$\hat{y}_2 = E(y_2|\mathbf{y}_1) \doteq [\{\Phi(\mathbf{w}_{2k}'\theta_2)\}]_{\theta_2=\theta_2^*} \,. \tag{17.101a}$$

$$\text{var}(y_2\text{-}\hat{y}_2) = \text{var}(y_2|\mathbf{y}_1) \doteq [\mathbf{D}'\text{var}(\theta_2|\mathbf{y}_1)\mathbf{D}]_{\theta_2=\theta_2^*} \,, \tag{17.101b}$$

where θ_2^* is the θ_2 component of the mode of $p(\theta_1,\theta_2|\mathbf{y}_1)$

$$\mathbf{D} = \text{Diag}\{\phi(\mathbf{w}_{2k}'\theta_2)\}$$

and var($\theta_2|\mathbf{y}_1$) is approximated from the θ_2 part in the inverse of minus the Hessian of the log-posterior density as in (17.61). Analogous formulae can be obtained for the method of Gilmour et al. (1985). However, these are just rough approximations, and more research is needed in this area, especially as far as comparison of models is concerned.

References

Agresti A (1984) Analysis of ordinal categorical data. Wiley, New York

Anderson TW (1984) An introduction to multivariate statistical analysis. Wiley, New York

Basu D, Pereira CA de B (1982) On the Bayesian analysis of categorical data: the problem of non-response. J Stat Plann Inf 6:345-362

Beitler P, Landis JR (1985) A mixed-effects model for categorical data. Biometrics 41:991-1000

Benzecri JP (1973) L'analyse des données. I. La taxinomie, II. L'analyse des correspondances. Dunod, Paris

Berger JO (1985) Statistical decision theory and Bayesian analysis. 2nd edn. Springer-Verlag, Berlin Heidelberg New York Tokyo

Berger PJ, Freeman AE (1978) Prediction of sire merit for calving difficulty. J Dairy Sci 61:1146-1150

Bishop YMM, Fienberg SE, Holland PW (1975) Discrete multivariate analysis. MIT Press, Cambridge

Bock RD (1975) Multivariate statistical methods in behavioral research. McGraw-Hill, New York

Box GEP, Tiao GC (1973) Bayesian inference in statistical analysis. Addison-Wesley, Reading

Brier SS (1980) Analysis of contingency tables under cluster sampling. Biometrika 67:591-596

Bulmer MG (1980) The mathematical theory of quantitative genetics. Clarendon Press, Oxford

Cady RA, Burnside EB (1982) Evaluating dairy bulls for dystocia and mortality of their progeny. 2nd World Congr Genet Appl Livest Prod 7:139-144. Editorial Garsi, Madrid

Cox DR, Hinkley DV (1974) Theoretical statistics. Chapman and Hall, London

Curnow RN (1972) The multifactorial model for the inheritance of liability to disease and its implication for relatives at risk. Biometrics 28:931-946

Danell O (1985) Theoretical aspects in the estimation of breeding values for all-or-none traits. 36th annu meet EAAP, Kallithea, Kassandra, Greece, Sept 30-Oct 3 1985, Mimeo

Daudin JJ, Meyniel L (1980) Notice d'utilisation du programme de regression qualitative requin. INA-PG, Dep math, Mimeo

Dempfle L (1977) Relation entre BLUP et estimateurs Bayesiens. Ann Genet Sel Anim 9:27-32

Dempster ER, Lerner IM (1950) Heritability of threshold characters. Genetics 35:212-236

Dempster AP, Laird NM, Rubin RB (1977) Maximum likelihood estimation from incomplete data via the EM algorithm. J R Stat Soc B 30:1-20

Djemali M (1985) The economic importance of dystocia in dairy cattle and the development of an ordered categorical analysis procedure to evaluate sires for calving ease. PhD thesis. Iowa State Univ, Ames, Iowa

Djemali M, Berger PJ, Freeman AE, Van Raden PM (1986) Ordered categorical sire evaluation for dystocia in Holsteins. J Dairy Sci 69 suppl 1 (Abstr):125

Ducrocq V, Colleau JJ (1986) Interest in quantitative genetics of Dutt's and Deak's methods for numerical computation of multivariate normal probability integrals. Genet Sel Evol 18:447-474

Edwards JH (1969) Familial predisposition in man. Br Med Bull 24:52-76

Elston RC (1977) Estimating heritability of a dichotomous trait. Biometrics 33:232-233

Falconer DS (1965) The inheritance of liability to certain diseases estimated from the incidence among relatives. Ann Hum Genet 29:51-76

Fernando RL, Gianola D (1984) Optimal properties of the conditional mean as a selection criterion. J Anim Sci 59 suppl 1 (Abstr):177

Fienberg SE (1980) The analysis of cross-classified categorical data. 2nd edn. MIT Press, Cambridge

Foulley JL, Elsen JM (1987) Posterior probability of the sire's genotype based on progeny results. 38th ann meet EAAP, Lisbon, Portugal

Foulley JL, Gianola D (1984) Estimation of genetic merit from bivariate "all-or-none" responses. Genet Sel Evol 16:285-306

Foulley JL, Gianola D (1986) Sire evaluation for multiple binary responses when information is missing on some traits. J Dairy Sci 69:2681-2695

Foulley JL, Gianola D, Thompson R (1983) Prediction of genetic merit from data on categorical and quantitative variates with an application to calving difficulty, birth weight and pelvic opening. Genet Sel Evol 15:407-424

Foulley JL, Gianola D, Planchenault D (1987a) Sire evaluation with uncertain paternity. Genet Sel Evol 19:83-102

Foulley JL, Im S, Gianola D, Hoschele I (1987b) Empirical Bayes estimation of parameters for n polygenic binary traits. Genet Sel Evol 19:197-224

Foulley JL, Gianola D, Im S (1987c) Genetic evaluation for traits distributed as Poisson-binomial with reference to reproductive traits. Theor Appl Genet 73:870-877

Gianola D (1982) Theory and analysis of threshold characters. J Anim Sci 54:1079-1096

Gianola D (1986) On selection criteria and estimation of parameters when the variance is heterogeneous. Theor Appl Genet 72:671-677

Gianola D, Fernando RL (1986) Bayesian methods in animal breeding theory. J Anim Sci 63:217-244

Gianola D, Foulley JL (1982) Non-linear prediction of latent genetic liability with binary expression: an empirical Bayes approach. 2nd World Congr Genet Appl Livest Prod 7:293-303, Editorial Garsi, Madrid

Gianola D, Foulley JL (1983a) Sire evaluation for ordered categorical data with a threshold model. Genet Sel Evol 15:201-224

Gianola D, Foulley JL (1983b) New techniques of prediction of breeding value for discontinuous traits. Proc 32nd Annu Nat Breed Roundtable. St Louis, Missouri

Gianola D, Foulley JL, Fernando RL (1986) Prediction of breeding values when variances are not known. Genet Sel Evol 18:485-498

Gill JL (1986) Review of analysis of contingency tables with dichotomous responses. J Anim Breed Genet 103:1-25

Gilmour AR (1983) The estimation of genetic parameters from categorical data. PhD thesis. Massey Univ, Palmerston North

Gilmour AR, Anderson RD, Rae AL (1985) The analysis of binomial data by a generalized linear mixed model. Biometrika 72:593-599

Gilmour AR, Anderson RD, Rae AL (1987) Variance components on an underlying scale for ordered multiple threshold categorical data using a generalized linear model. J Anim Breed Genet 104:149-155

Goffinet B, Elsen JM (1984) Critere optimal de selection: quelques resultats generaux. Genet Sel Evol 16:307-318

Goux JM (1970) L'héritabilité des caractères à seuil. Sciences agronomiques, Rennes

Griffiths DA (1973) Maximum likelihood estimation for the beta binomial distribution and an application to the household distribution of the total number of cases of a disease. Biometrics 29:637-648

Grizzle JE, Starmer CF, Koch GG (1969) Analysis of categorical data by linear models. Biometrics 25:489-504

Guiard V, Herrendorfer G, Tuchscherer A (1985) Variance component estimation for dichotomous characters and its use for estimating heritability. Biom J 27:653-658

Haberman SJ (1978) Analysis of qualitative data. 1. Introductory topics. Academic Press, New York

Haberman SJ (1979) Analysis of qualitative data. 2. New developments. Academic Press, New York

Harvey WR (1982) Least-squares analysis of discrete data. J Anim Sci 54:1067-1071

Harville DA (1974) Bayesian inference for variance components using only error contrasts. Biometrika 61:383-385

Harville DA, Mee RW (1984) A mixed model procedure for analyzing ordered categorical data. Biometrics 40:393-408

Henderson CR (1973) Sire evaluation and genetic trends. In: Proc Anim Breed Genet Symp in Honor of Dr. J.L. Lush. ASAS and ADSA, Champaign, Illinois pp 10-41

Henderson CR (1984) ANOVA, MIVQUE, REML and ML algorithms for estimation of variances and covariances. In: David HA, David HT (eds). Statistics: an appraisal. pp 257-280, Iowa State Univ Press, Ames

Henderson CR (1986) Recent developments in variance and covariance estimation. J Anim Sci 63:208-216

Henderson CR, Kempthorne O, Searle SR, von Krosigk CN (1959) Estimation of environmental and genetic trends from records subject to culling. Biometrics 1:192-218

Hill WG, Smith C (1977) Estimating heritability of a dichotomous trait. Biometrics 33:231-236

Hoschele I (1986) Estimation of breeding values and variance components with quasi-continuous data. Doctoral Thesis, Univ Hohenheim, Stuttgart

Hoschele I, Gianola D, Foulley JL (1987) Estimation of variance components with quasi-continuous data using Bayesian methods. J Anim Breed Genet 104:334-349

Im S (1982) Contribution a l'etude des tables de contingence a parametres aleatoires: utilisation en biometrie. Doctoral Thesis. Univ Paul Sabatier, Toulouse

Im S (1985) Analyse d'une variable binaire et de plusieurs variables continues. Rev Stat 33:15-28

Im S (1986) Mutual independence in a mixed effects model. Biometrics 42:997

James JW, McGuirk BJ (1982) Regression of offspring on parent for all-or-none traits. Z Tierz Züchtungsbiol 99:308-314

Jensen J (1986) Sire evaluation for type traits with linear and non-linear procedures. Livest Prod Sci 15:165-171

Judge GG, Griffiths WE, Carter Hill R, Lutkepohl H, Lee TC (1985) The theory and practice of econometrics. 2nd edn. Wiley, New York

Kackar RN, Harville DA (1981) Unbiasedness of two-stage estimation and prediction procedures for mixed linear models. Commun Stat Theor Meth A 10:1249-1261

Kakwani NC (1967) The unbiasedness of Zellner's seemingly unrelated regression equations estimators. J Am Stat Assoc 82:141-142

Kendall MG, Stuart A (1961) The advanced theory of statistics. Vol 2. Hafner, New York

Koch GG, Imrey PB, Reinfurt DW (1972) Linear model analysis of categorical data with incomplete response vectors. Biometrics 28:633-692

Kolakowski D, Bock RD (1981) A multivariate generalization of probit analysis. Biometrics 37:541-551

Laird NM, Lange N, Stram D (1987) Maximum likelihood computation with repeated measures: application to the EM algorithm. J Am Stat Assoc 82:97-105

Lalouel JM, Rao DC, Morton NE, Elston RC (1983) A unified model for complex segregation analysis. J Hum Genet 35:816-826

Landis JR, Koch GG (1977) A one-way components of variance model for categorical data. Biometrics 33:671-679

Lavergne C (1984) Contribution a l'etude des modeles a effets aleatoires dans l'analyse des donnees qualitatives. Doctoral Thesis. Univ Paul Sabatier, Toulouse

Lefort G (1980) Le modele de base de la selection: justifications et limites. In: Legay JM et al. (eds) Biometrie et Genetique 4:1-14, Societe Francaise de Biometrie, INRA, Dep biometrie

Lush JL, Lamoreux WF, Hazel LN (1948) The heritability of resistance to death in the fowl. Poult Sci 27:375-388

Malecot G (1947) Les criteres statistiques et la subjectivite de la connaissance scientifique. Ann Univ Lyon 10:43-74

McCullagh P (1980) Regression models for ordinal data. J R Stat Soc Ser B 42:109-142

McCullagh P (1986) Testing in industrial experiments with ordered categorical data. Technometrics 28:307

McCullagh P, Nelder JA (1983) Generalized linear models. Chapman and Hall, London

Meijering A (1985a) Practical aspects in the estimation of parameters and breeding values for all-or-none traits. 36th annu meet EAAP, Kallithea, Kassandra, Greece, Mimeo 27pp

Meijering A (1985b) Sire evaluation for calving traits by best linear unbiased prediction and non-linear methodology. Z Tierz Züchtungsbiol 102:95-105

Meijering A, Gianola D (1985a) Linear versus nonlinear methods of sire evaluation for categorical traits: a simulation study. Genet Sel Evol 17:115-132

Meijering A, Gianola D (1985b) Observations on sire evaluation with categorical data using heteroscedastic mixed linear models. J Dairy Sci 68:1126-1232

Millier C (1977) Codage de variables qualitatives: une application a des notations de tavelure sur des varietes de poires. INRA. Mimeo 15pp

Morton NE, MacLean CJ (1974) Analysis of family resemblance. III Complex segregation of quantitative traits. Am J Hum Genet 26:489-503

Nair VN (1986) Testing in industrial experiments with ordered categorical data. Technometrics 28:283-311

O'Hagan A (1980) Likelihood, sufficiency and ancilarity: reply to the discussion. In: Bernando JM, De Groot MH, Lindley DV, Smith AFM (eds) Bayesian statistics. pp 185-203, Univ Press, Valencia

Plackett RL (1981) The analysis of categorical data. 2nd edn. Griffin, London

Pollak EJ, Freeman AE (1976) Parameter estimation and sire evaluation for dystocia and calf size in Holsteins. J Dairy Sci 59:1817-1824

Quaas RL, Van Vleck LD (1980) Categorical trait sire evaluation by best linear unbiased prediction of future progeny categories frequencies. Biometrics 36:117-122

Quaas RL, Zhao Y, Pollak EJ (1988) Describing interactions in dystocia scores with a threshold model. J Anim Sci 66:396-399

Razungles J (1977) Héritabilité des caractères discrets: étude bibliographique critique. Ann Genet Sel Anim 9:43-61

Robertson A, Lerner IM (1949) The heritability of all-or-none traits: viability of poultry. Genetics 34:395-411

Ronningen K (1971) Some properties of the selection index derived by "Henderson's mixed model method" Z Tierz Züchtungsbiol 88:186-193

Rutledge JJ, Gunsett FC (1982) Analysis of categorical data in the animal sciences. J Anim Sci 54:1072-1078

Schaeffer LR (1979) Estimation of variance and covariance components for average daily gain and backfat thickness in swine. In: Searle SR, Van Vleck LD (eds.). Variance components and animal breeding. Proc conf in Honor of CR Henderson. pp 123-137, Cornell Univ, Ithaca

Schaeffer LR, Wilton JW (1976) Methods of sire evaluation for calving ease. J Dairy Sci 59:544-551

Schmidt P (1976) Econometrics. Dekker, New York

Searle SR (1971) Linear models. Wiley, New York

Searle SR, Speed FM, Henderson HV (1981) Some computational and model equivalences in analyses of variance of unequal subclass numbers data. Am Stat 35:16-33

Shah AK, Claypool PL (1984) Analysis of binary data for the balanced one-way classification. Commun Stat Theor Meth 13:1375-1390

Smith C (1970) Heritability of liability and concordance in monozygous twins. Ann Hum Genet 34:85-91

Smith SP, Allaire FR (1986) Analysis of failure times measures on dairy cows: theoretical considerations in animal breeding. J Dairy Sci 69:217-227

Smith SP, Graser H-U (1986) Estimating variance components in a class of mixed models by restricted maximum likelihood. J Dairy Sci 60:1156-1165

Stiratelli R, Laird N, Ware JH (1984) Random effects models for serial observations with binary response. Biometrics 40:961-971

Tallis GM (1962) The maximum likelihood estimation of correlation from contingency tables. Biometrics 18:342-353

Thompson R (1972) The maximum likelihood approach to the estimate of liability. Ann Hum Genet 36:221-231

Thompson R (1979) Sire evaluation. Biometrics 35:339-353

Thompson R, Cameron ND (1986) Estimation of genetic parameters. In: Dickerson GE, Johnson RK (eds) 3rd World Congr Genet Appl Livest Prod. Agric Commun, Univ Nebraska, Lincoln, Nebraska, XII:371-381

Thompson R, McGuirk BJ, Gilmour AR (1985) Estimating the heritability of all-or-none and categorical traits by offspring-parent regression. Z Tierz Züchtungsbiol 102:342-354

Tukey J (1962) The future of data analysis. Ann Math Stat 33:1-67

Van Raden PM, Freeman AE (1986) Computing restricted maximum likelihood estimates of variances and covariances. J Dairy Sci 69 suppl 1:208

Van Vleck LD, Karner PJ (1980) Sire evaluation by best linear unbiased prediction for categorically scored type traits. J Dairy Sci 63:1328-1333

Van Vleck LD, Edlin KM (1984) Multiple trait evaluation of bulls for calving ease. J Dairy Sci 67:3025-3033

Williams DA (1975) The analysis of binary responses from toxicological experiments involving reproduction and teratogenicity. Biometrics 31:949-952

Woolson RF, Clarke WR (1984) Analysis of categorical incomplete longitudinal data. J R Stat Soc Ser A 147:87-99

Wright S (1934a) An analysis of variability in number of digits in an inbred strain of guinea pig. Genetics 19:506-536

Wright S (1934b) The results of crosses between inbred strains of guinea pigs differing in number of digits. Genetics 19:537-551

Wright S (1968) Evolution and the genetics of populations. 1. Genetic and biometric foundations. Univ Chicago Press, Chicago

Zellner A, Rossi PE (1984) Bayesian analysis of dichotomous quantal response models. J Econometrics 25:365-393

APPENDIX 17.A

17.A.1 A General Linear Mixed Model for Categorical Data

The purpose of this Appendix is to extend the methodology of Beitler and Landis (1985) to more than one fixed factor, and to address the case of several categories as in Quaas and Van Vleck (1980).

Let k designate an elementary subclass formed by any (non-empty) combination of levels of fixed factors involved in the design. For $\mathbf{y}_{kl}$, the lth observation vector on c categories (one being deleted) in subclass k, one can write the model:

$$\mathbf{y}_{kl} = \mathbf{X}_k\mathbf{b}+\mathbf{Z}_k\mathbf{u}+\mathbf{e}_{kl}, \tag{17.A1}$$

where $\mathbf{y}_{kl}=(y_{kl1},y_{kl2},\ldots,y_{klc})'$ with $y_{klc}=1$ if response occurs in category c and 0 otherwise; $\mathbf{b}=(\mathbf{b}_1',\mathbf{b}_2',\ldots,\mathbf{b}_n',\ldots,\mathbf{b}_R')'$ is a column vector of fixed effects pertaining to R factors, each of them defined as $\mathbf{b}_r=(\mathbf{b}_{r1}',\mathbf{b}_{r2}',\ldots,\mathbf{b}_{rh}',\ldots)$, with categories ordered within levels such that $\mathbf{b}_{rh}=(\mathbf{b}_{rh1},\mathbf{b}_{rh2},\ldots,\mathbf{b}_{rhc})'$; $\mathbf{u}=(\mathbf{u}_1',\mathbf{u}_2',\ldots,\mathbf{u}_i',\ldots,\mathbf{u}_m')$ is an mc x 1 vector of effects of a random factor, with $\mathbf{u}_i=[\mathbf{u}_{i1},\mathbf{u}_{i2},\ldots,\mathbf{u}_{ic}]'$, and $\mathbf{e}_{kl}$ is a residual. For simplicity, just one random factor will be considered, but extension of the model to several factors does not raise any difficulty, except in notation.

As previously, the basic assumptions for the model are the following:

1. conditionally to $\mathbf{u}$, $y_{k.}=\Sigma y_{kl}$ has a multinomial distribution with parameters n_{k+}, π_k;
2. π_k can be decomposed into a linear combination of the effects of factors involved in the design. In short:

$$\mathbf{y}_{k.}|\mathbf{u} \sim M(n_{k+},\pi_k) \tag{17.A2}$$

$$\pi_k = \mathbf{X}_k\mathbf{b}+\mathbf{Z}_k\mathbf{u}. \tag{17.A3}$$

Therefore, the residuals $\mathbf{e}_{kl}$ are conditionally independent with expectation $E(\mathbf{e}_{kl}|\mathbf{u})=\mathbf{0}$ and variance $var(\mathbf{e}_{kl}|\mathbf{u})=Diag(\pi_k)-\pi_k\pi_k'$. Unconditionally, we have:

$$E(\mathbf{e}_{kl}) = \mathbf{0} \tag{17.A4a}$$

$$var(\mathbf{e}_{kl}) = Diag\ (\mathbf{X}_k\mathbf{b})-\mathbf{X}_k\mathbf{b}\mathbf{b}'\mathbf{X}_k'-\mathbf{Z}_k var(\mathbf{u})\mathbf{Z}_k' \tag{17.A4b}$$

$$E(\mathbf{y}_{kl}) = \mathbf{X}_k\mathbf{b} \tag{17.A5a}$$

$$var(\mathbf{y}_{kl}) = Diag(\mathbf{X}_k\mathbf{b})-\mathbf{X}_k\mathbf{b}\mathbf{b}'\mathbf{X}_k'. \tag{17.A5b}$$

Moreover, for any k and k', it turns out that:

$$\mathrm{Cov}(\mathbf{y}_{kl},\mathbf{y}'_{k'l'}) = \mathrm{E}(\mathbf{y}_{kl}\mathbf{y}'_{k'l'})-\mathbf{X}_k\mathbf{b}\mathbf{b}'\mathbf{X}'_{k'}.$$

In $\mathrm{E}(\mathbf{y}_{kl}\mathbf{y}'_{k'l'})$ the elements are interpretable as

$$\Pr(y_{klm} = y_{k'l'm'} = 1) \text{ for } m,m' = 1,2,...,c$$

and these can be expressed as

$$\mathrm{E}(\mathbf{y}_{kl}\mathbf{y}'_{k'l'}) = \mathbf{X}_k\mathbf{b}\mathbf{b}'\mathbf{X}'_{k'}+\mathbf{Z}_k\mathrm{var}(\mathbf{u})\mathbf{Z}'_{k'}$$

so that

$$\mathrm{Cov}(\mathbf{y}_{kl},\mathbf{y}'_{k'l'}) = \mathbf{Z}_k \ \mathrm{var}(\mathbf{u})\mathbf{Z}'_{k'}. \tag{17.A6}$$

Results given in (17.A4b), (17.A5b), and (17.A6) imply that

$$\mathrm{Cov}(\mathbf{e}_{kl},\mathbf{e}'_{k'l'}) = \mathbf{0}$$

$$\mathrm{Cov}(\mathbf{u},\mathbf{e}'_{kl}) = \mathbf{0}.$$

Let

$$\mathbf{p}_k = \sum_l \mathbf{y}_{kl}/n_{k+} \text{ for } k=1,2,...,k \tag{17.A7a}$$

$$\mathbf{p}' = (\mathbf{p}'_1,\mathbf{p}'_2,...,\mathbf{p}'_k,...,\mathbf{p}'_K) \tag{17.A7b}$$

$$\mathbf{X}' = (\mathbf{X}'_1,\mathbf{X}'_2,...,\mathbf{X}'_k,...,\mathbf{X}'_K) \tag{17.A7c}$$

$$\mathbf{Z}' = (\mathbf{Z}'_1,\mathbf{Z}'_2,...,\mathbf{Z}'_k,...,\mathbf{Z}'_K) \tag{17.A7d}$$

$$\bar{\mathbf{e}}' = (\bar{\mathbf{e}}'_1,\bar{\mathbf{e}}'_2,...,\bar{\mathbf{e}}'_k,...,\bar{\mathbf{e}}'_K). \tag{17.A7e}$$

The linear mixed model (17.A1) may be written using proportions as

$$\mathbf{p} = \mathbf{X}\mathbf{b}+\mathbf{Z}\mathbf{u}+\bar{\mathbf{e}} \tag{17.A8}$$

and is such that

$$E(\mathbf{p}) = \mathbf{Xb} \tag{17.A9a}$$

$$\text{var}(\mathbf{p}) = \mathbf{Z}\ \text{var}(\mathbf{u})\mathbf{Z}' + \text{var}(\bar{\mathbf{e}}) \tag{17.A9b}$$

$$\text{Cov}(\mathbf{p},\mathbf{u}') = \mathbf{Z}\ \text{var}(\mathbf{u}), \tag{17.A9c}$$

where

$$\text{var}(\mathbf{u}) = \mathbf{A} * \text{G}_o. \tag{17.A10}$$

Above, **A** is the relationship matrix among levels of **u**, and $\mathbf{G}_o$ a cxc matrix of "**u**" components of variance-covariance (e.g., additive genetic variances and covariances). Further:

$$\text{var}(\bar{\mathbf{e}}) = \mathbf{R}. = \bigoplus_{k=1}^{K} n_{k+}^{-1}\mathbf{R}_{o,k} \tag{17.A11}$$

with $\mathbf{R}_{o,k}$ being the unconditional residual variance-covariance matrix in (17.A4b). From a formal point of view, assuming $\mathbf{G}_o$ and $\mathbf{R}_{ok}$ (k=1,2,...,K) are known, the properties of (17.A8) given in (17.A9a), (17.A9b), and (17.A9c) allow the derivation of BLUP of **u** and GLS of **b** using the mixed model equations of Henderson:

$$\begin{bmatrix} \mathbf{X}'\mathbf{R}.^{-1}\mathbf{X} & \mathbf{X}'\mathbf{R}.^{-1}\mathbf{Z} \\ \mathbf{Z}'\mathbf{R}.^{-1}\mathbf{X} & \mathbf{Z}'\mathbf{R}.^{-1}\mathbf{Z}+\mathbf{A}^{-1}*\mathbf{G}_o^{-1} \end{bmatrix} \begin{bmatrix} \hat{\mathbf{b}} \\ \hat{\mathbf{u}} \end{bmatrix} = \begin{bmatrix} \mathbf{X}'\mathbf{R}.^{-1}\mathbf{p} \\ \mathbf{Z}'\mathbf{R}.^{-1}\mathbf{p} \end{bmatrix} \tag{17.A12}$$

In practice, one will have to replace $\mathbf{R}_{o,k}$ in **R**., and $\mathbf{G}_o$ by some estimates. Because the number of distinct residual dispersion matrices can be very large, the procedures proposed by Landis and Koch (1977) for one population, and by Beitler and Landis (1985) for one fixed factor to estimate these components will be rarely feasible in practice. A more realistic procedure might be based on using the alternative unknowns $\mathbf{G}_o$ and **b** as following:

1. Starting with $\mathbf{b}^{[o]}$=OLS estimate as in Grizzle et al. (1969) and a "prior" $\mathbf{G}_o^{[o]}$ for $\mathbf{G}_o$, use these values to compute $\mathbf{R}_{o,k}$, solve then (17.A12) and get $\mathbf{b}^{[1]}$ and $\mathbf{u}^{[1]}$.

2. Compute an "estimate" $\mathbf{G}_o^{[1]}$ as in the MINQUE procedure by equating the bilinear form $\mathbf{u}'^{[1]}_f\mathbf{A}^{-1}\mathbf{u}^{[1]}_g$ (where f and g refer to two different categories) to its expectation as described in Henderson (1984, 1986).
3. Use the values $\mathbf{b}^{[1]}$ and $\mathbf{G}_o^{[1]}$ to update $\mathbf{R}_{o,k}$, and solve (17.A12), then returning to (1). This algorithm can be applied iteratively proceeding to (2) and (3), and so on. This is an empirical procedure with unknown statistical properties. The results of Kakwani (1967) and Kackar and Harville (1981) do not hold for a linear model applied to categorical data because residuals in (A17.A1) are not symmetrically distributed.

It should be mentioned that $\hat{\mathbf{b}}$ must satisfy the constraint that the matrix in (17.A5b) must be positive-definite for any subpopulation k defined as a combination of levels of fixed factors. When the kth subclass is defined at an elementary level, in particular within the ith level of the random factor, formula (17.A4b) becomes

$$\mathrm{Diag}(\mathbf{X}_k\mathbf{b})-\mathbf{X}_k\mathbf{b}\mathbf{b}'\mathbf{X}_k'-\mathbf{G}_o\,. \tag{17.A13}$$

Because this is a variance-covariance matrix, $\mathbf{G}_o$ must not only be positive-definite but also such that (17.A13) is positive-definite for any "population" k. Likewise, estimates $\hat{\mathbf{b}}$ and $\hat{\mathbf{G}}_o$ should be such that this condition is satisfied.

Discussion Summary
PART V: PREDICTION AND ESTIMATION IN NON-LINEAR MODELS

A.R. Gilmour[1]

14 Generalized Linear Models and Applications to Animal Breeding

Generalized linear models were developed for analyzing data from distributions belonging to the exponential family. The procedure has been extended to the quasi-likelihood case where the distribution is not stated but the mean and variance of the observations are specified. Extension of the method to analysis of ranks is not possible because the mean and variance of the observations are not specified and the transformation to ranks is not done via a function.

The probit transformation is motivated by two considerations. Primarily, it is used to extend the range of the variable analyzed from (0,1) to the real line. A consequence of this transformation is that the variance is constant over the range of fixed effects (on the liability scale). This feature is important when it comes to drawing inferences about sire values. The model provides a natural way to predict probabilities for comparing sires in specific situations. There is no obvious consistent procedure for doing this if prediction is on the probability scale.

Simulation studies comparing "BLUP" with "pseudo-BLUP" on the liability scale have shown a 4 to 12% increase in efficiency from using the liability scale, even though correlations between ranks are generally extremely high. The explanation seems to be that while changes in rank are not large, they tend to be quite significant in the extremes when selecting a small proportion of the extreme animals.

15 Analysis of Linear and Non-Linear Growth Models with Random Parameters

When observations are missing in a non-random manner in a repeated measures analysis, care is required in setting up the model so that these are taken into account in a satisfactory way. Otherwise, the estimates of effects in the model will be biased. Also, in modelling the

[1] Agricultural Research and Veterinary Centre, Department of Agriculture, Orange, Australia

variance-covariance structure with multiple random variables, the covariance terms are important in generating a reasonable structure.

In Dempster et al. (1977), a formula is given for the convergence rate of the EM algorithm. This expression is difficult to evaluate because it involves expectations of second differentials. It might be useful in seeking ways to speed up convergence.

16 Survival, Endurance and Censored Observations in Animal Breeding

The result that genetic correlations under transformation of scale should be equal to 1, was challenged as "begging the question". It was noted that rank regression and Bayesian methods have the property of median invariance. In the rank regression procedures described, ranking was within strata-contemporary groups. The loss of information in rank regression is dependent on group size, and "proliferation" of strata (groups) to model fixed effects will exacerbate the loss of information. This corresponds to the introduction of a small sample bias in Cox's regression model.

17 Genetic Evaluation for Discrete Polygenic Traits in Animal Breeding

When validating the threshold probit model for genetic analysis, it has been common to generate data under this model and then compare the performance of linear and threshold procedures. This gives a bias in the direction of favoring the threshold model. Work needs to be done on generating discrete data using other distributions, to then assess the relative performance of the two procedures.

The threshold mixed models have been developed in the context of a sire model. In principle, there is no problem with using an animal model when variances are assumed known. Variance estimation will, in general, not be feasible in an animal model because the (0,1) data would lead to an infinite genetic variance.

It is necessary to validate the multiple threshold model for multinomial data. Calving difficulty is a trait for which the multiple threshold model has fitted well. One validation procedure proposed would be to look for a genetic correlation of 1 between categories of the trait. Another approach would be to fit a more general multinomial model and test for a significantly better fit.

Caution is also required in specifying the linear model for liability. For example, if herd-year-season effects are modelled, this may generate cell sizes that are quite small. Any cell where all observations are in one of the two classes must be completely dropped unless the factor is treated as random. There needs to be a careful consideration of this aspect of modelling, and avoidance of overspecification would seem to be quite important.

Part VI: Selection and Non-Random Mating

18 Accounting for Selection and Mating Biases in Genetic Evaluations

C.R. Henderson[1]

When selection has occurred, the means and variances of random variables are different from those under the usual assumptions of mixed linear models. Consequently, the solution to regular mixed model equations sometimes gives biased estimators and predictors. Given multivariate normal distributions and a vector variable used to make selection and mating decisions, the mixed model equations can be modified to yield unbiased estimators and predictors. This modification is described and several examples are given of its application, including cow culling, selection on records not available for analysis, estimation of genetic and environmental trends, association between sire and herd values, genetic groups, differential treatment, and assortative mating with and without selection.

18.1 Introduction

An interesting enigma arises in the application of linear models to the estimation of genetic parameters and the prediction of breeding values. Often, the most frequently involved model is one which is seldom applicable. This model, for which I probably deserve the major credit or blame, is

$$\mathbf{y} = \mathbf{X}\boldsymbol{\beta}+\mathbf{Z}\mathbf{u}+\mathbf{e}, \tag{18.1}$$

where $\boldsymbol{\beta}$ is fixed and unknown, $\mathbf{X}$ and $\mathbf{Z}$ are fixed and known, and $\mathbf{u}$, $\mathbf{e}$ are random variables, not observable, with null means, var($\mathbf{u}$)=$\mathbf{G}$, var(e)=$\mathbf{R}$, and cov($\mathbf{u}$, $\mathbf{e}'$)=$\mathbf{0}$.

Generally $\mathbf{G}$ and $\mathbf{R}$ have some known structure such that the number of parameters is considerably fewer than n(n+1)/2 and q(q+ 1)/2, where n and q are the dimensions of $\mathbf{R}$ and $\mathbf{G}$, respectively. These parameters, variances and covariances, are sometimes assumed known and in other cases need to be estimated. In many applications $\mathbf{G}$ or a submatrix of $\mathbf{G}$ is regarded as $\mathbf{A}\sigma_a^2$, where $\mathbf{A}$ is the numerator relationship matrix, and σ_a^2 is additive genetic variance. Also, $\mathbf{R}$ is often regarded as $\mathbf{I}\sigma_e^2$. For two-trait problems $\mathbf{G}$ might be

[1] Cornell University, Ithaca and University of Illinois, Urbana, Illinois, USA

$$\begin{bmatrix} \mathbf{A}g_{11} & \mathbf{A}g_{12} \\ \mathbf{A}g_{12} & \mathbf{A}g_{22} \end{bmatrix}, \text{ where} \begin{bmatrix} g_{11} & g_{12} \\ g_{12} & g_{22} \end{bmatrix}$$

is the additive genetic covariance matrix, and **R** might be

$$\begin{bmatrix} \mathbf{I}r_{11} & \mathbf{I}r_{12} \\ \mathbf{I}r_{12} & \mathbf{I}r_{22} \end{bmatrix}, \text{ where} \begin{bmatrix} r_{11} & r_{12} \\ r_{12} & r_{22} \end{bmatrix}$$

is the environmental covariance matrix. The form of the distribution is not always stated, but for the results on bias to hold as described in this paper we need to invoke the multivariate normal distribution.

18.2 Effect of Selection on u, e, G and R

Most data available to animal breeders arise either from selection experiments or from producers' herds. In either case selection has occurred and also possibly assortative mating. Then it makes no sense to regard **u** as having null means. If selection has been effective, we should expect the means of breeding values to increase generation by generation. Best linear unbiased prediction and selection index regress the predictions toward **0**. Consequently, are the later, and better, generations underevaluated relative to the earlier generations? It would appear so, but as we shall see subsequently, this is not necessarily always the case. Also the means of **e** can be altered by certain kinds of selection. Further, depending upon our points of view regarding the number of loci, we cannot assert that var(**a**)=$\mathbf{A}\sigma_a^2$. Generally we would expect the diagonals of var(**a**) to be smaller than σ_a^2, and also that its off-diagonal elements are no longer proportional to the off-diagonal elements of **A**. A further complication is that **R** may no longer be $\mathbf{I}\sigma_e^2$ and cov(**u**, **e**')$\neq$**0**.

What are the consequences of computing best linear unbiased estimators (BLUE) and best linear unbiased predictors (BLUP) using the assumptions of the model with no selection (18.1)? That is, what are the properties of $\mathbf{K}'\boldsymbol{\beta}°$ and $\hat{\mathbf{u}}$ computed from (18.2) below?

$$\begin{bmatrix} \mathbf{X}'\mathbf{R}^{-1}\mathbf{X} & \mathbf{X}'\mathbf{R}^{-1}\mathbf{Z} \\ \mathbf{Z}'\mathbf{R}^{-1}\mathbf{X} & \mathbf{Z}'\mathbf{R}^{-1}\mathbf{Z}+\mathbf{G}^{-1} \end{bmatrix} \begin{bmatrix} \boldsymbol{\beta}° \\ \hat{\mathbf{u}} \end{bmatrix} = \begin{bmatrix} \mathbf{X}'\mathbf{R}^{-1}\mathbf{y} \\ \mathbf{Z}'\mathbf{R}^{-1}\mathbf{y} \end{bmatrix}. \quad (18.2)$$

Are $\mathbf{K'\beta^\circ}$ and $\hat{\mathbf{u}}$ unbiased, where $\mathbf{K'\beta}$ is estimable? This paper gives a general answer to this question under the assumption of multivariate normality, **G** and **R** known to proportionality, and with **X** and **Z** fixed in conceptual repeated sampling, characteristic of classical statistics. Also, $\boldsymbol{\beta}$ is regarded strictly as fixed in repeated sampling. This is in contrast to the description of the model in the paper by Fernando and Gianola, Chapter 19, this Volume, in this symposium. Consequently, our results are not strictly comparable, but I think that our conclusions about how to evaluate animals are similar, and in some cases identical.

My paper deals entirely with elimination of biases due to selection. Nevertheless, I do not intend to suggest that we should always use unbiased predictors for selection. In fact, I conjecture in my earlier paper in this symposium (Henderson, Chap. 1, this Vol.) that probably always a biased predictor exists that is a better selection criterion than BLUP. Unfortunately, with unknown $\boldsymbol{\beta}$, even though **G** and **R** are known to proportionality, we do not know what this biased criterion really is. Therefore, should we restrict ourselves to unbiased predictors? I do not know the answer to this, but I think it is "no".

18.3 Means and Covariances Conditional on Selection Functions

Throughout this paper I shall use a result due to Pearson (1903) on means and covariances of random variables conditional upon a set of linear selection functions. Multivariate normality and model (18.1) is assumed. The matrix formulation of Pearson's result is due to Lawley (1943). The notation used in this paper is the same as in Henderson (1975).

Suppose that in repeated sampling selection is based on a random vector, **w**. This vector is selected in such a manner that its mean and covariance matrix are altered from what they would be had there been no selection. This formulation is different from the more common approach in statistics in which **w** is regarded as fixed in repeated samples. It is assumed, of course, that **w** is correlated with **u** and **e**, with **u** only, or with **e** only. In the unconditional (no selection model)

$$E\begin{bmatrix}\mathbf{y}\\ \mathbf{u}\\ \mathbf{e}\\ \mathbf{w}\end{bmatrix}=\begin{bmatrix}\mathbf{X\beta}\\ \mathbf{0}\\ \mathbf{0}\\ \mathbf{d}\end{bmatrix}, \qquad (18.3)$$

and

$$\operatorname{var}\begin{bmatrix}\mathbf{y}\\ \mathbf{u}\\ \mathbf{e}\\ \mathbf{w}\end{bmatrix} = \begin{bmatrix}\mathbf{ZGZ'+R} & \mathbf{ZG} & \mathbf{R} & \mathbf{B_y}\\ & \mathbf{G} & \mathbf{0} & \mathbf{B_u}\\ \text{Symmetric} & & \mathbf{R} & \mathbf{B_e}\\ & & & \mathbf{H}\end{bmatrix}, \tag{18.4}$$

where $\mathbf{B_y}=\mathbf{ZB_u}+\mathbf{B_e}$.

Now suppose that $\mathbf{w}$ is selected so that in repeated sampling $E(\mathbf{w})=\mathbf{s}\neq\mathbf{d}$, and $\operatorname{var}(\mathbf{w})=\mathbf{H_s}\neq\mathbf{H}$. Then by Pearson's result, the conditional means for

$$\begin{bmatrix}\mathbf{y}\\ \mathbf{u}\\ \mathbf{e}\\ \mathbf{w}\end{bmatrix} \text{ are } \begin{bmatrix}\mathbf{X\beta}+\mathbf{B_y H^{-1}(s-d)}\\ \mathbf{B_u H^{-1}(s-d)}\\ \mathbf{B_e H^{-1}(s-d)}\\ \mathbf{s}\end{bmatrix}. \tag{18.5}$$

Note that these are the same results as for fixed $\mathbf{w}=\mathbf{s}$. The conditional variance-covariance matrix for

$$\begin{bmatrix}\mathbf{y}\\ \mathbf{u}\\ \mathbf{e}\\ \mathbf{w}\end{bmatrix} \text{ is } \begin{bmatrix}\mathbf{ZGZ'+R-B_yH_oB_y'} & \mathbf{ZG-B_yH_oB_u'} & \mathbf{R-B_yH_oB_e'} & \mathbf{B_yH^{-1}H_s}\\ & \mathbf{G-B_uH_oB_u'} & \mathbf{B_uH_oB_e'} & \mathbf{B_uH^{-1}H_s}\\ \text{Symmetric} & & \mathbf{R-B_eH_oB_e'} & \mathbf{B_eH^{-1}H_s}\\ & & & \mathbf{H_s}\end{bmatrix}, \tag{18.6}$$

where $\mathbf{H_o}=\mathbf{H^{-1}(H-H_s)H^{-1}}$.

Note that if $\mathbf{w}$ is fixed, then $\mathbf{H_s}=\mathbf{0}$ and $\mathbf{H_o}=\mathbf{H^{-1}}$. This gives the traditional conditional covariance matrix of multivariate normal distributions.

We illustrate conditional means and variances with the following example.

	Animals					
Period	1	2	3	4	5	6
1	x	x	x	o	o	o
2	o	o	o	x	x	x

x denotes a record, o no record

The model is: $y_{ij}=p_i+a_j+e_{ij}$

$$\text{Var}(\mathbf{a}) = \begin{bmatrix} 1. & 0 & 0 & 0.5 & 0.5 & 0 \\ & 1. & 0 & 0 & 0 & 0.5 \\ & & 1. & 0 & 0 & 0 \\ & & & 1. & 0.25 & 0 \\ & & & & 1. & 0 \\ \text{symmetric} & & & & & 1. \end{bmatrix} \text{ and var}(\mathbf{e}) = 2\mathbf{I}.$$

In repeated sampling $y_{11}>y_{12}>y_{13}$ in all samples. Unconditional means are $\mathbf{u}=\mathbf{0}$, $\mathbf{e}=\mathbf{0}$; $\mathbf{y}'=(7, 7, 7, 9, 9, 9)$. Thus $\mathbf{p}=(7, 9)'$. Then, conditional

$$\text{var}\begin{bmatrix} y_{11} \\ y_{12} \\ y_{13} \end{bmatrix} \text{ is } \begin{bmatrix} 1.678 & 0.827 & 0.495 \\ & 1.346 & 0.827 \\ \text{sym.} & & 1.678 \end{bmatrix} \text{ and the conditional mean is } \begin{bmatrix} 8.466 \\ 7.0 \\ 5.534 \end{bmatrix}.$$

Unconditional var=diag(3, 3, 3).

In this case $\mathbf{w}$ can be written in two ways, and these give the same conditional means and covariances. These are

$$\begin{bmatrix} 1 & 0 & 0 & 0 & 0 & 0 \\ 0 & 1 & 0 & 0 & 0 & 0 \\ 0 & 0 & 1 & 0 & 0 & 0 \end{bmatrix} \mathbf{y}, \text{ and } \begin{bmatrix} 1 & -1 & 0 & 0 & 0 & 0 \\ & & & & & \\ 0 & 1 & -1 & 0 & 0 & 0 \end{bmatrix} \mathbf{y}.$$

This matrix pre-multiplying **y** will be referred to as **L**'. In the second case

$$\mathbf{H} = \begin{bmatrix} 6 & -3 \\ -3 & 6 \end{bmatrix}, \mathbf{H}_s = \begin{bmatrix} 1.370 & -0.187 \\ -0.187 & 1.370 \end{bmatrix}$$

and **s**-**d**=(1.4658, -1.4658)'.

In the first case **H**=diag(3, 3, 3),

$$\mathbf{H}_s = \begin{bmatrix} 1.678 & 0.827 & 0.495 \\ & 1.346 & 0.827 \\ \text{symmetric} & & 1.678 \end{bmatrix}$$

and **s**-**d**=(1.4658, 0, -1.4658)'.

For both **L**', conditional **u**=(0.4886, 0, -0.4886, 0.2443, -0.2443, 0)', conditional **e**=(0.9772, 0, -0.9772, 0, 0, 0)',

$$\text{conditional var}(\mathbf{a}) = \begin{bmatrix} 0.853 & 0.092 & 0.055 & 0.427 & 0.427 & 0.046 \\ & 0.816 & 0.092 & 0.046 & 0.046 & 0.408 \\ & & 0.853 & 0.027 & 0.027 & 0.046 \\ & & & 0.963 & 0.213 & 0.023 \\ & & & & 0.963 & 0.023 \\ & \text{symmetric} & & & & 0.954 \end{bmatrix},$$

$$\text{conditional var}(\mathbf{e}) = \begin{bmatrix} 1.413 & 0.368 & 0.220 & 0 & 0 & 0 \\ & 1.265 & 0.368 & 0 & 0 & 0 \\ & & 1.413 & 0 & 0 & 0 \\ & & & 2 & 0 & 0 \\ & & & & 2 & 0 \\ & \text{symmetric} & & & & 2 \end{bmatrix},$$

$$\text{and conditional cov}(\mathbf{a}, \mathbf{e}') = \begin{bmatrix} -0.294 & 0.184 & 0.110 & 0 & 0 & 0 \\ 0.184 & -0.368 & 0.684 & 0 & 0 & 0 \\ 0.110 & 0.184 & -0.294 & 0 & 0 & 0 \\ -0.147 & 0.092 & 0.055 & 0 & 0 & 0 \\ -0.147 & 0.092 & 0.055 & 0 & 0 & 0 \\ 0.092 & -0.184 & 0.092 & 0 & 0 & 0 \end{bmatrix}.$$

In spite of these alterations in means and covariances, regular mixed model equations written as though **G**=**A**, **R**=2**I**, E(**u**) and E(**e**) null yield, on an average in repeated sampling, $\hat{\mathbf{p}}$=(7, 9)', $\hat{\mathbf{u}}$= same as conditional means. This is because, as we shall see, **w**=**L'y** such that **L'X**=**0** for the second **L'**, correct σ_a^2 and σ_e^2 were used, and all data were employed.

18.4 BLUE and BLUP in a Selection Model

Henderson (1975) showed that modification of the regular mixed model equations in (18.2) yields BLUE and BLUP in the selection model with means of (18.5) and covariances of (18.6). These modified equations are in (18.7):

$$\begin{bmatrix} \mathbf{X'R^{-1}X} & \mathbf{X'R^{-1}Z} & \mathbf{X'R^{-1}B_e} \\ \mathbf{Z'R^{-1}X} & \mathbf{Z'R^{-1}Z+G^{-1}} & \mathbf{Z'R^{-1}B_e-G^{-1}B_u} \\ \mathbf{B_e'R^{-1}X} & \mathbf{B_e'R^{-1}Z-B_u'G^{-1}} & \mathbf{B_e'R^{-1}B_e+B_u'G^{-1}B_u} \end{bmatrix} \begin{bmatrix} \boldsymbol{\beta}^\circ \\ \mathbf{u}^\circ \\ \mathbf{t}^\circ \end{bmatrix}$$

$$= \begin{bmatrix} \mathbf{X'R^{-1}y} \\ \mathbf{Z'R^{-1}y} \\ \mathbf{B_e'R^{-1}y} \end{bmatrix}. \qquad (18.7)$$

Zero superscripts are used on the solution vector because the solution may not be unique. Then we need to find interesting and unique linear functions of $\boldsymbol{\beta}^\circ$ and $\mathbf{u}^\circ$. A remarkable feature of these equations is that we do not need $\mathbf{H}_s$ and **s**-**d**, both related to selection intensity. These equations were derived by finding those functions of **y** to predict

$\mathbf{K'\beta+M'u}$ that have expectation $\mathbf{K'\beta+M'B_u\ H^{-1}(s\text{-}d)}$, and using the conditional mean and covariance of $\mathbf{y}$.

In the same paper it was proved that the prediction error covariance of $\mathbf{K'\beta^\circ+M'(u^\circ-u)}$, provided $\mathbf{K'\beta+M'u}$ is predictable, is

$$[\mathbf{K'} : \mathbf{M'}]\begin{bmatrix} \mathbf{C}_{11} & \mathbf{C}_{12} \\ \mathbf{C}_{12}' & \mathbf{C}_{22} \end{bmatrix}\begin{bmatrix} \mathbf{K} \\ \mathbf{M} \end{bmatrix}, \qquad (18.8)$$

where some symmetric g-inverse of the coefficient matrix of (18.7) is:

$$\begin{bmatrix} \mathbf{C}_{11} & \mathbf{C}_{12} & \mathbf{C}_{13} \\ \mathbf{C}_{12}' & \mathbf{C}_{22} & \mathbf{C}_{23} \\ \mathbf{C}_{13}' & \mathbf{C}_{23}' & \mathbf{C}_{33} \end{bmatrix}. \qquad (18.9)$$

18.5 Estimability and Predictability

When the solution is not unique, we need to check estimability and predictability. $\mathbf{K'\beta^\circ}$ is BLUE if $\mathbf{K'(C_{11} : C_{12} : C_{13})F=(K': 0)}$,

$$\text{where } \mathbf{F} = \begin{bmatrix} \mathbf{X'R^{-1}X} & \mathbf{X'R^{-1}B_y} \\ \mathbf{Z'R^{-1}X} & \mathbf{Z'R^{-1}B_y} \\ \mathbf{B_e'R^{-1}X} & \mathbf{B_e'R^{-1}B_y} \end{bmatrix}.$$

$\mathbf{M'u^\circ}$ is BLUP of $\mathbf{M'u}$ if $\mathbf{M'[C_{12}' : C_{22} : C_{23}]F=(0 : M')}$. $\mathbf{THt}^\circ$ is BLUE of $\mathbf{s\text{-}d}$ if $\mathbf{T'(C_{13}' : C_{23}' : C_{33})F=(0 : T')}$. Also $\mathbf{K'\beta^\circ+M'u^\circ}$ is BLUP if:

$$[\mathbf{K'} : \mathbf{M'}]\begin{bmatrix} \mathbf{C}_{11} & \mathbf{C}_{12} & \mathbf{C}_{13} \\ \mathbf{C}_{12}' & \mathbf{C}_{22} & \mathbf{C}_{23} \end{bmatrix}\mathbf{F} = [\mathbf{K'} : \mathbf{M'}].$$

The remainder of this paper illustrates applications of these general results. The issue of whether we should or should not use unbiased predictors as selection criteria is discussed in my earlier paper in this symposium (Henderson Chap.1, this Vol.).

18.6 Cow Culling

The first example of selection bias is the one that forced me to think about this problem. When Robertson and I were studying with Dr. Lush in 1947, we looked at a thesis in which genetic and environmental trends were estimated by least-squares in the cow culling model,

$$y_{ij} = p_i + g_j + c_{jk} + e_{ijk},$$

where **p** represented periods, **g** represented generations, and c_{jk} represented real producing abilities of cows within generations. Fixed elements in the model are **p** and **g**. The observations were adjusted for age, but of course estimation of such factors in a culling model is not trivial. This problem was discussed later by Lush and Shrode (1950). In the thesis mentioned above, period differences were estimated from $p_i^\circ - p_j^\circ$. Genetic differences between generations were estimated by $g_i^\circ + \bar{c}_{i.}^\circ - g_j^\circ - \bar{c}_{j.}^\circ$.

Robertson and I were quite certain that this analysis yielded biased estimators under the assumption that culling on production had occurred. We did not know how to solve this problem conveniently, but we did realize that account needed to be taken of repeatability being less than one.

When I derived my mixed model equations (18.2) in 1948, I thought that perhaps a solution to these equations would solve the problem. Some small-scale simulation suggested that this was true. These results were reported at the 1949 ADSA meetings (Henderson 1949). Nothing more came of this for several years except for some tests of it by McGilliard at Michigan, Harvey at Idaho, and Plum at Nebraska. While I was in New Zealand, I derived with Searle's assistance the BLUE and BLUP properties of the mixed model equations. Also, proof was derived which showed that if selection is by translation invariant linear functions, and the distribution is multivariate normal, these equations yield unbiased estimators of $\boldsymbol{\beta}$ and predictors of **u**. The proofs that these were BLUE and BLUP came later and were published in Henderson (1975).

Kempthorne became interested in the culling problem and derived a result similar to mine except that he estimated σ_e^2/σ_c^2 and used this rather than the parameter value. This led to a joint paper, Henderson et al. (1959). Kempthorne's part of this paper anticipated a result of Gianola et al. (1986), which suggested estimation of variances by restricted maximum likelihood (REML), and then using these estimates in place of parameter values in the mixed model equations.

We illustrate with a very simple example.

	Cows		
Period	1	2	3
1	x	x	x
2	x	x	o

Above, x represents a record so cow 3 was culled.
The model is $y_{ij}=p_i+c_j+e_{ij}$

$$\text{Var}\begin{bmatrix} \mathbf{c} \\ \mathbf{e} \end{bmatrix} = \begin{bmatrix} \mathbf{I}\sigma_c^2 & \mathbf{0} \\ \mathbf{0} & \mathbf{I}\sigma_e^2 \end{bmatrix}, \text{ and } E\begin{bmatrix} \mathbf{u} \\ \mathbf{e} \end{bmatrix} = \begin{bmatrix} \mathbf{0} \\ \mathbf{0} \end{bmatrix},$$

ignoring culling, and $\sigma_c^2=\sigma_e^2=1$, that is r(repeatability)=0.5.

In repeated sampling $\bar{y}_{11}>\bar{y}_{12}>\bar{y}_{13}$. Consequently, we write **w** as **L'y**, where:

$$\mathbf{L}' = \begin{bmatrix} 1 & -1 & 0 & 0 & 0 \\ 0 & 1 & -1 & 0 & 0 \end{bmatrix}.$$

We simulated with $\bar{y}_{11}-\bar{y}_{12}=\bar{y}_{12}-\bar{y}_{13}=1$ and with **p**'=(6, 8). Conditional **c**=(0.5, 0, -0.5). Conditional **e**=(0.5, 0, -0.5, 0, 0). Regular mixed model equations gave an average of solutions, $\hat{\mathbf{p}}$=(6, 8)' and $\hat{\mathbf{c}}$=conditional **c**, unbiased.

Now let us change the model to include a covariate with values (2, 1, 1, 1, 2)', and associated regression, say b. In repeated sampling, b was ignored and culling was based on unadjusted y_{ij}. Suppose we write **L'** as before and again let **s**-**d**=(1, 1)'. Conditional **c** is as before, but now the mixed model equations gave, on average, the solution, $\hat{\mathbf{p}}$=(5.385, 7.346)', $\hat{\mathbf{b}}$=1.462, $\hat{\mathbf{c}}$=(0.461, -0.039, -0.423)'. These are biased. To explain the consequences of ignoring b in culling we look at the modified equations of (18.7). Now **w**=**L'y**. Then $\mathbf{X'R}^{-1}\mathbf{B}_e=\mathbf{X'L}$, $\mathbf{Z'R}^{-1}\mathbf{B}_e-\mathbf{G}^{-1}\mathbf{B}_u=\mathbf{0}$, $\mathbf{B}_e'\mathbf{R}^{-1}\mathbf{B}_e+\mathbf{B}_u\mathbf{G}^{-1}\mathbf{B}_u=\mathbf{L}'(\mathbf{ZGZ}'+\mathbf{R})\mathbf{L}$, and $\mathbf{B}_e'\mathbf{R}^{-1}\mathbf{y}=\mathbf{L'y}$. Then it is obvious that if **X'L**=**0**, the regular mixed model equations yield BLUE and BLUP even though selection has occurred. In our first example **L'X**=**0**, but in the second, it does not because of the covariates.

When the regular mixed model equations were modified to account for $\mathbf{L'X \neq 0}$, unbiased estimators and predictors were obtained.

One would hope that in practice adjustments are made for fixed effects, and then usually $\mathbf{L'X}$ would be null so we will not have the troublesome problem of having to write $\mathbf{L}$ and modifying the mixed model equations. When it is necessary to modify the equations to obtain unbiased predictors, it is not certain that the modified solution is a better predictor than the unmodified one in the sense of minimizing mean squared errors of prediction.

18.7 Translation Invariant Functions of Records Used in Selection Plus Other Unknown Selection Functions

Suppose selection is by

$$\mathbf{w} = \begin{bmatrix} \mathbf{L'y} \\ \mathbf{v} \end{bmatrix},$$

where $\mathbf{L'X=0}$ and $\mathbf{v}$ is not observable. Let cov($\mathbf{u}$, $\mathbf{v'}$)=$\mathbf{C_u}$, cov($\mathbf{e}$, $\mathbf{v'}$)=$\mathbf{C_e}$, cov($\mathbf{y}$, $\mathbf{v'}$)=$\mathbf{C_y}$, var($\mathbf{v}$)=$\mathbf{V}$. Then the modified equations are:

$$\begin{bmatrix} \mathbf{X'R^{-1}X} & \mathbf{X'R^{-1}Z} & \mathbf{0} & \mathbf{X'R^{-1}C_e} \\ \mathbf{Z'R^{-1}X} & \mathbf{Z'R^{-1}Z+G^{-1}} & \mathbf{0} & \mathbf{Z'R^{-1}C_e - G^{-1}C_u} \\ \mathbf{0} & \mathbf{0} & \mathbf{L'(ZGZ'+R)L} & \mathbf{L'C_y} \\ \mathbf{C_e'R^{-1}X} & \mathbf{C_e'R^{-1}Z - C_u'G^{-1}} & \mathbf{C_y'L} & \mathbf{V} \end{bmatrix} \begin{bmatrix} \boldsymbol{\beta}^{\circ} \\ \mathbf{u}^{\circ} \\ \boldsymbol{\theta}_1^{\circ} \\ \boldsymbol{\theta}_2^{\circ} \end{bmatrix}$$

$$= \begin{bmatrix} \mathbf{X'R^{-1}y} \\ \mathbf{Z'R^{-1}y} \\ \mathbf{L'y} \\ \mathbf{C_e'R^{-1}y} \end{bmatrix}. \qquad (18.10)$$

Now if cov(**u**, **v**') and cov(**e**, **v**') are null, so is cov(**y**, **v**'), and the regular mixed model equations yield BLUE and BLUP. In some applications it may be the case that **v** is nearly uncorrelated with **u**, **e**, and **y**. For example, it might represent death loss or infertility. Then we can use regular mixed model equations.

An example is presented next in which **v** represents linear functions of records that are not available. Then if these records were made by animals related to those with records, **v** is certainly correlated with **u** and **y**, and some account must be taken of this if the predictors are to be unbiased.

18.8 Selection on Previous Records Not Available for Analysis

Suppose that an experiment is conducted to compare two subpopulations of sires with respect to their progeny obtained from matings with a random sample of females. We assume that the sires were obtained by some random process from the same population, but records were used in such a way in making the selections that we suspect that the average genetic merits of the two groups differ. We illustrate with the following example. The three sires of group 1 have

$$\mathbf{A} = \begin{bmatrix} 1. & 0.25 & 0 \\ & 1. & 0 \\ \text{sym.} & & 1. \end{bmatrix},$$

and the three sires of group 2 have

$$\mathbf{A} = \begin{bmatrix} 1. & 0.5 & 0 \\ & 1. & 0 \\ \text{sym.} & & 1. \end{bmatrix}.$$

The two groups are unrelated. We obtain 20 progeny from each sire. Let the model be $\mu+s_i+e_{ij}$, and we simulate with Var(**s**)=**A**, var(**e**)=10**I**, μ=8.

We assume that in repeated sampling $\sum_{i=1}^{3} s_i - \sum_{i=4}^{6} s_i$ has average value=2.

Then we can describe **w** as **M's** where **M'**=[1 1 1 -1 -1 -1]. Then conditional s=[0.333, 0.333, 0.267, -0.4, -0.4, -0.267]', and conditional $\bar{e}_{i\cdot}=\mathbf{0}$.

Applying regular mixed model equations, the average solution is $\hat{\mu}$=7.995, $\hat{s}$'=[0.242, 0.242, 0.181, -0.296, -0.296, -0.174], biased. Mixed model equations modified for **M's** type selection are:

$$\begin{bmatrix} X'R^{-1}X & X'R^{-1}Z & 0 \\ Z'R^{-1}X & Z'R^{-1}Z+G^{-1} & -M \\ 0 & -M' & M'GM \end{bmatrix} \begin{bmatrix} \beta^\circ \\ u^\circ \\ \theta^\circ \end{bmatrix} = \begin{bmatrix} X'R^{-1}y \\ Z'R^{-1}y \\ 0 \end{bmatrix}. \tag{18.11}$$

Now we can easily absorb $\theta°$ to obtain

$$\begin{bmatrix} X'R^{-1}X & X'R^{-1}Z \\ Z'R^{-1}X & Z'R^{-1}Z+G^{-1}-M(M'GM)^{-1}M' \end{bmatrix} \begin{bmatrix} \beta^\circ \\ u^\circ \end{bmatrix} = \begin{bmatrix} X'R^{-1}y \\ Z'R^{-1}y \end{bmatrix}. \tag{18.12}$$

Because $\mathbf{M(M'GM)^{-1}M'}$ is positive-definite, the **u°** are shrunken less than in unmodified equations. Using Eqs. (18.12) in our example, we obtain μ=8, $\hat{s}$=same as conditional **s**, unbiased. Looking at (18.12) it can be seen that if **M=I**, these equations reduce to GLS equations for fixed **u**. In many cases treating **u** as fixed gives unbiased estimators and predictors, but these are not necessarily BLUE and BLUP.

We could attempt to handle the above **M's** example by adding g_1, g_2 to the equations and then computing with the regular mixed model equations. This gave the following average of solutions,

$$\hat{g} = (8.309, 7.653)'$$
$$\hat{s} = (0.018, 0.018, -0.028, -0.040, -0.40, 0.53)'$$

$\hat{g}_i+\hat{s}_{ij}$ does not give the same result as $\hat{\mu}+\hat{s}_{ij}$ of the **M's** modified equations, but $(\hat{g}_1+\hat{s}_{1.})-(\hat{g}_2+\hat{s}_{2.})$ of the second solution is the same as $\hat{s}_{1.}-\hat{s}_{2.}$ of the modified equations. Both of these are unbiased estimators of the difference in merit between the two groups.

18.9 Mixed Model Equations to Estimate Genetic and Environmental Trends

If selection is by translation invariant functions, all data are used, the distribution is multivariate normal, the model is additive genetic, and **G** and **R** are known to proportionality, the mixed model equations yield unbiased estimators of genetic and environmental trends.

Let the model be

$$y_{ij} = x_{ij}'\boldsymbol{\beta} + p_i + a_j + e_{ij},$$

where

- **p** represents fixed period effects;
- $\boldsymbol{\beta}$ represents other fixed effects;
- **a** represents additive genetic effects;

$$\text{Var}\begin{bmatrix}\mathbf{a}\\ \mathbf{e}\end{bmatrix} = \begin{bmatrix}\mathbf{A}\sigma_a^2 & \mathbf{0}\\ \mathbf{0} & \mathbf{I}\sigma_e^2\end{bmatrix},\quad E\begin{bmatrix}\mathbf{a}\\ \mathbf{e}\end{bmatrix} = \mathbf{0}$$

prior to selection.

We illustrate with a simple example by simulation. We have six base population parents: 1,...,6 with **A**=**I**. Then some of these produce the following progeny in generation 2.

Progeny	Parent
7	1-4
8	1-5
9	2-4
10	2-5

Generation 2 progeny produce these progeny for generation 3:

Progeny	Parent
11	7-9
12	7-9
13	7-10

Generations and periods correspond, so $\mathbf{X}\boldsymbol{\beta}$ is

$$\begin{bmatrix} 1 & 1 & 1 & 1 & 1 & 1 & 0 & 0 & 0 & 0 & 0 & 0 & 0 \\ 0 & 0 & 0 & 0 & 0 & 0 & 1 & 1 & 1 & 1 & 0 & 0 & 0 \\ 0 & 0 & 0 & 0 & 0 & 0 & 0 & 0 & 0 & 0 & 1 & 1 & 1 \end{bmatrix}' \begin{bmatrix} p_1 \\ p_2 \\ p_3 \end{bmatrix}.$$

For simulation, $\sigma_a^2 = \sigma_e^2 = 1$.

It is assumed that in repeated sampling $\bar{y}_1 > \bar{y}_3$, $\bar{y}_2 > \bar{y}_3$, $\bar{y}_4 > \bar{y}_6$, $\bar{y}_5 > \bar{y}_6$, $\bar{y}_7 > \bar{y}_8$, $\bar{y}_9 > \bar{y}_{10}$. This can be expressed as **L'y** selection, where

$$\mathbf{L}' = \begin{bmatrix} 1 & 0 & -1 & 0 & 0 & 0 & 0 & 0 & 0 & 0 & 0 & 0 & 0 \\ 0 & 1 & -1 & 0 & 0 & 0 & 0 & 0 & 0 & 0 & 0 & 0 & 0 \\ 0 & 0 & 0 & 1 & 0 & -1 & 0 & 0 & 0 & 0 & 0 & 0 & 0 \\ 0 & 0 & 0 & 0 & 1 & -1 & 0 & 0 & 0 & 0 & 0 & 0 & 0 \\ 0 & 0 & 0 & 0 & 0 & 0 & 1 & -1 & 0 & 0 & 0 & 0 & 0 \\ 0 & 0 & 0 & 0 & 0 & 0 & 0 & 0 & 1 & -1 & 0 & 0 & 0 \end{bmatrix}.$$

We simulated with $(\mathbf{s}-\mathbf{d})'$=(0.5, 0.5, 0.6, 0.6, 0.8, 0.9). Then conditional $\mathbf{u}'$ is [0.083, 0.083, -0.167, 0.221, -0.021, -0.200, 0.265, -0.082, 0.282, -0.099, 0.274, 0.274, 0.083].

The regular mixed model equations gave average solution, $\hat{\mathbf{p}}'$= (6, 8, 9), $\hat{\mathbf{u}}$=conditional **u**, unbiased. The $\hat{p}_i$ estimate unbiasedly environmental trend, and the means of the generational $\hat{a}_i$ represent unbiased estimates of genetic trend. The latter are (0, 0.0917, 0.2103). This illustrates that BLUP estimates genetic values relative to the base population.

If incorrect σ_e^2/σ_a^2 is used we obtain biased estimators. If $\sigma_e^2/\sigma_a^2=2$ is used, we get

$$\hat{\mathbf{p}}' = (6.000, 8.031, 9.067)$$
$$\hat{\mathbf{a}} = (0, 0.061, 0.144).$$

Environmental trend is overestimated and genetic trend is underestimated. The opposite is true when we use $\sigma_e^2/\sigma_a^2=0.5$. Then

$$\hat{\mathbf{p}}' = (6.000, 7.970\ 8.931)$$
$$\hat{\mathbf{a}} = (0, 0.122, 0.279).$$

The phenotypic trend is the same for all σ_e^2/σ_a^2 used.

18.10 The Problem of Association Between Sire Values and Herd Merits in Sire Evaluations

It is quite probable that the better AI sires tend to be used disproportionally in the better herds. Can bias be avoided if this is the case? The standard procedure in BLUP is to regard herds (or herd-year-seasons) as fixed for purposes of computation. This was first done in the northeast sire evaluation beginning in 1970 as described in the Lush Symposium (Henderson 1973). This is illustrated by the following simulation.

Three sires were tested in herds 1 and 2 with these numbers of progeny:

	Herds	
Sires	1	2
1	10	5
2	5	12
3	5	8

The model is $y_{ijk}=\mu+s_i+h_j+e_{ijk}$

$$\text{Var}\begin{bmatrix}\mathbf{s}\\ \mathbf{h}\\ \mathbf{e}\end{bmatrix}=\begin{bmatrix}\mathbf{I}\sigma_s^2 & \mathbf{0} & \mathbf{0}\\ & \mathbf{I}\sigma_h^2 & \mathbf{0}\\ \text{symmetric} & & \mathbf{I}\sigma_e^2\end{bmatrix}, \text{ with } \sigma_s^2 = 1,\ \sigma_h^2 = 6,\ \sigma_e^2 = 12.$$

It is assumed that in repeated sampling $\bar{h}_1-\bar{h}_2$ has mean, 3. The sires were evaluated in these herds by treating herds as fixed. Then

$$\begin{bmatrix}\hat{s}_1-\hat{s}_2\\ \hat{s}_2-\hat{s}_3\end{bmatrix}=\begin{bmatrix}0.290 & 0.246 & -0.234 & -0.319 & -0.056 & 0.073\\ 0.009 & -0.023 & 0.018 & 0.374 & -0.192 & -0.351\end{bmatrix}\begin{bmatrix}\bar{y}_{11.}\\ \bar{y}_{12.}\\ \bar{y}_{21.}\\ \bar{y}_{22.}\\ \bar{y}_{31.}\\ \bar{y}_{32.}\end{bmatrix}.$$

Suppose that in repeated sampling the means of these functions are (0.5, 0.9). That is, the sires rank on average 1, 2, 3. On a second set of progeny, we have data as follows in herds 3, 4

	Herds	
Sires	3	4
1	10	3
2	2	6
3	0	0

Note that the best sire has more progeny than the second best, and that the poorest was culled. Further, sires 1 and 2 have disproportionate numbers of progeny in the two herds. We let $\bar{h}_3$-$\bar{h}_4$=4 in repeated sampling. Then we have in this case

$$\mathbf{w} = \begin{bmatrix} \mathbf{L'y} \\ \mathbf{M'u} \end{bmatrix},$$

where $\mathbf{L}'$ is derived from the function of $\bar{y}_{1i.}$ described above, and

$$\mathbf{M'u} = \begin{bmatrix} 0 & 0 & 0 & 1 & -1 & 0 & 0 \\ 0 & 0 & 0 & 0 & 0 & 1 & -1 \end{bmatrix} \begin{bmatrix} \mathbf{s} \\ \mathbf{h} \end{bmatrix}.$$

Samples generated under these assumptions produced conditional

$$\mathbf{s} = [0.633,\ 0.133,\ -0.767]'$$
$$\mathbf{h} = [1.5,\ -1.5,\ 2.0,\ -2.0]'$$

with μ set to 8 in repeated samples. Regular mixed model equations with $\mathbf{h}$ random gave these averages:

$$\hat{u} = 8.007$$
$$\hat{s} = [0.709,\ 0.068,\ -0.778]'$$
$$\hat{h} = [1.340,\ -1.377,\ 1.664,\ -1.627]', \text{ biased.}$$

When $\hat{h}$ was treated as fixed the results were

$$\hat{\mu} = 8, \hat{s} \text{ and } \hat{h} = \text{conditional values, unbiased.}$$

In our example ranking is, on the average, not different for the biased as compared to the unbiased predictions. However, if $h_4 > h_3$ by a sufficient amount, $\hat{s}_2$ will on average be larger than $\hat{s}_1$ even though $s_1 > s_2$. If it is true that better sires tend to be used on better herds, treating herds as random magnifies the predicted differences for sires, but these may not affect ranking appreciably. However, this might tend to cause unfair pricing practices that discourage use of the better sires.

18.11 The Problem of Grouping in Sire Evaluations

Sire evaluation by BLUP routinely includes groups in the model with the supposed purpose of accounting for genetic trend and, in some cases, of accounting for stud differences. If all data used in selection of sires for sampling in artificial insemination (AI) were available and utilized, and $\mathbf{A}^{-1}$ were incorporated, grouping would accomplish nothing. However, in most situations only progeny records of sires and progeny records of their AI relatives are utilized. The relationships are usually calculated from sire and maternal grandsire rather than from sire and dam. Since the merits of the relatives with records not used are presumably increasing from generation to generation, failure to group could handicap sires of the later generations. In a sire and maternal grandsire model, a logical grouping system would be by generations with four classes of sires, namely sire and maternal grandsire known, sire known, maternal grandsire known, neither known.

Unfortunately, from the standpoint of bias, selection is by group + sire solutions $(\hat{g}+\hat{s})$ and, if g is fixed, the selection functions are not translation invariant. Should we therefore use mixed model equations modified for $\mathbf{L'X \neq 0}$? This is not being done at present.

With modification for $\mathbf{L'X \neq 0}$, the prediction error variances are increased over the unmodified equations. However, in the latter we have squared biases that inflate MSE. If we want to minimize MSE we cannot know which is better when **g** is regarded as fixed and unknown. If we could assume some prior distribution on **g** with known first and second moments, the problem would be simple. Unfortunately, the critical g_i is the most recent one, and we have no prior data on it.

18.12 The Problem of Differential Treatments

One of the unsolved problems in prediction of breeding values is the presence of unknown preferential treatment of individuals, e.g., the progeny of certain sires. This can be described as **w**=**L'e**. We illustrate with simulation. Consider:

	Sires		
Environments	1	2	3
1	3	1	2
2	1	4	2

The tabular values are numbers of progeny. It is thought that $\bar{\mathbf{e}}$ in the two environments differ. This can be expressed as **L'e**, where **L'**=[7 7 7 7 7 7 -6 -6 -6 -6 -6 -6 -6]. Further,

$$y_{ij} = \mu+s_i+e_{ij}$$

$$\text{Var}\begin{bmatrix} \mathbf{s} \\ \mathbf{e} \end{bmatrix} = \begin{bmatrix} \mathbf{I} & \mathbf{0} \\ \mathbf{0} & 8\mathbf{I} \end{bmatrix}.$$

Simulating with μ=8 and average of **L'e**=9, the conditional means are **s'**=[0, 0, 0], **e'**=[0.115 ..., 0.115, -0.099 ..., -0.099]. Regular mixed model equations gave average solutions, μ=8.002, $\hat{\mathbf{s}}$= [0.020, -0.022, 0.002], reflecting bias due to disproportionality of the selected **e**. When the equations were modified as follows, unbiased predictions were obtained:

$$\begin{bmatrix} \mathbf{X'R^{-1}X} & \mathbf{X'R^{-1}Z} & \mathbf{X'L} \\ \mathbf{Z'R^{-1}X} & \mathbf{Z'R^{-1}Z+G^{-1}} & \mathbf{Z'L} \\ \mathbf{L'X} & \mathbf{L'Z} & \mathbf{L'RL} \end{bmatrix} \begin{bmatrix} \boldsymbol{\beta}^{\circ} \\ \mathbf{u}^{\circ} \\ \theta^{\circ} \end{bmatrix} = \begin{bmatrix} \mathbf{X'R^{-1}y} \\ \mathbf{Z'R^{-1}y} \\ \mathbf{L'y} \end{bmatrix}.$$

In this example **X'L**=**0**, but not **Z'L**.

Another approach to this problem is simply to model as $y_{ijk}=\mu+ t_i+s_j+e_{ijk}$, where **t** refers to fixed environmental type. The average of solutions for this model were $\mu^\circ=0$, $\mathbf{t}^\circ=[8.11538,\ 7.90110]'$, $\hat{\mathbf{s}}=\mathbf{0}$. Note that $(6t_1^\circ+7t_2^\circ)/13=8$, the value of μ used in simulation. It must be obvious that if every progeny of a sire has a unique set of selected **e**, we cannot differentiate among sires. If we use either an **L'e** model or a treatment model, we obtain 0 as the evaluation of all sires regardless of their relative merits. This should not be confused with our simulation result in which the *average* for each sire in repeated sampling was 0.

It should be pointed out that if

$$\mathbf{w} = \begin{bmatrix} \mathbf{L'y} \\ \mathbf{M'e} \end{bmatrix},$$

L'y creates difficulty even when **L'X=0**. The modified equations for this case are:

$$\begin{bmatrix} \mathbf{X'R^{-1}X} & \mathbf{X'R^{-1}Z} & \mathbf{0} & \mathbf{X'M} \\ \mathbf{Z'R^{-1}X} & \mathbf{Z'R^{-1}Z+G^{-1}} & \mathbf{0} & \mathbf{Z'M} \\ \mathbf{0} & \mathbf{0} & \mathbf{L'VL} & \mathbf{L'RM} \\ \mathbf{M'X} & \mathbf{M'Z} & \mathbf{M'RL} & \mathbf{L'RL} \end{bmatrix} \begin{bmatrix} \boldsymbol{\beta}^\circ \\ \mathbf{u}^\circ \\ \boldsymbol{\theta}_1^\circ \\ \boldsymbol{\theta}_2^\circ \end{bmatrix} = \begin{bmatrix} \mathbf{X'R^{-1}y} \\ \mathbf{Z'R^{-1}y} \\ \mathbf{L'y} \\ \mathbf{L'y} \end{bmatrix},$$

where **V**=Var(**y**).

Because of the existence of **L'RM**, **L'** affects the solution. This is analogous to the problem of

$$\mathbf{w} = \begin{bmatrix} \mathbf{L'y} \\ \mathbf{M'u} \end{bmatrix}$$

discussed earlier.

18.13 The Problem of Assortative Mating

We need to deal with two related problems, assortative mating without selection and assortative mating accompanied by selection. It turns out that if the conditions described earlier for regular mixed model equations hold and base population parameters are used to compute BLUP, then assortative mating creates no problems. We illustrate with the following example of assortative mating without selection. There are six animals, where 1 and 2 are males and 3, 4, 5, 6 are females. The following progeny are obtained

Progeny	Parents
7	1,3
8	1,4
9	2,5
10	2,6

The model is $y_{ij}=p_i+a_j+e_{ij}$ where p_1 is period effect for parents, and p_2 for progeny,

$$\text{Var}\begin{bmatrix} \mathbf{a} \\ \mathbf{e} \end{bmatrix} = \begin{bmatrix} \mathbf{A} & \mathbf{0} \\ \mathbf{0} & 2\mathbf{I} \end{bmatrix}.$$

Animals 1-6 have $\mathbf{A}=\mathbf{I}$.

We simulated with $\mathbf{p}'=(5, 8)$, and under the assumption that in repeated sampling $\bar{y}_1>\bar{y}_2$ and $\bar{y}_3>\bar{y}_4>\bar{y}_5>\bar{y}_6$. Thus, on average, the better sire was mated to the two best females and the poorer sire to the two poorest females. We can write $\mathbf{L}'$ of $\mathbf{L}'\mathbf{y}$ as:

$$\begin{bmatrix} 1 & -1 & 0 & 0 & 0 & 0 & 0 & 0 & 0 & 0 \\ 0 & 0 & 1 & -1 & 0 & 0 & 0 & 0 & 0 & 0 \\ 0 & 0 & 0 & 1 & -1 & 0 & 0 & 0 & 0 & 0 \\ 0 & 0 & 0 & 0 & 1 & -1 & 0 & 0 & 0 & 0 \end{bmatrix}.$$

Simulation used $\mathbf{s}$-$\mathbf{d}$=[1.954, 1.268, 1.029, 1.268]'. This gave average values of $\mathbf{a}$=[0.326, -0.326, 0.594, 0.171, -0.171, -0.594, 0.460, 0.249, -0.249, -0.460]'. Note that the expected mean of the four progeny is 0 as would be expected without selection. The regular mixed model equations produced average values of $\hat{\mathbf{p}}=(5, 8)'$, and $\hat{\mathbf{a}}$=expected values above. These are unbiased as would be expected because $\mathbf{L}'\mathbf{X}=\mathbf{0}$.

As an example of assortative mating with selection, suppose we have four potential male parents, 1-4, and four potential female parents, 5-8. In repeated sampling $\bar{y}_1>\bar{y}_2>\bar{y}_3>\bar{y}_4$ and $\bar{y}_5>\bar{y}_6>\bar{y}_7>\bar{y}_8$. Then the following progeny were produced:

Progeny	Parents
9	1-5
10	2-6

Note that the best male was mated with the best female and the second best male with the second best female based on their records. The two worst males and two worst females were culled:

$$\mathbf{L}' = \begin{bmatrix} 1 & -1 & 0 & 0 & 0 & 0 & 0 & 0 & 0 & 0 \\ 0 & 1 & -1 & 0 & 0 & 0 & 0 & 0 & 0 & 0 \\ 0 & 0 & 1 & -1 & 0 & 0 & 0 & 0 & 0 & 0 \\ 0 & 0 & 0 & 0 & 1 & -1 & 0 & 0 & 0 & 0 \\ 0 & 0 & 0 & 0 & 0 & 1 & -1 & 0 & 0 & 0 \\ 0 & 0 & 0 & 0 & 0 & 0 & 1 & -1 & 0 & 0 \end{bmatrix}.$$

We assume the same model as in the preceding example. With **s**-**d**=[1.268, 1.029, 1.268, 1.268, 1.029, 1.268]', the average **a** is [0.594, 0.171, -0.171, -0.594, 0.594, 0.171, -0.171, -0.594, 0.594, 0.171]'.

Simulation with regular mixed model equations produced average values of $\hat{\mathbf{p}}$=(5, 8)' and $\hat{\mathbf{a}}$=conditional values. Note that the mean of a_i of the parental generation is 0 and that of the progeny generation is 0.382, reflecting selection.

18.14 Discussion

This paper deals exclusively with methods for obtaining unbiased predictors of breeding values. Should we, in fact, always want unbiased predictors? This question is addressed in my first paper in this symposium. Fernando and Gianola, Chapter 19, this Volume, also discuss the problem of finding optimum selection criteria. Let us assume for the present that we do want unbiased predictors. It should be understood that my results are based entirely on a multivariate normal distribution for $(\mathbf{u}, \mathbf{e}, \mathbf{y})'$ prior to selection. Also, we have these assumptions:

1. Changes in first and second moments due to selection follow Pearson's result.
2. **G** and **R** are known at least to proportionality.
3. Selection decisions are based on linear functions of the records, expressed by **L'y** or possibly by **L'y** and non-observable variables uncorrelated with $(\mathbf{u}, \mathbf{e}, \mathbf{y})$.

Then, if all records are used in the analysis, the solution to regular mixed model equations yields BLUE and BLUP under the selection model provided $\mathbf{L'X=0}$. If $\mathbf{L'X \neq 0}$, the mixed model equations can be modified so that BLUE and BLUP are obtained. This requires extensive computation as **L'** has many rows, and also raises the troublesome question of how to write **L**. One possibility would be to write it as **[I:0]** where the records are ordered by selected parents and culled prospective parents, and then unselected progeny. The matrix, **I**, has order equal to the number of selected and prospective parents. A better alternative is to have a "1" and a "-1" in each row of the subset referring to actual and prospective parents. Then some of the columns of **L'X** may be null. The second type of **L'** results in modified equations using all of the data, whereas the first type uses only certain elements of **y** for some of the elements of $\boldsymbol{\beta}^{\circ}$ and $\hat{\mathbf{u}}$.

Can we expect **L'X=0** in most applications? With the exception of the g+s problem, we should expect selection based on evaluations produced by universities and governmental agencies to produce predictors that have means=0 under a no selection model. That is, adjustments are made for fixed effects. As more evaluations by the animal model are developed, there will be little or no need for grouping and then the remaining problems in unbiased prediction will be:

1. Inaccurate $\hat{\mathbf{G}}$, $\hat{\mathbf{R}}$ used instead of **G**, **R**.
2. Data used for selection but not available or not used in the evaluation. This paper outlines an **M'u** approximation.
3. Preferential treatment. This is an essentially unsolved problem. If this becomes bad enough in AI, it might require contract herds for sampling AI sires with some control over how the progeny are managed.

With regard to unknown **G** and **R**, with the consequence of biased predictors if wrong proportionalities are used, this emphasizes the importance of using the best possible estimates. Unbiased predictors by the methods of this paper require that **G** and **R** prior to selection must be used in mixed model equations. Consequently we need a method for estimation of variances and covariances that estimates these parameters even though data resulting from selection have been used. It appears from work by Curnow (1961), Thompson (1973), and Rothschild et al. (1979) that REML estimates these parameters free of bias due to selection.

Fernando and Gianola, Chapter 19, this Volume, also discuss this problem in detail. I have also found with simulation that MIVQUE computed using quadratics in $\hat{\mathbf{u}}$ and $\hat{\mathbf{e}}$ from, if necessary, modified mixed model equations, yields unbiased estimators of variances and covariances provided the prior value used is correct to proportionality. This is, of course, not a very useful result since the proportionality is all that is needed in the mixed model equations. However, if we use proportionalities near the correct ones, we obtain "almost unbiased" estimators.

As a final statement, unbiased prediction is not an optimum solution to the selection problem. Also categorical and censored traits require more complicated techniques, as discussed by others in this volume.

References

Curnow RN (1961) The estimation of repeatability and heritability from records subject to culling. Biometrics 17:553-566

Gianola D, Foulley JL, Fernando RL (1986) Prediction of breeding values when variances are not known. In: Dickerson GE, Johnson RK (eds) Proc 3rd World Congr Genet Appl Livest Prod. Agric Commun, Univ Nebraska, Lincoln, Nebraska XII:356-370

Henderson CR (1949) Estimates of changes in herd environment. J Dairy Sci (Abstr) 32:706

Henderson CR (1973) Sire evaluation and genetic trends. In: Proc Anim Breed Genet Symp in Honor of Dr. J.L. Lush. ASAS and ADSA, Champaign, Illinois, pp 10-41

Henderson CR (1975) Best linear unbiased estimation and prediction under a selection model. Biometrics 31:423-447

Henderson CR, Kempthorne O, Searle SR, Von Krosigk CM (1959) The estimation of environmental and genetic trends from records subject to culling. Biometrics 15:192-218

Lawley DN (1943) A note on Karl Pearson's selection formulae. Proc R Soc Edinburgh A 62:28-36

Lush JL, Shrode RR (1950) Changes in milk production with age and milking frequency. J Dairy Sci 33:338-357

Pearson KC (1903) Mathematical contributions to the theory of evolution. XI. On the influence of natural selection on the variability and correlation of organs. Phil Trans R Soc London A 200:1-66

Rothschild MF, Henderson CR, Quaas RL (1979) Effects of selection on variances and covariances of simulated first and second lactations. J Dairy Sci 62:996-1002

Thompson R (1973) The estimation of variance and covariance components when records are subject to culling. Biometrics 29:527-550

19 Statistical Inferences in Populations Undergoing Selection or Non-Random Mating

R.L. Fernando and D. Gianola[1]

Data available to animal breeders often come from populations undergoing selection and non-random mating (SNRM). Unless this is taken into consideration, inferences based on such data may be misleading. Using a Bayesian setting, it is shown that when the information used to make breeding decisions is available, posterior densities constructed taking into consideration SNRM are identical to those constructed ignoring SNRM. Thus, methods of inference based on posterior densities including all information used to make breeding decisions, can be used ignoring complications due to SNRM. Properties of methods such as maximum likelihood (ML) and best linear unbiased prediction (BLUP) are examined under the assumption of multivariate normality in data from populations undergoing SNRM. Although expressions for ML estimators are identical with or without SNRM, their sampling properties are affected by SNRM. In the presence of SNRM, the matrices appearing in the mixed model equations cannot be considered fixed. However, it is shown under multivariate normality, that the usual expressions lead to BLUP, provided that SNRM decisions are based on translation invariant functions of the data available for calculation of BLUP.

19.1 Introduction

Data used for estimation of variance components, genetic and environmental trends, and prediction of breeding values often come from populations undergoing selection and non-random mating. Much effort has been devoted to develop methods for the proper use of such data in animal breeding.

Henderson (1949), based on simulation results, recommended maximum likelihood for estimating environmental trends in populations undergoing selection. Later (Henderson et al. 1959), maximum likelihood ignoring selection was used to estimate genetic and environmental trends from records subject to culling. The justification of this procedure was given in later publications (Henderson 1965, 1973, 1975, 1979, 1982). Best linear unbiased prediction (BLUP) has been shown to take into account selection when this is based on

[1] Department of Animal Sciences, University of Illinois, Urbana, Illinois 61801, USA

translation invariant functions of the data (Henderson 1975, 1982). Expressions for BLUP in more general types of selection have also been developed (Henderson 1975).

Henderson et al. (1959), Curnow (1961), and Thompson (1973) have shown that maximum likelihood (ML) equations under multivariate normality, for several situations, are identical with and without selection. Computer simulations have suggested that ML of variance and covariance components estimates the parameters prior to selection (Rothschild et al. 1979; Meyer and Thompson 1984).

Goffinet (1983) considered selection with selected records. This author showed that selecting candidates based on the conditional mean of the merit variable given the data was optimal. Fernando and Gianola (1984) extended this result to the situation where records and candidates available for evaluation are a selected subset.

The objective of this paper is to elaborate on a general result (Gianola and Fernando 1986) for making inferences from data subject to selection and non-random mating, using a Bayesian approach.

19.2 Dynamics of a Breeding Population

In a breeding population, phenotypic, classification, and relationship information is used to make breeding decisions. These decisions may involve the following: (1) choice and number of base individuals culled; these become unavailable for production or breeding. (2) Mating decisions concerning individuals kept for breeding; for example, sires with good evaluations will be used with greater frequency and may be mated to better cows. (3) Choice and number of new individuals added to the breeding program; typically, these individuals are offspring of the better sires and dams. These decisions result in individuals with varying amounts of information.

19.2.1 Mathematical Representation of a Breeding Population

Let $\mathbf{T}_0$ represent the breeding values, and $\mathbf{x}_0$ represent the phenotypic, classification, and relationship information of individuals in the base population. Based on $\mathbf{x}_0$, culling and mating decisions are made, and new individuals are added to the breeding program. This also produces new information. This process is represented as follows: when $\mathbf{x}_0 \varepsilon R_i^0$, let $\mathbf{T}_{1i}$ represent the merit of individuals in the population and $\mathbf{x}_{1i}$ represent the additional information; (R_i^0 i=1,...s) are mutually exclusive and exhaustive subsets of values $\mathbf{x}_0$ can take. This process can continue during several stages. Decisions based on $\mathbf{x}_0$ and $\mathbf{x}_{1i}$ give rise to more information and to a new set of individuals. When $\mathbf{x}_0 \varepsilon R_i^0$, and $\mathbf{x}_{1i} \varepsilon R_j^{1i}$, let $\mathbf{T}_{2ij}$ be the merit variables of the individuals in the population and $\mathbf{x}_{2ij}$ be the additional phenotypic information; R_j^{1i} (j=1,...,s_i) are mutually exclusive and exhaustive subsets of the values $\mathbf{x}_{1i}$, for i=1,...,s. Representation of further cycles of selection is straightforward but cumbersome.

Additional notation is needed for the ensuing development. Let $\mathbf{T}$ be a vector containing the unique elements of $\mathbf{T}_0$, $\mathbf{T}_{1i}$, (i=1, ...s), and $\mathbf{T}_{2ij}$ (j=1,...s_i and i=1,...s); similarly, $\mathbf{x}_1$ contains the unique elements of $\mathbf{x}_{1i}$, i=1,...s, and $\mathbf{x}_2$ contains the unique elements of $\mathbf{x}_{2ij}$, j=1,...s_i and i=1,...s.

Consider now the hypothetical situation where selection and non-random mating do not take place and where $\mathbf{T}$, $\mathbf{x}_0$, $\mathbf{x}_1$, and $\mathbf{x}_2$ are always realized. Let

$$f(\mathbf{T},\mathbf{x}_0,\mathbf{x}_1,\mathbf{x}_2,\theta) \tag{19.1}$$

be the joint density of these random variables; θ is a vector of unknowns such as location and dispersion parameters. In the Bayesian approach θ is treated as a random variable. In this population with no selection and non-random mating, the joint density of $\mathbf{T}_{2ij},\mathbf{x}_0,\mathbf{x}_{1i},\mathbf{x}_{2ij}$ and θ,

$$f(\mathbf{T}_{2ij},\mathbf{x}_0,\mathbf{x}_{1i},\mathbf{x}_{2ij},\theta) \tag{19.2}$$

is defined by integrating out the appropriate elements from (19.1). Then, the corresponding joint density with selection and non-random mating is

$$g(\mathbf{T}_{2ij},\mathbf{x}_0,\mathbf{x}_{1i},\mathbf{x}_{2ij},\theta) = f(\mathbf{T}_{2ij},\mathbf{x}_0,\mathbf{x}_{1i},\mathbf{x}_{2ij},\theta)/\text{Prob}(\mathbf{x}_0 \varepsilon R_i^0 \text{ and } \mathbf{x}_{1i} \varepsilon R_j^{1i}) \tag{19.3}$$

for $\mathbf{x}_0 \varepsilon R_i^0$ and $\mathbf{x}_{1i} \varepsilon R_j^{1i}$.

To illustrate the above, consider a population with two males m_1 and m_2 and two females f_1 and f_2, with one record each. Let y_{01},...y_{04} be the phenotypes of m_1,m_2,f_1,f_2, respectively; o_1,...o_4 be offspring from matings between m_i and f_j, j=1,2 and i=1,2; y_{11}, ...,y_{14} be the phenotypes of o_1,...o_4, respectively; and T_{11},...T_{18} the unobservable merits of m_1,m_2,f_1,f_2,o_1,...o_4, respectively. The vector $\mathbf{T}$ contains the elements T_{11},...T_{18}, $\mathbf{y}_0$ has elements y_{01},... y_{04}, and $\mathbf{y}_1$ contains y_{11},...y_{14}.

A possible scheme of non-random mating and selection would be to mate the male and female with the largest phenotypic values and cull the remaining individuals, i.e., m_i is mated to f_j, and m_i' and f_j' are culled when $y_{0i}>y_{0i}'$ for i,i'=1,2 and $y_{0j}>y_{0j}'$ for j,j'=3,4. For example, when $y_{01}>y_{02}$ and $y_{04}>y_{03}$, m_1 is mated to f_2 to produce o_2 and m_2 and f_1 are culled. Also, the phenotypic value y_{12} is realized. Now, with one generation of non-random mating and selection, the joint density of T_{11},T_{14},T_{16},$\mathbf{y}_0$,y_{12} and θ is:

$$g(T_{11},T_{14},T_{16},\mathbf{y}_0,y_{12},\theta)$$

$$= f(T_{11},T_{14},T_{16},\mathbf{y}_0,y_{12},\theta)/\text{Prob}(y_{01}>y_{02},y_{04}>y_{03}) \tag{19.4}$$

for $y_{01}>y_{02}$ and $y_{04}>y_{03}$, where $f(T_{11},T_{14},T_{16},\mathbf{y}_0,y_{12},\theta)$ is obtained by integrating out $y_{11},y_{13},y_{14},T_{12},T_{13},T_{15},T_{17}$, and T_{18} from the joint density of $\mathbf{T},\mathbf{y}_0,\mathbf{y}_1$, and θ.

19.3 Making Inferences in a Population Undergoing Non-Random Mating and Selection

In the Bayesian approach, the posterior density of the variable of interest given all the information is used to make inferences (Box and Tiao 1973). First, consider the problem of making inferences about θ. Using the notation introduced in the previous section, the joint density of θ and the available information after two generations of non-random mating and selection is

$$g(\mathbf{x}_0,\mathbf{x}_{1i},\mathbf{x}_{2ij},\theta) = f(\mathbf{x}_0,\mathbf{x}_{1i},\mathbf{x}_{2ij},\theta)/\text{Prob}(\mathbf{x}_0\varepsilon R_i^0,\ \mathbf{x}_{1i}\varepsilon R_j^{1i}) \tag{19.5}$$

for $\mathbf{x}_0\varepsilon R_i^0$ and $\mathbf{x}_{1i}\varepsilon R_j^{1i}$, where $f(\mathbf{x}_0,\mathbf{x}_{1i},\mathbf{x}_{2ij},\theta)$ is obtained by integrating out $\mathbf{T}_{2ij}$ from (19.2). The posterior density of θ with non-random mating and selection is:

$$g(\theta|\mathbf{x}_0,\mathbf{x}_{1i},\mathbf{x}_{2ij}) = g(\mathbf{x}_0,\mathbf{x}_{1i},\mathbf{x}_{2ij},\theta)/g(\mathbf{x}_0,\mathbf{x}_{1i},\mathbf{x}_{2ij})\ , \tag{19.6}$$

where $g(\mathbf{x}_0,\mathbf{x}_{1i},\mathbf{x}_{2ij})$ is the marginal density of the data under non-random mating and selection, given by:

$$g(\mathbf{x}_0,\mathbf{x}_{1i},\mathbf{x}_{2ij}) = f(\mathbf{x}_0,\mathbf{x}_{1i},\mathbf{x}_{2ij})/\text{Prob}(\mathbf{x}_0\varepsilon R_i^0,\ \mathbf{x}_{1i}\varepsilon R_j^{1i}) \tag{19.7}$$

and $f(\mathbf{x}_0,\mathbf{x}_{1i},\mathbf{x}_{2ij})$ is obtained integrating $\mathbf{T}_{2ij}$ and θ from (19.2). Thus:

$$g(\theta|\mathbf{x}_0,\mathbf{x}_{1i},\mathbf{x}_{2ij}) = (19.5)/(19.7) = f(\mathbf{x}_0,\mathbf{x}_{1i},\mathbf{x}_{2ij},\theta)/f(\mathbf{x}_0,\mathbf{x}_{1i},\mathbf{x}_{2ij}) \tag{19.8}$$

because the denominators of (19.5) and (19.7) are identical. Note that (19.8)=$f(\theta|\mathbf{x}_0,\mathbf{x}_{1i},\mathbf{x}_{2ij})$ is the posterior density of θ given the information, in the absence of non-random mating and selection. Thus, if the information used to make the selection decisions is available, inferences about θ can be made using (19.8) ignoring selection.

If $\theta'=[\boldsymbol{\beta}',\sigma']$, where $\boldsymbol{\beta}$ and σ are vectors of location and dispersion parameters, (19.8) would lead to joint inferences on $\boldsymbol{\beta}$ and σ. Marginal inferences about $\boldsymbol{\beta}$ and σ can be obtained from

$$g(\boldsymbol{\beta}|\mathbf{x}_0,\mathbf{x}_{1i},\mathbf{x}_{2ij}) = g(\mathbf{x}_0,\mathbf{x}_{1i},\mathbf{x}_{2ij},\boldsymbol{\beta})/g(\mathbf{x}_0,\mathbf{x}_{1i},\mathbf{x}_{2ij}) \tag{19.9}$$

and

$$g(\sigma|\mathbf{x}_0,\mathbf{x}_{1i},\mathbf{x}_{2ij}) = g(\mathbf{x}_0,\mathbf{x}_{1i},\mathbf{x}_{2ij},\sigma)/g(\mathbf{x}_0,\mathbf{x}_{1i},\mathbf{x}_{2ij}), \quad (19.10)$$

respectively. But,

$$g(\mathbf{x}_0,\mathbf{x}_{1i},\mathbf{x}_{2ij},\boldsymbol{\beta}) = f(\mathbf{x}_0,\mathbf{x}_{1i},\mathbf{x}_{2ij},\boldsymbol{\beta})/\text{Prob}(\mathbf{x}_0 \varepsilon R_i^0, \mathbf{x}_{1i} \varepsilon R_j^{1i}), \quad (19.11)$$

where $f(\mathbf{x}_0,\mathbf{x}_{1i},\mathbf{x}_{2ij},\boldsymbol{\beta})$ is obtained by integrating out σ and $\mathbf{T}_{2ij}$ from (19.2). Thus:

$$g(\boldsymbol{\beta}|\mathbf{x}_0,\mathbf{x}_{1i},\mathbf{x}_{2ij}) = (19.11)/(19.7) = f(\mathbf{x}_0,\mathbf{x}_{1i},\mathbf{x}_{2ij},\boldsymbol{\beta})/f(\mathbf{x}_0,\mathbf{x}_{1i},\mathbf{x}_{2ij}), \quad (19.12)$$

which is the posterior density of $\boldsymbol{\beta}$ in the absence of non-random mating and selection. Similarly:

$$g(\sigma|\mathbf{x}_0,\mathbf{x}_{1i},\mathbf{x}_{2ij}) = f(\mathbf{x}_0,\mathbf{x}_{1i},\mathbf{x}_{2ij},\sigma)/f(\mathbf{x}_0,\mathbf{x}_{1i},\mathbf{x}_{2ij}) = f(\sigma|\mathbf{x}_0,\mathbf{x}_{1i},\mathbf{x}_{2ij}). \quad (19.13)$$

Thus, when the information used to make selection decisions is available, inferences about $\boldsymbol{\beta}$ and σ can be made using (19.12) and (19.13) ignoring selection.

Inferences about the merit variables, $\mathbf{T}_{2ij}$, are made ideally using the marginal posterior density of $\mathbf{T}_{2ij}$, using the available information. In a population undergoing non-random mating and selection, the posterior density of $\mathbf{T}_{2ij}$ is:

$$g(\mathbf{T}_{2ij}|\mathbf{x}_0,\mathbf{x}_{1i},\mathbf{x}_{2ij}) = g(\mathbf{T}_{2ij},\mathbf{x}_0,\mathbf{x}_{1i},\mathbf{x}_{2ij})/g(\mathbf{x}_0,\mathbf{x}_{1i},\mathbf{x}_{2ij}). \quad (19.14)$$

Now,

$$g(\mathbf{T}_{2ij},\mathbf{x}_0,\mathbf{x}_{1i},\mathbf{x}_{2ij}) = f(\mathbf{T}_{2ij},\mathbf{x}_0,\mathbf{x}_{1i},\mathbf{x}_{2ij})/\text{Prob}(\mathbf{x}_0 \varepsilon R_i^0 \text{ and } \mathbf{x}_{1i} \varepsilon R_j^{1i}), \quad (19.15)$$

where $f(\mathbf{T}_{2ij},\mathbf{x}_0,\mathbf{x}_{1i},\mathbf{x}_{2ij})$ is obtained by integrating out θ from (19.2). Then, (19.14) can be written as

$$g(\mathbf{T}_{2ij}|\mathbf{x}_0,\mathbf{x}_{1i},\mathbf{x}_{2ij}) = (19.15)/(19.7) = f(\mathbf{x}_0,\mathbf{x}_{1i},\mathbf{x}_{2ij},\mathbf{T}_{2ij})/f(\mathbf{x}_0,\mathbf{x}_{1i},\mathbf{x}_{2ij}), \quad (19.16)$$

which is the posterior density of $\mathbf{T}_{2ij}$ in the absence of selection. Thus, when the information used to make the breeding decisions is available, the posterior density of $\mathbf{T}_{2ij}$ ignoring selection can be used to make inferences.

The results of this section can be summarized as follows: when the information used to make breeding decisions is available, joint and marginal posterior densities constructed taking into consideration non-random mating and selection are identical to those constructed ignoring selection. Thus, if inferences are based on posterior densities and all information used to make breeding decisions is used to construct such densities, inferences can be made ignoring the complications due to non-random mating and selection. This will hold for any distribution, any type of selection, or non-random mating based on the information available and any method of inference based on a posterior density. Although these results have been demonstrated only for two generations of selection, it is easy to see that the same would be true about posterior densities for any number of generations of selection and non-random mating, provided that all the information employed used for the selection decisions is employed when constructing the posterior distributions.

19.4 Making Inferences with Incomplete Information

In the previous section we considered the situation when all information used to make breeding decisions was available. However, often this is not the case.

For simplicity we will consider only one generation of non-random mating and selection. Assume that breeding decisions are based on information represented by $\mathbf{x}_0$ and $\mathbf{v}$. These decisions will define the set of individuals present in the population, and will also give rise to more information. This process can be represented mathematically as follows: when $\mathbf{x}_0 \varepsilon R_i^0$ and $\mathbf{v} \varepsilon R_j^v$, let $\mathbf{T}_{1ij}$ be the breeding values of the individuals in the population, and let $\mathbf{x}_{1ij}$ be the new information; R_i^0 (i=1,...s), are mutually exclusive and exhaustive subsets of the values $\mathbf{x}_0$ can take, and R_j^v (j=1,...s$_v$), are mutually exclusive and exhaustive subsets of values $\mathbf{v}$ can take. Let $\mathbf{T}$ contain the unique elements of $\mathbf{T}_0$ and $\mathbf{T}_{1ij}$ (i=1,...,s) and (j=1,...,s$_v$), and let x_1 contain the unique elements of $\mathbf{x}_{1ij}$ (i=1,...,s and j=1,...,s$_v$). Now, in a hypothetical population without selection and with random mating where $\mathbf{T}$, $\mathbf{x}_0$, $\mathbf{x}_1$, and $\mathbf{v}$ are always realized, the joint density of these variables and some other variables of interest, θ, will be denoted as:

$$f(\mathbf{T},\mathbf{x}_0,\mathbf{x}_1,\mathbf{v},\theta). \tag{19.17}$$

Then the joint density of $\mathbf{x}_0$, $\mathbf{v}$, $\mathbf{x}_{1ij}$, $\mathbf{T}_{1ij}$, and θ,

$$f(\mathbf{T}_{1ij},\mathbf{x}_0,\mathbf{v},\mathbf{x}_{1ij},\theta) \tag{19.18}$$

is obtained integrating out the appropriate elements from (19.17).

We will now consider making inferences about θ when non-random mating and selection have taken place, but $\mathbf{v}$ is not available for analysis. Inferences will be based on the posterior density of θ given $\mathbf{x}_0$ and $\mathbf{x}_{1ij}$ constructed under the non-random mating and selection scheme described above. The joint density of $\mathbf{x}_0,\mathbf{x}_{1ij}$, and θ under selection is:

$$g(\mathbf{x}_0,\mathbf{x}_{1ij},\theta) = \int_{R_j^v} f(\mathbf{x}_0,\mathbf{x}_{1ij},\mathbf{v},\theta)d\mathbf{v}/\text{Prob}(\mathbf{x}_0 \varepsilon R_i^0 \text{ and } \mathbf{v} \varepsilon R_j^v) \tag{19.19}$$

for $\mathbf{x}_0 \varepsilon R_i^0$, where $f(\mathbf{x}_0,\mathbf{x}_{1ij},\mathbf{v},\theta)$ is obtained by integrating $\mathbf{T}_{1ij}$ from (19.18). To obtain the posterior density of θ we need to divide (19.19) by the density of $\mathbf{x}_0$ and $\mathbf{x}_{1ij}$. This density under non-random mating and selection is:

$$g(\mathbf{x}_0,\mathbf{x}_{1ij}) = \int_{R_j^v} f(\mathbf{x}_0,\mathbf{x}_{1ij},\mathbf{v})d\mathbf{v}/\mathrm{Prob}(\mathbf{x}_0 \varepsilon R_i^0 \text{ and } \mathbf{v}\varepsilon R_j^v) \tag{19.20}$$

for $\mathbf{x}_0 \varepsilon R_i^0$, where $f(\mathbf{x}_0,\mathbf{x}_{1ij},\mathbf{v})$ is obtained by integrating out $\mathbf{T}_{1ij}$ and θ from (19.18). Because the denominators in (19.19) and (19.20) are identical, the posterior density of θ under selection is:

$$g(\theta|\mathbf{x}_0,\mathbf{x}_{1ij}) = \int_{R_j^v} f(\mathbf{x}_0,\mathbf{x}_{1ij},\mathbf{v},\theta)d\mathbf{v}/ \int_{R_j^v} f(\mathbf{x}_0,\mathbf{x}_{1ij},\mathbf{v})d\mathbf{v}. \tag{19.21}$$

The numerator of (19.21) can be written as

$$\int_{R_j^v} f(\mathbf{v}|\mathbf{x}_0,\mathbf{x}_{1ij},\theta).f(\mathbf{x}_0,\mathbf{x}_{1ij},\theta)d\mathbf{v} = \mathrm{Prob}(\mathbf{v}\varepsilon R_j^v|\mathbf{x}_0,\mathbf{x}_{1ij},\theta).f(\mathbf{x}_0,\mathbf{x}_{1ij},\theta) \tag{19.22}$$

and, similarly, the denominator of (19.21) is

$$= \mathrm{Prob}(\mathbf{v}\varepsilon R_j^v|\mathbf{x}_0,\mathbf{x}_{1ij}).f(\mathbf{x}_0,\mathbf{x}_{1ij}). \tag{19.23}$$

Thus, (19.21) is

$$g(\theta|\mathbf{x}_0,\mathbf{x}_{1ij}) = (19.22)/(19.23)$$

$$= \left[\mathrm{Prob}(\mathbf{v}\varepsilon R_j^v|\mathbf{x}_0,\mathbf{x}_{1ij},\theta)/\mathrm{Prob}(\mathbf{v}\varepsilon R_j^v|\mathbf{x}_0,\mathbf{x}_{1ij})\right].f(\theta|\mathbf{x}_0,\mathbf{x}_{1ij}) . \tag{19.24}$$

When

$$\mathrm{Prob}(\mathbf{v}\varepsilon R_j^v|\mathbf{x}_0,\mathbf{x}_{1ij},\theta) = \mathrm{Prob}(\mathbf{v}\varepsilon R_j^v|\mathbf{x}_0,\mathbf{x}_{1ij}), \tag{19.25}$$

it follows that

$$g(\theta|\mathbf{x}_0,\mathbf{x}_{1ij}) = f(\theta|\mathbf{x}_0,\mathbf{x}_{1ij}).$$

The condition in (19.25) is met when

$$f(\mathbf{v},\theta|\mathbf{x}_0,\mathbf{x}_{1ij}) = f(\mathbf{v}|\mathbf{x}_0,\mathbf{x}_{1ij}).f(\theta|\mathbf{x}_0,\mathbf{x}_{1ij}), \tag{19.26}$$

i.e., given $\mathbf{x}_0$ and $\mathbf{x}_{1ij}$, $\mathbf{v}$ and θ are independent. This can be seen by writing $f(\mathbf{v}|\mathbf{x}_0,\mathbf{x}_{1ij})$ as:

$$f(\mathbf{v}|\mathbf{x}_0,\mathbf{x}_{1ij},\theta) = f(\mathbf{v},\theta|\mathbf{x}_0,\mathbf{x}_{1ij})/f(\theta|\mathbf{x}_0,\mathbf{x}_{1ij})\ . \tag{19.27}$$

Now, substituting (19.26) in the numerator of (19.27) gives

$$f(\mathbf{v}|\mathbf{x}_0,\mathbf{x}_{1ij},\theta) = f(\mathbf{v}|\mathbf{x}_0,\mathbf{x}_{1ij}) \tag{19.28}$$

and (19.25) follows. Thus, if the missing information ($\mathbf{v}$) and θ are independent, given the data, the posterior density under non-random mating and selection is identical to the posterior density in the absence of selection. However, in general, when some of the information used to make selection decisions is not available, calculation of the posterior density (19.24) requires knowledge of the selection process.

By a similar reasoning, the posterior density of the breeding values can be shown to be

$$g(\mathbf{T}_{1ij}|\mathbf{x}_0,\mathbf{x}_{1ij})$$

$$= \left[\mathrm{Prob}(\mathbf{v}\varepsilon R_j^{\mathbf{v}}|\mathbf{x}_0,\mathbf{x}_{1ij},\mathbf{T}_{1ij})/\mathrm{Prob}(\mathbf{v}\varepsilon R_j^{\mathbf{v}}|\mathbf{x}_0,\mathbf{x}_{1ij})\right].f(\mathbf{T}_{1ij}|\mathbf{x}_0,\mathbf{x}_{1ij}). \tag{19.29}$$

This implies that when some of the information ($\mathbf{v}$) used to make breeding decisions is not available, the posterior distribution of breeding values under selection cannot be constructed unless $\mathbf{v}$ and $\mathbf{T}_{1ij}$ are conditionally (given the data) independent.

19.5 Multivariate Normality

In many animal breeding applications, multivariate normality is a reasonably approximation. For example, most continuous traits tend to follow a normal distribution (Falconer 1960). This is attributed to the genotypic and environmental components of the phenotype being determined by a large number of relatively small effects (e.g., Pirchner 1969). In this section we will examine the properties of a few well known methods of estimation and prediction, under the assumption of multivariate normality, when the data are from a population undergoing non-random mating and selection.

Consider a single trait sire evaluation problem. The usual mixed linear model is

$$\mathbf{y} = \mathbf{X}\boldsymbol{\beta}+\mathbf{Z}\mathbf{u}+\mathbf{e}, \tag{19.30}$$

where $\mathbf{X}$ and $\mathbf{Z}$ are matrices relating $\boldsymbol{\beta}$ and $\mathbf{u}$ to $\mathbf{y}$; $\mathbf{u}$ is the vector of transmitting abilities of the sires and $\mathbf{e}$ is a vector of residuals. In the absence of selection,

$$E(\mathbf{y}|\boldsymbol{\beta}) = \mathbf{X}\boldsymbol{\beta},\ E(\mathbf{u}) = \mathbf{0},\ E(\mathbf{e}) = \mathbf{0} \tag{19.31}$$

$$\text{Var}(\mathbf{u}|\sigma_u^2) = \mathbf{A}\sigma_u^2 \text{ and } \text{Var}(\mathbf{e}|\sigma_e^2) = \mathbf{R}\sigma_e^2. \tag{19.32}$$

Further, it is usual to assume that Cov($\mathbf{u}$,$\mathbf{e}'$)=$\mathbf{0}$ and that $\mathbf{u}$ and $\mathbf{e}$ follow a multivariate normal distribution with the location and dispersion parameters given above.

A single generation of selection and non-random mating will be considered. At the first stage, let $\mathbf{y}_0$ be the phenotypic information available, $\mathbf{u}_0$ be a vector of transmitting abilities of the sires evaluated, and $\mathbf{e}_0$ be a vector of residuals. Also,

$$E(\mathbf{y}_0|\boldsymbol{\beta}) = \mathbf{X}_0\boldsymbol{\beta},\ E(\mathbf{u}_0) = \mathbf{0},\ E(\mathbf{e}_0) = \mathbf{0} \tag{19.33}$$

$$\text{Var}(\mathbf{u}_0|\sigma_u^2) = \mathbf{A}_0\sigma_u^2 \text{ and } \text{Var}(\mathbf{e}_0|\sigma_e^2) = \mathbf{R}_0\sigma_e^2. \tag{19.34}$$

Evaluations based on the information available are used to make decisions such as evaluating additional sires so that more information is obtained. This is represented as follows: when $\mathbf{y}_0 \varepsilon R_i$,

$$\begin{aligned} \mathbf{y}' &= [\mathbf{y}_0', \mathbf{y}_i'] \\ \mathbf{X}' &= [\mathbf{X}_0', \mathbf{X}_i'] \\ \mathbf{Z}' &= [\mathbf{Z}_0', \mathbf{Z}_i'] \\ \mathbf{u}' &= [\mathbf{u}_0', \mathbf{u}_i']. \end{aligned} \tag{19.35}$$

Above, $\mathbf{y}$ is the phenotypic information available after non-random mating and selection, and $\mathbf{u}$ is the vector of breeding values of the sires to be evaluated after selection; $\mathbf{X}$ and $\mathbf{Z}$ are incidence matrices relating $\boldsymbol{\beta}$ and $\mathbf{u}$ to $\mathbf{y}$; R_i (i=1,...,s) are mutually exclusive and exhaustive subsets of the values $\mathbf{y}_0$ can take. The vectors $\mathbf{y}$ and $\mathbf{u}$ are no longer multivariate normal but inferences can be made using the posterior density as shown in the previous section.

19.5.1 Maximum Likelihood Estimation

We will consider first joint estimation of $\boldsymbol{\beta}$, σ_u^2, and σ_e^2. Let $\theta'=[\boldsymbol{\beta}', \sigma_u^2, \text{ and } \sigma_e^2]$. The posterior density of θ, in the absence of selection, can be written as:

$$f(\theta|\mathbf{y},\mathbf{X},\mathbf{Z},\mathbf{A},\mathbf{R}) = f(\mathbf{y}|\mathbf{X},\mathbf{Z},\mathbf{A},\mathbf{R},\theta).f(\theta|\mathbf{X},\mathbf{Z},\mathbf{A},\mathbf{R})/f(\mathbf{y}|\mathbf{X},\mathbf{Z},\mathbf{A},\mathbf{R}). \tag{19.36}$$

Because $f(\mathbf{y}|\mathbf{X},\mathbf{Z},\mathbf{A},\mathbf{R})$ is not a function of θ,

$$f(\theta|\mathbf{y},\mathbf{X},\mathbf{Z},\mathbf{A},\mathbf{R}) \propto f(\mathbf{y}|\mathbf{X},\mathbf{Z},\mathbf{A},\mathbf{R},\theta).f(\theta|\mathbf{X},\mathbf{Z},\mathbf{A},\mathbf{R}). \tag{19.37}$$

From the assumptions made here, in the absence of selection, $f(\mathbf{y}|\mathbf{X}, \mathbf{Z},\mathbf{A},\mathbf{R},\theta)$ has a multivariate normal distribution with mean $\mathbf{X}\boldsymbol{\beta}$ and variance $(\mathbf{ZAZ}'\sigma_u^2+\mathbf{R}\sigma_e^2)$. If $f(\theta|\mathbf{X},\mathbf{Z},\mathbf{A},\mathbf{R})$ in (19.37) is taken to be a constant,

$$f(\theta|\mathbf{y},\mathbf{X},\mathbf{Z},\mathbf{A},\mathbf{R}) \propto f(\mathbf{y}|\mathbf{X},\mathbf{Z},\mathbf{A},\mathbf{R},\theta). \tag{19.38}$$

Maximizing (19.38) with respect to the elements of θ gives the familiar maximum likelihood estimates of $\boldsymbol{\beta}$ and of the variances (e.g., Searle 1979). In Bayesian terminology, this is the posterior mode of θ or the most likely value of θ given the data.

With non-random mating and selection, **X**,**Z**,**A**, and **R** cannot be considered fixed. However, the posterior distribution of θ is constructed conditionally on all the information, including the observed **X**,**Z**,**A**, and **R** matrices, Then, in the absence of selection, if $f(\mathbf{y}|\mathbf{X},\mathbf{Z},\mathbf{A},\mathbf{R},\theta)$ is the density of a multivariate normal distribution with mean $\mathbf{X}\boldsymbol{\beta}$ and variance $(\mathbf{ZAZ}'\sigma_u^2+\mathbf{R}\sigma_e^2)$ and $f(\theta|\mathbf{X},\mathbf{Z},\mathbf{A},\mathbf{R})$ is constant, the posterior density of θ under non-random mating and selection, is identical to (19.38) because the information used to make breeding decisions is a subset of that used to construct the posterior density. Thus, the posterior mode which is also the ML estimate of θ, should be as in the absence of selection.

Without selection marginal inferences about $\sigma=[\sigma_u^2, \sigma_e^2]$ can be made from

$$f(\sigma|\mathbf{y},\mathbf{X},\mathbf{Z},\mathbf{A},\mathbf{R}) = \int_{R_\beta} f(\theta|\mathbf{y},\mathbf{X},\mathbf{Z},\mathbf{A},\mathbf{R})\, d\boldsymbol{\beta}. \tag{19.39}$$

Because the information used to make breeding decisions has been used to construct (19.39), this can also be used for making inferences even in the presence of selection. Given the assumptions made here, it has been shown (Harville 1974; Gianola et al. 1986) that the mode of (19.39) gives restricted maximum likelihood (REML) estimates of the variances (Patterson and Thompson 1971). Thus, if all information used to make breeding decisions is available, REML estimates can be calculated as if selection had not taken place.

Similar results on ML and REML, under normality, have been reported by others (Henderson et al. 1959; Curnow 1961; Thompson 1973). The results presented in this section are special cases of the general ones given earlier, which hold for any distribution. It must be pointed out, however, that the sampling properties of the ML or REML estimators in the presence of selection as described here, may differ from those holding when selection does not occur. A simple example is given below to illustrate this point. Define

$$\mathbf{y}_0' = [y_{01}, \ldots, y_{05}]$$

with

$$y_{0i} \sim N(0,\sigma^2)$$

and the mean known but σ^2 unknown. Sampling is such that y_0 is realized first, and the REML estimate of $\hat{\sigma}^2$ is calculated. When this estimate is smaller than 1, n_1 observations ($\mathbf{y}_1$) from the same distribution are taken and when $\hat{\sigma}^2$ is larger or equal than 1, n_2 observations ($\mathbf{y}_2$) are taken. Now, the REML estimate of σ^2 using $\mathbf{y}_0$ and $\mathbf{y}_1$ can be calculated as if no selection had taken place because the information used for selection, $\mathbf{y}_0$, is a subset of the information used to obtain REML. Now, if σ^2=1, n_1=0, and n_2>0, the REML estimator will be biased downwards. To demonstrate this, let e_1 be the estimator of σ^2 based on $\mathbf{y}_0$, and e_2 be the estimator based on all the data. We have:

$$e_1 = \sum_{j=1}^{5} y_{0j}^2/5$$

with expectation σ^2=1. This expected value can be written as:

$$E(e_1) = E(e_1|e_1<1).p_1+E(e_1|e_1>1).p_2 = \sigma^2 = 1, \qquad (19.40)$$

where p_1 is the probability that e_1 is less than 1, and p_2=1-p_1. If n_1=0, e_2 is:

$$e_2 = e_1 \qquad \text{if } e_1 < 1$$

$$e_2 = (5e_1+\sum_{j=1}^{n_2} y_{2j}^2)/(5+n_2) \qquad \text{if } e_1 > 1$$

and the expected value of e_2 can be written as

$$E(e_2) = E(e_1|e_1<1).p_1+E(e_2|e_1>1).p_2. \tag{19.41}$$

The first terms in (19.40) and (19.41) are identical and $E(e_2|e_1>1)$ is smaller than $E(e_1|e_1>1)$. This can be seen by writing the ratio of these two expected values as:

$$E(e_2|e_1>1)/E(e_1|e_1>1) = \{[5E(e_1|e_1>1)+n_2]/(5+n_2)\}/E(e_1|e_1>1)$$

$$= [5+n_2/E(e_1|e_1>1)]/(5+n_2). \tag{19.42}$$

Because $E(e_1|e_1>1)>1$, (19.42)<1. Thus (19.41) is smaller than (19.40), i.e., e_2 is biased downwards due to selection. If the decision to sample n_1 or n_2 observations were made at random, the REML estimator would be unbiased.

19.5.2 Best Linear Prediction

Consider the model given in (19.30). In the absence of selection, the best linear predictor (BLP) of **u** is:

$$BLP(\mathbf{u}) = \mathbf{C'V^{-1}(y\text{-}X\beta)}, \tag{19.43}$$

where $\mathbf{C'}$=cov($\mathbf{u}$,$\mathbf{y'}$) and $\mathbf{V}$=var($\mathbf{y}$) (Henderson 1973). Note that calculation of BLP requires knowledge of θ. When the joint density of **u** and **y** given θ,**X**,**Z**,**A**, and **R** is multivariate normal, (19.43) is identical to

$$E(\mathbf{u}|\mathbf{y},\mathbf{X},\mathbf{Z},\mathbf{A},\mathbf{R},\theta), \tag{19.44}$$

Because (19.44) is the mean of the posterior density of **u** given θ and all the information used for selection, the posterior mean with or without selection is identical to (19.43). It is well known that (19.44) minimizes mean squared error of prediction regardless of the distribution and the sample space of **y**. However, under normality (19.44) (e.g., Henderson 1973) is a linear function of **y** so it follows that (19.43) is also BLP of **u** when non-random mating and selection take place. Henderson (1979) showed that under normality, BLP with selection is identical to BLP without selection provided this is based on some linear function of $\mathbf{y}_0$. Henderson's (1979) result is extended here as follows: (1) the selection can be based on any function of $\mathbf{y}_0$. (2) This result is valid for more than one generation of selection provided all the information used to make selection decisions is used in calculating BLP. (3) The result also holds for random **X**,**Z**,**A**, and **R** matrices as shown in the following section in a more general setting.

19.5.3 Best Linear Unbiased Prediction

When $\boldsymbol{\beta}$ is unknown, BLP of **u** cannot be calculated. However, if σ is known, predictions of **u** can be made using the best linear unbiased predictor (BLUP) (Henderson 1963). In the absence of selection, BLUP of **u** is:

$$\text{BLUP}(\mathbf{u}) = \mathbf{C}'\mathbf{V}^{-1}(\mathbf{y}-\mathbf{X}\boldsymbol{\beta}^0), \tag{19.45}$$

where $\mathbf{X}\boldsymbol{\beta}^0$ is the generalized least-squares estimator of $\mathbf{X}\boldsymbol{\beta}$ (Henderson 1963). It can be shown that under normality (19.45) is identical to

$$E(\mathbf{u}|\mathbf{w},\mathbf{X},\mathbf{Z},\mathbf{A},\mathbf{R},\sigma), \tag{19.46}$$

where **w** is any set of n-r linearly independent translation invariant functions of **y**; n is the order of **y**, and r is the rank of **X** (Goffinet 1983; Fernando and Gianola 1986).

Consider the situation where breeding decisions at the first stage are not based directly on $\mathbf{y}_0$ but on some subset of **w**. Because selection decisions are based on a subset of the information used to construct the posterior mean (19.46), under normality this can be calculated using (19.45) as if selection had not taken place. With non-random mating and selection, **X**,**Z**,**A**, and **R** cannot be considered fixed. It is shown below that under normality (19.45) minimizes prediction error variance among all linear unbiased predictors even when **X**,**Z**,**A**, and **R** are random.

Let v be some linear function of **u**:

$$v = \ell'\mathbf{u}$$

and $\varepsilon = v - \hat{v}$, where $\hat{v}$ is some unbiased predictor of v. Note that any unbiased predictor depends on **y** only through **w**. Now, the mean squared error (prediction error variance in this case) of ε can be written as:

$$E(\varepsilon^2|\sigma) = E\{\{E[(v-\hat{v})^2|\mathbf{w},\mathbf{X},\mathbf{Z},\mathbf{A},\mathbf{R},\sigma]\}|\sigma\} = E\{\{\text{Var}(v|\mathbf{w},\mathbf{X},\mathbf{Z},\mathbf{A},\mathbf{R},\sigma)$$

$$+ [E(v|\mathbf{w},\mathbf{X},\mathbf{Z},\mathbf{A},\mathbf{R},\sigma)-\hat{v}]^2\}|\sigma\}. \tag{19.47}$$

The first term of (19.47) is not a function of $\hat{v}$ and, the second term is minimized by taking $\hat{v}$=E(v|**w**,**X**,**Z**,**A**,**R**,σ). If non-random mating and selection were based on a subset of **w**,**X**,**Z**,**A**, and **R**, E(v|**w**,**X**,**Z**,**A**,**R**,σ) can be calculated as if non-random mating and selection had not taken place, so under normality,

$$\hat{v} = \boldsymbol{\ell}'\mathbf{C}'\mathbf{V}^{-1}(\mathbf{y}-\mathbf{X}\boldsymbol{\beta}^0). \tag{19.48}$$

Thus, under normality, BLUP as defined by Henderson (1963) minimizes prediction error variance among all linear unbiased predictors, even when **X**,**Z**,**A**, and **R** are not fixed. This is provided that breeding decisions are based on translation invariant functions of the data. Further, because under normality, BLUP= (19.46), selection based on BLUP will maximize the mean of the selected individuals among all translation invariant selection rules (Goffinet 1983; Fernando and Gianola 1986).

When selection is based on some linear function of the data, **L'y**, with **L'X**=0, Henderson (1975) showed that under normality, BLUP can be calculated as if selection had not taken place. The presentation here extends Henderson's finding to include selection based on any translation invariant functions of the data and accommodates random **X**,**Z**,**A**, and **R** matrices. The result holds beyond a single generation of selection.

When $\boldsymbol{\beta}$ is an unknown constant, E(**u**|**y**,**X**,**Z**,**A**,**R**,σ) cannot be calculated using principles of classical statistics. However, as shown above, E(**u**|**w**,**X**,**Z**,**A**,**R**,σ) can be calculated under normality, provided that selection decisions are based on translation invariant functions of the data. Consider the situation where prior breeding decisions are based on a subset of **y**. The data vector **y** can be written as:

$$\mathbf{y} = \mathbf{X}\hat{\boldsymbol{\beta}}+\mathbf{w}, \tag{19.49}$$

where

$$\mathbf{w} = \mathbf{y}-\mathbf{X}\hat{\boldsymbol{\beta}} \tag{19.50}$$

and

$$\hat{\boldsymbol{\beta}} = (\mathbf{X}'\mathbf{X})^{-1}\mathbf{X}'\mathbf{y}. \tag{19.51}$$

Now, because selection is based on a subset of **y**, the posterior density constructed using only **w** will not be in general the same as when selection has not taken place. As shown earlier, calculation of the posterior density in this situation requires knowledge of the selection process. However, when selection is based on **L'y** and **L'X**≠0, Henderson (1975) has shown how to obtain BLUP. In this situation, BLUP has not been demonstrated to have the property of maximizing the mean of the selected individuals.

With the Bayesian approach, $E(\mathbf{u}|\mathbf{y},\mathbf{X},\mathbf{Z},\mathbf{A},\mathbf{R},\sigma)$ can be calculated even when $\boldsymbol{\beta}$ is unknown (e.g., Gianola and Fernando 1986). Here, uncertainty about $\boldsymbol{\beta}$ is expressed in the form of a prior density. It must be emphasized, however, that this Bayesian conditional mean has a different interpretation than the one defined in classical statistical theory. When the prior density of $\boldsymbol{\beta}$ is taken to be a constant, $E(\mathbf{u}|\mathbf{y},\mathbf{X},\mathbf{Z},\mathbf{A},\mathbf{R},\sigma)$=(19.46). Also, if selection is based on $\mathbf{y}$, $E(\mathbf{u}|\mathbf{y},\mathbf{X},\mathbf{Z},\mathbf{A},\mathbf{R},\sigma)$ is calculated ignoring selection.

Optimality properties of the conditional mean, derived using classical principles, may not be applicable to the conditional mean calculated using Bayesian arguments. In order to examine this issue, genetic evaluation using BLUP in a population undergoing selection based on $\mathbf{L'y}$ with $\mathbf{L'X}\neq\mathbf{0}$ (Henderson 1975) was compared to evaluation using (19.45), by computer simulation. Data were generated for four sires in two genetic groups, using the model:

$$y_{ij} = g_i+s_{ij}+e_{ijk} , \tag{19.52}$$

where s_{ij}~N(0,1), e_{ijk}~N(0,σ_e^2), and σ_e^2=1,3,7, or 15. These values of σ_e^2 correspond to "heritabilities" (h^2) of 2,1,0.5, and 0.25, respectively. The difference between the genetic groups was $\sigma_e/2$. Data were generated in two stages. The subclass numbers at the first stage were:

n_{11}	n_{12}	n_{21}	n_{22}
2	3	1	2

,

where n_{ij} is the number of records from sire j in group i. Ten more records were then obtained from the sire that had the highest mean in the first sampling. Evaluations were obtained using all data with Henderson's modified mixed model equations (MME), giving BLUP (Henderson 1975), and with the usual mixed model equations (UME), or expression (19.45). Criteria for comparison were calculated from 1000 replicates of the simulation.

Evaluations for the sire sampled a second time are given in Table 19.1. As expected, evaluations obtained with MME were unbiased while those obtained with UME were biased. The mean squared error of prediction was, however, lower for UME than for MME. The mean of the highest ranking sire for each method of evaluation is given in Table 19.2. For each situation considered, UME had the higher mean. It must be emphasized that these results do not prove that UME is always better than MME for genetic evaluation in populations undergoing an $\mathbf{L'X}\neq\mathbf{0}$ type of selection. Other sampling schemes may give different results.

Table 19.1 Comparison of evaluations for the sire sampled a second time

	Mixed model equations				Modified mixed model equations			
	Bias		MSE[a]		Bias		MSE	
h^2	s	g+s	s	g+s	s	g+s	s	g+s
2.0	0.329	-0.021	0.643	0.090	0.084	-0.001	3.592	0.110
1.0	0.283	-0.043	0.757	0.229	0.031	0.021	1.907	0.296
0.5	0.178	-0.122	0.827	0.527	0.024	-0.004	1.400	0.731
0.25	0.059	-0.244	0.901	1.227	-0.094	-0.051	1.250	1.608

[a] MSE: mean squared error

Table 19.2 Mean for the highest ranking sire, by method of evaluation

	Sire			Group and Sire		
h^2	Mean[a]			Mean		
	UME	MME	SED	UME	MME	SED
2.0	0.708	0.412	0.031	1.246	1.103	0.017
1.0	0.540	0.320	0.032	1.323	1.162	0.022
0.5	0.370	0.193	0.033	1.450	1.270	0.028
0.25	0.266	0.060	0.035	1.910	1.612	0.036

[a] UME: regular mixed model equations; MME: modified mixed model equations; SED: standard error of the difference between means.

References

Box GEP, Tiao GC (1973) Bayesian inference in statistical analysis. Addison-Wesley, Reading, MA

Curnow RN (1961) The estimation of repeatability and heritability from records subject to culling. Biometrics 17:553-566

Falconer DS (1960) Introduction to quantitative genetics. Ronald Press, New York

Fernando RL, Gianola D (1984) Optimal properties of the conditional mean as a selection criterion. J Anim Sci 59 suppl 1 (Abstr.):177

Fernando RL, Gianola D (1986) Optimal properties of the conditional mean as a selection criterion. Theor Appl Genet 72:822-825

Gianola D, Fernando RL (1986) Bayesian methods in animal breeding theory. J Anim Sci 63:217-244

Gianola D, Foulley JL, Fernando RL (1986) Prediction of breeding values when variances are not known. In: Dickerson GE, Johnson RK (eds) Proc 3rd World Congr Genet Appl Livest Prod. Agric Commun, Univ Nebraska, Lincoln, Nebraska XII:356-370

Goffinet B (1983) Selection on selected records. Genet Sel Evol 15:91-98

Harville DA (1974) Bayesian inference for variance components using error contrasts. Biometrika 61:383-385

Henderson CR (1949) Estimation of changes in herd environment. J Dairy Sci (Abstr) 32:706

Henderson CR (1963) Selection index and expected genetic advance. Nat Acad Sci - Nat Res Coun Publ No 982:141-163

Henderson CR (1965) A method for determining if selection will bias estimates. J Anim Sci (Abstr) 24:849

Henderson CR (1973) Sire evaluation and genetic trends. In: Proc Anim Breed Genet Symp in Honor of Dr. J.L. Lush. ASAS and ADSA, Champaign, Illinois, pp 10-41

Henderson CR (1975) Best linear unbiased prediction under a selection model. Biometrics 31:423-447

Henderson CR (1979) Using estimates of variances in predictions of breeding values under a selection model. In: Van Vleck LD, Searle SR (eds) Proc Conf in Honor of CR Henderson. pp 217-227

Henderson CR (1982) Best linear unbiased prediction in populations that have undergone selection. In: Barton RA, Smith WC (eds) Proc World Congr Sheep Beef Cattle Breed. 1:191-200, Dunmore Press, Palmerston North

Henderson CR, Kempthorne O, Searle SR, Von Krosigk CM (1959) The estimation of environmental and genetic trends from records subject to culling. Biometrics 15:192-218

Meyer K, Thompson R (1984) Bias in variance and covariance component estimators due to selection on a correlated trait. Z Tierz Züchtungsbiol 101:33-50

Patterson HD, Thompson R (1971) Recovery of interblock information when block sizes are unequal. Biometrika 58:545-554

Pirchner F (1969) Population genetics in animal breeding. WH Freeman, San Francisco

Rothschild MF, Henderson CR, Quaas RL (1979) Effects of selection on variances and covariances of simulated first and second lactations. J Dairy Sci 62:996-1002

Searle SR (1979) Notes on variance component estimation: a detailed account of maximum likelihood and kindred methodology. Biometrics Unit, Cornell Univ, Ithaca, New York

Thompson R (1973) The estimation of variance and covariance components when records are subject to culling. Biometrics 29:527-550

20 Problems in the Use of the Relationship Matrix in Animal Breeding

L. Dempfle[1]

In animal breeding the phenotypic variance-covariance matrix of observations and the variance-covariance matrix of underlying unobservable breeding values need to be specified. In more complicated settings, e.g., crossing schemes, it is very important to specify precisely the population with respect to which breeding values are defined, because these are not invariant to the populations in which they are tested. The variance-covariance matrix of breeding values (additive genetic values) can be expressed as the numerator relationship matrix (NRM) times the additive genetic variance. The NRM can be obtained in a purely probabilistic manner as twice the coancestries, or it can be derived from the linear relationship between the breeding values of parents and offspring. Whereas the NRM is well defined, the additive genetic variance may be much more critical when we connect observations gathered in time periods far apart, or in different herds. The construction of NRM is outlined and several applications are discussed. It is illustrated that an increase in the correlation between estimated and true breeding value is not easily related to increased genetic progress. With regard to partially known parentage, a modification of the NRM is suggested and its calculation is outlined. The modified NRM also has uses in other cases, e.g., artificially mixed semen (as in poultry), or naturally mixed semen (e.g., honey bees). The effect of mutation is discussed briefly.

20.1 Introduction

To characterize any observable trait in a population we can use the first two moments: mean and variance. If the trait is normally distributed, then we have a very precise description of its distribution, although for traits that are not normal the description is much less useful. In animal breeding, observable traits are usually called phenotypes and, thus, if we have a sample of animals we can describe their variation and relationships among them by the phenotypic variance-covariance matrix.

1 Institut für Tierwissenschaften, der Technische Universität München, D-8050 Freising-Weihenstephan, FRG

Instead of looking at phenotypes we often have to work with genotypic values or with breeding values. However, in order to deal with breeding values we need to specify precisely the populations to which the animals and their offspring belong. Usually these two populations are identical but in modern animal breeding, especially in poultry and pigs, they are often different. Even in dairy cattle we often deal, strictly speaking, with two populations separated in time.

Mainly due to estimability problems, we measure breeding values as deviations so that their expected value is zero. For a single trait model, the variance-covariance matrix of breeding values is the NRM matrix times the additive genetic variance. There is a purely probabilistic interpretation of the NRM. Whereas these probabilities are well defined in random mating populations, the additive genetic variance depends much more on the circumstances.

20.2 The Numerator Relationship Matrix

There are two slightly different interpretations of the NRM. The first is a straightforward probability interpretation called the coancestry (Falconer 1981). It is the probability that two gametes, one chosen at random from individual I and the other from individual J, carry alleles which are identical by descent. The coancestry is also the inbreeding coefficient (F- the probability that two uniting gametes carry identical alleles) of an offspring. The coancestry of an animal with itself (probability that two randomly chosen gametes of an individual carry alleles which are identical by descent) is 1/2(1+F). Defined in this way we can assemble all coancestries between the individuals in a matrix. Multiplying the matrix by two we obtain the numerator relationship matrix.

The second interpretation of the NRM is related more to quantitative genetic models, especially to the infinitesimal model. A detailed exposition of it is given by Dempfle (1983). Assume that in the base population we have breeding values $A_{01},\ldots,A_{0n}$ which, by definition, are assumed to pertain to unrelated and non-inbred individuals. Thus

$$E\,(A_0) = 0; \quad Var(A_0) = \mathbf{I}\sigma_A^2.$$

Then, for individuals of the next generation we use the simple relationship:

$$A_{1i} = \frac{A_{0s}}{2} + \frac{A_{0d}}{2} + Aw_{1i},$$

where:

A_{0s} is the breeding value of the sire;

A_{0d} is the breeding value of the dam; and

Aw_{1i} is the deviation of individual i from the mid-parent value (deviation from the true family mean).

More generally, and putting t'=t-1, we have:

$$A_{ti} = \begin{cases} Aw_{ti} & \text{if no parent is known;} \\ \frac{A_{t's}}{2} + Aw_{ti} & \text{if the sire is known;} \\ \frac{A_{t'd}}{2} + Aw_{ti} & \text{if the dam is known;} \\ \frac{A_{t's}}{2} + \frac{A_{t'd}}{2} + Aw_{ti} & \text{if both parents are known.} \end{cases}$$

In this way we can write breeding values of individuals (sorted by generation) as:

$$\mathbf{A} = \mathbf{LAw},$$

where **L** is a lower triangular matrix and **Aw** is a vector of deviations. Each row of **L** pertains to one individual and has unity on the diagonal and 1/2 in the columns pertaining to the known parents. The variance of A is:

$$\mathrm{Var}(\mathbf{A}) = \mathbf{L}\mathrm{Var}(\mathbf{Aw})\mathbf{L}'.$$

If parents are known, the **Aw** within a family are independent. In other cases they are assumed to be independent (taken to be independent due to lack of knowledge); thus, Var(**Aw**) is a diagonal matrix.

If no animal is inbred we have:

$$\mathrm{Var}(Aw_i) = \begin{cases} \sigma_A^2 & \text{if neither parent is known;} \\ 3/4\ \sigma_A^2 & \text{if only one parent is known;} \\ 1/2\ \sigma_A^2 & \text{if both parents are known.} \end{cases}$$

The last case is the well-known result that within full-sib families, we have half of the additive genetic variance, given that parents are not inbred.

This variation within full-sib families is caused by Mendelian segregation at segregating loci. By symmetry, half of it is caused by the male (variability between gametes of the sire around their mid-value), and half by the female. Since this variability caused by a non-inbred male is on average $\sigma_A^2/4$, it is changed by a factor $(1-F_s)$ if the sire is inbred with coefficient F_s. This is because a fraction F_s of loci are occupied by alleles which are identical by descent and, thus, do not contribute to the variability due to segregation. Thus, the general expression is

$$\mathrm{Var}(Aw_i) = \left[\frac{1-F_s}{4} + \frac{1-F_d}{4}\right]\sigma_A^2 = 1/4[4-(1+F_s)-(1+F_d)]\sigma_A^2.$$

If pedigrees are completely known, apart from those individuals in the base population, Aw represents the deviation from the true full-sib mean. Also, its interpretation as a random variable with expected value equal to zero, with any two Aw being independent, are well justified assumptions. With incomplete pedigrees the situation is more complicated.

Because **L** is a lower triangular matrix and Var(**A**w) is a diagonal matrix, very efficient computer programs can be developed to calculate Var(**A**) and its inverse:

$$[\mathrm{Var}(\mathbf{A})]^{-1} = \mathbf{L}'^{-1}[\mathrm{Var}(\mathbf{A}\mathrm{w})]^{-1}\mathbf{L}^{-1}$$

as described, e.g., by Henderson (1976a,b) and Quaas (1976). Another point is also worth noting. By writing

$$\mathbf{A} = \mathbf{L}\mathbf{A}\mathrm{w},$$

we partition any breeding value as a weighted sum of within-family deviations. This is an extension of the usual partition into within- and between-family components.

20.3 Additive Genetic Variance

To simplify the treatment, a locus with only two alleles is assumed. Using the terminology of Falconer (1981), with -a, d, and a as values of genotypes A_2A_2, A_2A_1, and A_1A_1, respectively, and p_T, q_T, p_S, q_S as allele frequencies of A_1 and A_2 in the target(T) and in the source(S) populations, we get the effect of allele A_1 as described below.

The mean value of genotypes in population T receiving allele A_1 from the source population and the other allele at random from the target population is:

$$m_1 = p_T a + q_T d.$$

The total mean is (one gamete from S and one from T):

$$m = p_S p_T a + p_S q_T d + q_S p_T d - q_S q_T a = (p_S p_T - q_S q_T)a + (p_S p_T + q_S q_T)d.$$

The additive effect of A_i is thus

$$\alpha_i = m_i - m,$$

and the breeding value of an individual A_iA_j is:

$$BV_{ij} = \alpha_i + \alpha_j.$$

The additive genetic variance in the source population with respect to the target population is

$$\sigma_A^2 = p_S^2 BV_{11}^2 + 2p_S q_S BV_{12}^2 + q_S^2 BV_{22}^2 = E(BV_{ij}^2).$$

The additive genetic variance of the source population with respect to a different target population (T') is easily obtained by replacing p_T with $p_{T'}$. The genetic covariance of two breeding values is

$$\sigma_{A_{TT'}} = E(BV_{ijT} BV_{ijT'}) = p_S^2 BV_{11T} BV_{11T'} + 2p_S q_S BV_{12T} BV_{12T'} + q_S^2 BV_{22T} BV_{22T'},$$

and the genetic correlation is thus:

$$f_{A_{TT'}} = \frac{\sigma_{A_{TT'}}}{\sigma_{A_T} \sigma_{A_{T'}}} .$$

Since breeding values are linear measures, the above correlation is either 1, 0, or -1 for a single locus with two alleles. In the example above, 0 or -1 are obtained only with overdominance.

If we look at several independent loci then

$$\sigma^2_{A_T} = \sum_i \sigma^2_{A_{Ti}} \qquad \sigma^2_{A_{T'}} = \sum_i \sigma^2_{A_{Ti'}} \qquad \sigma_{A_{TT'}} = \sum_i \sigma_{(A_{TTi})'}$$

and $f_{A_{TT'}}$ can take any value. In addition, and very importantly, $\sigma^2_{A_T}$ will not be equal to $\sigma^2_{A_{T'}}$.

This is illustrated in Table 20.1, where the additive genetic variance of the source population with respect to two different target populations is given. Using NRM.σ^2_A as the variance-covariance matrix of breeding values implies that the breeding values of all individuals considered have the same additive genetic variance. This implies that individuals are only from one source population (only one p_S), and that there is only one target population (one p_T). For individuals which are contemporary and are tested within the same population, this is easily fulfilled. However, if we include animals from different generations tested with different target populations (e.g., animals of different generations undergoing selection), then the above formula is at best approximate.

Table 20.1 Additive genetic variances and covariances with different target populations

	Target population		
Source population	1 p_T=0.8	2 $p_{T'}$=0.9	
	$\sigma^2_{A_T}$	$\sigma^2_{A_{T'}}$	$\sigma_{A_{TT'}}$
a=1, d=0			
p=0.1	0.18	0.18	0.18
0.5	0.50	0.50	0.50
0.9	0.18	0.18	0.18
a=1, d=1			
p=0.1	0.029	0.007	0.014
0.5	0.080	0.020	0.040
0.9	0.029	0.007	0.014
a=1, d=1.25			
p=0.1	0.011	0.0	0.0
0.5	0.031	0.0	0.0
0.9	0.011	0.0	0.0
a=1, d=1.50			
p=0.1	0.002	0.007	-0.004
0.5	0.005	0.020	-0.010
0.9	0.002	0.007	-0.004

Suppose observations are means of sires' daughters, and means of their sons' daughters. In this case there are sires of different generations tested within different populations and we have

$$\mathrm{Var}(\bar{y}) = \begin{pmatrix} \mathbf{I}\sigma^2_{A_T} & 1/2\,\mathbf{I}\sigma^2_{A_{TT'}} \\ 1/2\,\mathbf{I}\sigma_{A_{TT'}} & \mathbf{I}\sigma^2_{A_{T'}} \end{pmatrix} + \mathbf{I}\sigma^2_e/k,$$

where k is the number of daughters. In general, however, we cannot write this in the form

$$\begin{pmatrix} \mathbf{I} & 1/2\,\mathbf{I} \\ 1/2\,\mathbf{I} & \mathbf{I} \end{pmatrix} \sigma^2_A + \sigma^2_e/k.$$

This situation is very similar to the one where there are two different traits with unequal additive genetic variances and the genetic correlation is less than unity. The consequences are that observations on contemporaneous relatives should get higher weights than observations on ancestors tested in earlier generations. Thus, including as many relationships as possible to animals of previous generations without modification as above may not increase precision as much as expected.

In this treatment a very conservative attitude is taken with respect to additive genetic variance. It is assumed that basic parameters (a and d) stay constant, and that we have only variation in allele frequencies. From work with laboratory species it is well known that this is not necessarily true. The simplest deviation would be scale effects. We also get an effect if there are genotype-environment interactions. The man-influenced environment of decades ago may have been quite different from today's, e.g., the introduction of vaccines and hormones may have changed the environment drastically for today's animals.

20.4 Examples and Applications

20.4.1 Use of the NRM in a Simple Sire Evaluation

Assume there is a large breeding farm with a random sample of cows, and that IJ bulls, which belong to I paternal half-sib groups, are tested. Each test bull has k daughters. The trait is measured only on daughters of the test bull.

The simplest model used is:

$$\mathbf{y} = \mathbf{1}\mu+\mathbf{Zu}+\mathbf{e}$$

$$E(\mathbf{y}) = \mathbf{1}\mu;\ Var(\mathbf{u}) = NRM\sigma_A^2/4;\ Var(\mathbf{e}) = \mathbf{I}(\sigma_P^2\text{-}1/4\sigma_A^2)/k,$$

where μ is an effect common to all observations, **Z** is a known matrix, **u** is the vector of transmitting abilities of test bulls, **e** are residuals, and σ_P^2 is phenotypic variance.

In this case we calculate precision with μ known or more realistically, with μ estimated. In addition, the NRM was utilized or disregarded in estimation. The properties of estimators are given in Table 20.2.

It must be stressed that sires of test bulls were assumed to be randomly chosen from an unselected population to do these calculations. As shown in Table 20.2, increase in $f_{\hat{u}u}$ due to including the NRM is quite small, and corresponds roughly to less than two additional daughters. On the other hand, there is a large increase in correlations among estimates within half-sib families.

Relating these changes in correlations to an increase in genetic progress is very difficult. The following example, which relates to an aspect of sire reference schemes, illustrates this. Suppose we estimate the breeding values of bulls by progeny testing using a BLUP sire evaluation procedure. The following parameters are assumed: σ_P^2=1; h^2=0.25; n=50. We have 12 random unrelated sires with no systematic differences among them.

Table 20.2 Properties of estimators of transmitting abilities, for the bulls of I=4 sib groups, each group having J=4 bulls and each bull having k=20 daughters, h^2=0.25

μ	NRM	$f_{\hat{u}u}$ [a]	$f_{\hat{u}\hat{u}w}$ [b]	$f_{\hat{u}\hat{u}b}$ [c]
Known	Used	0.766	0.362	0.0
Known	Not used	0.756	0.143	0.0
Estimated	Used	0.715	0.266	-0.150
Estimated	Not used	0.705	0.059	-0.098
		k very large		
Known	-	1.0	0.25	0.0
Estimated	-	0.944	0.158	-0.121

[a] $f_{\hat{u}u}$ Correlations among estimated and actual breeding values
[b] $f_{\hat{u}\hat{u}w}$ Correlations among two estimated breeding values of bulls from the same half-sib families
[c] $f_{\hat{u}\hat{u}b}$ Correlations among two estimated breeding values from unrelated bulls

The number of herds available for testing bulls is varied and bulls are strictly nested within herds with k bulls tested on a herd. The model is now:

$$\mathbf{y = Xh+Zu+e},$$

where **X** is a known matrix relating herd effects (**h**) to observations. Further:

$$\mathrm{Var}(\mathbf{u}) = \mathbf{I}\sigma_u^2 = \mathbf{G};\ \mathrm{Var}(\mathbf{e}) = \mathbf{I}(\sigma_P^2-\sigma_u^2)/n;$$

$$\mathrm{Var}(\bar{\mathbf{y}}) = \mathbf{I}[\sigma_u^2+(\sigma_P^2-\sigma_u^2)/n] = \mathbf{V}$$

$$\hat{\mathbf{u}} = \mathbf{GZ'V^{-1}[I-X(X'V^{-1}X)^{-}X'V^{-1}]\bar{y} = GZ'V^{-1}Q\bar{y}}$$

$$\mathrm{Cov}(\hat{\mathbf{u}},\mathbf{u}) = \mathrm{Var}(\hat{\mathbf{u}}) = \mathbf{GZ'V^{-1}QZG}$$

$$= \sigma_u^2 \frac{\sigma_u^2}{\sigma_u^2+\sigma_e^2/n} \ [\mathbf{I-X(X'X)^{-}X'}]$$

$$f_{\hat{u}u} = \frac{\sqrt{\mathrm{Var}(\hat{u})}}{\sqrt{\mathrm{Var}(u)}}$$

$$f_{\hat{u}_1\hat{u}_2} = \frac{\mathrm{Cov}(\hat{u}_1,\hat{u}_2)}{\mathrm{Var}(\hat{u}_1)} = -\frac{1}{k-1} \ (u_1 \text{ and } u_2 \text{ on the same herd})$$

$$\mathrm{Var}(\hat{u}) = \mathrm{Var}(\hat{u})_{\text{without X}} \cdot \frac{1}{1-f_{\hat{u}\hat{u}}}$$

Consider now several special cases:

1. **Known herd effect**

X empty

$\mathrm{Var}(\hat{\mathbf{u}}) = 0.0481$; $f_{\hat{u}u} = 0.877$; $f_{\hat{u}\hat{u}} = 0$.

2. **One herd with unknown effect**

X=1

$$\mathrm{Var}(\hat{u}) = 0.0481 \times \frac{11}{12}; f_{\hat{u}u} = 0.877\left[\frac{11}{12}\right]^{1/2}; f_{\hat{u}\hat{u}} = -\frac{1}{12\text{-}1}.$$

3. **Six herds with unknown effects**

$$X = \begin{pmatrix} 1 & & \\ 1 & & \\ & 1 & \\ & 1 & \\ & & 1 \\ & & . \end{pmatrix}$$

$$\mathrm{Var}(\hat{u}) = 0.0481 \times \frac{1}{2}; f_{\hat{u}u} = 0.877\left[\frac{1}{2}\right]^{1/2}; f_{\hat{u}\hat{u}} = -\frac{1}{2\text{-}1} = -1.$$

For selection of s bulls from the k bulls within a herd we have for the standardized superiority (Owen and Steck 1962)

$$E(I_s) = i_{s|k}\left[(1-f_{\hat{u}\hat{u}})\right]^{1/2}.$$

Thus

$$\Delta G = E(I_s)\, f_{\hat{u}u} \cdot \sigma_u$$

$$= i_{s|k}\,(1-f_{\hat{u}\hat{u}})^{1/2}\, f_{\hat{u}u\ \text{without X}}\, 1/[1-f_{\hat{u}\hat{u}}]^{1/2}\sigma_u$$

$$= i_{s|k}\, f_{\hat{u}u_{\text{without X}}}\, \sigma_u.$$

Let us consider the case of 50% selected, and the sires distributed over two herds, each herd having six sires. This calculation assumes that we select the best three out of each herd. However, in reality we could select, e.g., four sires from one herd and two sires from the other. If selection is practiced in this way, the reduction in genetic gain (ΔG) would be somewhat smaller than the results shown in Table 20.3.

From this example, it should be clear that in the general case of sire evaluation, using the NRM changes $f_{\hat{u}u}$, $f_{\hat{u}\hat{u}w}$, and $f_{\hat{u}\hat{u}b}$, but it is extremely difficult to determine the change in ΔG.

Table 20.3 Reduction in genetic gain (ΔG) with various selection intensities (same intensity in each herd)

Fraction selected %	Number of herds	ΔG relative to one herd %
50	1	100
	2	94
	3	89
	6	75
33.3	1	100
	2	94
	4	83
25	1	100
	3	87
16.6	1	100
	2	92

20.4.2 Use of the NRM when Sires of the Test Bulls are a Selected Group

We envisage a situation where many bulls are tested each year, from which we select I to produce the next generation of test bulls. Thus, as in the example without selection, there are I sires with J test bulls each and k daughters per bull.

The variance between the selected I sires is, on average, approximately

$$\sigma_A^2[1-f^2 i(i-x)]\frac{I-1}{I} = \sigma_A^2 k_0 \quad 0<k_0<1,$$

where f is the correlation between estimated and true breeding value, i is the intensity of selection, and x the truncation point.

Similarly, the variance between bull dams is:

$$\sigma_A^2[1-f^2 i(i-x)] = \sigma_A^2 k_1.$$

Thus, the variance covariance matrix of sires and test bulls is (I=J=4)

$$\text{Var}\begin{bmatrix} A_1 \\ A_2 \\ A_3 \\ A_4 \\ A_{11} \\ A_{12} \\ A_{13} \\ A_{14} \\ A_{21} \\ . \end{bmatrix} = \begin{bmatrix} k_0 & 0 & 0 & 0 & k_0/2 & k_0/2 & k_0/2 & k_0/2 & 0 & . \\ 0 & k_0 & 0 & 0 & 0 & 0 & 0 & 0 & k_0/2 & . \\ 0 & 0 & k_0 & 0 & 0 & 0 & 0 & 0 & 0 & . \\ 0 & 0 & 0 & k_0 & 0 & 0 & 0 & 0 & 0 & . \\ k_0/2 & 0 & 0 & 0 & k_2 & k_0/4 & k_0/4 & k_0/4 & 0 & . \\ k_0/2 & 0 & 0 & 0 & k_0/4 & k_2 & k_0/4 & k_0/4 & 0 & . \\ k_0/2 & 0 & 0 & 0 & k_0/4 & k_0/4 & k_2 & k_0/4 & 0 & . \\ k_0/2 & 0 & 0 & 0 & k_0/4 & k_0/4 & k_0/4 & k_2 & 0 & . \\ 0 & k_0/2 & 0 & 0 & 0 & 0 & 0 & 0 & k_2 & . \\ . & . & . & . & . & . & . & . & . & . \end{bmatrix} \sigma_A^2$$

where:

$$[0.5+k_0/4+k_1/4]\sigma_A^2 = k_2\sigma_A^2.$$

The results in Table 20.4 illustrate that the relationship matrix is not very useful if there is strong selection. This is not surprising because in this situation one can think of the breeding value of a test bull as being composed of two parts: half the value of his sire plus the

Table 20.4 Properties of estimators of breeding value for bulls of I=4 half-sib groups, each group having J=4 bulls, and each bull k=20 daughters; selection of bull sires is intense (k_0=0.2, k_1=0.8, k_2=0.75, h^2= 0.25)

μ	NRM	$f_{\hat{u}u}$	$f_{\hat{u}\hat{u}w}$	$f_{\hat{u}\hat{u}b}$
Known	Used	0.708	0.102	0.0
Known	Not used	0.707	0.033	0.0
Estimated	Used	0.679	0.022	-0.089
Estimated	Not used	0.678	-0.038	-0.074

deviation within a half-sib family. To estimate the latter part, we can only use daughters of test bulls. For the first part, we also use information on daughters of his half-sibs if the NRM is included. However, under intense selection there is not much variation left among sires of test bulls. In any event, if the NRM is not used, parameterization should be different to accommodate change in variance due to selection.

20.5 The NRM and Unknown Parentage

Quite often in a breeding scheme, the parentage of some animals is not known precisely. With very expensive methods, parentage can sometimes be determined but most of the time animals are regarded as having unknown pedigrees. If the NRM is used and selection has been practiced for several generations, then these animals are essentially lost from the breeding scheme. The fraction lost can be sizable, being sometimes as high as one-quarter of the animals. In practice, although parentage may not be known exactly, one might be able to restrict possible parents to a small number of individuals. This is true, e.g., if a cow is mated with two bulls or is inseminated with mixed semen. In the latter case, it may be possible to amend the NRM in such a way that these bulls are treated "fairly".

Assume that we start with a base population of Nm and Nf animals. The design shall be such that environmental effects can be estimated precisely. Phenotypic variance is taken as unity and h^2 as 0.6. Selection is based on the phenotypic values (no progeny testing) so in the base population we have $\mathbf{y}\sim N(\mathbf{0},\mathbf{I})$, $Cov(\mathbf{y},\mathbf{A})=\mathbf{I}h^2$. If we select the best animals then we have $\mathbf{y}_s$ and the corresponding $\hat{\mathbf{A}}=h^2\mathbf{y}_s$. Expected values of $\mathbf{y}_s$ and $\mathbf{A}_s$ can be defined by reference to repeated replication of the design. After selecting the best two individuals of each sex and mating them assortatively, within-family variation still remains. The second generation is obtained selecting the best animals without regard to family.

Expected values of ordered observations for males in generations zero and one are:

$$E\mathbf{y}' = [\quad y_0 \quad | \quad y_1 \quad]$$

$$E\mathbf{y}' = [-1.03\ -0.30\ 0.30\ 1.03 | -0.68\ -0.07\ 0.43\ 1.04 | -0.24\ 0.37\ 0.87\ 1.48]$$

$$E\hat{\mathbf{A}}' = [-0.62\ -0.18\ 0.18\ 0.62 | -0.19\ \ 0.07\ 0.28\ 0.55 | 0.25\ 0.51\ 0.72\ 0\ .99\ ]$$

$$E\hat{\mathbf{A}}' = [\quad A_0 \quad | \quad A_1 \quad]$$

$$\mathbf{y}_{01} = [\quad \mathbf{y}_0 \quad | \quad \mathbf{y}_1 \quad]\ ;\ \mathbf{A}_{01} = [\quad \mathbf{A}_0 \quad | \quad \mathbf{A}_1 \quad].$$

It can be verified that:

$$EA_0 = Cov(\mathbf{A}_0, \mathbf{Y}_0)\ Var(\mathbf{Y}_0)^{-1}E(\mathbf{Y}_0 - \mu) = Cov(\mathbf{A}_0, \mathbf{Y}_{01})Var(\mathbf{Y}_{01})^{-1}E(\mathbf{Y}_{01} - \mu)$$

and

$$EA_{01} = Cov(\mathbf{A}_{01}, \mathbf{Y}_{01})Var(\mathbf{Y}_{01})^{-1}E(\mathbf{Y}_{01} - \mu).$$

An animal of the second generation obtained from mating the best male to the best female will have an expected value of 0.99, namely, the mid-parent average of expected breeding values. This value is obtained before any information on the animal is available. If the parentage of the animal is unknown and we do not take into account any relationship, then the estimated breeding value would be zero with no data, and 0.59 given that the phenotypic value is the family average (0.99). Thus, this animal would have essentially no chance of being selected. Repeating the argument for the average individual from the second mating, then instead of 0.99 and 0.59 we have 0.72 and 0.43, respectively. Similarly, if we have a daughter from the best dam but do not know the sire, then our *a priori* estimate would be 0.49; given that the phenotypic value is 0.99 the estimate would be 0.76.

If the exact parentage is unknown but we do know that the animal is an offspring of the best dam and one of the two best males used, a much better prediction before having data would be

$$[0.99+(0.96+0.72)/2]/2 = 0.92.$$

This assumes an equal chance of the progeny having been sired by either of the two best males.

20.5.1 Modifications of the NRM to Handle Certain Kinds of Unknown Parentage

If we use the first definition of the NRM, then we can calculate probabilities in the usual way. We can assign an equal chance (or specific probabilities) for each potential sire. In principle, this calculation is staightforward but for large numbers of animals may become quite involved. However, the inverse of the NRM rather than the matrix itself is needed to solve the mixed model equations.

Using the second interpretation, it is possible to develop an algorithm which is as efficient as the one used for the standard case. There we need $\mathbf{L}$, $\mathbf{L}^{-1}$, and Var(Aw), which are given below.

The Variance of Aw. If the parents are known we have

$$\mathrm{Var}(Aw) = \sigma_A^2\left[\frac{1-F_s}{4} + \frac{1-F_d}{4}\right].$$

If several sires have equal probability of being the sire, the variability of the offspring increases due to variability between these sires. If these M sires are unrelated and equally inbred, then the variability between half the breeding values is:

$$E\left[\frac{1}{4M}\sum_{i=1}^{M}(A_i-\bar{A})^2\right] = \frac{\sigma_A^2}{4}(1+F_s)\left(\frac{M-1}{M}\right).$$

The same is of course true for F possible dams and thus:

$$\mathrm{Var}(Aw) = \frac{\sigma_A^2}{4}\left[(1-F_s)+(1-F_d)+(1+F_s)\left(\frac{M-1}{M}\right)+(1+F_d)\left(\frac{F-1}{F}\right)\right]$$

$$= \frac{\sigma_A^2}{4}\left[4-\frac{(1+F_s)}{M}-\frac{(1+F_d)}{F}\right].$$

If M is equal to one, the above formula reduces to the usual formula; if the sire is unknown, then this implies that there are many possible sires, which is equivalent to M tending to ∞. In that case, the above also reduces to the well-known formula.

If the sires are related, letting **VA**=Var(**A**) and **A** be the vector of breeding values of potential sires, we have:

$$E\frac{1}{4M}\sum_{i=1}^{M}(A_i-\bar{A})^2 = \frac{1}{4M}\left[\mathrm{tr}(\mathbf{VA}) - \frac{1}{M}\mathbf{1}'\mathbf{VA}\mathbf{1}\right].$$

If so defined, all **Aw** are independent.

Rows of L. If animal i is derived from animals j and k, we have:

$$A_i = \frac{A_j}{2} + \frac{A_k}{2} + Aw_i.$$

The ith row of **L** is, therefore:

$$\mathbf{L}_i = \left[\frac{1}{2} L_j + \frac{1}{2} L_k\right] + [\delta_{ii}],$$

where $[\delta_{ii}]$ is a row vector with unity in the ith position and zeros elsewhere. If there are several equally likely parents, we have:

$$A_i = \frac{1}{2M} \sum_{j=1}^{M} A_j + \frac{1}{2F} \sum_{k=1}^{F} A_k + Aw_i.$$

The ith row of L is then

$$\mathbf{L}_i = \left[\frac{1}{2M} \sum_{j=1}^{M} L_j + \frac{1}{2F} \sum_{k=1}^{F} L_k\right] + [\delta_{ii}].$$

Rows of L^{-1}. **L** is a lower triangular matrix such that:

$$\mathbf{A} = \mathbf{LAw}.$$

Thus, **Aw=L^{-1}A**. In the usual case of both parents known:

$$Aw_i = -\frac{A_j}{2} - \frac{A_k}{2} + A_i.$$

Generalizing this, we obtain:

$$Aw_i = -\frac{1}{2M} \sum_{j=1}^{M} A_j - \frac{1}{2F} \sum_{k=1}^{F} A_k + A_i.$$

Calculation of VA^{-1}.

$$VA^{-1} = L'^{-1}Var(Aw)^{-1}L^{-1}.$$

If we express the rows of L^{-1} as t_i' then:

$$VA^{-1} = \Sigma \frac{1}{Var(Aw_i)} t_i t_i' \ .$$

Calculation of VA.

$$VA = LVar(Aw)L'$$

and expressing the columns of L as L_j, we have

$$VA = \Sigma L_j L_j' \ Var(Aw_j).$$

20.5.2 Example

Assume the following situation: two candidates are available for selection (C_1, C_2), with known dam (d) and paternal grandsire (gs), but whose sire can be any one out of M half-sib males ($s_1, s_2, ..., s_M$). The numerator relationship matrix has the form:

$$NRM \begin{bmatrix} A_{gs} \\ A_d \\ A_{s1} \\ A_{s2} \\ . \\ . \\ . \\ A_{c1} \\ A_{c2} \end{bmatrix} = \begin{bmatrix} 1 & 0 & 0.5 & 0.5 & \cdots & 0.25 & 0.25 \\ 0 & 1 & 0 & 0 & \cdots & 0.5 & 0.5 \\ 0.5 & 0 & 1 & 0.25 & \cdots & b & b \\ 0.5 & 0 & 0.25 & 1 & \cdots & b & b \\ . & . & . & . & \cdots & . & . \\ . & . & . & . & \cdots & . & . \\ . & . & . & . & \cdots & . & . \\ 0.25 & 0.5 & b & b & \cdots & 1 & a \\ 0.25 & 0.5 & b & b & \cdots & a & 1 \end{bmatrix}$$

where

$$a = \frac{5}{16} + \frac{3}{16M}$$

and

$$b = \frac{1}{8} + \frac{3}{8M} \;.$$

In order to use the second definition of NRM, we have to derive

$$E\frac{1}{4M}\sum_{i=1}^{M}(A_i-\bar{A})^2 = \frac{1}{4M}\left[\operatorname{tr}(\mathbf{VA}) - \frac{1}{M}\mathbf{1}'\mathbf{V}\,\mathbf{A}\,\mathbf{1}\right]$$

$$= \frac{\sigma_A^2}{4M}\left[M - \frac{1}{M}M - \frac{M(M-1)}{4}\right]$$

$$= \frac{\sigma_A^2}{4M}\cdot\frac{3}{4}(M-1).$$

Thus:

$$\operatorname{Var}(Aw) = \frac{\sigma_A^2}{4}\left[2 + \frac{3}{4}\bullet\frac{M-1}{M}\right] = \sigma_A^2\left[\frac{11}{16} - \frac{3}{16M}\right] \;.$$

The elements of **L** are such that:

$$\begin{array}{ccc} \mathbf{A} & = \quad \mathbf{L} & \mathbf{Aw} \end{array}$$

$$\begin{bmatrix} A_{gs} \\ A_d \\ A_{s1} \\ A_{s2} \\ \cdot \\ A_{c1} \\ A_{c2} \end{bmatrix} = \begin{bmatrix} 1 & & & & & & \\ 0 & 1 & & & & & \\ 0.5 & 0 & 0 & & & & \\ 0.5 & 0 & 0 & 0 & & & \\ \cdot & \cdot & \cdot & \cdot & \cdot & & \\ 0.25 & 0.5 & 1/2M & 1/2M & . & 1 & \\ 0.25 & 0.5 & 1/2M & 1/2M & . & 0 & 1 \end{bmatrix} \begin{bmatrix} Aw_{gs} \\ Aw_d \\ Aw_{s1} \\ Aw_{s2} \\ \cdot \\ Aw_{c1} \\ Aw_{c2} \end{bmatrix}$$

Discussion Summary
PART VI: SELECTION AND NON-RANDOM MATING

M. Goddard[1]

18 Accounting for Selection and Mating Biases in Genetic Evaluation

For BLUP prediction of breeding values, the usual mixed model equations must include the whole selection history of the population (via the **A** matrix) back to an unselected base population, and estimates of **G** and **R** on the base population are required. Unbiased estimates of **G** and **R** can be obtained by REML, again provided that the data used includes the base population and the **A** matrix. However, commercial populations are constantly undergoing selection and so records on an unselected base population may not be available. Fortunately, in dairy cattle, little effective selection was practiced in the early years of artificial insemination so these populations may include a suitable base. When analyzing data spanning many years, changes over time in means and variances must be suitably modelled in the equations.

Even if many years of records on a population undergoing selection are available, it may be desirable for practical reasons to include only the more recent data in the analysis. This can be done if the correct probability distribution of the observations at the start of the included data is used.

When selection is carred out in two stages: first on one trait, then, after collecting additional data, on another trait, the usual multivariate mixed model equations can still be used if the traits are multivariate normal. However, if first-stage selection was on the basis of a categorical trait (e.g., dystocia) we do not know how well the mixed model equations would correct for the selection practiced.

The need to modify the mixed model equations when selection is not based on a translation invariant function ($\mathbf{L'X} \neq 0$) appears to contradict the likelihood principle. This principle states that inferences should be based only on the data. The explanation for this may be that the mixed model equations can be viewed as calculating $E(\mathbf{u}|\mathbf{w})$, where **u** are the random effects and **w** are error contrasts (see Chap. 19). However, if selection has occurred on elements of the data vector (**y**) not included in **w**, this will be ignored by the usual mixed model equations.

[1] Department of Agriculture and Rural Affairs, Melbourne, Australia

As an alternative approach to dealing with selection it was suggested that the incidence matrices **X** and **Z** in the statistical model could be treated as random. This would avoid the need to "re-label" data, for instance, $y_1>y_2$, as is done in Henderson's approach.

19 Statistical Inferences in Populations Undergoing Selection or Non-Random Mating

Unbiased estimation of the fixed treatment effects (b) means that $E(\hat{b}|b)=b$. However, if one selects the "treatment" with the highest estimated value (highest $\hat{b}$), its true value (b) will on average be less than $\hat{b}$, i.e., $E(b|\hat{b})\neq\hat{b}$. This would be the reason why selection not based on translation invariant functions leads to biased estimators in the usual mixed model equations (UME). In the absence of selection, the UME predict random effects (u) such that $E(u|\hat{u})=\hat{u}$. But $E(b+u|\hat{b}+\hat{u})\neq\hat{b}+\hat{u}$. This means that if bulls are selected based on estimates of fixed group + sire, then the true breeding value of the selected bulls will be less than predicted. This would be overcome if groups could be regarded as random effects with a known variance. Then $E(b+u|\hat{b}+\hat{u})=\hat{b}+\hat{u}$, so the mean true breeding value of selected bulls equals their mean predicted breeding value.

Although Bayesian estimates are sometimes biased they are generally consistent (i.e., they approach the true parameter value as sample size increases).

20 Problems in the Use of the Relationship Matrix in Animal Breeding

The advantage to be gained from a reference sire program depends on the genetic differences between herds. If these differences are large the advantage of the reference sire scheme is greater than Dempfle's paper suggests.

If one is hoping to select individuals carrying mutations of large rather than small effect, it may be better not to use the relationship matrix in evaluating animals.

The amount of old data which should be included in genetic evaluations is a complex compromise allowing for changes in genetic parameters over time, the cost of using old data, and the importance of accounting for past selection. In a tightly designed breeding program in which selection is among a group of contemporary animals there is less to be gained by including back data than when selection is among animals from different generations. In poultry breeding no more than three generations of data are used.

Part VII: Statistics and New Genetic Technology

21 Identification of Genes with Large Effects

W.G. Hill[1] and S. Knott[2]

Recent findings of genes having a large effect on production traits and the possibilities of cloning and inserting genes into farm animals have stimulated interest in methods for identifying individual genes of major effect. Various simple and more elaborate methods are reviewed. It is argued that methods based on likelihood calculations such as segregation analysis are most likely to be appropriate, despite heavy computation, for analysis of both crosses between populations and individual random mating populations. The maximum likelihood algorithms used for human genetic analyses need extension to deal with the family and repeat record data structure of farm animals and to cope with non-normal data. At the molecular level there seems to be more scope for insertion of genes identified directly by their effect on the animal's physiology or biochemistry than for genes identified from segregation of major effects within populations.

21.1 Introduction and Motivation

The continuous distribution of traits and the particulate nature of Mendelian inheritance are reconciled by a model with many genes having small effect, together with environmental deviations. Because the action of individual genes can usually only be inferred rather than observed directly, the genetic properties of traits are generally described in terms of summary parameters such as heritability, genetic correlations, and depression in performance per unit of inbreeding. For many applications in animal breeding these parameters serve us well, but are nevertheless limited. For example, heritability is used to predict selection response, but must alter to some degree as a result of selection; it is only because such changes are small that the parameter has utility. Indeed, the slowness of change in heritability is itself an indication that gene effects are small. If we could identify the genotype of all individuals at all loci we could obviously promote more rapid improvement, but that is unrealistic. Some benefit can, however, be obtained from knowledge of genes having a major effect on the trait (e.g. one

[1] Institute of Animal Genetics, University of Edinburgh, Edinburgh, Scotland
[2] AFRC Institute of Animal Physiology and Genetics Research, Edinburgh Research Station, Edinburgh, Scotland

phenotypic standard deviation, but no precise definition is necessary) and, fortunately, the larger the effect, the easier it is to detect the gene and the more useful the information. Such knowledge can be utilized within the framework of standard quantitative genetic theory by incorporating the genotype as if it were a metric trait in a selection index (Neimann-Sorensen and Robertson 1961; Smith 1967; Smith and Simpson 1986). However, it is likely that better use can be made of individual genes by direct manipulation in a breeding program, for example by introgression into another breed or by progeny testing if recessive. For a general discussion of the role of major genes in animal breeding see Smith and Webb (1981), Roberts and Smith (1982), Hanset (1982), and Smith and Simpson (1986).

21.1.1 Motivation

There has been substantial work in recent years by human geneticists to develop methods for identifying genes of large effect on continuous and discrete traits such as disease incidence, motivated by attempts to understand the basis of their inheritance for use in counselling, and for finding the basic lesion and a possible cure. There has been a more recent stimulus of interest in animal improvement for several reasons. Firstly, animal breeders have for a long time used single genes in the manipulation of overt traits such as color and polledness, but the identification of genes of large, direct effects on commercial traits which can be incorporated into a breeding program is more recent. Notable examples are the use of the dwarf gene in poultry (Merat and Ricard 1974), Booroola in sheep (Piper and Bindon 1982; Piper et al. 1985), double muscling in cattle (Rollins et al. 1972; Hanset and Michaux 1985a,b), and halothane sensitivity in pigs (Smith and Bampton 1977). Secondly, the demonstration by Piper and Bindon that much of the advantage of the Booroola sheep was due to a single gene has stimulated further analyses and reports of genes of large effect on litter size in sheep (Hanrahan and Owen 1985; Jonmundsson and Adalsteinsson 1985; Bradford et al. 1986), suggesting that such searches might be fruitful. Thirdly, the development of transgenic techniques now enables genes to be inserted from different species and in many copies, and thereby creates a need for genes worth cloning and transferring. Fourthly, identification of a single gene may lead to a better understanding of how the biological system works. Finally, methods for breeding value estimation for quantitative traits are becoming optimal, so further improvements in efficiency will have to come from novel sources.

21.1.2 Prior Information - Number of Genes

In farm animals we have little information about the numbers and effects of genes influencing quantitative traits, apart from the few major genes which have been identified. In laboratory animals, and particularly *Drosophila*, data are much more extensive. There are several questions that can be posed; for example: How many loci and genes influence the trait? What is the (Wright) effective number of loci? How much selection response can be attributed to one or a few loci? What proportion of the variation in a population is accounted for by segregation at the most important locus, the second most important locus, etc.? Most completely: what is the distribution of gene effects on the trait in the population, supplemented if possible by the distribution of gene frequencies? As background, the methods and results of estimates of gene number are briefly reviewed.

Effective Number of Genes. The classical estimate of effective number of genes ($range^2$/variance x 8) devised by Wright (Castle 1921; Wright 1952) has been very widely used and does, in principle, demonstrate the presence of genes of large effect if the estimate of number is very low. Various modifications have recently been suggested (Lande 1981; Comstock and Enfield 1981; Cockerham 1986), but the estimate is likely to be biased downwards by linkage of genes in coupling. Estimates from different species and experiments vary greatly (see, for example, Falconer 1981), and it is difficult to draw general conclusions from them.

Genotype Assay. The genotype assay method proposed by Jinks and Towey (1976) can be used in plants where rapid inbreeding can be practiced, and genes are identified by segregation within sublines drawn from lines having already had two or more generations of selfing following a cross.

Chromosomal and Intra-Chromosomal Analysis. The identification of effects of individual chromosomes has long been carried out in analyses of selected lines of *Drosophila* using marker and cross-over suppressor techniques (e.g., Mather and Jinks 1982). These analyses can be extended to analyze effects within chromosomes by forming recombinants against multiple marked chromosome stocks and, in principle, if continued with enough effort, leads to the mapping of all genes differentiating a pair of extreme lines (Thoday 1961).

The most complete data are from analyses of lines of *Drosophila* selected for bristle number, where chromosome manipulation can be used to attribute effects to chromosomes and to sites within chromosomes. Analyses made by Mather and others (see Mather and Jinks 1982) have shown that selected lines usually differ at all chromosomes. Analyses of recombinants within chromosomes have suggested, at least in some cases, that most selection response is due to very few (two or three) genes; Thompson and Thoday (1979) give a review. A frequency distribution of effects has been compiled by Shrimpton (1981). It shows gene effects on bristles of up to more than two standard deviations, but with an increasingly higher frequency of genes of smaller effect, down to about one-half of a standard deviation. Below that size of effect, the method used does not enable the genes to be ascertained separately. A model with an exponential form seems appropriate to describe the distribution of effects, but hard evidence for it is lacking. It seems likely, but by no means certain from such detailed analysis, that there are some genes of large effect segregating for most traits.

21.2 Methods Using Population Differences

21.2.1 Segregation in Crosses and Backcrosses

Perhaps the standard method for identifying genes of large effect is from the analysis of segregation in crosses and backcrosses among homozygous lines, particularly those which differ substantially for the metric trait of interest. Mendel started that way! However, if the lines are inbred, then nothing can be done until the F2 and backcrosses are obtained, when it is possible to search for non-normality, or perhaps bimodality using "bump-hunting" or other likelihood fitting routines. With outbreeding lines, a significant increase in the variance of the F2 and backcrosses may be an indication that a gene of large effect is segregating. In any event, the analysis has to be continued for at least one more generation to test putative genotypes. When family information on two or more generations is combined, then maximum likelihood techniques such as segregation analysis can be used. It is clear, however, that two or more closely linked genes in coupling cannot be distinguished from a single major gene in the early generations after line crossing.

21.2.2 Segregation Analysis

The use of maximum likelihood methods for analysis of data on inbred lines and their crosses follows the same principles as for analysis within segregating populations, which is discussed more fully in the next section. With an initial cross, however, the method has a little more power in that gene frequencies do not have to be estimated, but, like other methods dependent on only a few generations, is biased by linkage. The maximum likelihood methods show considerable robustness even when the assumption of a normally distributed environmental effect is violated. The use and development of the method was originally described by Elston and Stewart (1973), and later extended by Elston (1984, Chap. 3, this Vol.). The power is likely to increase as further generations after the F2 and backcrosses are taken, using methods discussed later for within population analysis.

21.2.3 Repeated Backcrossing and Selection

Wright (1952) suggested that genes of large effect could be identified by repeated backcrossing of, e.g., crosses of a high scoring line to a low scoring line, and selecting for high score. This method leads to a halving of allele frequency in the absence of selection, so only genes having large effect in the heterozygote, such that fitness of animals carrying them is effectively doubled, can be maintained against the backcrossing force. Thus, selection for prolificacy by the Seears brothers maintained the Booroola gene despite their use of only bought-in rams (Piper and Bindon 1982).

21.2.4 Use of Linked Markers

By use of marker genes, which are now becoming available in large numbers as restriction fragment length polymorphisms (RFLPs) and minisatellites (Jeffreys et al. 1985), the effects on quantitative traits of regions of chromosomes can be estimated in the F2 and beyond (e.g., Soller and Beckmann 1982, 1985; Elston Chap. 22, this Vol.). For markers associated with regions of large effect, further generations of crossing can be performed to detect whether the effect is due mostly or entirely to a single gene and is close to the marker. Of course, as more markers become available and associations with more traits can be tested, the greater the chance that spurious effects will be detected unless the power of individual tests is reduced. Markers such as the RFLPs and minisatellite probes to particular sites are likely to be more useful than the available minisatellites which hybridize to many sites. This is so because although the latter are hypervariable, their allelism is usually difficult to ascertain and bands of the same mobility may not correspond to the same locus.

21.2.5 Use of Physiological Markers

On the assumption that more basic traits, such as levels of a hormone or metabolite, are influenced by fewer genes than traits of commercial importance such as growth rate or egg production, then it may be useful in analyses of crosses to monitor such traits. Any evidence of discontinuity or bimodality may point to genes affecting the physiological trait, and their effect on the economic trait can then be estimated. Segregation analysis can be employed to increase efficiency.

21.3 Within Population Analysis

Many methods have been proposed for detection of genes of large effect that are segregating in populations, particularly for analysis of data on man where experimental methods cannot be employed. Most of the tests are based on finding departures from normal distributions; some involve simple computations but others require heavy computation of likelihoods. Their utility depends on their ease of use, power in detecting segregation of a major gene, and sensitivity to breakdown of assumptions, notably of normality of environmental and background genetic distributions.

21.3.1 Departures from Normality

If no major genes are segregating, many genes with small effect are acting additively on the trait, and environmental deviations are continuously distributed and additive to genetic effects, the central limit theorem implies that, after appropriate scaling, observations of traits on individuals and their relatives follow a multivariate normal distribution. Various tests have been suggested that are based on departures from multivariate normality: skewness or kurtosis of distributions of observations and of family means, non-linearity of regression of performance of progeny on parent, asymmetry of responses to high and low selection, heterogeneity of variance within families, and association between variance and level of performance. These will be reviewed only briefly, for all such methods use only part of the information contained in the data and, therefore, as computing power increases, are likely to be superseded by methods making fuller use of the data, notably maximum likelihood.

Heterogeneity of Variance. If a gene of large effect is segregating in the population, heterogeneity of variance within families is expected. Further, as Pearson (1904) was the first to point out, the variability of progeny about the parental value will be greater for intermediate scoring parents than for extremes (see also Felsenstein 1973; Smith et al. 1978), and the variability will be greater within families of intermediate than of extreme mean performance (Fain 1978). Similarly, Penrose (1969) showed how the differences between the full-sib and offspring-parent correlations depended on the number of genes and mean score. Matthysse et al. (1979) suggested that the correlation between the within-family variance and the family mean be used as a statistic. The efficiency and sensitivity of these methods have been analyzed by their proposers and by others (Mayo et al. 1983), who consider their power, or lack thereof, and sensitivity to assumptions. These include genes having a geometric distribution of effects (Matthysse et al. 1979), differences in environmental variance between homozygotes and heterozygotes (Mayo et al. 1980), dominance (Felsenstein 1973), and so on.

Skewness and Kurtosis. Segregation of genes of large effect leads to both skewness, except at specific gene frequencies (Fisher et al. 1932), and kurtosis, and the degree of kurtosis can be used as an estimator of gene number (O'Donald 1971). The method rests very strongly on the basic normal assumptions of, for example, the environmental error distribution, and is sensitive to heterogeneity of variance. Hammond and James (1972) demonstrate the lack of power using data from *Drosophila*.

These methods were extended by Merat (1968), who suggested that deviations from family mean in families of high and low variance should be pooled, and skewness and kurtosis checked on the deviations. The method was discussed and used on *Drosophila* data by Hammond and James (1970).

Non-Linearity of Regression of Progeny on Parent. Robertson (1977) analyzed the influence of genes of large effect and other factors such as skewed environmental distributions on departures from linearity of the regression of progeny on parent and, consequently, on asymmetry of response to high and low selection for the trait. Departures are largest for recessive genes at low frequency. Maki-Tanila (1982) discussed this further, and considered the use of sib on sib regression in addition to the offspring on parent regression. The power of tests for single genes using non-linearity has not been investigated.

21.3.2 Structured Exploratory Data Analysis

A simple and quite different method, subsequently called structured exploratory data analysis (SEDA), was proposed by Karlin et al. (1979), and has been extended to include a group of tests, which, it has been suggested, should be applied together to indicate the presence of a major gene. The three tests that have been most used are the major gene index (MGI), offspring between-parents function (OBP), and the mid-parental correlation coefficient (MPCC).

The MGI has received most attention and is based on the argument that if a major gene is segregating, the deviation of observations on an individual (O_i) would tend to be larger than the geometric mean of deviations from the individual parents (S_i and D_i). For example, for an arbitrary coefficient k (usually 0.5, 1, 2), and summation over all individuals and families:

$$MGI(k) = \Sigma\, [|0_i-(S_i+D_i)/2|^k] \,/\Sigma\, (|0_i-S_i|^{k/2}|0_i-D_i\,|^{k/2}).$$

Subsequently Famula (1986) suggested that, rather than look at the behavior of the statistic for each k, differences between them for different k values should be examined. He also showed how mixed model methods could be employed to estimate fixed effects, but ignored the problem of variance estimation using the methods.

OBP defines the proportion of offspring within an interval of defined length around the mid-parental value. A larger proportion should be nearer the mid-parental value with multifactorial compared with monogenic inheritance. MPCC is claimed to vary in a fairly predictable way depending on the mode of inheritance; this is so because the trait value of all the offspring from each family is correlated with the mid-parental value of the offspring's parents.

Other tests and graphical methods have also been suggested. Analyses of properties of the methods have been undertaken by Karlin et al. (1979, 1981), Karlin and Williams (1981), Mayo et al. (1983), Morton et al. (1982), Famula (1986), and Kammerer et al. (1984). SEDA is not based on normal assumptions, which should make it more robust. However, although quick and simple to apply, the method seems rather *ad hoc* and, as pointed out by Mayo et al. (1983), for models of unequal effects gives values similar to those for multifactorial models. Kammerer et al. (1984) found that although reasonably sensitive in detecting the presence of a major gene, SEDA lacked specificity and consistently classified polygenic traits as due to a major gene. It seems likely that SEDA will be superseded by more formal methods.

21.3.3 Complex Segregation Analysis

The most rigorous of such formal methods was originally proposed by Elston and Stewart (1971), and was based on transmission probabilities. An alternative method, the mixed model, was developed by Morton and MacLean (1974), and included a major locus, a polygenic component, random and common environmental effects. Basically, for both methods, the likelihood of the data is maximized for a set of genetic models and compared over models. For example, to detect a major gene, Elston and Stewart fit a model allowing for Mendelian transmission probabilities (0, 1/2, and 1), and compare the likelihood with that of a non-genetic model, having equal transmission probabilities. A significant increase in likelihood, and no further significant increase in the likelihood for a general model with unrestricted transmission probabilities, indicates the presence of a major gene. Morton and MacLean fit a multivariate normal model in which genetic and environmental variances are estimated, and then the same model plus parameters for frequency, effect, and degree of dominance of a gene. A significant increase in likelihood indicates that a major gene is segregating. More recently, the approaches of Elston and Stewart and of Morton and MacLean have been unified by including terms for both continuous genetic variation and non-Mendelian transmission probabilities (Lalouel et al. 1983).

Complex segregation analysis is obviously more powerful than other methods based on normality because all the information contained in the data is used. In considering its efficiency, account has to be taken both of the possibility of detecting individual genes which are not actually present (false positives) and of missing genes of large effect that are segregating (false negatives). False positives could be caused by non-normality of the data, for example by skewness of the distribution, particularly when continuous genetic, individual and family environmental distributions, and a segregating locus are fitted (Elston 1979; Eaves 1983). Power transformations of the data (MacLean et al. 1976) can be used to transform to normality allowing for one, two, or three underlying distributions, and to reduce the chances of false detection. In any case, if two distributions, whatever their shape, fit better than one, this is suggestive of a major gene. There have been no analyses of the power of complex segregation analysis in data structures relevant to animal breeding, the usual structure being that of nuclear families of man, comprising parents and full-sibs. For such a structure, results have been given by Go et al. (1978) and MacLean et al. (1975) and these are not presented in a form readily transferable to the animal breeding context. The power of the model is, of course, dependent on the effect and frequency of the major gene and on family size, and is greater for continuously distributed than for all-or-none traits because more information is present in the data. Simulation results of MacLean et al. (1975) serve as an illustration of the power. They assumed 500 families with records on each of the father, mother, and four sibs on a trait with normally distributed genetic and environmental distributions. For an additive gene with frequency 0.1 and effect, expressed as the difference between homozygotes, of 0.5, 1.0, and 1.5 residual phenotypic standard deviations, the power of detecting its presence was, respectively, negligible, about one-half, and close to one.

A major problem is the very laborious computation of the likelihood for large pedigrees and considerable ingenuity has gone into designing efficient algorithms (for example, Cannings et al. 1978; Lalouel and Morton 1981). MacCluer et al. (1983) have compared the ability of different computer packages to determine the mode of inheritance. Although tests were based on only 40 nuclear families, the results obtained were in fairly good agreement, with some programs giving parameter estimates very similar to those simulated. As computing costs fall and methods of programming become more sophisticated, it is likely that the use of maximum likelihood will become more widespread.

Recently Bonney (1984,1986) has proposed the use of regressive models in which the phenotypes of relatives are fitted as covariates in computing the likelihood under different genetic models. These methods may lead to very substantial improvements in computing efficiency, and are discussed further by Elston, Chapter 3, this Volume.

The likelihood methods were proposed and developed for data on man, both for continuous and for all-or-none traits such as familial diseases. Extension appropriate for the typical data structure in farm animals seems necessary and it would seem best to concentrate on the model originally proposed by Morton and MacLean, with continuous genetic variation and additional variation due to a single locus.

21.3.4 Miscellanea

Non-Normally Distributed Traits. Segregation analysis is appropriate for traits which, in the absence of segregation of a major gene, are normally distributed or can be transformed to normality, and for all-or-none characters in which a normal-threshold model can be assumed. Some traits in farm animals, however, such as litter size in pigs or in prolific breeds of sheep, have discrete distributions with many classes; these traits are likely to be close enough to normal for crude but not for fine-scale analysis. Other traits have notably non-normal distributions: particularly egg number in poultry and body size in fish after rearing under competitive conditions. These distributions are skewed and not normalized by any standard transformation. It is clear that developments in the formal methods are needed, but it is not clear how these should proceed. Nevertheless, *ad hoc* methods can be used successfully. For example, Piper and Bindon (1982) adopted an arbitrary cut-off at three lambs born, and were thereby able to demonstrate major gene inheritance of litter size in the Booroola strain of sheep. Also, Hanrahan and Owen (1985) used the fact that the repeatability of litter size is likely to be much higher in sheep populations in which a major gene is segregating. Neither of these authors tested alternative hypotheses, however, and the high repeatability was associated with a high mean and heritability.

Linked Markers. The association of segregating markers with performance of the quantitative trait provides a method of identifying chromosomal regions with major effect. This method has been discussed for farm animals by Soller and Beckmann (1982, 1985) and is reviewed further in this volume by Elston, Chapter 22, this Volume.

Physiological Variables. Although in analyses of differences between populations recording basic traits such as hormone levels as a way of finding genes of large effects might be rewarding, there seems little point in doing so when there is no prior information within a segregating population. If, however, such data were collected for other purposes there might be benefit in analyzing them.

Breeding from Extreme Animals. In view of the low power of detection of single genes by assessing departures from normality, it may be worthwhile to breed from extreme animals on a regular basis, as suggested by Roberts and Smith (1982). In this way more data is accumulated for testing, and if indeed a major gene is present, progress would be made towards its identification.

21.4 Use of Selected Populations

If widely divergent populations for analysis through crossing and backcrossing methods as discussed previously are not available, it may be possible to create these by selecting high and low individuals from some base. Then, any genes of large effect which are segregating in the population will contribute a large part of the high-low difference. The increase in frequency is proportional to the effect of the gene in standard units, the selection intensity and, to a first approximation, generation number (t). The variance of change is proportional to t and to the inverse of effective population size. Thus, the efficiency, measured as the ratio of expected change in gene frequency to its standard error, increases as the square root of t. A discussion of the design of selection experiments on quantitative traits intended to give efficient estimates of gene frequency change at loci affecting them is given by Hill (1971).

Such a scheme can only utilize variants segregating initially and so it would seem to have little to offer for gene identification over an analysis using maximum likelihood directly in the base population. There is the potential benefit that in the cross between high and low lines the genes may be at intermediate frequency, but the disadvantage that in the crosses the effects of different genes are correlated by linkage disequilibrium.

Selection has more promise as a technique for identifying genes through their effects on other physiological or structural variables. If the trait is expensive to measure then use of divergent lines is more efficient than analysis of the base population. For example, assume selection is for high and low growth rate. As Bulfield (1985) has suggested, two-dimensional (2-D) gel electrophoresis can be used to identify differences in amount and structure of proteins in several tissues. If such differences are found, the protein can, in principle, be identified and ultimately cloned. Such analyses can give us information about the nature of genes which are associated with genetic changes in the quantitative trait, and have the potential of enabling subsequent manipulation using the cloned gene. Although 2-D gel analyses offer the possibility of screening very large numbers of loci at one time, unless gels are of very high quality differences in protein positions and intensity are hard to detect.

It is clear that a genetic change in Y, for example, the amount of a protein in the liver, the activity of a metabolic enzyme, or the level of a circulating hormone, consequent on a genetic change from selection in a production trait X, such as growth rate, implies a genetic covariance between X and Y and, therefore, that changing Y would itself change X. Thus selection among animals on the variable Y should be effective to some degree in increasing X. It does not follow, however, that inserting the structural gene or altering the level of expression of the gene producing Y will lead to the same responses because the genetic and metabolic system is complex. In statistical terms, the multivariate associations among the traits within the population do not enable predictions of what will happen when a variable or variables are moved outside the original parameter space.

At the level of continuous underlying variables, such as hormones or enzyme activities, analysis of segregation of crosses of selected lines offers an efficient method for detection of genes of large effect. If there is little or no difference in these variables between the selected lines, there is no point in continuing. Further, if there is a difference between the lines but there is no increased variance or bimodality in the F2, there is also no point in continuing the analysis of correlated variables as potential indicators of commercial traits. Only genes producing very large differences in the indicator trait are likely to be of interest because the consequent difference in the commercial trait is likely to be smaller.

21.5 Molecular Manipulation

The discussion has concentrated largely on statistical methods for identifying loci which are already polymorphic within or between populations, perhaps with the intention of cloning them for subsequent transgenic manipulation. Molecular methods have potential both for identifying and creating new variation.

21.5.1 Transposon Tagging

Although new variants can be produced by mutation, these are not likely to be useful unless they occur with high frequency and can be identified. Thus, chemical or radiation-induced mutagenesis together with strong selection may provide specific new variants in microorganisms, for example the ability to grow on some deficient medium. In farm animals such selection pressures cannot be practiced so as to isolate any mutant form should it occur. Straightforward directional selection for a metric trait with concurrent use of these mutagenic agents might turn up a mutant of large effect, but unless it had some overt phenotype, it would not be noticed and located.

In principle, transposable elements or retroviruses are potentially more useful mutagenic agents in that the mutant genes can be identified and cloned by tagging with the element (Mackay 1985). Thus, Mackay selected for abdominal bristle numbers in *Drosophila* in P-M dysgenic crosses and found in low lines a classical mutant gene; using the P-element probe it should be possible to clone this gene. Nevertheless, there are limitations to the method. If the transposition rate is low, it is unlikely that any mutants affecting the trait substantially will occur; if the rate is high, relevant mutants may occur, but it will be difficult to identify which is the relevant insertion in the company of many others. Although *in situ* hybridization may be feasible in *Drosophila* , it is not in other species, and so RFLPs using the transposon as probe would have to be studied. It might, however, be possible to analyze these in the same way as if they were linked markers, but knowing that the mutant gene itself is being located (Soller and Beckmann 1985). With large numbers of inserts, high quality Southern blotting techniques would be needed.

A potential problem with the use of transposable elements or retroviruses is that they may be unstable either for mutation or gene insertion, for if they entered a site they may also leave again. It may be possible to overcome this problem by use of vectors which are immobilized. As yet, transposable element mutagenesis has not been achieved in farm animals.

Although of primary interest as a means of incorporating specific genes or constructs, DNA inserts may also act as mutagens at their site of landing; there are, for example, instances in mice of developmental mutants (Palmiter and Brinster 1985). This seems unlikely to be useful for the improvement of metric traits, because most constructs appear to go into a single site as long head to tail sequences and, therefore, the mutation rate is too low per unit effort.

21.5.2 Transgenics

The site of incorporation and/or copy number of most, perhaps all, transgenic animals is likely to be different, so every transgene is effectively a new gene. If the inserted construct codes for some biochemically or physiologically active product, and is present in many copies, it is likely that each new gene will have large, certainly measurable, effect on one or many traits. Thus, a supply of new genes, each identifiable, can be created, the numbers being limited only by the labor input, facilities, and success of the insertion techniques. Smith et al. (1987) have considered how these lines might be evaluated for production traits; it is clear that extensive resources will be required for testing, and the utility will depend largely on choosing useful constructs.

Methods have been discussed above for finding genes by 2-D gel electrophoresis, for example, for subsequent cloning. Such methods rely on polymorphism of the genes, technology for running gels, protein identification, and gene cloning, but not on knowledge of the physiology of the trait. Genes for insertion can be chosen, however, without *a priori* polymorphism. For example, the growth hormone was inserted in the classic experiment of Palmiter et al. (1982) because a clone was available and because it was likely to have large physiological effects. The polymorphism is created in the population by the transgenic procedure. There are, of course, potential problems with a transgene in contrast with an already present segregant, perhaps with fitness of the animals or perhaps because the insert is not stable; however, there is the chance to make either more of the same or others with different promotors or with a modified promotor or structural region. There is also the problem of choice of gene, but studies on exogenous hormones, for example, are likely to be a useful guide. It is not, at this stage, obvious that transgenic techniques are going to be important ways of improving farm animals for commercial traits, and particularly for aggregate economic value. However, they are certainly a potential source of major genes and seem to us a more likely source than identification of segregants in present populations.

21.6 Discussion

A large number of methods for identifying genes of large effect have been reviewed, but a more quantitative analysis and discussion is still needed. Several workers have investigated the efficiency of individual methods, but it is necessary to bring these together and compare their powers and limitations for a range of models. For example, how powerful are they at detecting a single gene when all the rest are of infinitesimal effect, or when the rest have a distribution of effects? How sensitive are the methods to heteroscedasticity of environmental variance, or to non-normality of environmental deviations? How readily can they cope with sex-limited traits and nearly continuous traits such as litter size?

The formal methods, notably segregation analysis, for detection of genes within populations have been developed for data on man. Their extension to farm animal populations is clearly necessary and the data available are generally more suitable. For livestock there are repeated records, large family sizes, data on individuals from previous generations recorded at the same age, and large contemporary environmental groups. If human geneticists can get anything from segregation analysis, then surely so can the animal breeders. There are problems, for example, of selection; in principle, however, maximum likelihood techniques can handle these, so there is obviously room for a lot of work and for a very large computer. It is also important to consider the power of the methods and the size and design of experiments to provide data for such analyses. It seems likely to be of little benefit, except in an exploratory sense, to pursue simple methods based on departures from normality. Karlin's SEDA procedure would seem to have the benefit of being less dependent on normality of underlying variation, but has only an *ad hoc* foundation. There is clearly a need to adapt maximum likelihood methods to non-normal data, for example, on egg production of poultry.

There are obvious benefits in being able to attribute variation between and within populations to single genes. This is illustrated by the use in breeding practice of the dwarf, halothane, double muscling, and Booroola genes in poultry, pigs, cattle, and sheep, respectively. Fortunately, the larger the effect of the gene, the easier it is to find it but the task is not likely to be easy. Ironically, the Booroola gene has too large an effect on litter size for many management systems, and even its inheritance was not clarified until many years after the Booroola flock was known.

Acknowledgements. We are grateful to the Agricultural and Food Research Council for financial support through a research grant (to WGH) and studentship (to SK), and to colleagues, particularly Chris Haley and Patricia Simpson, for helpful comments.

References

Bonney GE (1984) On the statistical determination of major gene mechanisms in continuous human traits: regressive models. Am J Med Genet 18:731-749

Bonney GE (1986) Regressive logistic models for familial disease and other binary traits. Biometrics 42:611-625

Bradford GE, Quirke JF, Sitorus P, Inounu, I Tiesnamurti, Bell FL, Fletcher IC, Torell DT (1986) Reproduction in Javanese sheep: evidence for a gene with large effect on ovulation rate and litter size. J Anim Sci 63:418-431

Bulfield G (1985) The potential for improvement of commercial poultry by genetic engineering techniques. In: Hill WG, Manson JM, Hewitt D (eds) Poultry genetics and breeding. Brit Poult Sci Ltd, Longman, Harlow, pp 37-46

Cannings C, Thompson EA, Skolnick MH (1978) Probability functions on complex pedigrees. Adv Appl Probab 10:26-61

Castle WE (1921) An improved method of estimating the number of genetic factors concerned in cases of blending inheritance. Science 54:223

Cockerham CC (1986) Modifications in estimating the number of genes for a quantitative character. Genetics 114:659-664

Comstock RE, Enfield FD (1981) Gene number estimation when multiplicative genetic effects are assumed - growth in flour beetles and mice. Theor Appl Genet 59:373-379

Eaves LJ (1983) Errors of inference in the detection of major gene effects on psychological test scores. Am J Hum Genet 35:1179-1189

Elston RC (1979) Major locus analysis for quantitative traits. Am J Hum Genet 31:655-661

Elston RC (1984) The genetic analysis of quantitative trait differences between two homozygous lines. Genetics 108:733-744

Elston RC, Stewart J (1971) A general model for the genetic analysis of pedigree data. Hum Hered 21:523-542

Elston RC, Stewart J (1973) The analysis of quantitative traits for simple genetic models from parental, F1 and backcross data. Genetics 73:695-711

Fain PR (1978) Characteristics of simple sibship variance tests for the detection of major loci and application to height, weight and spatial performance. Ann Hum Genet 42:109-120

Falconer DS (1981) Introduction to quantitative genetics, 2nd edn. Longman, London

Famula TR (1986) Identifying single genes of large effect in quantitative traits using best linear unbiased prediction. J Anim Sci 63:68-76

Felsenstein J (1973) Estimation of number of loci controlling variation in a quantitative character. Genetics 74 suppl part 2 :s78-s79

Fisher RA, Immer FR, Tedin O (1932) The genetical interpretation of statistics of the third degree in the study of quantitative inheritance. Genetics 17:107-124

Go RCP, Elston RC, Kaplan EB (1978) Efficiency and robustness of pedigree segregation analysis. Am J Hum Genet 30:28-37

Hammond K, James JW (1970) Genes of large effect and the shape of the distribution of a quantitative character. Aust J Biol Sci 23:867-876

Hammond K, James JW (1972) The use of higher degree statistics to estimate the number of loci which contribute to a quantitative character. Heredity 28:146-147

Hanrahan JP, Owen JB (1985) Variation and repeatability of ovulation rate in Cambridge ewes. Anim Prod (Abstr.) 40:529

Hanset R (1982) Major genes in animal production, examples and perspectives: cattle and pigs. Proc 2nd World Congr Genet Appl Livest Prod. Garsi, Madrid 6:439-453

Hanset R, Michaux C (1985a) On the genetic determinism of muscular hypertrophy in the Belgian White and Blue cattle breed. I. Experimental data. Genet Sel Evol 17:359-368

Hanset R, Michaux C (1985b) On the genetic determinism of muscular hypertrophy in the Belgian White and Blue cattle breed. II. Population data. Genet Sel Evol 17:369-386

Hill WG (1971) Design and efficiency of selection experiments for estimating genetic parameters. Biometrics 27:293-311

Jeffreys AJ, Wilson V, Thein SL (1985) Hypervariable 'minisatellite' regions in human DNA. Nature 314:67-73

Jinks JL, Towey P (1976) Estimating the number of genes in a polygenic system by genotype assay. Heredity 37:69-81

Jonmundsson JV, Adalsteinsson S (1985) Single genes for fecundity in Icelandic sheep. In: Land RB, Robinson DW (eds) Genetics of reproduction in sheep. Butterworths, London, pp 159-168

Kammerer CM, MacCluer JW, Bridges JM (1984) An evaluation of three statistics of structured exploratory data analysis. Am J Hum Genet 36:187-196

Karlin S, Williams PT (1981) Structured exploration data analysis (SEDA) for determining mode of inheritance of quantitative traits. II. Simulation studies on the effect of ascertaining families through high-valued probands. Am J Hum Genet 33:282-292

Karlin S, Carmelli D, Williams R (1979) Index measures for assessing the mode of inheritance of continuously distributed traits. 1. Theory and justifications. Theor Popul Biol 16:81-106

Karlin S, Williams PT, Carmelli D (1981) Structured exploratory data analysis (SEDA) for determining mode of inheritance of quantitative traits. I. Simulation studies on effect of background distributions. Am J Hum Genet 33:262-281

Lalouel JM, Morton NE (1981) Complex segregation analysis with pointers. Hum Hered 31:312-321

Lalouel JM, Rao DC, Morton NE, Elston RC (1983). A unified model for complex segregation analysis. Am J Hum Genet 35:816-826

Lande R (1981) The minimum number of genes contributing to quantitative variation between and within populations. Genetics 99:541-553

MacCluer JW, Wagener DK, Spielman RS (1983) Genetic analysis workshop. I: Segregation analysis of simulated data. Am J Hum Genet 35:784-792

Mackay TFC (1985) Transposable element-induced response to artificial selection in *Drosophila melanogaster*. Genetics 111:351-374

MacLean CJ, Morton NE, Lew R (1975) Analysis of family resemblance. IV. Operational characteristics of segregation analysis. Am J Hum Genet 27:365-384

MacLean CJ, Morton NE, Elston RC, Yee S (1976) Skewness in commingled distributions. Biometrics 32:695-699

Maki-Tanila A (1982) The validity of the heritability concept in quantitative genetics. PhD Thesis, Univ Edinburgh

Mather K, Jinks JL (1982) Biometrical genetics, 3rd edn. Chapman and Hall, London

Matthysse S, Lange K, Wagener DK (1979) Continuous variation caused by genes with graduated effects. Proc Nat Acad Sci USA 76:2862-2865

Mayo O, Hancock TW, Baghurst PA (1980) Influence of major genes on variance within sibships for a quantitative trait. Ann Hum Genet 43:419-421

Mayo O, Eckert SR, Nugroho WH (1983) Properties of the major gene index and related functions. Hum Hered 33:205-212

Merat P (1968) Distributions de frequences, interpretation du determinisme genetique des characteres quantitatifs et recherche de "genes majeurs". Biometrics 24:277-293

Merat P, Ricard FH (1974) Etude d'un gene de nanisme lie aus sexe chez la poule: importance de l'etat d'engraissement et gain de poids chez l'adulte. Ann Genet Sel Anim 6:211-217

Morton NE, MacLean CJ (1974) Analysis of family resemblance. III. Complex segregation of quantitative traits. Am J Hum Genet 26:489-503

Morton NE, Williams WR, Lew R (1982) Trials of structured exploratory data analysis. Am J Hum Genet 34:489-500

Neiman-Sorensen A, Robertson A (1961) The association between blood groups and several production characters in three Danish cattle breeds. Acta Agric Scand 11:163-196

O'Donald P (1971) The distribution of genotypes produced by alleles segregating at a number of loci. Heredity 26:233-241

Palmiter RD, Brinster RL (1985) Transgenic mice. Cell 41:343-345

Palmiter RD, Brinster RL, Hammer RE, Trumbauer ME, Rosenfeld MG, Birnberg NC, Evans RM (1982) Dramatic growth of mice that develop from eggs microinjected with metallothionein-growth hormone fusion genes. Nature 300:611-615

Pearson K (1904) Mathematical contributions to the theory of evolution. XII. On a generalised theory of alternative inheritance, with special reference to Mendel's laws. Phil Trans R Soc London A203:53-86

Penrose LS (1969) Effects of additive genes at many loci compared with those at a set of alleles at one locus in parent-child and sib correlations. Ann Hum Genet 33:15-21

Piper LR, Bindon BM (1982) Genetic segregation for fecundity in Booroola Merino sheep. In: Barton RA, Smith WC (eds) Proc World Conf Sheep Beef Cattle Breed. Dunmore Press, Palmerston North, New Zealand. Vol 1, pp 395-400

Piper LR, Bindon BM, Davis GH (1985) The single gene inheritance of the high litter size of the Booroola Merino. In: Land RB, Robinson DW (eds) Genetics of reproduction in sheep. Butterworths, London, pp 115-125

Roberts RC, Smith C (1982) Genes with large effects - theoretical aspects in livestock breeding. Proc 2nd World Congr Genet Appl Livest Prod. Garsi, Madrid 6:420-438

Robertson A (1977) The non-linearity of offspring-parent regression. In: Pollak E, Kempthorne O, Bailey TB, Jr (eds) Proc Int Conf Quant Genet. Iowa State Univ Press, Ames, pp 297-304

Rollins WC, Tanaka M, Nott CFG, Thiessen RB (1972) On the mode of inheritance of double-muscled conformation in bovines. Hilgardia 41:433-456

Shrimpton AE (1981) The isolation of polygenic factors controlling bristle score in *Drosophila melanogaster*. PhD Thesis, Univ Edinburgh

Smith C (1967) Improvement of metric traits through specific genetic loci. Anim Prod 9:349-358

Smith C, Bampton PR (1977) Inheritance of reaction to halothane anaesthesia in pigs. Genet Res 29:287-292

Smith C, Simpson SP (1986) The use of genetic polymorphisms in livestock improvement. Z Tierz Züchtungsbiol 103:205-217

Smith C, Webb AJ (1981) Effects of major genes on animal breeding strategies. Z Tierz Züchtungsbiol 98:161-169

Smith C, Meuwissen THE, Gibson JP (1987) On the use of transgenes in livestock improvement. Anim Breed Abstr 55:1-10

Smith CAB, Loesch DZ, Bener A (1978) Search for heterozygosis in quantitative characters. Ann Hum Genet 42:121-128

Soller M, Beckmann JS (1982) Restriction fragment length polymorphisms and genetic improvement. Proc 2nd World Congr Genet Appl Livest Prod. Garsi, Madrid 6:396-404

Soller M, Beckmann JS (1985) Restriction fragment length polymorphisms and animal genetic improvement. In: Leng, RA, Barker, JSF, Adams DB, Hutchinson KJ (eds) Reviews in rural science, 6. Biotechnology and recombinant DNA Tech Anim Prod Indust pp 10-18

Thoday JM (1961) Location of polygenes. Nature 191:368-370

Thompson JN, Thoday JM (1979) Synthesis: polygenic variation in perspective. In: Thompson JN, Thoday JM (eds) Quantitative genetic variation. Academic Press, New York, pp 295-301

Wright S (1952) The genetics of quantitative variability. In: Reeve ECR, Waddington CH (eds) Quantitative inheritance. Her Majesty's Stationery Office, London, pp 5-41

22 A General Linkage Method for the Detection of Major Genes

R.C. Elston[1]

Extending a method previously proposed specifically for sib-pair data, a general method is suggested to detect linkage between a trait locus and a marker locus when the genetic mechanism underlying the trait is unknown. The method consists essentially of regressing the squared trait difference between a pair of relatives on the proportion of genes they share (or are estimated to share) identical by descent at the marker locus. This regression is expected to be $-2(1-2\theta)^2\sigma_a^2$, where θ is the recombination fraction between the two loci and σ_a^2 is the additive genetic variance due to the trait locus. It is suggested that a transformation be used to better approximate normality, and the likelihood ratio criterion used to test the null hypothesis that no linked trait locus exists. The method is extended to dichotomous traits and disease traits with variable age of onset, and to linkage with multiple markers. The method is computationally fast, but has little power if θ is not close to 0 or if σ_a^2 is small.

22.1 Introduction

Most early work on statistical methods to detect linkage treated the case in which the traits being studied are qualitative and Mendelizing, i.e., they display simple monogenic inheritance. Penrose (1938) was the first to propose a test of linkage for the case when one or both traits are quantitative and determined by an unknown mode of inheritance, using data on independent sib pairs. Jayakar (1970) and Hill (1975) investigated testing for linkage between a known marker locus and a locus for a quantitative trait by comparing, within families, trait variability among marker genotypes to variability within marker genotypes. These ideas were extended by Gelderman (1975) and Smith (1975). Early experimental studies to locate quantitative trait loci on the genome were conducted in chicken (Lowry and Schultz 1959), *Drosophila* (Thoday 1961), and wheat (Law 1966), while more recent experimental studies have been conducted in tomato (Tanksley et al. 1982). The recent discovery that at the DNA level there exists an enormous number of polymorphic markers, the so-called restriction fragment length polymorphisms (Botstein et al. 1980) will undoubtedly

[1] Department of Biometry and Genetics, Louisiana State University Medical Center, New Orleans, Louisiana, USA

give further impetus to such studies for plant and animal improvement (Tanksley 1983; Beckman and Soller 1983).

If the genetic mechanism underlying the variability of a trait is known, likelihood methods can be used to detect and estimate linkage relationships between that trait and marker loci (Ott 1985). Here, however, I shall be concerned with the case in which the underlying genetic mechanism is unknown, and linkage analysis is used to detect it. I shall indicate a very general and computationally fast method of doing this, based on the method of linkage analysis first proposed by Haseman and Elston (1972) for data collected on sib pairs sampled from a random mating population. I shall indicate how the method can be extended to dichotomous traits or to disease traits with variable age of onset, and how certain statistical techniques can be incorporated to generalize the method even further. For simplicity, however, I shall assume only additive genetic variance in the trait being investigated, that the population is at panmictic equilibrium with respect to the loci being considered, and that we have data on a random sample of independent pairs of relatives from such a population. I shall indicate in the discussion to what extent some of these assumptions may be relaxed.

22.2 A Generalization of Haseman and Elston's (1972) Method

Suppose we have a random sample of n pairs of relatives and let x_{ij} (i=1,2; j=1,...,n) denote the value of the quantitative trait for the i-th member of the j-th pair. We assume a model of the form

$$x_{ij} = \mu + g_{ij} + e_{ij},$$

where μ is an overall mean, g_{ij} is the genetic effect of a locus linked to a particular marker, and e_{ij} is a residual "environmental" effect that includes any contribution to x_{ij} not due to the single additive trait locus under consideration, and may in fact include a genetic component due to loci unlinked to the marker. It is assumed that g_{ij} and $e_j = e_{1j} - e_{2j}$ are random variables with means 0 and variances σ_a^2 and σ_e^2, respectively. We need not assume that the environmental effects e_{1j} and e_{2j} are independent, but we shall assume that the difference between them, e_j, is uncorrelated to the difference $g_{1j} - g_{2j}$. It follows from these assumptions, if we define $Y_j = (x_{1j} - x_{2j})^2$ and use E to denote expectation and Cov to denote covariance, that

$$E(Y_j) = 2\sigma_a^2 - 2\mathrm{Cov}(g_{1j}, g_{2j}) + \sigma_e^2.$$

Now the value of $\mathrm{Cov}(g_{1j}, g_{2j})$ in a random mating population at equilibrium depends only on σ_a^2 and on how many genes the pair of relatives shares identical by descent (i.b.d.) at the trait locus - 0, 1, or 2. Consider a particular type of relative pair for whom the probabilities of sharing 0, 1, or 2 genes i.b.d. at an autosomal locus are respectively c_0, c_1, and c_2. Table 22.1 gives the values of these probabilities for different types of relative pairs. For such a type of relative pair we have

$$E(Y_j | \text{ the pair share i genes i.b.d. at the trait locus})$$

$$= (2-i)\sigma_a^2+\sigma_e^2 \quad (22.1)$$

and

$$E(Y_j) = \underset{i}{E}(2-i)\sigma_a^2+\sigma_e^2$$

$$= 2\sigma_a^2+\sigma_e^2-(c_1+2c_2)\sigma_a^2 \quad (22.2)$$

Table 22.1 Probabilities, c_i, that a pair of relatives shares i genes identical by descent at an autosomal locus

Type of relative pair	i=0	i=1	i=2
Sibs	1/4	1/2	1/4
Parent-offspring	0	1	0
First cousins	3/4	1/4	0
Grandparent-grandchild, half-sibs, uncle/aunt, nephew/niece	1/2	1/2	0

The coefficient c_1+2c_2 is twice the average proportion of genes the particular type of relative pair shares i.b.d. and, as can be seen from Table 22.1, it can be different for different types of relatives. Thus, we could regress values of Y_i for different types of relative pairs on the corresponding values of c_1+2c_2 and, since this regression is expected to be $-\sigma_a^2$, detect the presence of an additive genetic effect by testing whether or not the empirical regression coefficient is significantly less than 0. This, however, implicitly assumes that σ_e^2 is the same for the different types of relative pairs. Differences in within-pair environmental correlations among the types of relative pairs, which are to be expected unless special precautions are taken to randomize all environmental influences, motivate us to develop an analogous test **within** each type of relative pair.

Let π_{jt} be the proportion of genes shared i.b.d. at the trait locus by the j-th pair of relatives. Then, analogous to (22.1) and (22.2), we can write

$$E(Y_j|\pi_{jt}) = 2\sigma_a^2+\sigma_e^2-2\sigma_a^2\pi_{jt}. \quad (22.3)$$

Thus, ideally, we should like to measure the proportion of genes shared i.b.d. at the trait locus by the j-th pair, and regress Y_j on it, doing this separately for each type of relative pair. It should be noted, however, that parent-offspring pairs cannot be used to estimate such a regression, because there is no variability in the proportion of genes shared i.b.d. by parent-offspring pairs; all such pairs share exactly half their genes i.b.d. at every autosomal locus. The basic idea behind the approach proposed by Haseman and Elston (1972) is to regress Y_j on π_{jm}, the proportion of genes shared i.b.d. at a marker locus, instead of on π_{jt}. Provided the trait locus and the marker locus are linked, this regression is expected to be negative; if the two loci are unlinked, or if $\sigma_a^2=0$, the regression is expected to be 0. This can be seen by noting that

$$E(Y_j|\pi_{jm}) = \underset{\pi_{jt}}{E} (Y_j|\pi_{jm},\pi_{jt})P(\pi_{jt}|\pi_{jm})$$

$$= \underset{\pi_{jt}}{E} (Y_j|\pi_{jt})P(\pi_{jt}|\pi_{jm}) ,$$

assuming that the marker locus does not itself have a pleiotropic effect on the quantitative trait. From this, (22.3) and Table 22.2, which is obtained from the joint distribution of π_{jt} and π_{jm} derived by Haseman and Elston (1972), we arrive at

$$E(Y_j|\pi_{jm}) = 2[1\text{-}2\theta(1\text{-}\theta)]\sigma_a^2+\sigma_e^2\text{-}2(1\text{-}2\theta)^2\sigma_a^2\pi_{jm}, \qquad (22.4)$$

where θ is the recombination fraction between the two loci. Thus, the regression coefficient $-2(1-2\theta)^2\sigma_a^2$ is 0 if either $\theta=1/2$ or $\sigma_a^2= 0$, and is negative if both $\theta<1/2$ and $\sigma_a^2>0$.

Table 22.2 Conditional distribution $P(\pi_{jt}|\pi_{jm})$ as a function of $\psi=\theta^2+(1+\theta)^2$, where θ is the recombination fraction between the trait and marker loci

π_{jt}	$\pi_{jm}=0$	$\pi_{jm}=1/2$	$\pi_{jm}=1$
0	ψ^2	$\psi(1-\psi)$	$(1-\psi)^2$
1/2	$2\psi(1-\psi)$	$1-2\psi(1-\psi)$	$2\psi(1-\psi)$
1	$(1-\psi)^2$	$\psi(1-\psi)$	ψ^2
Total	1	1	1

In some cases there is no difficulty in obtaining π_{jm} from the marker phenotypes available. Suppose, for example, the marker locus is very polymorphic and we have a pair of sibs whose parents' genotypes are known to be A_1A_2 x A_3A_4; in this case, π_{jm} is simply half the number of alleles the sibs have in common at the marker locus. In other cases π_{jm} may not be known with certainty, and then it can be replaced by its Bayes' estimate based on all the marker information available; it will still be true that the regression coefficient (provided it is estimable) is zero if, and only if, both $\theta \neq 1/2$ and $\sigma_a^2 > 0$. The Bayes' estimate of π_{jm} is

$$\hat{\pi}_{jm} = f_{2j} + f_{1j}/2,$$

where f_{ij} (i=0, 1, or 2) is the posterior probability, based on all the marker information available, that the two relatives share i genes i.b.d. at the marker locus.

The use of Bayes' theorem to calculate the probabilities f_{ij} is most easily illustrated for the case in which all alleles at the marker locus are codominant and marker information is available only on the two relatives under consideration. Denote the marker alleles A_j, with corresponding gene frequencies p_j. Then, no matter how polymorphic the marker locus is, there are at most seven pair types that need to be considered. In this connection, the term "pair type" is used in a broad sense that classifies genotypically analogous but different pairs as belonging to the same type. The two pairs A_1A_1-A_1A_1 and A_2A_2-A_2A_2, for example, are of the same type (Type I) because they both involve a pair of identical homozygotes. Table 22.3 indicates, for each of the seven pair types possible, the probability of its occurrence conditional on the pair sharing 0, 1, or 2 genes i.b.d.

Table 22.3 Probability of occurrence of each of the seven pair types conditional on the pair sharing i genes i.b.d. at a marker locus with alleles A_j, and corresponding gene frequences p_j (all alleles codominant). It is assumed that A_j, A_k, A_l, and A_m are all distinct alleles

Pair types	i=0	i=1	i=2
I. A_jA_j - A_jA_j	p_j^4	p_j^3	p_j^2
II. A_jA_j - A_kA_k	$2p_j^2p_k^2$	0	0
III. A_jA_j - A_jA_k	$4p_j^3p_k$	$2p_j^2p_k$	0
IV. A_jA_j - A_kA_l	$4p_j^2p_kp_l$	0	0
V. A_jA_k - A_jA_k	$4p_j^2p_k^2$	$p_jp_k(p_j+p_k)$	$2p_jp_k$
VI. A_jA_k - A_jA_l	$8p_j^2p_kp_l$	$2p_jp_kp_l$	0
VII. A_jA_k - A_lA_m	$8p_jp_kp_lp_m$	0	0

Denoting these probabilities P(marker information on pair|i genes i.b.d.), we calculate for the j-th pair, i=1, 2:

$$f_{ij} = \frac{c_i P(\text{marker information on j-th pair}|i \text{ genes i.b.d})}{\sum_k c_k P(\text{marker information on j-th pair}|k \text{ genes i.b.d})}$$

where c_i is the prior probability that the pair of relatives share i genes i.b.d. at the marker locus (Table 22.1).

When marker data on other relatives are also available, they can (and, if informative, should) be used. In the case of sibs, for example, marker data on their parents will often be informative. The general method of incorporating such information, and of allowing for dominance at the marker locus, is discussed by Haseman and Elston (1972) for the case of full sibs. The procedure is analogous, but somewhat more complicated, in the case of other types of relative pairs.

22.3 Transformations to Approximate Normality

The usual t-statistic for testing the null hypothesis $\beta=0$ in the regression model

$$E(Y_j|\hat{\pi}_j) = \alpha+\beta\hat{\pi}_j \tag{22.5}$$

assumes that the residuals are normally distributed. Because this is unlikely to be the case, it is appropriate to use a transformation, such as a power transformation of the type suggested by Box and Cox (1964), John and Draper (1980), or George and Elston (1988). The last of these can remove skewness and kurtosis and induce normality from a much broader class of distributions than can the other two, and is defined as

$$h(y) = \begin{cases} \operatorname{sgn}(y-\delta)\left\{\dfrac{(|y-\delta|+1)^{\lambda}-1}{\lambda}\right\} & \text{if } \lambda > 0 \\ \operatorname{sgn}(y-\delta)\,\ln(|y-\delta|+1) & \text{if } \lambda = 0. \end{cases}$$

The parameters of the transformation, λ and δ, are estimated simultaneously with the other parameters of the regression model (22.5), α and β. In order to preserve the model, however, both sides of (22.5) are transformed (Carroll and Ruppert 1984; George and Elston

1988). In other words, maximum likelihood estimates of λ, δ, α, β and σ^2 are obtained on the assumption that the residuals $h(y_j)-h(\alpha+\beta\hat{\pi}_j)$ are normally distributed with variance σ^2, using numerical maximization of the likelihood surface. To test the null hypothesis $\beta=0$, the likelihood ratio criterion can be conveniently used. A second maximization is performed assuming the residuals $h(y_i)-h(\alpha)$ are normally distributed with variance σ^2, the maximization now being over λ, δ, α, and σ^2. In each case, the maximum likelihood is calculated and must be standardized by dividing by the Jacobian of the transformation:

$$J = \prod_{j=1}^{n} (|y_j-\delta|+1)^{(\lambda+1)}.$$

The corresponding estimates of δ and λ are used in calculating J to standardize each likelihood. Under the null hypothesis, twice the difference in the standardized ℓn likelihoods is then expected to be approximately distributed as chi-square with one degree of freedom, in large samples. Provided the estimate of β is negative, the appropriate significance level is taken to be half the usual tabulated value for chi-square, which corresponds to a two-sided test.

Blackwelder and Elston (1982) simulated samples of 20 and 30 sib pairs and found, for samples even as small as these, that the original Haseman and Elston test, with no transformation used, is fairly robust (though slightly liberal). They warned, however, that in their simulations cases of discrete distributions or extreme outliers were not included. Interestingly, they found that if Y_j is power-transformed to minimize the skewness of the residuals in model (22.5), the test becomes slightly more liberal, i.e., slightly more likely to reject the null hypothesis $\beta=0$ than it should, based on the nominal significance level. This is presumably due to the fact that the power transform was being estimated to make the residuals from the model (including regression on $\hat{\pi}$) approximately normally distributed, and then kept fixed at that value when testing the null hypothesis $\beta=0$. The bias that results from such a procedure should disappear if the fact that the power transform is being estimated is allowed for in the manner indicated above.

22.4 Dichotomous Traits and Disease Traits with Variable Age of Onset

If the trait x_{ij} is a dichotomy, to which we can without loss of generality assign the values 0 and 1, Y_j also takes on the two values 0 and 1, corresponding to concordance or discordance of the j-th relative pair for the disease. Haseman and Elston's test is then identical to testing whether the mean proportion of genes i.b.d. is the same for concordant and discordant pairs (Elston et al. 1973). In this case it may be helpful to perform a transformation of the

fraction of 0. But spurious "linkage" may also be detected in the absence of any pleiotropic effect, if the two loci are unlinked but their allelic frequencies are not independent. Cockerham and Weir (1983) have proposed a method of analysis, for sibships of size three or more, that is not affected by such disequilibria. The assumption of random mating is probably not otherwise critical, provided the marker genotype frequencies approximately follow Hardy-Weinberg equilibrium proportions, though it may be necessary to allow for pedigree structures with inbreeding loops when determining the prior probabilities c_0, c_1, and c_2 for a particular pair of relatives (Nadot and Vaysseix 1973).

It is not critical to have a completely random sample of relative pairs. It is sufficient that the pairs be random with respect to either the marker or trait loci. Provided the markers have no selective effect, this implies that it can be advantageous to select the sample on the basis of the trait measures; the larger the additive genetic variance of the trait **in the sample**, the more powerful will be the test of the null hypothesis $\beta=0$. Dependencies among the relative pairs, however, will affect the outcome and could conceivably give rise to spurious results. It is my belief that the effect of such dependencies will be minimal in large samples, however, in view of the findings by Blackwelder and Elston (1982; 1985) for sibships of size greater than two. Further studies are needed to investigate this point.

Finally, I should emphasize that my aim has been to describe a very general, computationally fast, linkage method for detecting major genes, i.e., genes that account for an appreciable amount of the trait variance in the sample being studied. Careful consideration should be given to power considerations before conducting any study based on these methods, since it is known that enormous samples may be required if the trait locus has only a small additive genetic variance (Blackwelder and Elston 1982; Soller and Genizi 1978; Soller and Brody 1976; McMillan and Robertson 1974).

Acknowledgements. This work was supported in part by a Public Health Service Research Grant (GM 28356) from the National Institute of General Medical Sciences.

References

Beckmann JS, Soller M (1983) Restriction fragment length polymorphisms in genetic improvement: methodologies, mapping and costs. Theor Appl Genet 67:35-43

Blackwelder WC, Elston RC (1982) Power and robustness of sib-pair linkage tests and extension to larger sibships. Commun Stat Theor Meth 11:449-484

Blackwelder WC, Elston RC (1985) A comparison of sib-pair linkage tests for disease susceptibility loci. Genet Epid 2:85-98

Botstein D, White RL, Skolnick MH, Davis RW (1980) Construction of a genetic linkage map in man using restriction fragment length polymorphisms. Am J Hum Genet 32:314-331

Box GEP, Cox DR (1964) An analysis of transformations. J R Stat Soc B 26:211-252

Carroll RJ, Ruppert D (1984) Power transformation when fitting theoretical models to data. J Am Stat Assoc 79:321-328

Cockerham CC, Weir BS (1983) Linkage between a marker locus and a quantitative trait of sibs. Am J Hum Genet 35:263-273

Elston RC, Kringlen E, Namboodiri KK (1973) Possible linkage relationships between certain blood groups and schizophrenia or other psychoses. Behav Genet 3:101-106

Gelderman H (1975) Investigations on inheritance of quantitative traits in animals by gene markers. I. Methods. Theor Appl Genet 46:319-330

George VT, Elston RC (1988) Generalized modulus power transformations. Commun Stat Theor Meth 17:2933-2952

Haseman JK, Elston RC (1972) The investigation of linkage between a quantitative trait and a marker locus. Behav Genet 2:3-19

Hill AP (1975) Quantitative linkage: a statistical procedure for its detection and estimation. Ann Hum Genet 38:439-449

Jayakar SD (1970) On the detection and estimation of linkage between a locus influencing a quantitative character and a marker locus. Biometrics 26:451-464

John JA, Draper NR (1980) An alternative family of transformations. Appl Stat 29:190-197

Law CN (1966) The location of genetic factors affecting a quantitative character in wheat. Genetics 53:487-498

Lowry DC, Schultz FT (1959) Testing association of metric traits and marker genes. Ann Hum Genet 23:83-90

McMillan I, Robertson A (1974) The power of methods for the detection of major genes affecting quantitative characters. Heredity 32:349-356

Nadot R, Vaysseix G (1973) Apparentement et identité. Algorithme du calcul des coefficients d'identité. Biometrics 29:347-349

Ott, J (1985) Analysis of human genetic linkage. Johns Hopkins Univ Press, Baltimore

Penrose LS (1938) Genetic linkage in graded human characters. Ann Eugen 8:233-237

Smith CAB (1975) A non-parametric test for linkage with a quantitative character. Ann Hum Genet 38:451-460

Soller M, Brody T (1976) On the power of experimental designs for the detection of linkage between marker loci and quantitative loci in crosses between inbred lines. Theor Appl Genet 47:35-39

Soller M, Genizi A (1978) The efficiency of experimental designs for the detection of linkage between a marker locus and a locus affecting a quantitative trait in segregating populations. Biometrics 34:47-55

Tanksley SD (1983) Molecular markers in plant breeding. Plant Molec Biol Rep 1:3-8

Tanksley SD, Medina-Filho H, Rick CM (1982) Use of naturally-occurring enzyme variation to detect and map genes controlling quantitative traits in an interspecific backcross of tomato. Heredity 49:11-25

Thoday JM (1961) Location of polygenes. Nature 191:368-370

23 Reproductive Technology and Genetic Evaluation

B.W. Kennedy and L.R. Schaeffer[1]

The effects of use of embryo transfer and embryo splitting, and potential applications of embryo and semen sexing, chimeric and polyploid animals and gene transfer on genetic evaluation of dairy cattle are considered. Procedures for joint cow and bull evaluation under an animal model to accommodate embryo transfer in the population are reviewed. Genetic evaluation models based on data involving animals of identical genotype, as created through embryo splitting or other forms of cloning, are proposed. Approaches to genetic evaluation for traits influenced by both transgenes and polygenes are also suggested. Evaluation of cytoplasmic effects in embryo transfer programs is considered, and of dominance effects in populations where considerable use is made of embryo transfer and splitting. Problems of preferential treatment of genotypes produced through expensive reproductive technology are addressed.

23.1 Introduction

Historically, implementation of new reproductive technology has provided impetus for improved methods of genetic evaluation. Growth and expansion of artificial insemination (AI) in dairy cattle breeding during the 1950's created the need for genetic evaluation of bulls across herds, and prompted the replacement of daughter-dam comparisons with contemporary or herd-mate comparison methods (Henderson et al. 1954; Robertson and Rendel 1954). Early AI programs were based on fresh semen and the dairy farmer had little choice in the bull used. As a result, bulls were used randomly across herds and the contemporary comparison worked satisfactorily. The transition to frozen semen in the early 1960's and the success of AI for genetic improvement further complicated genetic evaluation. The dairy farmer could now pick and choose between bulls, so old and young bulls could be used simultaneously. The contemporary comparison did not account effectively for genetic changes over time nor for the non-random use of sires. This led to its replacement beginning in the early 1970's with the sire comparison method (Henderson 1966, 1973) based on best linear unbiased

---

[1] Centre for Genetic Improvement of Livestock, University of Guelph, Guelph, Canada

prediction (BLUP) principles. The sire comparison method has become the international standard for genetic evaluation of dairy bulls, and it, or an approximation to it, is used in most developed countries today.

Application of reproductive technology is continuing to create needs for newer methods of genetic evaluation. The sire comparison method is based largely on the assumption that the progeny of a sire constitute a group of half-sisters, each with an equal additive genetic relationship to each other and to the sire. However, use of embryo transfer (ET), for example, can produce full-sib families in addition to half-sib families, and both ET and the continued widespread use of AI have contributed to increasingly complicated relationship structures among animals. Some sophistication has been added to the sire comparison method by accounting for relationships among sires (e.g., Kennedy and Moxley 1975), and for the maternal grandsire of the cow (e.g., Quaas et al. 1979), but these modifications do not account fully for the complex breeding structure of the population. This can bias genetic evaluation if selection has operated on animals whose relationships are not fully identified as might occur, for example, if greater use of ET is made with genetically superior cows. Other reproductive techniques, such as embryo splitting, which is now done commercially (Baker and Shea 1985), further complicate genetic evaluation.

This paper reviews current and developing techniques in reproductive biology in the context of their effect on genetic evaluation, principally of dairy cattle. Advances in reproductive technology seem to have the most rapid and widespread application in cattle, particularly dairy cattle; however, the principles given are applicable also to other classes of livestock. This paper is concerned only with genetic evaluation, and use of the new techniques in reproductive biology to increase rate of genetic improvement (e.g., Nicholas and Smith 1983) are not considered, as this has been dealt with elsewhere (e.g., Van Vleck 1981).

23.2 Reproductive Technology and Evaluation for Additive Genetic Merit

For this section assume generally that we are predicting additive genetic values of animals for traits controlled by a large number of additive loci. Non-additive inheritance is considered later. The effects of embryo transfer, embryo splitting and other forms of cloning, embryo and semen sexing, androgenesis including self-fertilization, and creation of chimeric and polyploid animals on evaluation for additive genetic merit are considered. Also, consideration is given to genetic evaluation for traits influenced by a major gene in the context of implementation of gene transfer.

23.2.1 Embryo Transfer

Applications of embryo transfer to cattle have been reviewed by Seidel (1981, 1984). In North America, close to 50% of young Holstein bulls sampled in AI are from ET (Seidel 1984), but only about 2% of animal registrations in the Holstein breed are the result of ET, at least in Canada.

Unquestionably, the use of ET complicates genetic evaluation, but this has received little attention. Powell (1981) suggested some *ad hoc* modifications to cow and sire evaluation systems used in the United States to accommodate increased use of ET, but a general consideration of the problem has not been provided. Most attention has been given to genetic improvement through ET (Van Vleck 1981; Nicholas and Smith 1983; Everett 1984), not to genetic evaluation in populations in which ET is used. Nicholas and Smith (1983), in their innovative proposal for use of embryo transfer and other reproductive techniques in a nucleus herd to produce bulls for use in a larger AI bred population, considered a selection index based on dam, full-sib, and half-sib information. However, such indices would lose efficiency in a small nucleus herd under selection over a period of time, because of complications caused by finite population size and selection.

Use of ET produces increasingly complex genetic relationships among animals, a greater intensity of selection, increased inbreeding, and ET is usually associated with assortative mating. All of these should be accommodated in genetic evaluation such as through analysis of the data with an "animal" model (e.g., Meyer and Burnside 1988). Unlike sire evaluation models, where a predefined family structure is assumed, the animal model can account for any and all additive genetic relationships among animals in the population. Its application leads to joint evaluation of both cows and bulls.

Consider the model

$$\mathbf{y} = \mathbf{Xb}+\mathbf{Za}+\mathbf{e}, \tag{23.1}$$

where $\mathbf{y}$, $\mathbf{b}$, $\mathbf{a}$, and $\mathbf{e}$ represent vectors of observations, fixed effects, additive genetic values of animals, and random (environmental) residuals, respectively. $\mathbf{X}$ and $\mathbf{Z}$ are incidence matrices, and in the absence of selection $E(\mathbf{y})=\mathbf{Xb}$ and $Var(\mathbf{y})=\mathbf{ZAZ}'\sigma_a^2+\mathbf{I}\sigma_e^2=\mathbf{V}$. Evaluation of additive genetic value is

$$\hat{\mathbf{a}} = \mathbf{C}'\mathbf{V}^{-1}(\mathbf{y}-\mathbf{X}\tilde{\mathbf{b}}), \tag{23.2}$$

where $\mathbf{C}=Cov(\mathbf{y},\mathbf{a}')=\mathbf{ZA}\sigma_a^2$ and $\tilde{\mathbf{b}}$ is a generalized least-squares solution to $\mathbf{b}$. In practice, Henderson's mixed model equations are easier to solve computationally;

$$\begin{bmatrix} \tilde{\mathbf{b}} \\ \hat{\mathbf{a}} \end{bmatrix} = \begin{bmatrix} \mathbf{X'X} & \mathbf{X'Z} \\ \mathbf{Z'X} & \mathbf{Z'Z+A^{-1}}\lambda \end{bmatrix}^{-} \begin{bmatrix} \mathbf{X'y} \\ \mathbf{Z'y} \end{bmatrix} \tag{23.3}$$

for $\lambda=\sigma_e^2/\sigma_a^2=(1-h^2)/h^2$, which leads to the same result as (23.2).

The matrix of additive genetic relationships, **A**, is composed of off-diagonal elements a_{ij} equal to the numerator of Wright's (1922) coefficient of relationship between the ith and j^{th} animals, and diagonal elements a_{ii} equal to $1+F_i$, where F_i is the coefficient of inbreeding of the ith animal. The essential elements of the animal model are in Henderson (1963), but practical application was not possible, except for small data sets, until the development of computationally easy methods to obtain the inverse of **A** (Henderson 1976; Quaas 1976). The animal model given in (23.3) is appropriate for the situation where single records on a single trait are available, although the trait need not be recorded on all animals. Extension to repeated records on animals and multiple trait evaluation is straightforward (Henderson 1984), but can be more demanding computationally.

We use (23.3) with a small numerical example. Suppose bull 1 is mated to cow 2 and through ET two daughters (full-sibs) 5 and 6 are produced. Additionally, bull 3 is mated to cow 4 in a normal fashion, and a single daughter 7 is produced. Cows 2, 4, 5, 6, and 7 have phenotypic records of 11, 7, 9, 8, and 8, respectively. Cow 2 was chosen over cow 4 for ET because she had a better record. The bulls do not have phenotypic records. If the animals are ordered by number, the relationship matrix, assuming sires and dams are unrelated and not inbred, is:

$$\mathbf{A} = \begin{bmatrix} 1 & 0 & 0 & 0 & 1/2 & 1/2 & 0 \\ 0 & 1 & 0 & 0 & 1/2 & 1/2 & 0 \\ 0 & 0 & 1 & 0 & 0 & 0 & 1/2 \\ 0 & 0 & 0 & 1 & 0 & 0 & 1/2 \\ 1/2 & 1/2 & 0 & 0 & 1 & 1/2 & 0 \\ 1/2 & 1/2 & 0 & 0 & 1/2 & 1 & 0 \\ 0 & 0 & 1/2 & 1/2 & 0 & 0 & 1 \end{bmatrix}.$$

If $h^2=0.5$ and assuming, for simplicity, that the only fixed effect is the mean (μ), then (23.3) becomes

$$\begin{bmatrix} \hat{\mu} \\ \hat{a}_1 \\ \hat{a}_2 \\ \hat{a}_3 \\ \hat{a}_4 \\ \hat{a}_5 \\ \hat{a}_6 \\ \hat{a}_7 \end{bmatrix} = \begin{bmatrix} 5 & 0 & 1 & 0 & 1 & 1 & 1 & 1 \\ 0 & 2 & 1 & 0 & 0 & -1 & -1 & 0 \\ 1 & 1 & 3 & 0 & 0 & -1 & -1 & 0 \\ 0 & 0 & 0 & 3/2 & 1/2 & 0 & 0 & -1 \\ 1 & 0 & 0 & 1/2 & 5/2 & 0 & 0 & -1 \\ 1 & -1 & -1 & 0 & 0 & 3 & 0 & 0 \\ 1 & -1 & -1 & 0 & 0 & 0 & 3 & 0 \\ 1 & 0 & 0 & -1 & -1 & 0 & 0 & 3 \end{bmatrix}^{-1} \begin{bmatrix} 43 \\ 0 \\ 11 \\ 0 \\ 7 \\ 9 \\ 8 \\ 8 \end{bmatrix} = \begin{bmatrix} 8.519 \\ -0.284 \\ 1.099 \\ -0.037 \\ -0.778 \\ 0.432 \\ 0.099 \\ -0.444 \end{bmatrix}$$

All animals are evaluated. Note that the average $\hat{a}_i$ for the parental (unrelated base population) animals is zero, i.e., $1/4(\hat{a}_1+\hat{a}_2+\hat{a}_3+\hat{a}_4)=0$, but the average $\hat{a}_i$ of the offspring generation is greater than zero, i.e., $1/3(\hat{a}_5+\hat{a}_6+\hat{a}_7)=0.029$. This reflects genetic change in the offspring generation through selecting the better dam to have relatively more progeny through ET. Also, note that the evaluations of the sires are adjusted for the estimated genetic contributions of their mates, i.e., $\hat{a}_1=1/2[(\hat{a}_5+\hat{a}_6)-\hat{a}_2]= -0.284$ and $\hat{a}_3=1/3(2\hat{a}_7-\hat{a}_4)=-0.037$. The fact that bull 1 was mated to a selected ET donor cow does not bias its genetic evaluation. Consider now an evaluation with the usual sire model. Records on the mates would be ignored and the two progeny of bull 1, cows 5 and 6, would be treated as half-sibs rather than full-sibs. The equations would be

$$\begin{bmatrix} \hat{\mu} \\ \hat{s}_1 \\ \hat{s}_3 \end{bmatrix} = \begin{bmatrix} 3 & 2 & 1 \\ 2 & 9 & 0 \\ 1 & 0 & 8 \end{bmatrix}^{-1} \begin{bmatrix} 25 \\ 17 \\ 8 \end{bmatrix} = \begin{bmatrix} 8.32 \\ 0.04 \\ -0.04 \end{bmatrix},$$

resulting in $\hat{a}_1=2\hat{s}_1=0.08$ and $\hat{a}_3=2\hat{s}_3=-0.08$. Now bull 1 is incorrectly estimated to be genetically better than bull 3.

As mentioned previously, use of ET and of other reproductive technology complicates genetic evaluation because many complex additive genetic relationships are created in the population and selection is intensified. However, if **A** includes all relationships back to a base population prior to selection, and heritability in the base population is known, then use of the animal model gives best linear unbiased predictors (BLUP) of additive genetic values, under selection, as long as the model is correctly specified, selection is on a linear translation invariant function of **y**, and **a** and **y** are multivariate normal (Henderson 1975).

In practice, these conditions are not likely to be met exactly. Discussion of genetic assumptions implicit in (23.3) and consequences of violating these assumptions are in Sorensen and Kennedy (1986) and Kennedy, Chapter 5, this Volume. There are some other practical problems with implementation of the animal model, the consequences of which are not completely understood. For example, all animals in the population may not have been derived from the same initial or base population, and there may be different subpopulations with respect to time or origin, as would occur, for example, if ET were used to introduce new breeding stock derived from an external source unrelated to the existing population. Animals would need to be identified as to population of origin and grouping strategies would be required to accommodate different subpopulations. How to do this when groups and relationships are to be used together is not perfectly understood (Thompson 1979; Dempfle 1982). Westell and Van Vleck (1984) assigned groups only to animals that had unidentified parents. Additionally, the relationship matrix may not be completely accurate, either as a result of missing information or errors in identification. This reduces the accuracy of evaluation and can lead to bias in selected populations. Little can be done about errors in identification, except by taking fairly elaborate precautions through blood testing (e.g., Geldermann et al. 1986). Presumably, increased use of embryo transfer might reduce unintentional identification errors because there likely would be better record keeping associated with it, although accidents can occur. Missing information can be handled by judicious grouping to account for selection not accounted for by the relationship matrix (Pollak and Quaas 1983).

One of the impediments to implementation of animal models has been computational problems. The order of the equations to be solved exceeds the number of records. Fortunately, developments with computers, both in hardware and software, have kept pace with those in reproductive technology. Some animal breeders can now access "super" computers. Also, computing algorithms which reduce the computational load, are being developed. Quaas and Pollak (1980) suggested the innovative "reduced animal model" or gametic model whereby equations need only be formed explicitly for parent animals. Equations for progeny who never become parents are, in effect, absorbed into the parent equations. Solutions for progeny, if desired, are obtained by backsolution. Evaluations are identical to those obtained with the animal model. Schaeffer and Kennedy (1986a,b) illustrated an alternative procedure that iterates on the model (23.1), without setting up the coefficient matrix (23.3). In their application this was more efficient than the reduced animal model. These and other algorithms have been applied to routine genetic evaluation under an animal model for populations of pigs (e.g., Hudson and Kennedy 1985; Schaeffer and Kennedy 1986b) and beef cattle (e.g., Benyshek et al. 1986), and applications to dairy cattle are under development (e.g., Meyer and Burnside 1988).

Lastly, although a normal distribution of genotypes is assumed implicitly with the animal model under selection, the phenotypic distribution need not be normal. For example, for binary or ordered categorical traits, a threshold model can be invoked if the underlying distribution of genotypic values is normal (Falconer 1981). Procedures for genetic analysis of such traits have been developed and applied to sire evaluation (Gianola and Foulley 1983; Harville and Mee 1984; Foulley and Gianola 1986), and application to cow and bull evaluation under an animal model is straightforward, although computationally more difficult. Some difficulties, however, remain in obtaining unbiased estimates of heritability for categorical traits (Gilmour et al. 1985).

In summary, continued and increasing use of ET, along with continued use of AI, will provide an impetus for replacing current sire evaluation and cow indexing procedures by joint cow and bull evaluation under an animal model. Computational difficulties will be less of a barrier to implementation because of improved computers and computing algorithms.

23.2.2 Embryo Splitting (Cloning)

Splits of embryos in two are now produced commercially (Baker and Shea 1985), and higher order splits have been made experimentally (Polge 1986). Splitting embryos is a form of cloning and animals resulting from a split are identical genetically. The use of split embryos offers considerable potential for genetic improvement (Nicholas and Smith 1983). Split embryos complicate genetic evaluation in the same way as do naturally occurring identical twins; however, the frequency of such occurrences will presumably be increased. The major complication with identical genotypes is that $\mathbf{A}$ is singular and $\mathbf{A}^{-1}$ does not exist. Henderson (1984) has suggested avoiding $\mathbf{A}^{-1}$ by solving

$$\begin{bmatrix} \tilde{\mathbf{b}} \\ \tilde{\boldsymbol{\alpha}} \end{bmatrix} = \begin{bmatrix} \mathbf{X'X} & \mathbf{X'ZG} \\ \mathbf{GZ'X} & \mathbf{GZ'ZG} + \mathbf{G}\sigma_e^2 \end{bmatrix}^{-} \begin{bmatrix} \mathbf{X'y} \\ \mathbf{GZ'y} \end{bmatrix}, \tag{23.4}$$

where $\mathbf{G}=\mathbf{A}\sigma_a^2$. These are similar to equations Henderson (1973) suggested for an animal model (with non-singular $\mathbf{A}$) prior to the development of easy methods for obtaining $\mathbf{A}^{-1}$. Additive genetic values are then estimated as $\hat{\mathbf{a}}=\mathbf{G}\tilde{\boldsymbol{\alpha}}$. Note that despite a singular $\mathbf{A}$, $\hat{\mathbf{a}}$ is unique. Also, for a family of t clones, $\hat{a}_1=\hat{a}_2=...=\hat{a}_t$, that is, all clones of the same genotype will have identical estimated genetic values, as they should.

A limitation of (23.4) is its computational difficulty. For a large number of animals, **A** is more difficult to compute than $\mathbf{A}^{-1}$. A simpler alternative would be to treat records on clones as repeated observations on the same genotype. For example, consider a young bull that is progeny tested but his frozen clone, the result of the other half of the split embryo, is subsequently thawed and used for production of semen. Treating the records of daughters of both clones as being from progeny of the same sire with a first and second crop of daughters would lead to the same genetic evaluation. Even with an animal model, it is simpler to define **A** such that genetically identical animals are considered to be the same individual rather than using (23.4). This is so because there is one row and column in **A** per genotype rather than per animal; then, **Z** can account for repeated measures on the same genotype, and evaluation is carried out with (23.3). The results are identical but (23.3) is computationally simpler than (23.4) because it involves $\mathbf{A}^{-1}$ rather than **A**. This approach holds provided non-additive genetic effects are unimportant.

We illustrate with a simple example. Suppose bull 1 is mated with cow 2 and the resulting female embryo is split to produce clones 3 and 4. Records on 2, 3, and 4 are 11, 10, and 9, respectively. Again assume h^2=0.5 and that the only fixed effect is the mean. The relationship matrix, ordered by animal number, is

$$\mathbf{A} = \begin{bmatrix} 1 & 0 & 1/2 & 1/2 \\ 0 & 1 & 1/2 & 1/2 \\ 1/2 & 1/2 & 1 & 1 \\ 1/2 & 1/2 & 1 & 1 \end{bmatrix}.$$

Note that the rows and columns for animals 3 and 4 are identical, because they have identical genotypes. For simplicity, set $\sigma_a^2=\sigma_e^2=1$, so (23.4) is:

$$\begin{bmatrix} \hat{\mu} \\ \tilde{\alpha}_1 \\ \tilde{\alpha}_2 \\ \tilde{\alpha}_3 \\ \tilde{\alpha}_4 \end{bmatrix} = \begin{bmatrix} 3 & 1 & 2 & 2\,1/2 & 2\,1/2 \\ 1 & 1\,1/2 & 1/2 & 1\,1/2 & 1\,1/2 \\ 2 & 1/2 & 2\,1/2 & 2 & 2 \\ 2\,1/2 & 1\,1/2 & 2 & 3\,1/4 & 3\,1/4 \\ 2\,1/2 & 1\,1/2 & 2 & 3\,1/4 & 3\,1/4 \end{bmatrix}^{-} \begin{bmatrix} 30 \\ 9.5 \\ 20.5 \\ 24.5 \\ 24.5 \end{bmatrix}.$$

Equations for α_3 and α_4 are identical, but setting $\tilde{\alpha}_4=0$ leads to a solution

$$\begin{bmatrix} \hat{\mu} \\ \tilde{\alpha} \end{bmatrix} = (10.1 \quad 0 \quad 0.6 \quad -0.6 \quad 0)'.$$

The estimated additive genetic values of the four animals are

$$\begin{bmatrix} \hat{a}_1 \\ \hat{a}_2 \\ \hat{a}_3 \\ \hat{a}_4 \end{bmatrix} = \begin{bmatrix} 1 & 0 & 1/2 & 1/2 \\ 0 & 1 & 1/2 & 1/2 \\ 1/2 & 1/2 & 1 & 1 \\ 1/2 & 1/2 & 1 & 1 \end{bmatrix} \begin{bmatrix} 0 \\ 0.6 \\ -0.6 \\ 0 \end{bmatrix} = \begin{bmatrix} -0.3 \\ 0.3 \\ -0.3 \\ -0.3 \end{bmatrix}.$$

Note that $\hat{a}_3=\hat{a}_4=-0.3$.

A simpler alternative is to delete the row and column for animal 4 from **A** at the start, and evaluate with (23.3). The equations are now

$$\begin{bmatrix} \hat{\mu} \\ \hat{a}_1 \\ \hat{a}_2 \\ \hat{a}_3 \end{bmatrix} = \begin{bmatrix} 3 & 0 & 1 & 2 \\ 0 & 1\ 1/2 & 1/2 & -1 \\ 1 & 1/2 & 2\ 1/2 & -1 \\ 2 & -1 & -1 & 4 \end{bmatrix}^{-1} \begin{bmatrix} 30 \\ 0 \\ 11 \\ 19 \end{bmatrix} = \begin{bmatrix} 10.1 \\ -0.3 \\ 0.3 \\ -0.3 \end{bmatrix},$$

with solutions exactly as before, except that we now have only one solution ($\hat{a}_3$) per identical genotype. Note that the records of both 3 and 4 are included in the evaluation, but they are treated as repeated records on the same genotype.

With repeated records on animals, (23.1) and (23.3) can be expanded to accommodate permanent environmental and other contributions to covariances between records on the same animal in the usual manner. Solution of the equations results in one estimated genetic value for each genotype, but different estimated permanent environmental effects for each animal with records. The estimated additive genetic value of the jth clone of the ith genotype is simply $\hat{a}_i$, but its "real producing ability" is estimated as $\hat{a}_i$ plus $\hat{p}_{ij}$, where $\hat{p}_{ij}$ is the estimated permanent environmental effect peculiar to the ij^{th} clone.

If identical genotypes are present, it is essential that recording and registration programs identify them as such, and that **A** be structured accordingly, either in full as in (23.4), or with replicate genotypes deleted. Failure to do so will result in identical genotypes being treated as full-sibs under an animal model, or possibly as unrelated under some sire evaluation models, and genetic evaluations of identical genotypes will differ. This is less accurate and potentially embarrassing.

23.2.3 Embryo and Semen Sexing

Although a reliable method of sexing sperm has not been documented (Church et al. 1986), sexing of embryos can be done (Betteridge et al. 1981). The ability to sex semen and embryos can have some impact on genetic improvement (Van Vleck 1981), but semen sexing and embryo sexing would have little effect on genetic evaluation.

Sexing would likely result in non-random association of genotypes with sex. Conceptually, this does not differ from current practices (e.g., culling males on pedigree or performance) except that selection is prior to or shortly after conception, rather than postnatal. This presents no problem for genetic evaluation as long as the information (data and relationships) upon which selection is based is included in the evaluation process.

23.2.4 Androgenous Matings and Self-Fertilization

Androgenous mating, in which an oocyte is fertilized with two sperm and the female genetic material is removed, has been the subject of speculation (Seidel 1984). One approach would involve the microsurgical removal of the female pronucleus of the fertilized ovum immediately after fertilization, and replacement with a male pronucleus. The result would be the union of two sperm with the ovum acting simply as a carrier. Potential genetic improvement from producing bulls with only sires as parents, for example, is considerable (Van Raden and Freeman 1985). Twice as many male offspring as female offspring would be expected from androgenous matings, unless sexed semen were available. There is doubt as to whether successful androgenous matings are possible in mammals (McLaren 1984), and current knowledge is reviewed by Surani et al. (1987). In any event, having two "sires" as parents of an individual, rather than a sire and dam, does not complicate genetic evaluation for additive genetic merit (assuming the absence of sex-linked genes). Sex of the parent is immaterial to the computation of **A** and $\mathbf{A}^{-1}$.

Male matings where the same sire is used as both parents would be a form of self-fertilization. The first generation of selfed individuals would be 50% inbred, and the fifth generation of repeated selfing would be 97% inbred. Van Vleck (1981) has considered the use of selfing on accuracy of sire evaluation, and concluded that the gains in accuracy would likely not be enough to offset the biological and economic costs of selfing. In terms of the technical aspects of genetic evaluation for additive genetic merit, self-fertilization adds no complication if evaluation is under an animal model. This is because **A** accommodates the contributions of inbreeding to genetic relationships and changes in genetic variance from inbreeding (Sorensen and Kennedy 1984a). The possibility of inbreeding depression is a separate consideration, and is not relevant for traits for which all genetic variation is additive. Some discussion of inbreeding depression is given in a later section on evaluation for non-additive genetic merit.

23.2.5 Chimeras

Chimeras, both within and between species, have been produced experimentally. They may be of practical use only in an experimental context; nonetheless, the introduction of chimeras into livestock populations would create some interesting problems in genetic evaluation.

Chimeras can be produced by the aggregation of blastomeres from two embryos (Mintz 1971) or by injecting cells from one genetic source into the blastocyst of an embryo of another (Gardner 1978). The resulting animal has in effect four parents. The chimeric animal is made up of a mixture of two types of cells, representing the two parental embryos, but the individual cell is diploid. If the two types of cells are equally represented in producing tissue, cell numbers are not increased and their contribution to product is additive, then the additive genetic value of the chimera would be the average of the additive genetic values of the two contributing embryos. Assuming the two parental embryos are unrelated, then the additive genetic variance among chimeric animals would be half the usual additive genetic variance.

Because the relationship matrix, $\mathbf{A}\sigma_a^2$, is simply a matrix of additive genetic variances and covariances among animals, it could be modified to account for the presence of chimeric animals in the population. For example, the covariance between a normal parent and its chimeric offspring is one-quarter rather than the usual one-half, assuming no inbreeding. Other relationships can be more difficult to determine. A correctly modified **A** matrix could then be applied to the mixed model equations (23.3) for genetic evaluation if simple procedures could be developed for obtaining $\mathbf{A}^{-1}$, or to equations similar to (23.4) if not.

The preceding is based on unrealistic and simplistic assumptions on the expression and transmission of additive genetic merit in chimeric animals. For example, bovine chimeras from double-muscled and normal embryos have not shown double-muscled characteristics (Church et al. 1986), and an assumption of additive contributions of cells is not realistic in this case. Additionally, the parental cell types are usually not equally represented in tissue and the relative contribution of each cell type to the chimera may vary with age (Church et al. 1986). Also, the parental embryos may contribute differentially to reproductive tissue, or one parental embryo may not contribute at all. If chimeric animals are used for breeding purposes, this can complicate genetic evaluation further, because the genetic composition of tissue contributing to the trait being evaluated may differ from the genetic composition of gonadal tissue. In effect, an extra form of sampling is introduced in the transmission of genes from parent to offspring over and above the usual Mendelian sampling process. Until some of these questions and problems are resolved, little practical use could be made of chimeric animals in genetic evaluation.

23.2.6 Polyploidy

Polyploid embryos can be produced without great difficulty, at least in principle (Fechheimer 1985), and both triploid and tetraploid stocks of fish have been developed (Kinghorn 1983). The primary value of polyploid fish is in their retarded sexual development, which avoids problems of reduced growth rate and poor meat quality associated with sexual maturation. Polyploidy in fish is induced by subjecting fertilized eggs to heat or cold shock. In mice, triploidy has been induced by suppressing the second meiotic division of oogenesis or treatment of eggs at time of fertilization, and tetraploid embryos are induced by suppressing cytokinesis of the first cleavage division of normal fertilized eggs (Fechheimer 1985). However, polyploid mouse embryos apparently do not survive to term. Triploids are expected to be sterile but the possibility of fertile tetraploids exists (Fechheimer 1985).

Evaluation of diploid parent breeding stock on the basis of their polyploid progeny would require recalculation of additive genetic variances and covariances to accommodate the polyploid nature of the offspring. Genetic relationships and inbreeding coefficients have been modified for bees to account for their mixture of haploid and diploid individuals (Crow and Roberts 1950). One potential complication with triploids is that they may not be fully homozygous for the duplicated chromosomes, depending upon how the triploid is produced, due to crossing over in the oocyte at meiosis.

23.2.7 Gene Transfer

There is potential for application of gene transfer to genetic improvement of livestock if genes of major beneficial effect can be identified. A review is given by Smith et al. (1987). Considerable attention has been paid to methods for detecting genes of major effect (e.g., Famula 1986; Elston Chap. 22, this Vol.; Hill Chap. 21, this Vol.), but little work has been done on genetic evaluation for traits influenced by major genes. The situation with transferred genes in some situations could be analogous to evaluation for traits controlled by a gene with a major effect plus many polygenes, although the site of integration and the number of copies usually cannot be controlled for transgenes.

Application of gene transfer to livestock populations could complicate genetic evaluations considerably. If animals upon which observations are available can be readily classified as to whether or not they possess the transgene, BLUP of additive genetic merit ($\hat{\mathbf{a}}$) of animals for the polygenes influencing the trait could be obtained by treating the transgene as a fixed effect in (23.1) and (23.3), assuming the gene is expressed equally in all animals (which may be unrealistic). If the effect of the transgene on the trait of concern is additive or incompletely dominant, then hemizygotes (TO) would need to be distinguished from homozygotes (TT and OO).

To include the transgene in the estimated breeding value of an animal would be a little more complicated and would require knowledge of the genotype of the animal (TT, TO, or OO). Ideally, computation of estimated breeding value involves the frequency of the gene and the effect of its substitution in the population, expressed such that the population average is zero (Falconer 1981). Breeding value then changes as gene frequency changes. Operationally, if animals are to be evaluated on their combined value for the transgenotype and all other loci (**a**), it might be preferable to evaluate the ith animal simply as the sum of $\hat{a}_i$ plus the solution from (23.3) of the effect of its transgenotype, if inheritance of the transgene is additive. Of course, all this assumes that transgenotype is known, and only the effects of genotype need to be estimated. In this case, normality assumptions about **y**, **a**, and **e** are not altered.

If it is not practical or feasible to classify the individual as to transgenotype, then one could resort to the usual genetic evaluation system (23.3) whereby the transgene is simply considered as one of the many genes influencing the trait, and its effect is included as part of **a**. This, however, is not too satisfactory if the transgene has a large effect, and presumably there would be no incentive for the transfer if this were not the case. Multivariate normality could no longer be assumed legitimately, even prior to selection. The greater the effect of the transgene, the greater will be the departures from normality.

An alternative worth investigating is to consider the observations as arising from a mixture of two or more normally distributed populations. Mixture models, which are distinct from mixed models, were first considered by Pearson (1894) who used the method of moments to estimate parameters of a mixture of two normal densities; these were the means and variances of each population and the proportion of each population in the mixture. Maximum likelihood procedures have been developed for mixture models (Day 1969). Their application is demanding computationally, perhaps too demanding for application to large data sets. Computational requirements, however, are less if a common variance can be assumed for the populations. Applications have been made in the field of human genetics to distinguish between the effects of major genes and polygenes (e.g., Elston and Stewart 1971; Morton and MacLean 1974).

In the case of gene transfer data, a mixture of two (complete dominance) or three (incomplete dominance) populations, each normally distributed, could be hypothesized. Evaluation could be according to the model

$$\mathbf{y} = \mathbf{Pt}+\mathbf{Xb}+\mathbf{Za}+\mathbf{e}, \tag{23.5}$$

where **t** represents the unknown effect of transgenotype [t_1=homozygous for transferred gene (TT), t_2=hemizygous (TO) and t_3=absence of transferred gene (OO)]. If an additive model is assumed for the transgene, $\hat{\mathbf{t}}$ should be restricted such that $\hat{t}_2=0.5(\hat{t}_1+\hat{t}_3)$. What distinguishes (23.5) from (23.1) is that elements of **t** cannot be assigned to **y** with certainty by use of an incidence matrix. Rather $\hat{\mathbf{t}}$ and $\hat{\mathbf{p}}$, the proportion of the mixture arising from each population, would be estimated jointly with $\hat{\mathbf{a}}$, $\hat{\mathbf{b}}$, and $\hat{\mathbf{e}}$. Using normal distribution theory, one could then assign a vector of probabilities ($\hat{\pi}$) to each observation for each of the three populations on the basis of the solutions to (23.5). The matrix **P** is not a fixed incidence matrix of the usual form but is the collection of $\hat{\pi}$ for all observations. The value of **P** would be adjusted at each cycle of iteration. The estimated genetic value for the transgenotype of the animal that made the observation would at convergence be $\hat{\pi}'\hat{\mathbf{t}}$. The solution for $\hat{\mathbf{a}}$ could be added to estimate genetic merit over all loci. Development of equations for solution of the parameters in (23.5) is not an easy exercise. One avenue might be to estimate **b**, **a**, and **e**, and then **t**, **p**, and **P** in a two-stage procedure per iterate. Also, one would want to make use of elements of the relationship matrix in computing **P**, if possible.

23.3 Evaluation for Non-Additive Genetic Merit

Except for the systematic use of line and breed crossing, and perhaps selection on producing ability of individuals, little has been done to capitalize on non-additive gene action in animal breeding programs. Applications of techniques in reproductive biology such as ET and zygote splitting will result in the creation of many more non-additive genetic relationships between individuals than found in conventionally bred populations. This may provide opportunities to evaluate and "select" for non-additive and perhaps extranuclear genetic value. Jansen and Wilton (1985) have considered selection when genetic merit is a non-linear function of component traits and the same principles can be applied to selection for non-additive genetic merit for a single trait. In this section the implications of reproductive technology to genetic evaluation for cytoplasmic and dominance genetic effects are considered. Also, problems of preferential treatment are examined.

23.3.1 Cytoplasmic Inheritance

Bell et al. (1985) presented evidence of important effects of cytoplasmic inheritance on milk and fat production. Although it is likely they misinterpreted variance due to drift as cytoplasmic effects (Kennedy 1986), the possibility of inherited cytoplasmic effects through mitochondrial deoxyribonucleic acid (mtDNA) remains. If cytoplasmic effects were important, this would have implications with respect to the use of embryo transfer and genetic evaluation of cows. ET dams could be selected on their cytoplasmic genetic value in addition to additive genetic value.

To accommodate cytoplasmic effects in genetic evaluation, model (23.1) could be expanded to

$$\mathbf{y} = \mathbf{Xb}+\mathbf{Za}+\mathbf{Wc}+\mathbf{e},$$

where $\mathbf{c}$ is a vector of cytoplasmic genetic effects, $\mathbf{W}$ is an incidence matrix, and other terms are as defined previously. The structure of $\mathbf{W}$ and var($\mathbf{c}$) would depend upon the nature of the inheritance of mtDNA.

If mtDNA is transmitted from dam to offspring intact, with no alteration, then the elements of $\mathbf{c}$ would simply represent maternal sources of cytoplasm. Genetic evaluation could be through solution of a simple expansion of (23.3);

$$\begin{bmatrix} \tilde{\mathbf{b}} \\ \hat{\mathbf{a}} \\ \hat{\mathbf{c}} \end{bmatrix} = \begin{bmatrix} \mathbf{X'X} & \mathbf{X'Z} & \mathbf{X'W} \\ \mathbf{Z'X} & \mathbf{Z'Z}+\mathbf{A}^{-1}\lambda & \mathbf{Z'W} \\ \mathbf{W'X} & \mathbf{W'Z} & \mathbf{W'W}+\mathbf{I}\theta \end{bmatrix}^{-} \begin{bmatrix} \mathbf{X'y} \\ \mathbf{Z'y} \\ \mathbf{W'y} \end{bmatrix}, \tag{23.6}$$

where $\theta=\sigma_e^2/\sigma_c^2$, and σ_c^2 is the variance due to cytoplasmic effects. For populations with a history of pedigree recording, the number of sources of cytoplasm would likely be small relative to the number of cows. Accordingly the order of $\mathbf{W'W}$ would be relatively small, and evaluation for cytoplasmic effects would not necessarily complicate genetic evaluation greatly. Evaluation of total genetic merit of animals with records would be $\mathbf{W}\hat{\mathbf{c}}+\hat{\mathbf{a}}$.

This is illustrated with a small example. Consider bull 1, who is mated to cows 2 and 3, producing daughters 4 and 5, respectively. Cows 2, 3, 4, and 5 have records 10, 8, 9, 7, respectively. Assume that $\sigma_a^2=0.25\sigma^2$, $\sigma_c^2=0.25\sigma^2$, and $\sigma_e^2=0.50\sigma^2$, where σ^2 is the phenotypic variance, (i.e., $\lambda=\theta=2$). For this example, if animals are ordered by number,

$$\mathbf{A} = \begin{bmatrix} 1 & 0 & 0 & 1/2 & 1/2 \\ 0 & 1 & 0 & 1/2 & 0 \\ 0 & 0 & 1 & 0 & 1/2 \\ 1/2 & 1/2 & 0 & 1 & 1/4 \\ 1/2 & 0 & 1/2 & 1/4 & 1 \end{bmatrix},$$

$$\text{and } \mathbf{A}^{-1}\lambda = \begin{bmatrix} 4 & 1 & 1 & -2 & -2 \\ 1 & 3 & 0 & -2 & 0 \\ 1 & 0 & 3 & 0 & -2 \\ -2 & -2 & 0 & 4 & 0 \\ -2 & 0 & -2 & 0 & 4 \end{bmatrix}.$$

There are two cytoplasmic sources (cows 2 and 3), so

$$\mathbf{I}\theta = \begin{bmatrix} 2 & 0 \\ 0 & 2 \end{bmatrix}.$$

Assuming the only fixed effect is the mean, (23.6) is

$$\begin{bmatrix} \hat{\mu} \\ \hat{a}_1 \\ \hat{a}_2 \\ \hat{a}_3 \\ \hat{a}_4 \\ \hat{a}_5 \\ \hat{c}_1 \\ \hat{c}_2 \end{bmatrix} = \begin{bmatrix} 4 & 0 & 1 & 1 & 1 & 1 & 2 & 2 \\ 0 & 4 & 1 & 1 & -2 & -2 & 0 & 0 \\ 1 & 1 & 4 & 0 & -2 & 0 & 1 & 0 \\ 1 & 1 & 0 & 4 & 0 & -2 & 0 & 1 \\ 1 & -2 & -2 & 0 & 5 & 0 & 1 & 0 \\ 1 & -2 & 0 & -2 & 0 & 5 & 0 & 1 \\ 2 & 0 & 1 & 0 & 1 & 0 & 4 & 0 \\ 2 & 0 & 0 & 1 & 0 & 1 & 0 & 4 \end{bmatrix}^{-1} \begin{bmatrix} 34 \\ 0 \\ 10 \\ 8 \\ 9 \\ 7 \\ 19 \\ 15 \end{bmatrix} = \begin{bmatrix} 8.52 \\ -0.19 \\ 0.37 \\ -0.18 \\ 0.09 \\ -0.38 \\ 0.37 \\ -0.37 \end{bmatrix}$$

The evaluation of total genetic merit of an animal includes the cytoplasmic contribution. For example, the evaluation of cow 4 is $\hat{c}_1+\hat{a}_4=0.37+0.09=0.46$. Note that the whole cytoplasmic genetic value is transmitted from dam to daughter.

If there were some alteration of mtDNA from generation to generation, then the situation becomes more complex. If that alteration would occur at a constant rate from generation to generation, then the elements of **c** would represent a combination of cytoplasmic source and number of generations from that source. For example, var(**c**) could be assumed to have the following form

$$\mathbf{R}\sigma_c^2 = \begin{bmatrix} \mathbf{R}_1 & \mathbf{0} & \cdots & \mathbf{0} \\ \mathbf{0} & \mathbf{R}_2 & \cdots & \mathbf{0} \\ \cdot & \cdot & \cdot & \cdot \\ \cdot & \cdot & \cdot & \cdot \\ \cdot & \cdot & \cdot & \cdot \\ \mathbf{0} & \mathbf{0} & \cdots & \mathbf{R}_t \end{bmatrix} \sigma_c^2$$

for the t sources of cytoplasm or "cow families". The form of $\mathbf{R}_i$, if **c** is ordered by generation number within cytoplasmic source, is

$$\mathbf{R}_i = \begin{bmatrix} 1 & \rho & \rho^2 & \cdots & \rho^{n_i-1} \\ \rho & 1 & \rho & \cdots & \rho^{n_i-2} \\ \rho^2 & \rho & 1 & \cdots & \rho^{n_i-3} \\ \cdot & \cdot & \cdot & \cdot & \cdot \\ \cdot & \cdot & \cdot & \cdot & \cdot \\ \cdot & \cdot & \cdot & \cdot & \cdot \\ \rho^{n_i-1} & \rho^{n_i-2} & \rho^{n_i-3} & \cdots & 1 \end{bmatrix},$$

where ρ is one minus the alteration rate, and n_i is the largest number of generations from the cytoplasmic source in the ith cow family. For genetic evaluation, $\mathbf{R}^{-1}$ would be substituted for $\mathbf{I}$ in (23.6). Obtaining $\mathbf{R}^{-1}$ is not difficult, because of the block diagonal structure of $\mathbf{R}$. The maximum order of any $\mathbf{R}_i$ is the largest number of generations from cytoplasmic source, which in dairy cattle populations would be relatively small, probably less than 25. Because all $\mathbf{R}_i$ have the same general form and differ potentially only in number of generations, the number of different $\mathbf{R}_i^{-1}$ would be small and elements of each $\mathbf{R}_i^{-1}$ could be stored conveniently and accessed, rather than inverting $\mathbf{R}_i^{-1}$ as each cow family is processed. This is true because $\mathbf{R}_i^{-1}$ has tridiagonal form with predictable elements for all i.

Evaluation of cytoplasmic effects would require good estimates of both h^2 and σ_c^2, and of ρ if there is alteration of mtDNA. This is not a trivial matter. If "conventional" variance component estimation models and techniques are used, drift variance and cytoplasmic variance can be confounded (e.g., Bell et al. 1985). Use of the relationship matrix (**A**) can account for drift variance (Sorensen and Kennedy 1983), and estimation of σ_a^2 and σ_c^2 by MIVQUE or REML, where **A** is included in the estimation process (e.g., Sorensen and Kennedy 1984b) should provide reliable estimates given sufficient data. Nonetheless, obtaining such estimates would be demanding computationally.

In any event, evaluation of total genetic merit of the ij^{th} cow ($\hat{g}_{ij}$) would be $\hat{g}_{ij}=\hat{c}_i+\hat{a}_{ij}$ where $\hat{c}_i$ is the estimated cytoplasmic genetic value of the ith source or i^{th} source-generation, depending upon alteration of mtDNA. However, for purposes of selection, one might want to consider different weightings on $\hat{a}_{ij}$ and $\hat{c}_i$ to maximize genetic merit in future generations over specified time frames.

More complicated structures for **R** could also be envisaged whereby alteration of mtDNA is a continual process, and correlations between animals in the same generation for cytoplasmic effects are less than one. Accommodation of this would require more detailed knowledge about the alteration process than is currently available. However, if published estimates of base substitution rates of mtDNA are correct or anywhere near so (Upholt and David 1977; Brown et al. 1979), then the simple structure of transmission of mtDNA assumed in (23.6), where mtDNA is transmitted unaltered from generation to generation is a very good approximation to the real situation, even for populations spanning many generations. There is still no concrete evidence that cytoplasmic effects on milk production are important (Kennedy 1986), and if σ_c^2 is negligible, debate over the structure of **R** is academic.

23.3.2 Dominance Effects

Offspring from an embryo transfer flush are full-sibs, which in the absence of inbreeding have a dominance relationship of 0.25 between each other. The frequency of other types of relatives which also share dominance relationships, such as double first cousins (both sets of grandparents in common), is also increased through ET. Similarly, identical genotypes produced through zygote splitting have a dominance relationship of one. The net effect of use of these and other reproductive techniques is to increase greatly the number and magnitude of dominance relationships in the population, and to allow for genetic evaluation of animals for dominance effects. Indeed, if these are important, they should be included in the evaluation process. Selection for dominance effects might prove useful for improvement of some traits, particularly those for which dominance effects are large relative to additive effects. For example one could predict total genetic merit, including dominance merit, for all possible mating pairs and select on the outcome (e.g., Jansen and Wilton 1985).

Henderson (1984) reviewed in some detail computational procedures for BLUP of dominance genetic effects and estimation of dominance variance, and only an outline of the former will be given here. Evaluation of dominance effects simply requires an extension of (23.1) such that

$$\mathbf{y} = \mathbf{Xb}+\mathbf{Za}+\mathbf{Zd}+\mathbf{e}, \tag{23.7}$$

with $\mathbf{d}$ representing a vector of dominance genetic effects, var($\mathbf{d}$)= $\mathbf{D}\sigma_d^2$, where σ_d^2 is the dominance variance, and other elements are as defined previously. A solution for $\mathbf{d}$ can be obtained from solving:

$$\begin{bmatrix} \tilde{\mathbf{b}} \\ \hat{\mathbf{a}} \\ \hat{\mathbf{d}} \end{bmatrix} = \begin{bmatrix} \mathbf{X'X} & \mathbf{X'Z} & \mathbf{X'Z} \\ \mathbf{Z'X} & \mathbf{Z'Z}+\mathbf{A}^{-1}\lambda & \mathbf{Z'Z} \\ \mathbf{Z'X} & \mathbf{Z'Z} & \mathbf{Z'Z}+\mathbf{D}^{-1}\gamma \end{bmatrix}^{-} \begin{bmatrix} \mathbf{X'y} \\ \mathbf{Z'y} \\ \mathbf{Z'y} \end{bmatrix}, \tag{23.8}$$

with $\gamma=\sigma_e^2/\sigma_d^2$. Estimation of total genetic merit is $\hat{\mathbf{g}}=\hat{\mathbf{a}}+\hat{\mathbf{d}}$.

Consider a simple example where bull 1 is mated to unrelated cow 2 and two full-sib daughters, cows 3 and 4, result through ET. Records on cows 2, 3, and 4 are 10, 9, and 8, respectively. Assume that $\sigma_a^2=0.25\sigma^2$, $\sigma_d^2=0.25\sigma^2$, and $\sigma_e^2=0.5\sigma^2$, i.e. $\lambda=\gamma=2$. The additive relationship matrix is

$$\mathbf{A} = \begin{bmatrix} 1 & 0 & 1/2 & 1/2 \\ 0 & 1 & 1/2 & 1/2 \\ 1/2 & 1/2 & 1 & 1/2 \\ 1/2 & 1/2 & 1/2 & 1 \end{bmatrix}, \text{ with } \mathbf{A}^{-1}\lambda = \begin{bmatrix} 4 & 2 & -2 & -2 \\ 2 & 4 & -2 & -2 \\ -2 & -2 & 4 & 0 \\ -2 & -2 & 0 & 4 \end{bmatrix}.$$

The dominance relationship matrix is

$$\mathbf{D} = \begin{bmatrix} 1 & 0 & 0 & 0 \\ 0 & 1 & 0 & 0 \\ 0 & 0 & 1 & 1/4 \\ 0 & 0 & 1/4 & 1 \end{bmatrix}, \text{ with } \mathbf{D}^{-1}\gamma = \begin{bmatrix} 2 & 0 & 0 & 0 \\ 0 & 2 & 0 & 0 \\ 0 & 0 & 2.13 & -0.53 \\ 0 & 0 & -0.53 & 2.13 \end{bmatrix}.$$

Again, assuming that the mean is the only fixed effect, (23.8) is

$$\begin{bmatrix} \hat{\mu} \\ \hat{a}_1 \\ \hat{a}_2 \\ \hat{a}_3 \\ \hat{a}_4 \\ \hat{d}_1 \\ \hat{d}_2 \\ \hat{d}_3 \\ \hat{d}_4 \end{bmatrix} = \begin{bmatrix} 3 & 0 & 1 & 1 & 1 & 0 & 1 & 1 & 1 \\ 0 & 4 & 2 & -2 & -2 & 0 & 0 & 0 & 0 \\ 1 & 2 & 5 & -2 & -2 & 0 & 1 & 0 & 0 \\ 1 & -2 & -2 & 5 & 0 & 0 & 0 & 1 & 0 \\ 1 & -2 & -2 & 0 & 5 & 0 & 0 & 0 & 1 \\ 0 & 0 & 0 & 0 & 0 & 2 & 0 & 0 & 0 \\ 1 & 0 & 1 & 0 & 0 & 0 & 3 & 0 & 0 \\ 1 & 0 & 0 & 1 & 0 & 0 & 0 & 3.13 & -0.53 \\ 1 & 0 & 0 & 0 & 1 & 0 & 0 & -0.53 & 3.13 \end{bmatrix}^{-1} \begin{bmatrix} 27 \\ 0 \\ 10 \\ 9 \\ 8 \\ 0 \\ 10 \\ 9 \\ 8 \end{bmatrix} = \begin{bmatrix} 9.02 \\ -0.14 \\ 0.14 \\ 0.01 \\ -0.15 \\ 0 \\ 0.28 \\ -0.06 \\ -0.29 \end{bmatrix}.$$

Note that the estimated dominance effect for the sire ($\hat{d}_1$) is zero because the sire has no record or dominance relationship with any animal with a record. The total genetic merit of an animal is $\hat{a}_i+\hat{d}_i$. For example, the total genetic merit of cow 3 is $\hat{a}_3+\hat{d}_3$= 0.01-0.06=-0.05.

There are, however, some operational difficulties with (23.8). Computation of $\mathbf{D}^{-1}$, may be difficult and if the order of **D** is large, solution of (23.8) can also be difficult, if not impossible. A simple extension of equations suggested by Henderson (1984) can help to circumvent this problem by solving

$$\begin{bmatrix} \mathbf{X'X} & \mathbf{X'ZK} \\ \mathbf{Z'X} & \mathbf{Z'ZK+A^{-1}}\lambda \end{bmatrix} \begin{bmatrix} \tilde{\mathbf{b}} \\ \hat{\mathbf{a}} \end{bmatrix} = \begin{bmatrix} \mathbf{X'y} \\ \mathbf{Z'y} \end{bmatrix}, \tag{23.9}$$

where $\mathbf{K}=(\mathbf{I}+\mathbf{D}\mathbf{A}^{-1}\sigma_d^2/\sigma_a^2)$. The solution for $\mathbf{d}$ is obtained from $\hat{\mathbf{d}}=\mathbf{D}\mathbf{A}^{-1}\hat{\mathbf{a}}\sigma_d^2/\sigma_a^2$. The procedure requires computation of $\mathbf{D}$ and $\mathbf{A}^{-1}$, but avoids computation of $\mathbf{D}^{-1}$.

The structure of $\mathbf{D}$ in inbred and selected populations is not well understood, but Smith (1984) has proposed a method for computing $\mathbf{D}$ in inbred populations, based on a genomic table for computing relationships (Smith and Allaire 1985).

23.3.3 Preferential Treatment

Use of expensive reproductive technology on a differential basis will increase the likelihood of animals being treated preferentially, which can bias genetic evaluation. The consequences can be serious particularly with respect to evaluation and selection of bull dams, where reproductive technology (ET, zygote splitting, and sexing) can potentially reduce the number of dams selected greatly (Van Vleck 1981).

Animals may be treated differentially or preferentially with different interpretations and consequences. For example, cows are treated differentially when they are fed according to level of production, yet this is a routine management practice which we accept as a function of ability to produce. Preferential treatment occurs when some cows are provided special environmental conditions, box stalls for example, that are not in response to production but are to stimulate production in excess of that of other cows with similar production abilities. If preferential treatment can be identified and categorized, its effects can be corrected for by including it as a fixed effect in the model (23.1), and unbiased genetic evaluations can be obtained.

When preferential treatment cannot be identified, one approach is to use outlier theory to identify records on animals that have likely been influenced by preferential treatment. The usual method is to examine residuals associated with each observation (e.g., Anscombe 1960). However, with mixed models such as (23.1), the residual should include $\mathbf{a}$ as well as $\mathbf{e}$ because the preferential treatment effect will be apportioned between the two, with the amount to each depending upon heritability and the structure of relationships. An alternative, particularly if one suspects preferential treatment within family, is to examine the Mendelian sampling portions of $\hat{\mathbf{a}}$, which can be estimated simply as

$$\hat{\mathbf{a}}_m = \hat{\mathbf{a}}-1/2(\hat{\mathbf{a}}_s+\hat{\mathbf{a}}_d)\,,$$

where $\hat{\mathbf{a}}_s$ and $\hat{\mathbf{a}}_d$ represent vectors of sire and dam estimated additive genetic values, respectively. However, preferential treatment of progeny will affect $\hat{\mathbf{a}}_s$ and $\hat{\mathbf{a}}_d$ also, and a form of iterative analysis with suspect records rejected or modified might be required. Whether or not to reject suspect records outright is debatable, and there may be optimum procedures to minimize losses from both rejection of a valid record and retention of a preferentially treated record through regression of outliers towards the mean. Also, potential outliers should be examined closely if a major gene is possibly segregating for the trait of concern.

23.4 Conclusions

Advances in reproductive technology have historically complicated genetic evaluation and will continue to do so. Widespread use of artificial insemination and increasing use of embryo transfer have made current sire evaluation systems obsolete, and their replacement by joint cow and bull evaluation systems under an animal model is warranted. The introduction of zygote splitting requires that identical genotypes be identified as such in recording and registration programs and the presence of identical genotypes has to be accounted for in construction of the relationship matrix and mixed model equations. Applications of gene transfer will present new problems in genetic evaluation and will create a new field for study and development. Computations for genetic evaluation will become increasingly complex, but these problems will be offset by improved computer hardware and computing algorithms.

Increasing use of new techniques in reproductive technology will facilitate evaluation of non-additive genetic merit, particularly dominance effects, and will foster the development of mating schemes to exploit non-additive genetic variation more fully. Study is still required on estimation of dominance effects in inbred and selected populations.

Application of expensive techniques for reproduction will likely increase the incidence of preferential treatment of animals, and better schemes for recording preferential treatments are required as is the development of statistical techniques for the identification and correction of preferentially treated records.

Lastly we quote from Seidel (1984): "making copies of the best animals available will be a trivial exercise relative to making animals better than the best".

Acknowledgements. Helpful suggestions were made by A.M. Gibbins, K.J. Betteridge and C. Smith. Financial support was provided by the Canadian Association of Animal Breeders, the Ontario Ministry of Agriculture and Food, and IBM Canada.

References

Anscombe FJ (1960) Rejection of outliers. Technometrics 2:123-147

Baker RD, Shea BF (1985) Commercial splitting of bovine embryos. Theriogenology 23:3-12

Bell BR, McDaniel BT, Robison OW (1985) Effects of cytoplasmic inheritance on production traits of dairy cattle. J Dairy Sci 68:2038-2051

Benyshek L, Bertrand K, Johnson M, Little D (1986) Angus sire evaluation - the 1986 edition. Angus J, Oct. pp 116-118

Betteridge KJ, Hare WCD, Singh EL (1981) Approaches to sex selection in farm animals. In: Brackett BG, Seidel GE Jr, Seidel SM (eds) New technologies in animal breeding. Academic Press, New York, pp 109-125

Brown WM, George M Jr, Wilson AC (1979) Rapid evolution of animal mitochondrial DNA. Proc Nat Acad Sci, USA 76:1967-1971

Church RB, McRae A, McWhir J (1986) Embryo manipulation and gene transfer in livestock production. In: Dickerson GE, Johnson RK (eds) Proc 3rd World Congr Genet Appl Livest Prod, Agri Commun, Univ Nebraska, Lincoln, Nebraska XII:133-138

Crow JF, Roberts WC (1950) Inbreeding and homozygosis in bees. Genetics 35:612-621

Day NE (1969) Estimating the components of a mixture of normal distributions. Biometrika 56:463-474

Dempfle L (1982) Problems in estimation of breeding values. Proc 2nd World Congr Genet Appl Livest Prod. Garsi, Madrid V:104-118

Elston RC, Stewart J (1971) A general model for the genetic analysis of pedigree data. Hum Hered 21:523-542

Everett RW (1984) Impact of genetic manipulation. J Dairy Sci 67:2812-2818

Falconer DS (1981) Introduction to quantitative genetics. 2nd edn. Longman, London

Famula TR (1986) Identifying single genes of large effect in quantitative traits using best linear unbiased prediction. J Anim Sci 63:68-76

Fechheimer NS (1985) Prospects for genetic engineering in domestic animals. In: Chapman AB (ed) General and quantitative genetics. World Anim Sci A4, Elsevier, Amsterdam, pp 385-398

Foulley JL, Gianola D (1986) Sire evaluation for multiple binary responses when information is missing on some traits. J Dairy Sci 69:2681-2695

Gardner RL (1978) Production of chimeras by injecting cells or tissue into the blastocyst. In: Daniel JC Jr (ed) Methods in mammalian reproduction. Academic Press, New York, pp 137-165

Geldermann H, Pieper U, Weber EW (1986) Effect of misidentifcation on the estimation of breeding value and heritability in cattle. J Anim Sci 63:1759-1768

Gianola D, Foulley JL (1983) Sire evaluation for ordered categorical data with a threshold model. Genet Sel Evol 15:201-224

Gilmour AR, Anderson RD, Rae AL (1985) The analysis of binomial data by a generalized linear mixed model. Biometrika 72:593-599

Harville DA, Mee RW (1984) A mixed model procedure for analyzing ordered categorical data. Biometrics 40:393-408

Henderson CR (1963) Selection index and expected genetic advance. In: Hanson WD, Robinson WR (eds) Statistical genetics and plant breeding. National Academy of Sciences, Washington DC, Publication 982:141-163.

Henderson CR (1966) A sire evaluation method which accounts for unknown genetic and environmental trends, herd differences, season, age effects and differential culling. Proc Symp Estimating Breeding Values of Dairy Sires and Cows, Washington DC

Henderson CR (1973) Sire evaluation and genetic trends. In: Proc Anim Breed Genet Symp in Honor of Dr. J.L. Lush. ASAS and ADSA, Champaign, Illinois, pp 10-41

Henderson CR (1975) Best linear unbiased estimation and prediction under a selection model. Biometrics 31:423-447

Henderson CR (1976) A simple method for computing the inverse of a numerator relationship matrix used in prediction of breeding values. Biometrics 32:69-83

Henderson CR (1984) Applications of linear models in animal breeding. Univ Guelph Press, Guelph, Canada

Henderson CR, Carter HW, Godfrey JT (1954) Use of contemporary herd average in appraising progeny tests of dairy bulls. J Anim Sci (Abstr) 14:949

Hudson GFS, Kennedy BW (1985) Genetic evaluation of swine for growth rate and backfat thickness. J Anim Sci 61:83-91

Jansen GB, Wilton JW (1985) Selecting mating pairs with linear programming techniques. J Dairy Sci 68:1302-1305

Kennedy BW (1986) A further look at evidence for cytoplasmic inheritance of production traits in dairy cattle. J Dairy Sci 69:3100-3111

Kennedy BW, Moxley JE (1975) Comparison of genetic group and relationship methods for mixed model sire evaluation. J Dairy Sci 58:1507-1514

Kinghorn BP (1983) A review of quantitative genetics in fish breeding. Aquaculture 31:283-304

McLaren A (1984) Methods and success of nuclear transplantation in mammals. Nature 309:671-672

Meyer K, Burnside EB (1988) Joint sire and cow evaluation for conformation traits using an individual animal model. J Dairy Sci 71:1034-1049

Mintz B (1971) Allophenic mice of multi-embryo origin. In: Daniel JC Jr, (ed) Methods in mammalian embryology. Freeman WH, San Francisco, pp 184-214

Morton NE, MacLean CJ (1974) Analysis of family resemblance. III. Complex segregation of quantitative traits. Am J Hum Genet 26:489-503.

Nicholas FW, Smith C (1983) Increased rates of genetic change in dairy cattle by embryo transfer and splitting. Anim Prod 36:341-353

Pearson K (1894) Contributions to the mathematical theory of evolution. Phil Trans R Soc 185:71-110

Polge C (1986) Current and potential reproductive technology. In: Dickerson GE, Johnson RK (eds) Proc 3rd World Congr Genet Appl Livest Prod, Agric Commun, Univ Nebraska, Lincoln, Nebraska XII:81-85

Pollak EJ, Quaas RL (1983) Definition of group effects in sire evaluation models. J Dairy Sci 66:1503-1509

Powell RL (1981) Possible effects of embryo transfer on evaluation of cows and bulls. J Dairy Sci 64:2476-2483

Quaas RL (1976) Computing the diagonal elements and inverse of a large numerator relationship matrix. Biometrics 32:949-953

Quaas RL, Pollak EJ (1980) Mixed model methodology for farm and ranch beef cattle testing programs. J Anim Sci 51:1277-1287

Quaas RL, Everett RW, McClintock AE (1979) Maternal grandsire model for dairy sire evaluation. J Dairy Sci 62:1648-1654

Robertson A, Rendel JM (1954) The performance of heifers got by artificial insemination. J Agric Sci 44:184-192

Schaeffer LR, Kennedy BW (1986a) Computing strategies for solving mixed model equations. J Dairy Sci 69:575-579

Schaeffer LR, Kennedy BW (1986b) Computing solutions to mixed model equations. In: Dickerson GE, Johnson RK (eds) Proc 3rd World Congr Genet Appl Livest Prod, Agric Commun, Univ Nebraska, Lincoln, Nebraska XII:382-393

Seidel GE Jr (1981) Superovulation and embryo transfer in cattle. Science 211:351-358

Seidel GE Jr (1984) Applications of embryo transfer and related technologies to cattle. J Dairy Sci 67:2786-2796

Smith C, Meuwissen THE, Gibson JP (1987) On the use of transgenes in livestock improvement. Anim Breed Abstr 55:1-10

Smith SP (1984) Dominance relationship matrix and inverse for an inbred population. Unpublished mimeo, Dept Dairy Sci, Ohio State Univ, Columbus, Ohio

Smith SP, Allaire FR (1985) Efficient selection rules to increase non-linear merit: application in mate selection. Genet Sel Evol 17:387-406

Sorensen DA, Kennedy BW (1983) The use of the relationship matrix to account for genetic drift variance in the analysis of genetic experiments. Theor Appl Genet 66:217-220

Sorensen DA, Kennedy BW (1984a) Estimation of response to selection using least-squares and mixed model methodology. J Anim Sci 58:1097-1106

Sorensen DA, Kennedy BW (1984b) Estimation of genetic variances from unselected and selected populations. J Anim Sci 59:1213-1223

Sorensen DA, Kennedy BW (1986) Analysis of selection experiments using mixed model methodology. J Anim Sci, 68:245-258

Surani MAH, Barton SC, Norris ML (1987) Experimental reconstruction of mouse eggs and embryos: an analysis of mammalian development. Biol Reprod 36:1-16

Thompson R (1979) Sire evaluation. Biometrics 35:339-353

Upholt WB, David IB (1977) Mapping of mitochondrial DNA of individual sheep and goats: Rapid evolution in the D loop region. Cell 11:571-583

Van Raden PM, Freeman AE (1985) Potential gains from producing bulls with only sires as parents. J Dairy Sci 68:1425-1431

Van Vleck LD (1981) Potential genetic impact of artificial insemination, sex selection, embryo transfer, cloning, and selfing in dairy cattle. In: Brackett BG, Seidel GE Jr, Seidel SM (eds) New technologies in animal breeding. Academic Press, New York, pp 221-242

Westell RA, Van Vleck LD (1984) Simultaneous genetic evaluation of sires and cows under an animal model. J Anim Sci 59, suppl 1(Abstr.):175

Wright S (1922) Coefficients of inbreeding and relationship. Am Nat 56:330-338

Discussion Summary

PART VII: STATISTICS AND NEW GENETIC TECHNOLOGY

J.S.F. Barker[1]

21 Identification of Genes with Large Effects

This paper provided a comprehensive review of methods for the detection of genes with large effects, although it was emphasized that some of the methods are not applicable to livestock at this time.

No disagreement was expressed with the contention that complex segregation analysis (see Chap. 3) appears to be the most suitable method for use in livestock populations. However, as this was developed for human data, further work is necessary to consider effects of the typical data structures encountered in livestock on the application and power of this method.

Genes of large effect are likely to be at low frequency in the general population, and the power of the test is less at low than at intermediate gene frequencies. Two possibilities were considered for reducing this problem: (1) where divergent selection lines are available, basing analysis on a cross between the lines, and (2) where a major gene is suspected, by ascertaining families through "affected" individuals, so as to increase the frequency of the putative major gene in the sample to be analyzed.

Farm animal data banks may provide opportunities for detection of genes of large effect that are not available for human data. This is because records of performance typically contain information on quantitative traits for thousands of individuals, and progeny test information for some males. With such a large number of individuals, animals carrying a gene of large effect may occur as outliers in the distribution. While outliers can cause problems in data analysis (due either to recording errors, or to extreme environmental or genetic effects), such animals should be used in breeding tests to determine whether a gene of large effect is involved. Progeny test information may be used to study the distribution of breeding values, and non-normality would indicate genes of large effect.

It was emphasized that while segregation analysis may indicate a possible gene of large effect, progeny testing is necessary to provide final proof.

[1] Department of Animal Science, University of New England, Armidale, Australia

22 A General Linkage Method for the Detection of Major Genes

Discussion centered particularly on the assumptions of the model and problems of extension to farm animal data, in particular effects of dominance and the need to combine information from different types of relatives and from different populations. The assumption of allowing for only additive genetic variance would seem to be critical, and further work including dominance variance is essential.

In order to obtain large numbers of pairs of relatives, different types of relatives and samples from different populations will be necessary. These populations may differ in mean, and the question was raised as to how this could be allowed for. It was argued that analysis should be done within types of relative and within populations, on the basis that the regressions would be expected to have common slopes, but different intercepts. Some doubts were expressed, and further clarification appears necessary.

Finally, it was pointed out that in some farm animal data there are many more than two individuals per family, and fitting a linear model may be a useful approach.

23 Reproductive Technology and Genetic Evaluation

Some of the techniques from reproduction technology were recognized as not likely to have much impact in farm animal breeding in the near future (e.g., androgenous matings, chimeras, and polyploidy). Discussion centered on problems of evaluation of breeding value with embryo transfer and embryo splitting, and bias in evaluation from preferential treatment.

While treating records on clones as repeated observations on the same genotype appears to be the simplest alternative, it was noted that records from different individual clones may differ because of permanent environmental effects. However, the clones must have the same breeding value, so that expansion of the equations is necessary. In this context, and extrapolating from the very high heritabilities estimated from data on identical twins, it was argued that the additional records from clones may provide only limited further information.

Preferential treatment, if reported by breeders, could be handled by extension of the definition of management groups. However, if it is unknown to have occurred, there are very real problems. Discussion centered on whether there would be more noise or bias in an animal model or in a sire model, but no conclusion was reached.